# MXene-based Electrochemical Sensors

The scope of this book encompasses the design, fabrication, and characterization of MXene-based composite materials for various electrochemical sensing applications, including, but not limited to, environmental monitoring, biomedical diagnostics, and industrial process control.

*MXene-based Electrochemical Sensors* delves into the fundamental principles of electrochemical sensing, explores the synergistic effects of MXene composites, and highlights the recent advancements in sensor performance and sensitivity achieved through their integration. It comprehensively covers MXene-based electrochemical sensors, discusses strategies to fabricate MXene-based sensor devices, and includes functionalization of MXene, device fabrication, and electrochemical sensing of food additives and adulterants. It also reviews industrially relevant topics, including device fabrication using MXene-based composites, and explores the vast potential of MXenes in nanoelectronics, photonics, and flexible device technologies.

This book is aimed at researchers and graduate students in chemical engineering, materials science and engineering, chemistry, and electrochemistry.

# MXene-based Electrochemical Sensors

Edited by

P. Abdul Rasheed, Menon Ankitha,
and Ajith Mohan Arjun

CRC Press
Taylor & Francis Group
Boca Raton London New York

CRC Press is an imprint of the
Taylor & Francis Group, an **informa** business

Designed cover image: ©Ajith Mohan Arjun

First edition published 2027
by CRC Press
2385 NW Executive Center Drive, Suite 320, Boca Raton FL 33431

and by CRC Press
4 Park Square, Milton Park, Abingdon, Oxon, OX14 4RN

*CRC Press is an imprint of Taylor & Francis Group, LLC*



ISBN: 9781032812700 (hbk)
ISBN: 9781041232889 (pbk)
ISBN: 9781003736318 (ebk)

DOI: 10.1201/9781003736318

Typeset in Times
by codeMantra

# Contents

# About the Editors

**P. Abdul Rasheed** is a distinguished researcher in nanobiotechnology focussing on electrochemical sensing and drug delivery, currently serving as a Ramalingaswami Fellow at the Indian Institute of Technology Palakkad, India. He obtained his Ph.D. in Nanoscience and Technology from the National Institute of Technology Calicut in 2015, where his work focused on nanomaterial-based DNA sensors for highly sensitive gene detection. Following his doctoral studies, he carried out postdoctoral research at Korea University, South Korea, and the Qatar Environment and Energy Research Institute, Qatar, gaining expertise in optical biosensors, drug delivery, and 2D nanomaterials for environmental and biomedical applications. His research integrates nanomaterials, electrochemistry, and device engineering, with a strong emphasis on developing low-cost, point-of-care diagnostic technologies. His current work focuses on MXene-based sensor platforms for early detection of neurological and metabolic biomarkers, along with wearable and portable electrochemical devices. He has published over 80 international research articles in high-impact journals, with an h-index exceeding 30, reflecting the broad impact of his contributions. He is also an innovator with multiple patents, including four granted patents and several filed applications, particularly in wearable biosensors and electrochemical device fabrication. Recognized for his scientific impact, he has been listed among the world's top 2% scientists for three consecutive years. His research is supported by several competitive grants and is directed toward translating advanced nanotechnology into practical healthcare and environmental monitoring solutions.

**Menon Ankitha** is a Prime Minister's Research Fellow at the Indian Institute of Technology (IIT) Palakkad, India. She was awarded the prestigious Saroj Chandrasekhar Award 2025 for Women Scientists in Biosciences and Biotechnology, in recognition of research excellence. She graduated with the first rank for M.Sc. in Applied Chemistry from Mahatma Gandhi University, Kerala, in 2019. She also completed her bachelor's degree (B.Sc.) in Chemistry with the university rank in 2017. Menon Ankitha has received honors from Smriti Irani, Minister in the Union Cabinet of India, for her higher secondary academic performance. She joined as full-time research scholar at the Indian Institute of Technology research group in fall 2022. She is currently working on surface modification of MXene and has a general research interest in the field of synthesis of nanomaterial-based composites for electrochemical sensors and environmental remediation.

**Ajith Mohan Arjun** is a Development Scientist in industry, specializing in the architectural design and kinetic optimization of enzymatic sensing platforms. He holds a Ph.D. in Materials Science and Engineering and an M.Tech in Nanotechnology. His work focuses on engineering electrochemical signal transduction pathways for the quantification of analytes from unprocessed human matrices and environmental samples, enabling diagnostics from discrete samples such as a drop of blood.

His technical experience includes the development of high-reliability aptamer-based sensing platforms for non-invasive monitoring, utilizing metal-organic frameworks (MOFs) for high-density probe immobilization. He also oversees the design and surface functionalization of immunosensors using selective antibodies and the application of redox-active molecularly imprinted polymers (MIPs) for minimally invasive diagnostics. His engineering expertise encompasses a diverse range of material architectures, including transition metal carbides (MXenes) and carbon nanostructures, alongside the integration of microfluidic architectures for enhanced assay efficiency. He has authored numerous research articles in peer-reviewed scientific journals.

# Contributors

**Parvathy Anil**
School of Chemistry, Indian Institute of Science Education and Research Thiruvananthapuram, Maruthamala PO, Thiruvananthapuram, India

**Menon Ankitha**
Department of Chemistry, Indian Institute of Technology Palakkad, 678623, Kerala, India

**Aravind Kannoth Anilkumar**
School of Chemistry, Indian Institute of Science Education and Research Thiruvananthapuram, Maruthamala PO, Thiruvananthapuram, India

**Nishant Arya**
DBT-National Institute of Animal Biotechnology (NIAB), Hyderabad 500032, Telangana, India
DBT-Regional Centre for Biotechnology (RCB), Faridabad 121001, Haryana, India

**Jithin Thomas Chacko**
Department of Biological Sciences and Engineering, Indian Institute of Technology Palakkad, 678623, Kerala, India

**Shingirirai Chakoma**
Laboratory for Integrated Nano BioElectronics Innovation, The Henry Samueli School of Engineering, University of California, Irvine, CA 92697, USA
Department of Electrical Engineering and Computer Science, University of California, Irvine, CA 92697, USA

**Arghya Chakravorty**
Centre for Nanotechnology Research, Vellore Institute of Technology, Vellore 632014, Tamil Nadu, India

**Chiranjit Das**
Department of Biological Sciences and Engineering, Indian Institute of Technology Palakkad, 678623, Kerala, India

**Sudip Das**
Centre for Nanotechnology Research, Vellore Institute of Technology, Vellore 632014, Tamil Nadu, India

**Rahim Esfandyarpour**
Laboratory for Integrated Nano BioElectronics Innovation, The Henry Samueli School of Engineering, University of California, Irvine, CA 92697, USA
Department of Electrical Engineering and Computer Science, University of California, Irvine, CA 92697, USA

**Sonu Gandhi**
DBT-National Institute of Animal Biotechnology (NIAB), Hyderabad 500032, Telangana, India
DBT-Regional Centre for Biotechnology (RCB), Faridabad 121001, Haryana, India

**Andrews Nirmala Grace**
Centre for Nanotechnology Research, Vellore Institute of Technology, Vellore 632014, Tamil Nadu, India

**Keerthi Kodakat**
Department of Biological Sciences and Engineering, Indian Institute of Technology Palakkad, 678623, Kerala, India

**Mamta Kumari**
DBT-National Institute of Animal Biotechnology (NIAB), Hyderabad 500032, Telangana, India

**Rajendra Kurapati**
School of Chemistry, Indian Institute of Science Education and Research Thiruvananthapuram, Maruthamala PO, Thiruvananthapuram, India

**Lakavathu Manikrishna**
School of Chemistry, Indian Institute of Science Education and Research Thiruvananthapuram, Maruthamala PO, Thiruvananthapuram, India

**Anson T. Mathew**
Department of Biological Sciences and Engineering, Indian Institute of Technology Palakkad, 678623, Kerala, India

**Misheal Merlin V. S.**
Department of Chemistry, Christ University, Bangalore 560029, India

**Aarcha Appu Mini**
Centre for Nanotechnology Research, Vellore Institute of Technology, Vellore 632014, Tamil Nadu, India

**Ajith Mohan Arjun**
Department of Pharmacy and Therapeutics, Institute of Systems, Molecular and Integrative Biology, University of Liverpool, 6 W Derby St, Liverpool L7 8TX, UK

**Mufeeda Mundummal**
Department of Bioengineering, King Abdullah University of Science and Technology, Thuwal 23955-6900, Saudi Arabia

**Kannankuzhiyan Niyas**
Department of Biological Sciences and Engineering, Indian Institute of Technology Palakkad, 678623, Kerala, India

**Vimala Raghavan**
Centre for Nanotechnology Research, Vellore Institute of Technology, Vellore 632014, Tamil Nadu, India

**Jerome Rajendran**
Laboratory for Integrated Nano BioElectronics Innovation, The Henry Samueli School of Engineering, University of California, Irvine, CA 92697, USA
Department of Electrical Engineering and Computer Science, University of California, Irvine, CA 92697, USA

**Rijo Rajeev**
Department of Chemistry, Christ University, Bangalore 560029, India

**Niranjani Ramakrishnan**
DBT-National Institute of Animal Biotechnology (NIAB), Hyderabad 500032, Telangana, India

**P. Abdul Rasheed**
Department of Biological Sciences and Engineering, Indian Institute of Technology Palakkad, 678623, Kerala, India

**Jeethu Raveendran**
Department of Biological Sciences and Engineering, Indian Institute of Technology Palakkad, 678623, Kerala, India

**Bartholomew Richard**
Department of Biological Sciences and Engineering, Indian Institute of Technology Palakkad, 678623, Kerala, India

**Arka Prava Roy**
Centre for Nanotechnology Research, Vellore Institute of Technology, Vellore 632014, Tamil Nadu, India
Department of Bioengineering, University of Texas at Dallas, Richardson, TX 75083, USA

**Shivmuni Sarup**
DBT-National Institute of Animal Biotechnology (NIAB), Hyderabad 500032, Telangana, India

**Sanjiv Sharma**
Department of Pharmacy and Therapeutics, Institute of Systems, Molecular and Integrative Biology, University of Liverpool, 6 W Derby St, Liverpool L7 8TX, UK

**Korappath Swetha**
School of Chemistry, Indian Institute of Science Education and Research Thiruvananthapuram, Maruthamala PO, Thiruvananthapuram, India

**Anitha Varghese**
Department of Chemistry, Christ University, Bangalore 560029, India

# Preface

*Dear Readers,*

Welcome to the world of MXene-based systems for electrochemical sensor applications. It is with great pleasure and excitement that I present to you this comprehensive volume that explores the interface between cutting-edge materials science intersects with the vital field of electrochemistry. The journey into the potential of MXene-based systems as electrochemical sensors has been one of deep scientific curiosity, rigorous research, and unwavering dedication.

MXenes, a remarkable family of two-dimensional transition metal carbides, nitrides, and carbonitrides, have rapidly emerged as transformative materials in the field of sensing. Their exceptional electrical conductivity, rich surface chemistry, and large specific surface area make them uniquely suited for diverse electrochemical applications. As global research continues to uncover the versatile capabilities of MXenes, we find ourselves on the threshold of a technological revolution that is redefining how we detect and interact with our environment.

Electrochemical sensors play a pivotal role across sectors, including environmental monitoring, medical diagnostics, food safety, and industrial process control. The ability to achieve real-time, highly sensitive, and selective detection of critical analytes is essential to addressing many of the pressing challenges of our time. Through the integration of MXenes, sensor platforms are being enhanced in unprecedented ways, launching a new era of performance and miniaturization.

This book represents a concerted effort to document the current advancements, innovative approaches, and future directions in MXene-based electrochemical sensors. With contributions from leading researchers and interdisciplinary experts, each chapter sheds light on the synthesis, functionalization, and application of MXenes and their composites in the realm of electrochemical sensing. From fundamental mechanisms to applied technologies such as wearable, microfluidic, and self-powered sensors, this volume offers an expansive yet detailed narrative.

Whether you are a seasoned researcher, a graduate student, or an enthusiast of materials science and sensor technology, this book is crafted to inform, inspire, and engage. I extend my sincere gratitude to all the contributors who have enriched this work with their valuable insights, research, and vision. Their contributions have been instrumental in making this compilation a meaningful resource for the scientific community. As you turn these pages, I invite you to explore the potential that MXenes hold in transforming electrochemical sensing. May this book spark new ideas, foster collaboration, and motivate innovation as we collectively navigate toward a smarter, safer, and more sustainable future.

**With Warm Regards,**
**Dr. P. Abdul Rasheed, Dr. Ajith Mohan Arjun, and Ms. Ankitha Menon**

# 1 The Rise of MXenes – Synthesis and Properties

*Menon Ankitha, Ajith Mohan Arjun, and P. Abdul Rasheed*

## 1.1 INTRODUCTION

The journey of 2D materials started with the discovery of graphene, a single-atom-thick sheet of carbon organized in a hexagonal pattern (Geim and Novoselov 2007). This discovery led to a resurgence in interest due to remarkable properties associated with 2D materials, such as flexibility, strength, and electrical conductivity (Novoselov et al. 2016). The diverse family of 2D materials includes carbon nitrides ($g\text{-}C_3N_4$)(Liu, Wang, and Antonietti 2016), transition metal dichalcogenides (TMDs) like molybdenum disulfide ($MoS_2$) (Manzeli et al. 2017), hexagonal boron nitride (h-BN) (Zhang et al. 2017b), phosphorene (Carvalho et al. 2016), and transition metal carbides like $Ti_3C_2T_x$, $Nb_2CT_x$, and $V_4C_3T_x$ (Naguib, Barsoum, and Gogotsi 2021). A broad spectrum of electronic characteristics is observed in these materials, including insulating, semiconducting, and superconducting behavior, which enables their integration into technologies such as energy storage, electronics, and optoelectronics.

The initial focus of 2D material research was on mechanical exfoliation, using adhesive tape to peel off atomically thin layers from bulk crystals. However, this method was not scalable for industrial applications (Huang et al. 2020). Targeted toward large-scale industrial applications of these interesting materials, significant efforts have been devoted to developing alternative synthesis techniques like chemical vapor deposition (CVD) (Zhang et al. 2020), liquid-phase exfoliation (Dong, Zhang, and Feng 2018), and epitaxial growth (Dong et al. 2020). These advancements in the synthesis approaches have produced high-quality 2D materials on a larger scale, paving the way for their integration into various technologies. Furthermore, there have also been research efforts directed to the creation of heterostructures by stacking different 2D materials to develop new functionalities and customize their characteristics for targeted uses (Novoselov et al. 2016).

MXenes, one of the notable additions to the growing group of 2D materials, have grabbed significant interest in recent years. Discovered in 2011, they represent a class of transition metal carbides, nitrides, and carbonitrides produced by selectively removing the 'A' element from their MAX phase precursors (Naguib et al. 2014). Their unique combination of hydrophilicity, metallic conductivity, and tunable surface chemistry makes them promising candidates for a wide range of applications (VahidMohammadi, Rosen, and Gogotsi 2021). MXenes have demonstrated

DOI: 10.1201/9781003736318-1

exceptional performance in energy storage devices like supercapacitors and batteries, showing high capacitance and excellent rate capability (Naguib et al. 2013). Moreover, their potential extends to catalysis (Morales-García, Calle-Vallejo, and Illas 2020), electromagnetic interference shielding (Han et al. 2020), and even water purification (Rasool et al. 2019), highlighting their versatility and promising future in materials science and nanotechnology.

MXenes' high electrical conductivity guarantees quick electron transport, resulting in sensors that are responsive and sensitive (Arjun, Shinde, and Slaughter 2023). The large surface area inherent to their 2D structure provides ample space for analyte adsorption, increasing sensitivity and lowering detection limits (Abdul Rasheed et al. 2020). Furthermore, the rich surface chemistry of MXenes, with their diverse functional groups, allows for tailored modifications to selectively detect a wide range of analytes. This versatility, combined with their tunable properties, hydrophilicity, and biocompatibility, positions MXenes as promising candidates for applications in the next generation of electrochemical sensors. The rich surface chemistry of the material also makes them suitable for immobilizing biocatalysts, such as enzymes (Rakhi et al. 2016) and bioreceptors like aptamers (Zahra et al. 2022) and antibodies (Lorencova et al. 2019).

## 1.2 SYNTHESIS

### 1.2.1 HF Etching Method

Since their initial discovery, MXenes have traditionally been synthesized from MAX phases using hydrofluoric acid (HF) for more than ten years. In a pioneering study in 2011, Gogotsi and co-workers showed that treating $Ti_3AlC_2$ with 50% HF could selectively extract the aluminum layers, attributed to the strong affinity between fluoride ions and the aluminum layer of the precursor phase (Naguib et al. 2011). $Ti_3C_2T_x$ powder with an accordion-like structure was produced; the layers were kept together by van der Waals forces and hydrogen bonds between surface groups. The resulting $Ti_3C_2T_x$ powder shows similar stoichiometry and crystal structure to the parent MAX phase, except that the Al atomic layer is absent. Later, various Al-containing MAX phases, such as $V_4AlC_3$, $V_2AlC$, $Mo_2TiAlC_2$, $Ta_4AlC_3$, and $Nb_2AlC$, have been effectively etched using HF to produce their respective MXenes (Naguib et al. 2012; Zhou et al. 2019; Naguib et al. 2013; Anasori et al. 2015). Figure 1.1(C) displays the X-ray diffraction (XRD) patterns recorded before and after the etching process. Post-treatment with HF, the distinctive $Ti_3AlC_2$ peak at 39° vanished, and the (002) peak shifted toward a lower angle, indicating an expansion in interlayer spacing as described by Bragg's law. The scanning electron microscopy (SEM) image of the MAX phase are given in Figure 1.1(A), and the resulting MXene is shown in Figure 1.1(B). This work marked the introduction of a new two-dimensional transition metal carbide, paving the possibilities for further advancements in the field of 2D materials.

The two most important factors in converting MAX phases are temperature and etching time. When 10 and 25 wt% HF solutions were employed, the (002) diffraction peak of the produced $Ti_3C_2T_x$ shifted from roughly 9.8°, seen in $Ti_3AlC_2$, to around 7°, signifying an expanded interlayer distance due to the extraction of aluminum layers. In contrast, etching with 50 wt% HF resulted in a more prominent accordion-like morphology, although the (002) peak was observed at approximately

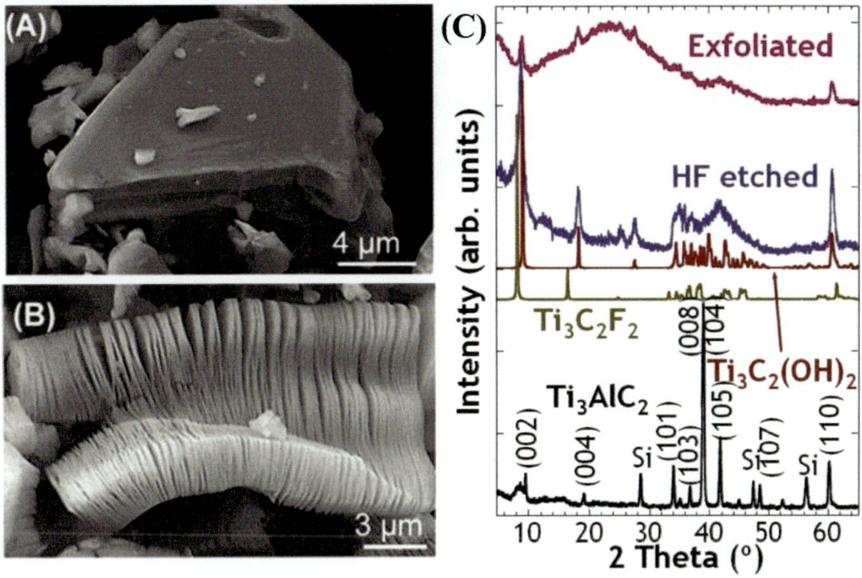


**FIGURE 1.1** SEM images of (A) $Ti_3AlC_2$, (B) $Ti_3C_2T_x$, and (C) XRD patterns of $Ti_3AlC_2$, simulated patterns of $Ti_3C_2(OH)_2$ and $Ti_3C_2F_2$. Reproduced with permission from Naguib et al. (2012) and Salim et al. (2019).

9.136°, suggesting reduced interlayer spacing compared to the lower-concentration HF treatments. This contraction could be explained by the MAX structure dissolving too much and the possible ternary fluoride production at high HF levels. As a result, extended etching with HF concentrations exceeding 30 wt% should be avoided to minimize over-etching effects (Mashtalir et al. 2013). Given the corrosive and toxic nature of hydrofluoric acid, using higher concentrations or extending the etching duration can introduce more structural defects and reduce the lateral dimensions of the resulting MXene flakes. Hence, the etching parameters should be optimized based on the intended functionality and target application. MXenes with fewer defects and larger flake areas are ideal for use in fields like electronics, optics, and electromagnetic devices, where gentle etching, such as lower HF levels and shorter reaction times, is preferred. However, for applications like gas sensing and catalysis, smaller flakes with numerous exposed edge sites are more effective, making harsher etching conditions more suitable in such cases.

Combining HF with another acid, such as HCl or $H_2SO_4$, has been used to lower the HF concentration while maintaining sufficient etching capability. For instance, by mixing 10 wt% HCl, the HF concentration required for etching $Mo_2TiC_2$ and $Mo_2Ti_2C_3$ from their respective $Mo_2TiAlC_2$ and $Mo_2Ti_2AlC_3$ phases precursors can be reduced from 50 wt% to 10 wt% (Zhou et al. 2019; Shahzad et al. 2016). Alhabeb et al. (2023) introduced a range of mixed etchants combining dilute HF with various oxidizing agents, including $H_2O_2$, $HNO_3/(NH_4)_2S_2O_8$, $KMnO_4$, and $FeCl_3$. These chemical mixtures facilitated the targeted extraction of silicon layers from the $Ti_3SiC_2$ precursor, leading to the production of multilayered $Ti_3C_2T_x$ MXene.

### 1.2.2 In Situ HF Etching Methods

In 2014, Ghidiu et al. (2023) were the first to introduce the use of a hydrochloric acid and lithium fluoride (HCl/LiF) mixture in 2014 to etch $Ti_3AlC_2$ at 40°C. This approach successfully yielded $Ti_3C_2T_x$ in the form of conductive clay. Other fluoride salts have been investigated as etching agents in addition to LiF. Liu and co-workers prepared a mixed solution by combining HCl with different fluoride salts, including NaF, KF, $NH_4F$, and LiF, to etch $Ti_3AlC_2$ effectively (Liu et al. 2017). Zou et al. used a solution of 2 M $H_2SO_4$ and 1.1 M LiF at 30°C for 30 minutes to successfully etch $Cr_2AlC$ and produce $Cr_2CT_x$ (Zou et al. 2021). In 2014, $NH_4HF_2$ was first reported as an etchant for removing layers from sputter-deposited $Ti_3AlC_2$ films at room temperature (Halim et al. 2014). Wu et al. (2020) introduced an innovative hybrid etching system composed of ammonium fluoride ($NH_4F$) and a low-eutectic solvent blend of choline chloride and oxalic acid. This etching was carried out via a hydrothermal method, with temperatures between 100°C and 180°C maintained for 24 hours to process the MAX phase.

### 1.2.3 Alkali Etching Methods

Although many of the aforementioned techniques depend on acidic conditions to extract the A-layer atoms, alkaline treatments have also been shown to selectively etch the MAX phase. Xie et al. (2014) proposed a two-stage etching strategy, where $Ti_3AlC_2$ was initially dispersed in 1 M NaOH solution for 100 hours, followed by exposure to a 1 M $H_2SO_4$ solution at 80°C for 2 hours (Figure 1.2(A)). When the concentration and temperature of the alkali solution are significantly increased, the reaction with the MAX phase undergoes a qualitative transformation. For instance, using 27.5 M NaOH at 270°C, the Al layer can be effectively removed from $Ti_3AlC_2$, resulting in $Ti_3C_2T_x$ with a high yield of 92% (Figure 1.2(B)) (Li et al. 2018). In a similar approach, Li et al. (2017) etched 100 mg of $Ti_3AlC_2$ by combining it with 350 mg of KOH in 50 μL of water, followed by hydrothermal treatment at 180°C for 24 hours. Highly hydrophilic MXenes with fluorine-free surface terminations have been successfully produced by MAX phase etching using concentrated alkaline solutions. The viability of this approach for large-scale synthesis is limited by the safety concerns associated with the use of strong alkali and high temperatures. Furthermore, the products usually exhibit an accordion-like, multilayered morphology and need to undergo further processes (exfoliation and intercalation) in order to produce single-layered MXene nanosheets.

### 1.2.4 Molten Salt Etching Methods

Urbankowski et al. (2016) employed a mixture of molten salts, LiF, KF, and NaF to etch $Ti_4AlN_3$, successfully producing nitride MXenes for the first time, namely $Ti_4N_3T_x$ (Figure 1.3(A)). This demonstrated the potential of molten salts to generate MXenes with high formation energy. Transition metal halides can react with the A-layer of the MAX phase due to their electron-accepting properties when they are in a molten state. In this regard, Huang et al. (Li et al. 2019) etched $Ti_3AlC_2$, $Ti_2AlC$, $Ti_2AlN$, and $V_2AlC$ MAX phases using a mixed $ZnCl_2$/NaCl/KCl molten salt system,

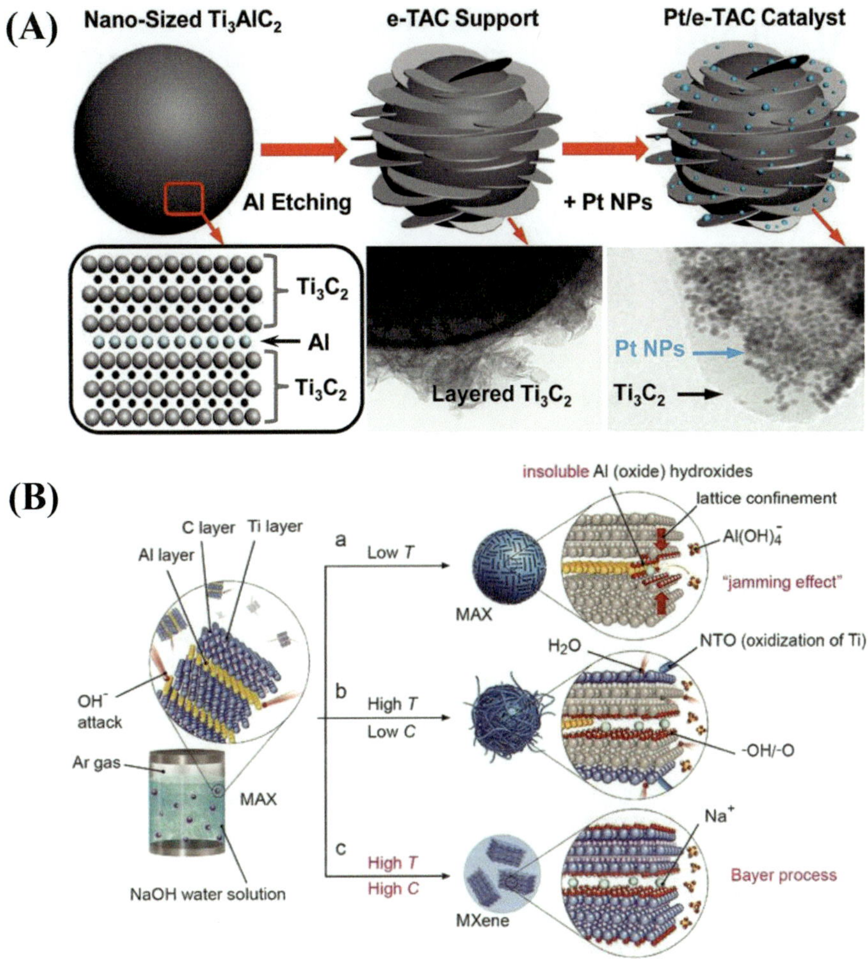


**FIGURE 1.2** (A) Schematic for the formation of layered $Ti_3C_2$. (B) The reaction between $Ti_3AlC_2$ and NaOH under different conditions. Reproduced with permission from Xie et al. (2014) and Li et al. (2018).

conducting the process in a nitrogen-protected environment. The variety of molten etching agents has grown to incorporate Lewis acid salts (Figure 1.3(B)), enabling the processing of both Al-based and non-Al MAX phases, including those with elements like Zn, Si, and Ga. In this approach, molten $CuCl_2$ was employed to etch the $Ti_3SiC_2$ MAX phase. This was made possible due to the higher redox potential of the $Cu^{2+}/Cu$ couple relative to that of $Si^{4+}/Si$, which facilitated the targeted extraction of the silicon layers, resulting in $Ti_3C_2T_x/Cu$ powder (Li et al. 2020). Notably, superconducting properties were first observed in MXenes by functionalizing $Nb_2CT_x$ with different terminations such as –Cl, –S, –NH, and –Se. This approach enabled the substitution or removal of different surface terminations on MXenes using molten inorganic salts, offering a novel pathway for tuning their properties.

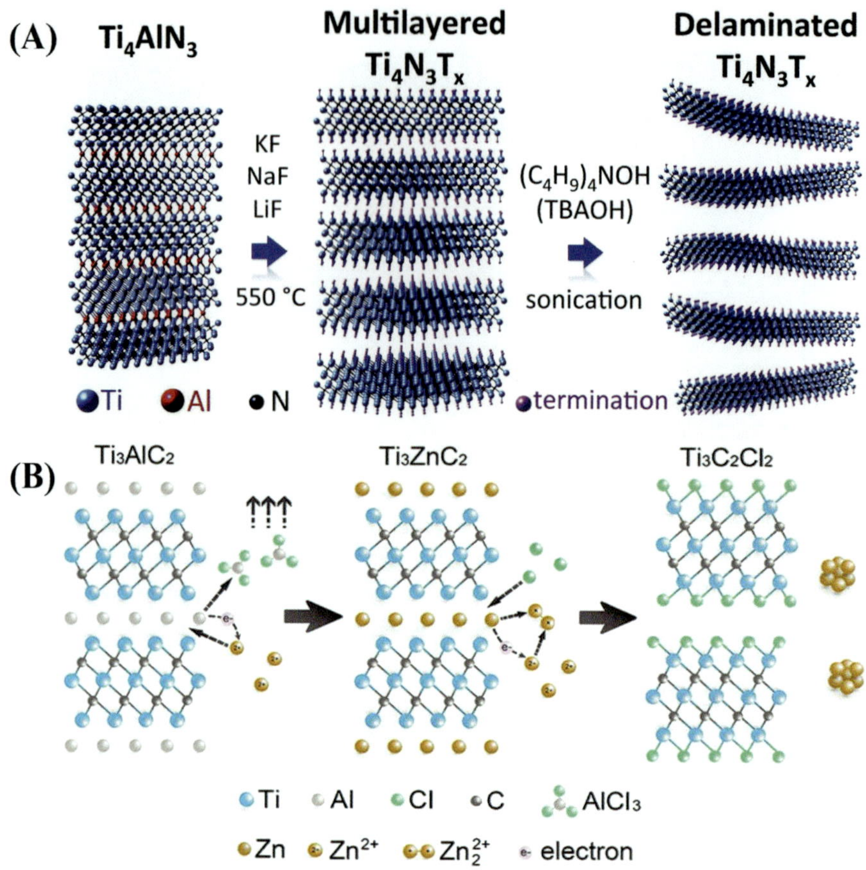


**FIGURE 1.3** Schematic illustration of (A) $Ti_4N_3T_x$ synthesis. (B) Top-down approach enabling the substitution of the late transitional element (Zn) at the A-site in the pre-existing MAX phase structure. Reproduced with permission from Li et al. (2019) and Urbankowski et al. (2016).

### 1.2.5 Electrochemical Etching Methods

By using a regulated voltage and the MAX phase as the working electrode, aluminum atomic layers can be selectively removed during electrochemical etching to create MXenes. By employing electrolytes such as NaCl, HCl, or HF, this method can also produce carbon produced from carbides from MAX precursors. Cyclic voltammetry is performed in a typical electrochemical process between 0 and 2.5 V, when the A-layer can separate from the MAX structure due to the breaking of M–A bonds (Lukatskaya et al. 2014). Electrochemical etching of the MAX phase was carried out at 0.6 V versus the Ag/AgCl electrode. In this setup, thin blocks of bulk $Ti_2AlC$ acted as the working electrode, with Ag/AgCl serving as the reference, platinum as the counter electrode, and hydrochloric acid as the electrolyte. This method offers notable benefits, including lower processing temperatures, decreased energy requirements, and limited reliance on aggressive acids. However, a common drawback is the simultaneous formation of a carbide-derived carbon (CDC) layer on the MXene surface, which often results in reduced overall yields (Sun et al. 2017). To address this, Yang et al. (2018) utilized a two-electrode system where $Ti_3AlC_2$ MAX pieces served as both the working and counter electrodes. Various electrolytes, including $H_2SO_4$, $HNO_3$, NaOH, $NH_4Cl$, and $FeCl_3$, were used to study their influence on the etching process. Hao et al. (Pang et al. 2019) introduced a novel thermo-assisted electrochemical etching method capable of achieving efficient etching without the need for intercalators. This process utilized a three-electrode system to etch various MAX phases, including $Ti_2AlC$, $V_2AlC$, and $Cr_2AlC$. The system employed 1 M HCl as the electrolyte, Pt as the counter electrode, and a calomel electrode as the reference electrode.

Electrochemical etching is considered a safer and more eco-friendly approach for synthesizing MXenes, characterized by low energy requirements. Nevertheless, issues like the formation of unwanted CDC layers and limited material output remain major limitations. Although the MAX phase electrode can be used repeatedly, the overall yield of MXene from a single etching cycle is typically low, posing challenges for scaling up the process to industrial levels.

### 1.2.6 Other Synthetic Approaches

To achieve efficient, green, and safe production of MXenes, researchers are continually developing and refining various synthesis approaches. Recently, several innovative methods have been introduced to advance this goal.

MAX phases have also been etched using halogen compounds as reactive agents. Shi et al. (2021) utilized iodine ($I_2$) as an etchant to process $Ti_3AlC_2$ in anhydrous acetonitrile at 100°C, yielding accordion-like $Ti_3C_2I_x$. The resulting material was then treated with 1 M HCl to eliminate $AlI_3$ generated during etching. Due to the instability of the surface terminations, the –I groups were subsequently replaced by −OH and −O groups in the acidic environment, leading to the formation of fluorine-free $Ti_3C_2T_x$.

Jawaid et al. (2021) further advanced this method by etching $Ti_3AlC_2$ in organic solvents using halogen and interhalogen compounds, resulting in MXenes with uniform –Cl, –Br, or –I terminations. These reactions were performed under inert conditions in a glove box, and the resulting materials were preserved in tetrahydrofuran to inhibit changes in surface functional groups. However, handling reactive interhalogen species like ICl and IBr posed certain difficulties. ICl etching was carried out in carbon disulfide ($CS_2$) at a controlled temperature of –78°C for 4 hours under constant agitation, whereas IBr treatment took place at ambient temperature with 8 hours of stirring. For elemental halogens such as $Br_2$ or $I_2$, etching was done in cyclohexane, either at room temperature or at 70°C, also with continuous stirring for 8 hours.

Experimental evidence suggests that algae can be used to exfoliate MXene nanosheets from their ternary layered sources. Zada et al. (2020) demonstrated the use of organic acids secreted by algae to remove Al from the MAX phase. The bioactive molecules produced by the algae served as intercalating agents, promoting both delamination and cleavage of the layered structure. After 24 hours of etching at ambient temperature, $V_2CT_x$ MXenes were obtained, featuring lateral sizes in the range of 50–100 nm with a thickness of 1.8 nm.

The bottom-up approach offers an alternative approach to the top-down method. Unlike the top-down approach, this method enables the fabrication of defect-free, large-area 2D monolayer crystals. For instance, Xu et al. (2015) reported the fabrication of ultrathin two-dimensional $Mo_2C$ crystals via CVD conducted at a high temperature of 1085°C. In this approach, methane functioned as the carbon precursor, while a copper–molybdenum foil served as the substrate and molybdenum source. The resulting $Mo_2C$ crystals featured a thickness of just a few nanometers and exhibited excellent structural stability under ambient conditions.

A recently developed class of MAX phases, termed 'i-MAX,' has emerged as a promising precursor for MXene fabrication. These in-plane chemically ordered structures adhere to the general formula '$(M_{2/3}M'_{1/3})_2AlC$,' where M and M′ are different transition metals arranged in an ordered configuration within the basal plane. Because the $M_2$ elements are situated between conventional M and A elements in the periodic table, they can be selectively removed along with the A-layer by fine-tuning the etching parameters. This strategy enables the synthesis of two-dimensional materials exhibiting in-plane chemical ordering or controlled vacancy patterns. For instance, Rosen et al. (Tao et al. 2017) successfully etched Al and Sc atoms from the chemically ordered $(Mo_{2/3}Sc_{1/3})_2AlC$ i-MAX phase using either 48 wt% HF or a LiF/HCl mixture.

### 1.2.7 Intercalation and Delamination

MXenes exhibit rheological behaviors comparable to those of clays like kaolinite. As a result, organic species commonly used as intercalators in clay materials show promise for MXenes as well. In 2013, Mashtalir et al. (Tao et al. 2017) demonstrated that multilayer $Ti_3C_2T_x$ could be exfoliated into single-layer MXene sheets by intercalating dimethyl sulfoxide (DMSO). In this process, MXenes were intercalated with DMSO molecules, which weakened the interlayer forces, making the material easier to delaminate. Beyond DMSO, various organic solvents including N,N-dimethylformamide (DMF), hydrazine monohydrate, and urea have been investigated for their effectiveness in intercalating and exfoliating multilayer $Ti_3C_2T_x$. Compared to DMSO, bulkier organic bases like hydroxyl choline, tetrabutylammonium hydroxide (TBAOH), and n-butylamine exhibit broader intercalation capabilities, making them more versatile in facilitating delamination.

These organic bases are effective in intercalating into the layers of MXenes, aiding in the exfoliation process (Naguib et al. 2015). LiCl has been employed as an intercalating agent for multilayer MXenes by introducing $Li^+$ ions between the layers, thereby increasing the interlayer spacing. This expansion weakens van der Waals interactions, enabling the separation of the layers into monolayer MXene sheets through ultrasonic treatment. However, this technique is primarily applicable to MXenes synthesized via etching with an HF/HCl mixture. For example, Ghidiu et al. (2016) employed a combination of LiCl and HF to etch $Ti_3AlC_2$, yielding a characteristic accordion-like morphology with $Li^+$ ions inserted between the layers. Despite successful intercalation, subsequent ultrasonic treatment failed to achieve delamination. Alternatively, mechanical exfoliation where longitudinal or transverse forces are applied to the surface has proven effective for separating nanosheets in layered materials such as graphene, $MoS_2$, and others.

However, MXenes, with their stronger interlayer interactions, present a challenge to complete exfoliation through mechanical stress alone. Xu et al. (Ghidiu et al. 2016) and Lai et al. (2015) used an adhesive tape delamination process to separate $Ti_3C_2T_x$ nanosheets onto a Si wafer. While this process successfully achieved partial delamination, it only resulted in a few-layer MXene nanosheets. Furthermore, the mechanical approach is not appropriate for large-scale production due to its low delamination yields. More recently, a method for delaminating MXenes that uses cyclic freezing and thawing has been documented (Huang and Wu 2020). By increasing interlayer space through the expansion of water molecules during freezing, this method reduces van der Waals forces and facilitates delamination without the use of intercalators.

## 1.3 PROPERTIES

### 1.3.1 Electronic Properties

It is established that MXenes possess high conductivity, and the conductivity depends on the synthesis and delamination methods (Wang et al. 2021a). MXenes generally display metallic properties, and the properties of MXene depend upon its constituents such as the transition metal M, the X atom, and the surface groups present in the MXenes (Wang et al. 2021a). The Density Functional Theory (DFT) studies by Xie et al. indicated that unmodified MXenes exhibit metallic behavior, and it turns to a semiconducting nature when MXenes are terminated with oxygen (Xie and Kent 2013). For example, $Ti_3C_2$ shows metallic behavior, while $Ti_3C_2F_2$ and $Ti_3C_2(OH)_2$ show semiconducting nature with band gaps of 0.1 eV and 0.05 eV, respectively (Naguib et al. 2023). MXenes such as $Sc_2CT_2$ (T = F, OH, and O) show semiconducting behavior with band gaps estimated to be 0.45, 1.03, and 1.8 eV for T = OH, F, and O, respectively. Similarly, $Ti_2CO_2$, $Zr_2CO_2$, and $Hf_2CO_2$ show semiconducting properties with band gaps of 0.24, 0.88, and 1 eV for $Hf_2CO_2$, respectively (Khazaei et al. 2017).

DC transport measurements were used to evaluate the electronic properties of MXene. Figure 1.4(A) shows the resistivity as a function of temperature measured on $Ti_3C_2T_x$ and the Mo-containing MXenes (Anasori et al. 2016b). The resistivities of the multilayered $Mo_2TiC_2T_x$ and $Mo_2Ti_2C_3T_x$ are higher than those of delaminated $Mo_2TiC_2T_x$ and $Mo_2Ti_2C_3T_x$ samples, However, the resistivity of delaminated Mo-containing MXenes is almost one order of magnitude higher than that of the multilayer counterparts. To understand the resistivity difference between the Mo-containing MXenes and $Ti_3C_2T_x$, the resistivity values of multilayered $Mo_2TiC_2T_x$ (red) and a multilayered $Ti_3C_2T_x$ (black) are plotted in Figure 1.4(B).

Nitride-based MXenes exhibit stronger metallic properties than carbide-based MXenes, as nitrogen contains one more electron compared to carbon. For example, $Ti_3N_2(OH)_2$ is more metallic than $Ti_3C_2(OH)_2$ (Peng et al. 2019a). Similarly, carbonitride MXenes have also shown stronger metal characteristics than carbide MXenes (Peng et al. 2019b). For example, $Ti_3C_2N_2(OH)_2$ exhibits stronger metallic properties than $Ti_3C_2(OH)_2$. However, other double transition metal MXene such as $Mo_2ZrC_2O_2$, $Mo_2HfC_2O_2$, and $W_2TiC_2O_2$ also exhibit semiconducting properties (Anasori et al. 2016a). In contrast, MXenes such as $W_2C_2O_2$, $Mo_2C_2O_2$, and $Cr_2C_2O_2$, which are projected to have a strong spin–orbit coupling (SOC) effect, show insulating properties with a higher SOC effect for $W_2C_2O_2$ (Weng et al. 2015).

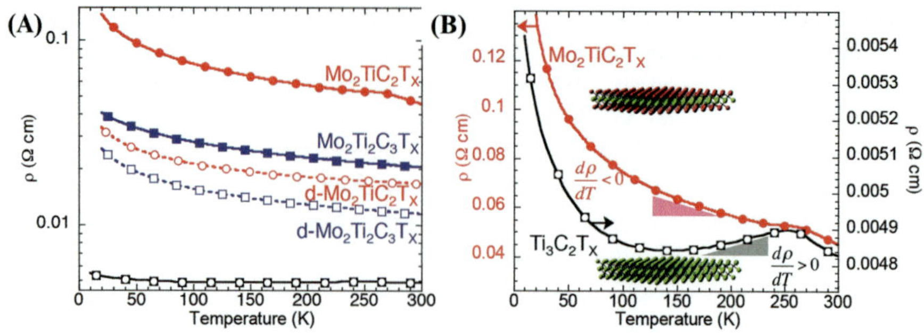


**FIGURE 1.4** (A) Temperature dependence of resistivity for a $Ti_3C_2T_x$ (black), $Mo_2Ti_2C_3T_x$ (blue), and $Mo_2TiC_2T_x$ (red). The dashed lines represent delaminated films, and the solid lines represent pressed multilayered powders. (B) The resistivity comparison of $Mo_2TiC_2T_x$ (red, left-side $y$-axis) and $Ti_3C_2T_x$ (black, right-side $y$-axis). The differences in $d\rho/dT$ are given as colored triangles (red and black) below each resistivity curve. Reproduced with permission from Anasori et al. (2016).

### 1.3.2 Electrochemical Activity

MXenes are considered as a suitable candidate for electrochemical applications such as energy storage media, catalysts, sensors, and components in electronic devices owing to their large surface area, high electrical conductivity, and variable surface chemistry. Similar to other properties, the electrochemical activity of MXene can be influenced by the morphology of MXene layers, as well as the surface terminations (Shekhirev et al. 2021). As per the comparison studies of MXenes with other materials, the surface functional groups play a critical role in deciding the electroactivity (Bilibana 2023). The conductivity of MXenes can be tuned by adjusting the surface functional groups, in which the –OH and –F terminations provide higher conductivity compared to –O as well as –OH/–F mixed terminations. This can be used for enhancing the ability of an electrochemical biosensor to detect certain biomolecules, which further enhances the affinity and selectivity of the biosensor. This tunability of surface functional groups also enables enhanced sensor performance toward a wide range of applications, from environmental monitoring to medical diagnostics.

The large surface area of MXenes offers plenty of sites for charge storage. Similarly, the high surface-to-volume ratio can enhance the accessibility of ion active sites. Overall, the electrochemical activity can be enhanced by enhancing the number of active sites which enables ion adsorption, and by making tunable surface functional groups. Additionally, the interlayer distance of MXenes can be changed by adding different intercalants which can accommodate bigger ions and enhance charge storage (Bilibana 2023).

### 1.3.3 Chemical Stability

MXene is usually unstable and likely to be oxidized under ambient conditions, which results in the degradation of structure integrity, which prevents the use of MXenes in future applications (Cao et al. 2022; Lipatov et al. 2016). To solve this issue, many strategies have been proposed for enhancing chemical stability by mitigating the oxidation of MXene. Researches have been done with different storage conditions, such as inert atmosphere protection (Lipatov et al. 2016) and cryogenic freezing (Chae et al. 2019), aiming to reduce the activity of oxygen molecules and water molecules in the aqueous MXene solution. It is already established that the surrounding media and surface functional groups of MXenes significantly influence the chemical stability of MXene (Lipatov et al. 2016). In addition, changing the surrounding environment of MXene and storing MXene in organic solvents has also been found to prevent MXene from oxidation (Wu et al. 2022; Zhang et al. 2021). Zhao et al found that the use of an antioxidant such as sodium L-ascorbate can prevent the oxidation of colloidal $Ti_3C_2T_x$ by protecting the edge of the MXene against the attack of water molecules (Zhao et al. 2019). Similarly, the presence of inorganic salts can also enhance the storage life of MXene aqueous solution (Wang, Wang, and Qiu 2021b). It has been established that the reaction between MXene and generated free radicals causes the degradation of MXene, and the use of molecules that can quench free radicals retards the oxidation of MXene (Wan et al. 2021). It has been reported that the oxidation of MXene starts from the edge of the flake and proceeds inward, and hence, covering the edges/defect sites with some molecules through electrostatic interaction can also prevent the oxidation of MXenes (Natu et al. 2019; Wang et al. 2017). However, the introduction of more insulating additives could deteriorate the electrical conductivity of MXene.

Another approach to increase the environmental stability of MXene is by functionalizing MXene in a covalent or non-covalent method (Cao et al. 2024). Similarly, the formation of MXene-based composites is another way to prevent the oxidation of MXenes and enhance the desired properties (Luo et al. 2022). Another method to improve the electronic properties and stability of MXene is the heat treatments such as hydrogen annealing or annealing under inert gas (Lee et al. 2020; Zhao et al. 2020).

### 1.3.4 Optical Properties

MXenes possess excellent optical properties, such as light scattering, light emission, nonlinear light absorption, broad spectral response range, and photoluminescence properties (Borah et al. 2024). MXenes show strong light absorption properties ranging from ultraviolet (UV) light to near infrared (NIR) light, and also thin films of MXenes have promising transmittance (Ayodhya 2023). For example, $Ti_3C_2T_z$ thin films show good absorbance at 300–500 nm light, and it shows high transmittance of 91.2% when the thickness of the film is 5 nm, and the transmittance can be varied by varying the thickness as well as by using different intercalation agents (Hantanasirisakul et al. 2016; Zhang et al. 2017a). By using a suitable thickness of thin film, higher absorbance at longer wavelengths (800 nm) has been observed and the absorbance is in linear correlation with film thickness (Borah et al. 2024).

Theoretical studies show that the optical properties of MXene can vary with different M atoms and functional groups present in the MXenes. MXenes with different M, $M_2C$ (M = Ti, Zr, Hf) showed varying band gaps of 0.92–1.75 eV, which lie in the visible spectrum (Zhang et al. 2016). In addition, F–and OH– groups in MXenes can reduce the optical absorption and reflectivity in the visible range, while they enhance the optical reflectivity in the UV region in comparison with unfunctionalized MXenes (Zhang et al. 2016). It was also found that the absorption property of

MXenes declines with the reduction of their lateral size (Wang et al. 2021a). Due to their strong reflectivity in the ultraviolet area, MXenes can be utilized to create anti-UV coating materials, and their good optical transparency and metal conductivity in the visible light region make them a promising choice for the production of flexible transparent electrode materials (Zhang et al. 2016). Additionally, the excellent photothermal conversion property of MXenes can be utilized for biomedical treatment applications (Lu et al. 2023).

## 1.4 CONCLUSION

To conclude, the distinctive synergy of properties within these 2D materials marks them as exceptionally promising, holding transformative potential for diverse technological frontiers. While challenges remain in terms of scalable production, stability, and controlled functionalization, ongoing research efforts are paving the way for overcoming these hurdles. As synthesis, characterization, and processing methods continue to progress, MXenes are poised to revolutionize various fields, including electronics, energy storage, catalysis, and sensing. As the current research is directed to develop a deep and fundamental understanding of MXenes, in addition to exploring their diverse functionalities, we can anticipate a future where these materials play a pivotal role in shaping technological advancements and addressing global challenges.

## 1.5 CHALLENGES AND FUTURE PERSPECTIVES

Several challenges remain to be addressed before the full potential of MXenes can be realized. One of the primary challenges is the scalability of production. Current synthesis methods especially etching methods involve harsh conditions, which limit their commercial viability. Additionally, stability in ambient conditions, particularly susceptibility to oxidation, is a concern that needs to be addressed for long-term applications. Another challenge is the controllable functionalization of MXenes. While various surface terminations can be introduced during synthesis or post-synthesis treatments, achieving precise control over the type and distribution of functional groups remains a challenge. Developing MXenes suited for specific applications necessitates a more comprehensive understanding of the connection between their structure and resulting properties. This includes understanding the effects of different synthesis methods, surface terminations, and defect structures on the resulting properties.

With ongoing research and developments that are addressing the existing challenges and exploring new boundaries, there is a wide range of potential uses for MXenes. One promising direction is the development of scalable and environmentally friendly synthesis methods, such as electrochemical etching and mechanochemical synthesis. These methods offer the potential for large-scale production with reduced environmental impact. Enhancing stability through techniques including surface passivation, encapsulation, and the creation of MXene compositions that are resistant to oxidation is another area of emphasis. Furthermore, advancements in controlled functionalization techniques are expected toward the development of MXenes with tailored properties for specific applications. Investigating new materials based on

MXene, such as composites and hybrid structures, is also a promising avenue for expanding their functionalities and applications. Looking ahead, the successful integration into advanced systems like flexible electronics, energy storage devices, and sensors holds the promise of accelerating innovation and enabling the emergence of next-generation technologies.

## ACKNOWLEDGMENTS

PAR acknowledges the Ramalingaswami fellowship (BT/RLF/Re-entry/75/2020) from the Department of Biotechnology (DBT), Government of India. The authors acknowledge the Indian Institute of Technology Palakkad for providing the facilities.

## REFERENCES

Abdul Rasheed, Pathath, Ravi P. Pandey, Tricia Gomez, Khadeeja A. Jabbar, Kaitlyn Prenger, Michael Naguib, Brahim Aïssa, and Khaled A. Mahmoud. 2020. "Nb-based MXenes for efficient electrochemical sensing of small biomolecules in the anodic potential." *Electrochemistry Communications* 119:106811. doi: 10.1016/j.elecom.2020.106811.

Alhabeb, Mohamed, Kathleen Maleski, Tyler S. Mathis, Asia Sarycheva, Christine B. Hatter, Simge Uzun, Ariana Levitt, and Yury Gogotsi. 2023. "Selective etching of silicon from $Ti_3SiC_2$ (MAX) to obtain 2D titanium carbide (MXene)." In *MXenes*, Edited by Yury Gogosti, 451–462. Jenny Stanford Publishing.

Anasori, Babak, Chenyang Shi, Eun J. Moon, Yu Xie, Cooper A. Voigt, Paul R. C. Kent, Steven J. May, Simon J. L. Billinge, Michel W. Barsoum, and Yury Gogotsi. 2016a. "Control of electronic properties of 2D carbides (MXenes) by manipulating their transition metal layers." *Nanoscale Horizons* 1 (3):227–234.

Anasori, Babak, Chenyang Shi, Eun J. Moon, Yu Xie, Cooper A. Voigt, Paul R. C. Kent, Steven J. May, Simon J. L. Billinge, Michel W. Barsoum, and Yury Gogotsi. 2016b. "Control of electronic properties of 2D carbides (MXenes) by manipulating their transition metal layers." *Nanoscale Horizons* 1 (3):227–234. doi: 10.1039/C5NH00125K.

Anasori, Babak, Yu Xie, Majid Beidaghi, Jun Lu, Brian C. Hosler, Lars Hultman, Paul R. C. Kent, Yury Gogotsi, and Michel W. Barsoum. 2015. "Two-dimensional, ordered, double transition metals carbides (MXenes)." *ACS Nano* 9 (10):9507–9516.

Arjun, Ajith Mohan, Megha Shinde, and Gymama Slaughter. 2023. "Application of MXene in the electrochemical detection of neurotransmitters: A review." *IEEE Sensors Journal* 23 (15):16456–16466. doi: 10.1109/JSEN.2023.3285594.

Ayodhya, Dasari. 2023. "A review of recent progress in 2D MXenes: Synthesis, properties, and applications." *Diamond and Related Materials* 132:109634. doi: 10.1016/j.diamond.2022.109634.

Bilibana, Mawethu P. 2023. "Electrochemical properties of MXenes and applications." *Advanced Sensor and Energy Materials* 2 (4):100080. doi: 10.1016/j.asems.2023.100080.

Borah, Ashish J., Varun Natu, Abhijit Biswas, and Anchal Srivastava. 2024. "A review on recent progress in synthesis, properties, and applications of MXenes." *Oxford Open Materials Science* 5 (1). itae017. Doi 10.1093/oxfmat/itae017 3/oxfmat/itae017.

Cao, Fangcheng, Ye Zhang, Hongqing Wang, Karim Khan, Ayesha K. Tareen, Wenjing Qian, Han Zhang, and Hans Ågren. 2022. "Recent advances in oxidation stable chemistry of 2D MXenes." *Advanced Materials* 34 (13):2107554.

Cao, Wei, Junli Nie, Ye Cao, Chengjie Gao, Mingsheng Wang, Weiwei Wang, Xiaoli Lu, Xiaohua Ma, and Peng Zhong. 2024. "A review of how to improve $Ti_3C_2T_x$ MXene stability." *Chemical Engineering Journal* 496:154097. doi: 10.1016/j.cej.2024.154097.

Carvalho, Alexandra, Min Wang, Xi Zhu, Aleksandr S. Rodin, Haibin Su, and Antonio H. Castro Neto. 2016. "Phosphorene: From theory to applications." *Nature Reviews Materials* 1 (11):16061. doi: 10.1038/natrevmats.2016.61.

Chae, Yoonjeong, Seon J. Kim, Soo-Yeon Cho, Junghoon Choi, Kathleen Maleski, Byeong-Joo Lee, Hee-Tae Jung, Yury Gogotsi, Yonghee Lee, and Chi W. Ahn. 2019. "An investigation into the factors governing the oxidation of two-dimensional $Ti_3C_2$ MXene." *Nanoscale* 11 (17):8387–8393.

Dong, Jichen, Leining Zhang, Xinyue Dai, and Feng Ding. 2020. "The epitaxy of 2D materials growth." *Nature Communications* 11 (1):5862. doi: 10.1038/s41467-020-19752-3.

Dong, Renhao, Tao Zhang, and Xinliang Feng. 2018. "Interface-assisted synthesis of 2D materials: Trend and challenges." *Chemical Reviews* 118 (13):6189–6235. doi: 10.1021/acs.chemrev.8b00056.

Geim, Andre Konstantin, and Konstantin Sergeevich Novoselov. 2007. "The rise of graphene." *Nature Materials* 6 (3):183–191. doi: 10.1038/nmat1849.

Ghidiu, Michael, Joseph Halim, Sankalp Kota, David Bish, Yury Gogotsi, and Michel W. Barsoum. 2016. "Ion-exchange and cation solvation reactions in $Ti_3C_2$ MXene." *Chemistry of Materials* 28 (10):3507–3514.

Ghidiu, Michael, Maria R. Lukatskaya, Meng-Qiang Zhao, Yury Gogotsi, and Michel W. Barsoum. 2023. "Conductive two-dimensional titanium carbide 'clay'with high volumetric capacitance." In *MXenes*, Edited by Yury Gogosti, 379–399. Jenny Stanford Publishing.

Halim, Joseph, Maria R. Lukatskaya, Kevin M. Cook, Jun Lu, Cole R. Smith, Lars-Åke Näslund, Steven J. May, Lars Hultman, Yury Gogotsi, and Per Eklund. 2014. "Transparent conductive two-dimensional titanium carbide epitaxial thin films." *Chemistry of Materials* 26 (7):2374–2381.

Han, Meikang, Christopher E. Shuck, Roman Rakhmanov, David Parchment, Babak Anasori, Chong M. Koo, Gary Friedman, and Yury Gogotsi. 2020. "Beyond $Ti_3C_2T_x$: MXenes for electromagnetic interference shielding." *ACS Nano* 14 (4):5008–5016. doi: 10.1021/acsnano.0c01312.

Hantanasirisakul, Kanit, Meng-Qiang Zhao, Patrick Urbankowski, Joseph Halim, Babak Anasori, Sankalp Kota, Chang E. Ren, Michel W. Barsoum, and Yury Gogotsi. 2016. "Fabrication of $Ti_3C_2T_x$ MXene transparent thin films with tunable optoelectronic properties." *Advanced Electronic Materials* 2 (6):1600050.

Huang, Xianwu, and Peiyi Wu. 2020. "A facile, high-yield, and freeze-and-thaw-assisted approach to fabricate MXene with plentiful wrinkles and its application in on-chip micro-supercapacitors." *Advanced Functional Materials* 30 (12):1910048.

Huang, Yuan, Yu-Hao Pan, Rong Yang, Li-Hong Bao, Lei Meng, Hai-Lan Luo, Yong-Qing Cai, Guo-Dong Liu, Wen-Juan Zhao, Zhang Zhou, Liang-Mei Wu, Zhi-Li Zhu, Ming Huang, Li-Wei Liu, Lei Liu, Peng Cheng, Ke-Hui Wu, Shi-Bing Tian, Chang-Zhi Gu, You-Guo Shi, Yan-Feng Guo, Zhi Gang Cheng, Jiang-Ping Hu, Lin Zhao, Guan-Hua Yang, Eli Sutter, Peter Sutter, Ye-Liang Wang, Wei Ji, Xing-Jiang Zhou, and Hong-Jun Gao. 2020. "Universal mechanical exfoliation of large-area 2D crystals." *Nature Communications* 11 (1):2453. doi: 10.1038/s41467-020-16266-w.

Jawaid, Ali, Asra Hassan, Gregory Neher, Dhriti Nepal, Ruth Pachter, W. Joshua Kennedy, Subramanian Ramakrishnan, and Richard A. Vaia. 2021. "Halogen etch of $Ti_3AlC_2$ MAX phase for MXene fabrication." *ACS Nano* 15 (2):2771–2777.

Khazaei, Mohammad, Ahmad Ranjbar, Masao Arai, Taizo Sasaki, and Seiji Yunoki. 2017. "Electronic properties and applications of MXenes: A theoretical review." *Journal of Materials Chemistry C* 5 (10):2488–2503. doi: 10.1039/C7TC00140A.

Lai, Shen, Jaeho Jeon, Sung K. Jang, Jiao Xu, Young J. Choi, Jin-Hong Park, Euyheon Hwang, and Sungjoo Lee. 2015. "Surface group modification and carrier transport properties of layered transition metal carbides ($Ti_2CT_x$, T: –OH,–F and–O)." *Nanoscale* 7 (46): 19390–19396.

Lee, Yonghee, Seon J. Kim, Yong-Jae Kim, Younghwan Lim, Yoonjeong Chae, Byeong-Joo Lee, Young-Tae Kim, Hee Han, Yury Gogotsi, and Chi W. Ahn. 2020. "Oxidation-resistant titanium carbide MXene films." *Journal of Materials Chemistry A* 8 (2):573–581.

Li, Gengnan, Li Tan, Yumeng Zhang, Binghan Wu, and Liang Li. 2017. "Highly efficiently delaminated single-layered MXene nanosheets with large lateral size." *Langmuir* 33 (36): 9000–9006.

Li, Mian, Jun Lu, Kan Luo, Youbing Li, Keke Chang, Ke Chen, Jie Zhou, Johanna Rosen, Lars Hultman, and Per Eklund. 2019. "Element replacement approach by reaction with Lewis acidic molten salts to synthesize nanolaminated MAX phases and MXenes." *Journal of the American Chemical Society* 141 (11):4730–4737.

Li, Tengfei, Lulu Yao, Qinglei Liu, Jiajun Gu, Ruichun Luo, Jinghan Li, Xudong Yan, Weiqiang Wang, Pan Liu, and Bin Chen. 2018. "Fluorine-free synthesis of high-purity $Ti_3C_2T_x$ (T = OH, O) via alkali treatment." *Angewandte Chemie International Edition* 57 (21):6115–6119.

Li, Youbing, Hui Shao, Zifeng Lin, Jun Lu, Liyuan Liu, Benjamin Duployer, Per O. Å. Persson, Per Eklund, Lars Hultman, and Mian Li. 2020. "A general Lewis acidic etching route for preparing MXenes with enhanced electrochemical performance in non-aqueous electrolyte." *Nature Materials* 19 (8):894–899. doi 10.1038/s41563-020-0657-0

Lipatov, Alexey, Mohamed Alhabeb, Maria R. Lukatskaya, Alex Boson, Yury Gogotsi, and Alexander Sinitskii. 2016. "MXene materials: Effect of synthesis on quality, electronic properties and environmental stability of individual monolayer $Ti_3C_2$ MXene flakes." *Advanced Electronic Materials* 2 (12), 1600255. doi 10.1002/aelm.201600255

Liu, Fanfan, Aiguo Zhou, Jinfeng Chen, Jin Jia, Weijia Zhou, Libo Wang, and Qianku Hu. 2017. "Preparation of $Ti_3C_2$ and $Ti_2C$ MXenes by fluoride salts etching and methane adsorptive properties." *Applied Surface Science* 416:781–789. doi 10.1016/j.apsusc.2017.04.239

Liu, Jian, Hongqiang Wang, and Markus Antonietti. 2016. "Graphitic carbon nitride "reloaded": emerging applications beyond (photo)catalysis." *Chemical Society Reviews* 45 (8):2308–2326. doi: 10.1039/C5CS00767D.

Lorencova, Lenka, Veronika Gajdosova, Stefania Hroncekova, Tomas Bertok, Jana Blahutova, Alica Vikartovska, Lucia Parrakova, Pavol Gemeiner, Peter Kasak, and Jan Tkac. 2019. "2D MXenes as perspective immobilization platforms for design of electrochemical nanobiosensors." *Electroanalysis* 31 (10):1833–1844. doi: 10.1002/elan.201900288.

Lu, Haozi, Junhua Wang, Huimin Li, Wei Zhou, Quan Yuan, and Song Liu. 2023. "Efficient photothermal conversion of MXenes and their application in biomedicine." *Materials Chemistry Frontiers* 7 (19):4372–4399.

Lukatskaya, Maria R., Joseph Halim, Boris Dyatkin, Michael Naguib, Yulia S. Buranova, Michel W. Barsoum, and Yury Gogotsi. 2014. "Room-temperature carbide-derived carbon synthesis by electrochemical etching of MAX phases." *Angewandte Chemie* 126 (19): 4977–4980.

Luo, Yijia, Wenxiu Que, Xiaoqing Bin, Chenji Xia, Bingshan Kong, Bowen Gao, and Ling B. Kong. 2022. "Flexible MXene-based composite films: Synthesis, modification, and applications as electrodes of supercapacitors." *Small* 18 (27):2201290.

Manzeli, Sajedeh, Dmitry Ovchinnikov, Diego Pasquier, Oleg V. Yazyev, and Andras Kis. 2017. "2D transition metal dichalcogenides." *Nature Reviews Materials* 2 (8):17033. doi: 10.1038/natrevmats.2017.33.

Mashtalir, Olha, Michael Naguib, Boris Dyatkin, Yury Gogotsi, and Michel W. Barsoum. 2013. "Kinetics of aluminum extraction from $Ti_3AlC_2$ in hydrofluoric acid." *Materials Chemistry Physics* 139 (1):147–152.

Morales-García, Ángel, Federico Calle-Vallejo, and Francesc Illas. 2020. "MXenes: New horizons in catalysis." *ACS Catalysis* 10 (22):13487–13503. doi: 10.1021/acscatal.0c03106.

Naguib, Michael, Joseph Halim, Jun Lu, Kevin M. Cook, Lars Hultman, Yury Gogotsi, and Michel W. Barsoum. 2013. "New two-dimensional niobium and vanadium carbides as promising materials for Li-ion batteries." *Journal of the American Chemical Society* 135 (43):15966–15969.

Naguib, Michael, Michel W. Barsoum, and Yury Gogotsi. 2021. "Ten years of progress in the synthesis and development of MXenes." *Advanced Materials* 33 (39):2103393. doi: 10.1002/adma.202103393.

Naguib, Michael, Murat Kurtoglu, Volker Presser, Jun Lu, Junjie Niu, Min Heon, Lars Hultman, Yury Gogotsi, and Michel W. Barsoum. 2011. "Two-dimensional nanocrystals: Two-dimensional nanocrystals produced by exfoliation of $Ti_3AlC_2$." *Advanced Materials* 23 (37):4207–4207.

Naguib, Michael, Murat Kurtoglu, Volker Presser, Jun Lu, Junjie Niu, Min Heon, Lars Hultman, Yury Gogotsi, and Michel W. Barsoum. 2023. "Two-dimensional nanocrystals produced by exfoliation of $Ti_3AlC_2$." In *MXenes*, Edited by Yury Gogosti, 15–29. Jenny Stanford Publishing.

Naguib, Michael, Olha Mashtalir, Joshua Carle, Volker Presser, Jun Lu, Lars Hultman, Yury Gogotsi, and Michel W. Barsoum. 2012. "Two-dimensional transition metal carbides." *ACS Nano* 6 (2):1322–1331.

Naguib, Michael, Vadym N. Mochalin, Michel W. Barsoum, and Yury Gogotsi. 2014. "25th anniversary article: MXenes: A new family of two-dimensional materials." *Advanced Materials* 26 (7):992–1005. doi: 10.1002/adma.201304138.

Naguib, Michael, Raymond R. Unocic, Beth L. Armstrong, and Jagjit Nanda. 2015. "Large-scale delamination of multi-layers transition metal carbides and carbonitrides "MXenes"." *Dalton Transactions* 44 (20):9353–9358.

Natu, Varun, James L. Hart, Maxim Sokol, Helen Chiang, Mitra L. Taheri, and Michel W. Barsoum. 2019. "Edge capping of 2D-MXene sheets with polyanionic salts to mitigate oxidation in aqueous colloidal suspensions." *Angewandte Chemie* 131 (36):12785–12790.

Novoselov, K. S., A. Mishchenko, A. Carvalho, and A. H. Castro Neto. 2016. "2D materials and van der Waals heterostructures." *Science* 353 (6298):aac9439. doi: 10.1126/science.aac9439.

Pang, Sin-Yi, Yuen-Ting Wong, Shuoguo Yuan, Yan Liu, Ming-Kiu Tsang, Zhibin Yang, Haitao Huang, Wing-Tak Wong, and Jianhua Hao. 2019. "Universal strategy for HF-free facile and rapid synthesis of two-dimensional MXenes as multifunctional energy materials." *Journal of the American Chemical Society* 141 (24):9610–9616.

Peng, Jiahe, Xingzhu Chen, Wee-Jun Ong, Xiujian Zhao, and Neng Li. 2019a. "Surface and heterointerface engineering of 2D MXenes and their nanocomposites: Insights into electro- and photocatalysis." *Chem* 5 (1):18–50. doi: 10.1016/j.chempr.2018.08.037.

Peng, Jiahe, Xingzhu Chen, Wee-Jun Ong, Xiujian Zhao, and Neng Li. 2019b. "Surface and heterointerface engineering of 2D MXenes and their nanocomposites: Insights into electro- and photocatalysis." *Chem* 5 (1):18–50.

Rakhi, Raghavan Baby, Pranati Nayak, Chuan Xia, and Husam N. Alshareef. 2016. "Novel amperometric glucose biosensor based on MXene nanocomposite." *Scientific Reports* 6 (1):36422. doi: 10.1038/srep36422.

Rasool, Kashif, Ravi P. Pandey, Pathath Abdul Rasheed, Samantha Buczek, Yury Gogotsi, and Khaled A. Mahmoud. 2019. "Water treatment and environmental remediation applications of two-dimensional metal carbides (MXenes)." *Materials Today* 30:80–102. doi: 10.1016/j.mattod.2019.05.017.

Shahzad, Faisal, Mohamed Alhabeb, Christine B. Hatter, Babak Anasori, Soon M. Hong, Chong M. Koo, and Yury Gogotsi. 2016. "Electromagnetic interference shielding with 2D transition metal carbides (MXenes)." *Science* 353 (6304):1137–1140.

Shekhirev, Mikhail, Christopher E. Shuck, Asia Sarycheva, and Yury Gogotsi. 2021. "Characterization of MXenes at every step, from their precursors to single flakes and assembled films." *Progress in Materials Science* 120:100757. doi: 10.1016/j.pmatsci.2020.100757.

Shi, Huanhuan, Panpan Zhang, Zaichun Liu, SangW. Park, Martin R. Lohe, Yuping Wu, Ali S. Nia, Sheng Yang, and Xinliang Feng. 2021. "Ambient-stable two-dimensional titanium carbide (MXene) enabled by iodine etching." *Angewandte Chemie International Edition* 60 (16):8689–8693.

Sun, Wanmei, S. A. Shah, Yexiao Chen, Z. Tan, H. Gao, T. Habib, M. Radovic, and M. J. Green. 2017. "Electrochemical etching of $Ti_2AlC$ to $Ti_2CT_x$ (MXene) in low-concentration hydrochloric acid solution." *Journal of Materials Chemistry A* 5 (41):21663–21668.

Tao, Q., M. Dahlqvist, J. Lu, S. Kota, R. Meshkian, J. Halim, J. Palisaitis, L. Hultman, M. W. Barsoum, P. O. Persson, and J. Rosen. 2017. "Two-dimensional $Mo_{1.33}C$ MXene with divacancy ordering prepared from parent 3D laminate with in-plane chemical ordering." *Nature Communications* 8:14949.

Urbankowski, Patrick, Babak Anasori, Taron Makaryan, Dequan Er, Sankalp Kota, Patrick L. Walsh, Mengqiang Zhao, Vivek B. Shenoy, Michel W. Barsoum, and Yury Gogotsi. 2016. "Synthesis of two-dimensional titanium nitride $Ti_4N_3$ (MXene)." *Nanoscale* 8 (22): 11385–11391.

VahidMohammadi, Armin, Johanna Rosen, and Yury Gogotsi. 2021. "The world of two-dimensional carbides and nitrides (MXenes)." *Science* 372 (6547):eabf1581. doi: 10.1126/science.abf1581.

Wan, Yan-Jun, Krishnamoorthy Rajavel, Xing-Miao Li, Xiao-Yun Wang, Si-Yuan Liao, Zhi-Qiang Lin, Peng-Li Zhu, Rong Sun, and Ching-Ping Wong. 2021. "Electromagnetic interference shielding of $Ti_3C_2T_x$ MXene modified by ionic liquid for high chemical stability and excellent mechanical strength." *Chemical Engineering Journal* 408:127303.

Wang, Bing, Shipeng Zhong, Ping Xu, and Han Zhang. 2021a. "Booming development and present advances of two dimensional MXenes for photodetectors." *Chemical Engineering Journal* 403:126336. doi: 10.1016/j.cej.2020.126336.

Wang, Xingyu, Zhiyu Wang, and Jieshan Qiu. 2021b. "Stabilizing MXene by hydration chemistry in aqueous solution." *Angewandte Chemie* 133 (51):26791–26795.

Zixing Wang, Vidya Kochat, Prafull Pandey, Sanjay Kashyap, Soham Chattopadhyay, Atanu Samanta, Suman Sarkar, Praveena Manimunda, Xiang Zhang, Syed Asif, Abhisek K. Singh, Kamanio Chattopadhyay, Chandra Sekhar Tiwary, Pulickel M. Ajayan. 2017. "Metal immiscibility route to synthesis of ultrathin carbides, borides, and nitrides." *Advanced Materials* 29, 1700364. doi: 10.1002/adma.201700364.

Weng, Hongming, Ahmad Ranjbar, Yunye Liang, Zhida Song, Mohammad Khazaei, Seiji Yunoki, Masao Arai, Yoshiyuki Kawazoe, Zhong Fang, and Xi Dai. 2015. "Large-gap two-dimensional topological insulator in oxygen functionalized MXene." *Physical Review B* 92 (7):075436. doi: 10.1103/PhysRevB.92.075436.

Wu, Junbiao, Yu Wang, Yaopeng Zhang, Hao Meng, Yan Xu, Yide Han, Zhuopeng Wang, Yanfeng Dong, and Xia Zhang. 2020. "Highly safe and ionothermal synthesis of $Ti_3C_2$ MXene with expanded interlayer spacing for enhanced lithium storage." *Journal of Energy Chemistry* 47:203–209.

Wu, Tao, Paul R. C. Kent, Yury Gogotsi, and De-en Jiang. 2022. "How water attacks MXene." *Chemistry of Materials* 34 (11):4975–4982.

Xie, Xiaohong, Yun Xue, Li Li, Siguo Chen, Yao Nie, Wei Ding, and Zidong Wei. 2014. "Surface Al leached $Ti_3AlC_2$ as a substitute for carbon for use as a catalyst support in a harsh corrosive electrochemical system." *Nanoscale* 6 (19):11035–11040.

Xie, Yu, and Paul. R. C. Kent. 2013. "Hybrid density functional study of structural and electronic properties of functionalized $Ti_{n+1}X_n$ (X = C, N) monolayers." *Physical Review B* 87 (23):235441. doi: 10.1103/PhysRevB.87.235441.

Xu, Chuan, Libin Wang, Zhibo Liu, Long Chen, Jingkun Guo, Ning Kang, Xiu-Liang Ma, Hui-Ming Cheng, and Wencai Ren. 2015. "Large-area high-quality 2D ultrathin $Mo_2C$ superconducting crystals." *Nature Materials* 14 (11):1135–1141.

Yang, Sheng, Panpan Zhang, Faxing Wang, Antonio G. Ricciardulli, Martin R. Lohe, Paul W. M. Blom, and Xinliang Feng. 2018. "Fluoride-free synthesis of two-dimensional titanium carbide (MXene) using a binary aqueous system." *Angewandte Chemie* 130 (47):15717–15721.

Zada, Shah, Wenhao Dai, Zhang Kai, Huiting Lu, Xiangdan Meng, Yiyi Zhang, Yaru Cheng, Fang Yan, Pengcheng Fu, and Xueji Zhang. 2020. "Algae extraction controllable delamination of vanadium carbide nanosheets with enhanced near-infrared photothermal performance." *Angewandte Chemie International Edition* 59 (16):6601–6606.

Zahra, Qurat ul Ain, Salim Ullah, Faisal Shahzad, Bensheng Qiu, Xiaona Fang, Ayesha Ammar, Zhaofeng Luo, and Shabi A. Zaidi. 2022. "MXene-based aptasensors: Advances, challenges, and prospects." *Progress in Materials Science* 129:100967. doi: 10.1016/j.pmatsci.2022.100967.

Zhang, Chuanfang, Babak Anasori, Andrés Seral-Ascaso, Sang-Hoon Park, Niall McEvoy, Aleksey Shmeliov, Georg S. Duesberg, Jonathan N. Coleman, Yury Gogotsi, and Valeria Nicolosi. 2017a. "Transparent, flexible, and conductive 2D titanium carbide (MXene) films with high volumetric capacitance." *Advanced Materials* 29 (36):1702678.

Haijun Zhang, Guang Yang, Xueqin Zuo, Huaibao Tang, Qun Yang, Guang Li . 2016. "Computational studies on the structural, electronic and optical properties of graphene-like MXenes ($M_2CT_2$, M = Ti, Zr, Hf; T = O, F, OH) and their potential applications as visible-light driven photocatalysts." *Journal of Materials Chemistry A* 4 (33):12913–12920. doi: 10.1039/c6ta04628b.

Zhang, Jing, Fan Wang, Vivek B. Shenoy, Ming Tang, and Jun Lou. 2020. "Towards controlled synthesis of 2D crystals by chemical vapor deposition (CVD)." *Materials Today* 40:132–139. doi: 10.1016/j.mattod.2020.06.012.

Zhang, Kailiang, Yulin Feng, Fang Wang, Zhengchun Yang, and John Wang. 2017b. "Two dimensional hexagonal boron nitride (2D-hBN): Synthesis, properties and applications." *Journal of Materials Chemistry C* 5 (46):11992–12022. doi: 10.1039/C7TC04300G.

Zhang, Qingxiao, Huirong Lai, Runze Fan, Peiyi Ji, Xueli Fu, and Hui Li. 2021. "High concentration of $Ti_3C_2T_x$ MXene in organic solvent." *ACS Nano* 15 (3):5249–5262.

Zhao, Xiaofei, Aniruddh Vashisth, Evan Prehn, Wanmei Sun, Smit A. Shah, Touseef Habib, Yexiao Chen, Zeyi Tan, Jodie L. Lutkenhaus, Miladin Radovic, and Micah J. Green. 2019. "Antioxidants unlock shelf-stable $Ti_3C_2T_x$ (MXene) nanosheet dispersions." *Matter* 1 (2):513–526. doi: 10.1016/j.matt.2019.05.020.

Zhao, Xiaofei, Dustin E. Holta, Zeyi Tan, Ju-Hyun Oh, Ian J. Echols, Muhammad Anas, Huaixuan Cao, Jodie L. Lutkenhaus, Miladin Radovic, and Micah J. Green. 2020. "Annealed $Ti_3C_2T_z$ MXene films for oxidation-resistant functional coatings." *ACS Applied Nano Materials* 3 (11):10578–10585.

Zhou, Jiafeng, Shuai Lin, Yanan Huang, Peng Tong, Bangchuan Zhao, Xuebin Zhu, and Yuping Sun. 2019. "Synthesis and lithium ion storage performance of two-dimensional $V_4C_3$ MXene." *Chemical Engineering Journal* 373:203–212.

Zou, Xinshu, Hao Liu, Hao Xu, Xueyun Wu, Xiaocang Han, Jianli Kang, and Kolan M. Reddy. 2021. "A simple approach to synthesis $Cr_2CT_x$ MXene for efficient hydrogen evolution reaction." *Materials Today Energy* 20:100668.

# 2 Evolution of MXenes and Derivatives
## *Synthesis and Characterization*

*Korappath Swetha, Aravind Kannoth Anilkumar, Lakavathu Manikrishna, Parvathy Anil, and Rajendra Kurapati*

## 2.1 INTRODUCTION

The discovery of graphene has led to the exploration of a large number of two-dimensional inorganic materials (2DMs), such as transition dichalcogenides, elemental 2D materials (Xenes), transition metal carbides ($M_2C$, M = Sc, Ti, V, Cr, Zr, Nb, Ta), nitrides ($M_2N$, M = Ti, Cr, Zr), and carbonitrides (MXenes, e.g., $Ti_3C_2T_x$) with surface functional groups including F, OH, and O groups (Kumar et al. 2024; Kurapati et al. 2016). In general, MXenes are represented by $M_{n+1}X_nT_x$, where $n$ = 1–4, M = transition metals (Ti, V, Ta, Nb, etc.), and X = C and/or N; $T_x$ is surface functional groups (–O, –OH, –F, –Cl, etc.) (Gogotsi and Anasori 2019). However, MXenes have been rapidly increasing and are part of a large family of 2DMs, which are mainly derived from the MAX phases with a general formula of $M_{n+1}AX_x$ where A represents a metal belonging to group 13 or 14, such as Al, Si, and Ga (He et al. 2022). Generally, 2D MXenes were obtained from the bulk-layered MAX phases using HF etching, electrochemical etching, water-free etching, fluoride salt etching, etc. (Naguib et al. 2011). In addition, various derivatives of MXenes or products of MXenes are usually obtained by oxidation, chlorination, and sulfuration (Tao et al. 2020). Some of the derivatives (e.g., $TiO_xN_y$) by oxidation of MXenes show better physicochemical properties than precursor MXenes, such as transition metal oxides and carbon, which are especially promising for energy storage applications (Tao et al. 2019; Tao et al. 2020). Due to the unique physicochemical properties of MXenes, numerous promising industrial and biomedical applications are increasingly reported (Chen et al. 2022b). MXenes are better than other inorganic 2DMs for biomedical applications concerning high photothermal-conversion ability and surface reactivity, including water dispersibility. Overall, this chapter will elaborately discuss the synthesis methods of MXenes from their respective MAX phases, MXenes carbides, MXenes nitrides, and MXenes carbonitrides, including various derivatives of MXenes.

DOI: 10.1201/9781003736318-2

## 2.2 MXENES SYNTHESIS METHODS

### 2.2.1 Top-Down Strategies

The top-down approach is used extensively for synthesizing the MXenes. Herein, bulk MAX phases are cleaved or etched to synthesize the MXenes. The defects found on the surface of the bulk material act as the initial sites for breaking down bulk materials into tiny particles (Jiang et al. 2022; Rahman et al. 2022). The commonly used top-down approaches for fabricating MXenes include different etching processes like acid (HF) treatment, electrochemical, water-free, and alkali etching methods. Intercalation of ions and molecules is also extensively used in producing MXenes. In this section, the top-down strategies for the MXenes synthesis are discussed in detail.

**HF etching**: Treatment with HF is one of the most commonly used methods. There are generally two ways for the etching process: i.e., directly treating with HF or in situ HF, as shown in Figure 2.1 (Alhabeb et al. 2017). HF acid displaces A layers from the MAX phase (where M stands for the transition metals, A represents the group III A or IVA elements, and X represents the C and N elements). In the $Ti_3AlC_2$-MAX phase, MXene (i.e., $Ti_3C_2T_x$) was obtained by extracting the Al from $Ti_3AlC_2$. To synthesize $Ti_3AlC_2$ powder, $Ti_3AlC$ and TiC powders were combined at a 1:1 molar ratio and subjected to ball milling for 24 hours using zirconia balls. The milled mixture was subsequently heated to 1350°C for 2 hours under an inert atmosphere. Post-heating, the resultant compound was manually crushed using a mortar and pestle and then treated with concentrated HF solution at room temperature for 2 hours. The suspension was then repeatedly washed with deionized water and centrifuged to isolate the powders. In some instances, to align the flakes and create free-standing disks, the treated powders were cold

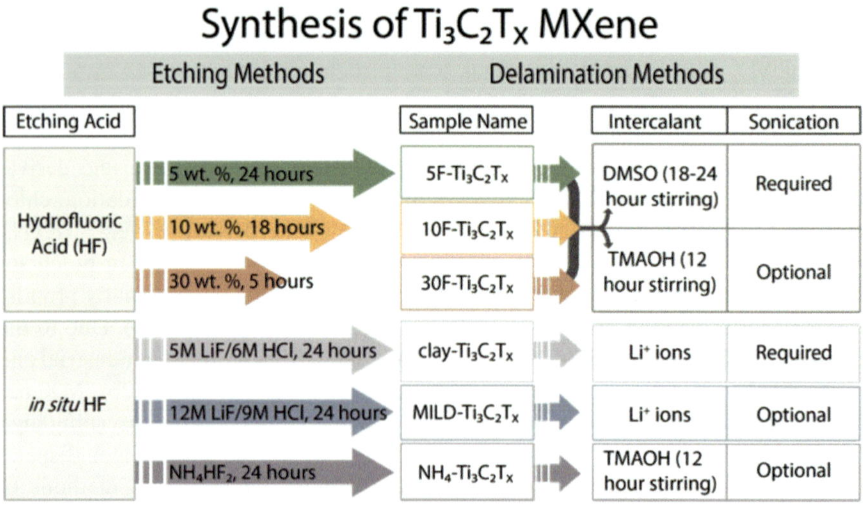


**FIGURE 2.1** General guidelines for synthesizing $Ti_3C_2T_x$ from different $Ti_3AlC_2$ MAX phases; this figure is adapted with permission from Alhabeb et al. (2017).

pressed at a pressure corresponding to 1 GPa in a steel die (Naguib et al. 2011). The etching process, duration, temperature, and concentration of fluoride ions are crucial for producing high-quality MXene layers. It is reported that HF treatment creates atomic vacancies in the obtained MXenes. However, HF etching often yields MXenes significantly, which is challenging to achieve with other acid solutions (Sang et al. 2016).

**Electrochemical etching**: The electrochemical-etching technique utilizes electrolytes to selectively remove a targeted metal or semiconductor by applying electrochemical principles. Liu et al. utilized carbon nanotubes (CNTs) and reduced graphene oxide (rGO) as carbon sources to prepare $Ti_2AlC$. This was followed by an in situ molten salt electrochemical-etching process to produce $Ti_2CT_x$ (MXene) (Liu et al. 2023).

The electrochemical-etching processes at the anode and cathode are given below:

At anode

$$3Cl^- + Ti_2AlC \text{ LiCl/KCl} \rightarrow Ti_2C + AlCl_3(g) + 3e^-$$

$$Ti_2C + xCl^- \text{ LiCl/KCl} \rightarrow E - Ti_2CCl_x + xe^-$$

At cathode

$$Li + K^+ + e^- \text{ LiCl/KCl} \rightarrow Li(K)$$

MXenes were also prepared in various aqueous electrolytes, such as LiOH/LiCl and HCl (Chen et al. 2022a; Sun et al. 2017). For example, Chen et al. synthesized F-free, Cl-containing $Ti_3C_2T_x$ in an aqueous solution at room temperature within 5 hours without using F-based reagents, as shown in Figure 2.2(A) (Chen et al. 2022a). A prior study reported that tetramethylammonium hydroxide (TMAOH) could etch $Ti_3AlC_2$ and delaminate $Ti_3C_2T_x$. However, an initial HF pretreatment was required, and the yield was relatively low (Xuan et al. 2016).

**Water-free etching**: It is a technique developed to overcome the limitations of using water as a solvent for water-sensitive applications of MXenes. Traditionally, water was used as a solvent in the acid etching to prepare MXenes, which selectively etch out only the A-atomic layers from the MAX phase and results in 2D MXenes. However, as shown in Figure 2.2(B), Michel et al. demonstrated the possibility of using solvents such as propylene carbonate (PC), acetonitrile (ACN), dioxane (DXN), N,N-dimethylformamide (DMF), dimethyl sulfoxide (DMSO), and N-methyl-2-pyrrolidone (NMP) along with $NH_4HF_2$ solution instead of water (Natu et al. 2020). The water-free etching method avoids hazardous HF and enhances the versatility of MXenes for various applications. Moreover, the organic solvent intercalation provides opportunities for further functionalization and use in advanced materials.

**Fluoride salt etching**: A combination of fluoride salts (such as NaF, LiF, KF, and $NH_4F$) and strong acids can also be used to etch MAX-phase precursors. The fluoride salts and acids selectively etch specific atoms, making it easier for cations ($K^+$, $Na^+$, $Li^+$, and $NH_4^+$) to intercalate in situ. $Ti_3AlC_2$ and $Ti_2AlC$ powders were immersed in four distinct etching solutions. These solutions were prepared by ultrasonically mixing specific fluoride salts (2.07 g LiF, 3.35 g NaF, 7.52 g KF·$2H_2O$, or 2.96 g $NH_4F$) with 20 mL of HCl solution (36–38 wt%) for 15 minutes. The varied masses of fluoride salts were calculated to obtain an etching solution of identical molar concentration of 4 mol/L. For the etching process, 0.5 g of $Ti_3AlC_2$ or $Ti_2AlC$ powder is immersed in 20 mL of the prepared solution, stirred for 1 minute, and then maintained for 12, 24, and 48 hours at different temperatures (30°C, 40°C, 50°C, and 60°C) with continuous magnetic stirring. Following etching, the samples underwent multiple washings with deionized water and diluted hydrochloric acid (2 mol/L), further washing with deionized water, and separation by centrifugation.

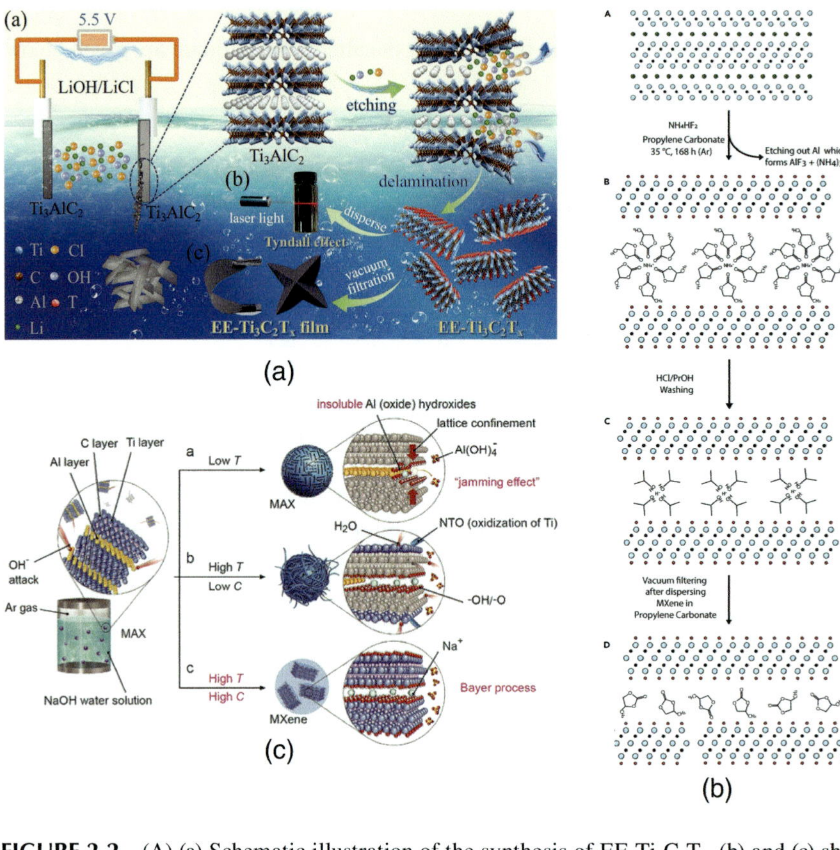


**FIGURE 2.2** (A) (a) Schematic illustration of the synthesis of EE-$Ti_3C_2T_x$. (b) and (c) show the photographs of EE-$Ti_3C_2T_x$ dispersion and films, respectively. This figure is adapted with permission from Chen et al. (2022). (B) Schematic of etching and washing steps: (a) starting $Ti_3AlC_2$ phase, (b) after etching with $NH_4HF_2$ in organic solvent, (c) after washing in HCl/2-propanol mixture, and (d) final state after filtering. This figure is adapted with permission from Natu et al. (2020). (C) $Ti_3AlC_2$ and NaOH water solution reaction under different conditions. (a) Al (oxide) hydroxides block the Al extraction at low temperatures. (b) Some Al (oxide) hydroxides dissolve in NaOH under high temperatures but low NaOH concentrations (c). This figure is adapted with permission from Li et al. (2018).

Finally, the samples were washed twice with ethanol until the supernatant reached a near-neutral pH. The resulting MXene powders were dried under a vacuum at 80°C (Liu et al. 2017).

**Alkali etching and other etching methods**: In an early attempt, MXenes were produced through alkali etching by immersing $Ti_3AlC_2$ MAX in a diluted NaOH solution at 80°C, as reported by Xie et al. In summary, $Ti_3AlC_2$ powder was soaked in NaOH and agitated for 100 hours at 80°C to eliminate the surface Al atoms between the unique $Ti_3C_2$ layers of the bulk $Ti_3AlC_2$. However, due to the low concentration of NaOH, only the outermost layer of MAX could be etched (Li et al. 2018b).

There are various alkali-assisted etching methods to obtain $Ti_3C_2T_x$ from $Ti_3AlC_2$. The pristine $Ti_3AlC_2$ powders are subjected to hydrothermal treatment using NaOH aqueous solutions. Various hydrothermal temperatures and NaOH concentrations were utilized. These treatments are performed under an argon atmosphere to minimize oxidation of the samples—one specific treatment involved using a 27.5 M NaOH aqueous solution at 270°C. Figure 2.2(C) shows that the formation of oxide/hydroxide layers like $Al_2O_3$ on the $Ti_3AlC_2$ surface hinders the production of MXenes via alkali attack. Even after removing this oxide layer with a 10% HF water solution, the efficiency in producing $Ti_3C_2T_x$ through repeated treatments with sodium hydroxide (NaOH) remained very poor. This suggested that $Al(OH)_3$ and $Al_2O_3$ might further hinder the desired Al extraction (Li et al. 2018b).

**Choice of intercalation method**: Intercalation refers to inserting ions or molecules between the layers of materials with weak bonds, such as graphite and transition metal dichalcogenides (TMDs) like $MoS_2$. Since MXenes have weak bonding between the MX layers, they can accommodate various ionic, organic, and inorganic species, which intercalate between the layers, weakening the interlayer bonds and increasing the interlayer spacing.

**Ionic intercalation**: MXenes can accommodate a wide range of multivalent cations, including $Na^+$, $Li^+$, $K^+$, $Mg^{2+}$, $NH_4^+$, and $Al^{3+}$. This capability is advantageous for applications in metal-ion batteries and supercapacitors. Figure 2.3(A) illustrates the intercalation of cations in $Ti_3C_2T_x$, one of the most studied MXenes (Lukatskaya et al. 2013).

**Organic molecule intercalation**: Polar organic molecules of varying sizes, such as tetrabutylammonium hydroxide (TBAOH), urea, hydrazine, isopropanol, and dimethyl sulfoxide (DMSO), can also be intercalated into MXenes to increase the interlayer spacing further (Salim et al. 2019). For example, when $Ti_3C_2T_x$ was intercalated with hydrazine, the lattice parameter increased from 19.5 to 25 Å. This intercalation was crucial for producing 'MXene paper,' which involved sonication in water and filtration of the resultant colloidal solution.

### 2.2.2 Characterization of MXenes Synthesized Using the Top-Down Approach

The characterization of MXenes synthesized through top-down strategies is important to understand their structural, morphological, and electrochemical properties. Techniques such as X-ray diffraction (XRD) were commonly used to study the structural changes in MXenes, which revealed the *c*-axis expansion due to the removal of

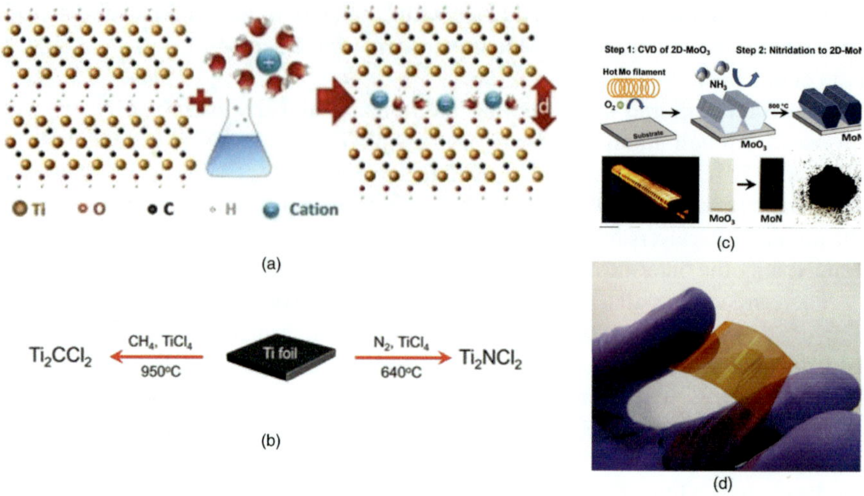


**FIGURE 2.3** (A) Intercalation of cations in $Ti_3C_2T_x$ MXene. This figure is adapted with permission from Lukatskaya et al. (2013). (B) Chemical vapor deposition to synthesize MXenes. This figure is adapted with permission from Wang et al. (2023). (C) Two-step process for creating 2D δ-MoN layers. Step 1: 2D $MoO_3$ layers are synthesized using hot-filament chemical vapor deposition, where oxygen reacts with a heated molybdenum filament. Step 2: Post-synthesis nitridation converts 2D $MoO_3$ to 2D MoN by exposing it to ammonia at 800°C. This figure is adapted with permission from Joshi et al. (2017). (D) PLD $MoS_2$ film (thickness ~30 nm) deposited on flexible plastic Kapton substrate. This figure is adapted with permission from Late et al. (2014).

Al layers and surface functionalization (–OH, –F, –O) (Naguib et al. 2012). Scanning electron microscopy (SEM) provided high-resolution images to examine the morphology of etched powders, and transmission electron microscopy (TEM) enabled visualization of exfoliated 2D flakes at atomic resolution. The surface composition analysis was conducted using X-ray photoelectron spectroscopy (XPS), which provided detailed information on the chemical states and termination groups introduced during the etching process. Zeta potential measurements were used to assess the colloidal stability in solvents, and contact angle measurements to quantify hydrophilicity changes (Natu et al. 2020). Electrical properties, such as sheet resistance, were evaluated using a four-probe technique (Naguib et al. 2012), while electrochemical characterization including cyclic voltammetry (CV) (Lukatskaya et al. 2013), galvanostatic cycling (Mashtalir et al. 2013), and electrochemical impedance spectroscopy provided an insight into charge storage behavior and ion intercalation kinetics. The techniques like Brunauer–Emmett–Teller (BET) for surface area analysis were used to measure adsorption properties, particularly for MXenes prepared via fluoride salt etching or alkali etching (Li et al. 2018b).

### 2.2.3 Bottom-Up Strategies

While the top-down approach involves the breakdown of bulk materials as precursors, the bottom-up approach uses small molecules as precursors. Bottom-up synthesis methods are emerging as promising alternatives as they offer the potential for high-quality, large-area films and the ability to create compositions not achievable through top-down techniques (Jiang et al. 2022; Rahman et al. 2022). This section discusses some of the most commonly used bottom-up strategies, such as chemical vapor deposition, plasma-enhanced pulsed laser deposition, and the template method.

**Chemical vapor deposition (CVD)**: CVD is a widely utilized technique in material science for depositing solid materials from a vapor phase. This process involves a chemical reaction on or near a heated substrate surface, forming a solid material. The substrate is typically maintained at an elevated temperature to facilitate the desired chemical reactions and promote the deposition of the material.

MXenes are synthesized using a CVD process on a titanium (Ti) surface, employing a gas mixture of $CH_4$ and $TiCl_4$ diluted in argon (Ar), as shown in Figure 2.3(B). The Ti foil was exposed to the $TiCl_4$ and $CH_4$ vapor at 950°C for 15 minutes. Similarly, $Zr_2CCl_2$ and $Zr_2CBr_2$ MXenes can be synthesized by exposing a Zr foil to $CH_4$ and either $ZrCl_4$ or $ZrBr_4$ vapor at 975°C. Figure 2.3(B) shows the production of nitride MXene $Ti_2NCl_2$ by reacting Ti foil with $TiCl_4$ and $N_2$ at temperatures exceeding 640°C (Wang et al. 2023). The CVD method is not always preferred because it typically produces very thin films instead of single layers. The CVD method enables the creation of MXenes with more accessible surfaces and exposed catalytically active edges, which are beneficial for applications like supercapacitors.

### 2.2.4 Template Method

Template method uses higher-dimensional templates (2D or 3D) to direct the growth of 2D materials, preventing growth in the *z*-direction (Tian, Wei, and Sun 2020). This method offers precise control over the size, shape, and configuration of materials, making it advantageous for applications like aqueous supercapacitors, zinc ion batteries, and zinc air batteries. Standard 2D templates include graphene and derivatives, which are widely used due to their high conductivity and mechanical robustness. The TMDs also provide confined spaces but are challenging to remove after synthesis. Metal-organic frameworks (MOFs) offer periodic structures and large surface areas, ideal for synthesizing various 2D nanosheets and layered double hydroxides, including metal oxides.

The 3D templates involve salt crystals, which facilitate the growth of various 2D materials through confined surface growth. Metal foams are used for their high catalytic activity and mechanical stability, though they may leave residuals. Another template involves liquid metal alloys, unique for their "self-limiting native-oxide skin," helpful in synthesizing atomically thin materials. This method's key challenges include template removal and maintaining material integrity post-synthesis (Tian, Wei, and Sun 2020). Joshi et al. synthesized 2D hexagonal molybdenum nitride (MoN) nanosheets using a template method (Figure 2.3(C)). Initially, 2D $MoO_3$ nanosheets were vertically grown on a substrate by reacting a hot Mo filament with $O_2$ (Joshi et al. 2017).

Subsequently, these $MoO_3$ nanosheets were transformed into 2D MoN nanosheets by heating them in $NH_3$ at 800°C. The MoN nanosheets retained the same morphology as the original $MoO_3$ nanosheets, standing vertically on the substrate.

### 2.2.5 Plasma-Enhanced Pulsed Laser Deposition (PEPLD)

Plasma-enhanced deposition grows thin films or coatings on substrates. It combines the advantages of two existing methods: plasma-enhanced chemical vapor deposition (PECVD) and pulsed laser deposition (PLD). In conventional PECVD systems, plasma is generated by ionizing gas molecules (such as methane) using an electric field. The active species in the plasma facilitate chemical reactions at lower temperatures than traditional thermal CVD. PLD is a method for depositing thin films using high-energy laser pulses to evaporate material from a solid target in a vacuum chamber. The vaporized material then condenses onto a substrate, creating a thin film several micrometers thick (Liu 2010). Figure 2.3(D) gives an example wherein the PLD method was used to grow $MoS_2$ films on flexible Kapton substrate to evaluate photo-response properties (Late et al. 2014).

In PEPLD, a pure metal target is ablated by a (radio frequency) RF inductively coupled plasma, which produces reactive oxygen species. These species interact with the laser-generated plasma, forming oxides that deposit onto a substrate. PEPLD was used in a recent study to grow large-area single-crystalline $Mo_2C$ films of different thicknesses on sapphire (0001) (Zhang et al. 2017). By incorporating a high-voltage electrode at the methane intake of the chamber, it was possible to modify a standard PLD setup and enable ionization of the gas upon entry into the PEPLD system. A power generator supplies a 500 V direct current voltage. Methane is pumped into the chamber at a pressure of 15 Pa before deposition, and glow discharge is used to ionize the methane and create plasma. A molybdenum (Mo) metal target was exposed to a 248 nm KrF laser (Lambda Physik COMPex 205). The pulse frequency was 5 Hz, and the energy density was roughly 5 J $mm^2$. For film deposition, the sapphire (0001) substrate was placed 3.5 cm away from the target and heated to 700°C. The sample was then allowed to cool down to room temperature.

PEPLD can be applied to various materials, including metals, oxides, and 2D materials. It offers a way to grow large-area films with better crystallinity and fewer defects. Researchers use PEPLD to explore novel materials for electronic devices, energy storage, and catalysis.

### 2.2.6 Characterization of MXenes Synthesized Using Bottom-Up Approach

The characterization of MXenes synthesized using bottom-up strategies, i.e., mainly CVD, template method, and PEPLD, involves different optical, microscopic, and spectroscopic techniques to analyze their structural, chemical, and physical properties. The morphology of the thin films and ultrathin crystals of MXene was analyzed mainly using optical microscopy, TEM, and SEM analyses. HR-TEM along with selected area electron diffraction (SAED) measurements also gave an insight into the chemical composition and complete crystalline structure of MXene (Geng et al. 2017; Xu et al. 2015). The thickness and roughness of the MXene sheets and films were analyzed using atomic force microscopy (AFM). Moreover, the Raman spectroscopy was also used in some cases to understand the chemical bonding and composition of MXene. Since these MXenes were largely used in electronic and energy storage applications, scanning tunneling microscopy and spectroscopy (STM and STS) (Joshi et al. 2017). were used to analyze the n-type character and *I–V* characteristic of the thin films on substrates (Late et al. 2014). Finally, XPS is widely used to understand the elemental composition, and the oxidation state of elements of the MXene films and crystals (Geng et al. 2017). These characterization methods are essential to understand the unique properties of MXene and optimizing their applications in energy storage, filtration, biosensing, and other fields.

## 2.3 MXENES DERIVATIVES

### 2.3.1 Completely Oxidized MXenes (MXene-Derived $TiO_2$, MXene-Derived Titanates, and Other MXene-Derived Transition Metal Compounds)

The primary method for preparing completely oxidized MXenes involves an oxidation reaction. In this process, the MXene is treated with potent oxidizing agents, such as hydrogen peroxide ($H_2O_2$) or nitric acid ($HNO_3$), depending on the extent of oxidation. Interestingly, the composition of the oxidation products depends on the degree of oxidation. For instance, as the name suggests, partial oxidation results in partially oxidized MXenes with $TiO_2$ and $Ti_3C_2$ residues, while complete oxidation leads to fully oxidized MXenes with $TiO_2$ and C hybrids. Herein, in this section, some of the techniques to completely oxidize the MXene are discussed. Naguib et al. found that $TiO_2$/C composites with a layered structure can be produced by oxidizing $Ti_3C_2$ in a $CO_2$ environment or using pressurized water. Similarly, Yuan et al. introduced a simple one-step $CO_2$ oxidation technique to fabricate 2D $TiO_2$/C hybrids from $Ti_3C_2$ (He et al. 2020). Figure 2.4(A) shows the schematic for synthesizing the Ti-peroxo complex derived from MXenes ($Ti_3C_2$) and further conjugation with reduced graphene oxide composites (M-$TiO_2$–RGO). MXene–$Ti_3C_2$ was prepared by HF etching of $Ti_3AlC_2$ at room temperature. The concentrated $H_2O_2$ (30 wt%) was mixed with $Ti_3C_2$ MXenes powder, resulting in a brownish-yellow mixture, releasing a large amount of gas and heat. After the reaction, the mixture turned to a transparent yellow solution, which indicated the formation of the Ti-peroxo complex ion (Tang et al. 2016). Figure 2.4(B) shows the MXene-derived $TiO_2$ nanoparticles/RGO composite as the anode material for lithium-ion and potassium-ion batteries (PIBs), which were fabricated via a one-step hydrothermal method by Fang et al (Fang et al. 2019). However, Tang et al. synthesized layered a-$TiO_2$ (anatase titania) using a two-stage oxidation process starting from HF-etched $Ti_3C_2T_x$, aimed at lithium-ion battery (LIB) applications as depicted in Figure 2.4(C). As compared to directly calcining MXene in air, their approach involved hydrothermal treatment followed by calcination. This method effectively reduced abnormal grain growth, resulting in a 2D morphology of a-$TiO_2$ nanoparticles. The presence of carbon during hydrothermal oxidation of MXene impeded mass diffusion across crystal boundaries during calcination, contributing to this beneficial outcome (Wang et al. 2018b).

### 2.3.2 Deeply Oxidized MXenes

Among the MXene derivatives, the transition metal oxides are the most common. However, low electronic conductivity makes them less applicable in the energy and storage field. Hence, the carbon material composites are used to improve their performance. Since MXene (carbide) has carbon in its composition, the controlled oxidation of MXene will result in the MXene-derived oxide/C composite, which enhances the electronic properties. There are several methods to control the degree of oxidation of MXene.

Flash oxidation is a commonly used technique to partially oxidize the MXene, wherein the MXene powders are annealed at a very high temperature around 1000°C for 30–60 seconds in the presence of air. In 2014, Naguib and co-workers used this method to synthesize $TiO_2$ nanocrystals entrapped in disordered graphitic carbon sheets (Naguib et al. 2014). Briefly, the multilayer MXene $Ti_3C_2T_x$ sheets were placed on the alumina holders. They were subjected to the furnace's hot area at around 1150°C for 30 seconds and immediately pushed out of the stove, resulting in $TiO_2$–C hybrid structures. They also proved the formation of distorted graphitic C using Raman spectroscopy and TEM analyses. They also demonstrated that these hybrid structures showed improved electrochemical properties compared to the anatase $TiO_2$. Similarly, Gao et al. also reported the flash oxidation of $Ti_3C_2T_x$ at different furnace temperatures to understand the effect of flash temperature on oxidation (Gao et al. 2015). Different temperatures from 200°C to 1000°C were used for 60 seconds in an oxygen atmosphere for oxidation. However, the MXene-derived oxide/C composites were formed at 280°C–400°C, which was confirmed using Raman spectroscopy and XRD. The Raman spectroscopic analyses showed a gradual increase in the D and G bands of the carbon material; however, at very high temperatures, these bands were missing, which indicated the complete oxidation of the MXene into the rutile phase of titanium oxide. Although this oxidation technique can synthesize MXene-derived oxide/C material, flash oxidation was not a controllable strategy as it involves a concise span of synthesis time.

The annealing technique was performed at comparatively low temperatures to overcome this hurdle. The hydrothermal technique to partially oxidize the MXene at temperatures around 800°C yielded lower amounts of amorphous carbon. Hence, to conserve the carbon content during the hydrothermal process, $CO_2$ was used in the calcination atmosphere. Yuan et al. reported the oxidation of $Ti_3C_2T_x$ MXene using $CO_2$ at different temperatures like 700°C, 800°C, and 900°C for 1 hour (Yuan et al. 2017). The hybrid structure of C/$TiO_2$ was obtained efficiently at a temperature of 700°C, which showed higher photocatalytic activity. At a temperature of 700°C, the MXene started to oxidize and convert $Ti_3C_2$ and $CO_2$ to C and $TiO_2$. At higher temperatures, C was oxidized in the presence of $CO_2$ to get CO. Similarly, Byeon et al. also reported the calcination of another MXene $Mo_2CT_x$ wherein the oxidation using $CO_2$ was used to acquire $MoO_2$/C composite, which was used for lithium storage (Byeon et al. 2017). In this case, the $Mo_2CT_x$ underwent oxidation using $CO_2$ at 500°C for 1 hour and 850°C for 30 min. In both cases, the oxidation of the MXene was confirmed using various characterization techniques like XRD and Raman spectroscopy. However, higher oxidation and more extensive amorphous C content were found at higher temperatures of 850°C.

Different methods like solvent thermal decomposition, microwave irradiation, and high-energy ball milling were exploited to produce MXene-derived oxide/C composites. Zou and co-workers developed the urchin-like rutile phase $TiO_2$/C nanocomposites using the in situ conversion of MXene to its deeply oxidized derivative. The mixture of MXene and $FeCl_3$ was made using isopropanol and ethylene glycol as solutions. This mixture was sonicated and stirred at room temperature, then heated to a high temperature of 180°C–300°C using Teflon autoclaves in an electric blast drying oven. The resulting precipitate was collected and washed with ethyl alcohol to remove the unreacted solution. The precipitate was then dried to obtain the urchin-shaped $TiO_2$/C nanocomposites (Zou et al. 2016). Following this work, Zou and co-workers also used solvent thermal decomposition to synthesize the $TiO_2$/C composite, which had a flower-shaped morphology. For the first time, alcoholysis of MXene was performed to synthesize the MXene derivative flower-shaped $TiO_2$/C. The composite was synthesized similarly to urchin-shaped rutile $TiO_2$/C composites. These derivatives were characterized using different microscopic and spectroscopic techniques to confirm the flower morphology and formation of the $TiO_2$/C MXene derivative (Zou et al. 2017). These complexes also showed excellent catalytic activity for the dehydrogenation of $NaAlH_4$. Another common oxidation process was using microwave irradiation. Briefly, conductive acetylene powder along with $Ti_3C_2T_x$ powder was ground and then heated using microwave irradiation for around 20 s at 800 W. The resulting powder of $TiO_2$/C nanocomposite was further characterized using TEM, XRD, and Raman spectroscopy. The unique structure of $TiO_2$/C composite can be used as an electrode material to increase ion diffusion and electron passage (Wang et al. 2018a). Later, MXene-derived oxides were used as substrates to grow carbon nanotubes (CNTs), which were used to test the efficiency of anodes in LIBs. The MXene and ferrocene were first mixed using acetone; then, the mixture was shifted to a crucible where carbon fibers were added to obtain CNTs on the surface of MXene nanosheets. This composite was irradiated with a microwave at 900 W for 40 s. This irradiation cycle was repeated three times and the final product of the hybrid structure with $TiO_2$/CNT/Fe was synthesized (Zheng et al. 2018). Li et al. and co-workers used the high-energy ball milling strategy using agate jars and agate balls. Different sizes of agate balls were filled in the agate jar and $Ti_3C_2T_x$ powder was added to it. The powder was milled in the atmosphere for 1.5 hours, resulting in the formation of $TiO_2$@C. Thus, the powder was cold pressed and further characterized using XRD, XPS and TEM (Li et al. 2018a).

### 2.3.3 Surface-Oxidized MXenes

MXenes, two-dimensional transition metal carbides with metallic conductivity, are promising for high-performance composite electrodes. Incorporating conductive carbon can mitigate the low electronic conductivity of MXene-derived transition metal oxides. The metallic conductivity of MXenes can surpass that of carbon, improving the conductivity of composite materials (Xu et al. 2021; Zhang et al. 2020; He et al. 2020).

The synthesis of composites that combine transition metal oxides with MXenes can be achieved through self-assembly or in situ growth. The self-assembly of metallic

$Ti_3C_2T_x$ MXene flakes with positively charged carbon nanotubes (CNTs) results in a unique 2D/1D composite material that forms a highly efficient porous structure (Wang et al. 2020). In situ technique facilitates the direct formation of metal oxides on MXene surfaces, improving their structural stability and electrochemical performance (Gogotsi and Huang 2021). A more interesting approach involves the partial oxidation of MXenes. This process allows for forming surface-oxidized MXenes, which retain a portion of their original structure while generating transition metal oxide derivatives on their surfaces.

### 2.3.4 Titanium-based MXene Composites with $TiO_2$

$TiO_2/Ti_2C$ composite synthesis involves a mild oxidation process conducted at room temperature using hydrogen peroxide ($H_2O_2$) as the oxidizing agent. This method allows for controlled oxidation of HF-etched $Ti_2C$ MXenes, where the concentration of $H_2O_2$ and the reaction time are critical parameters that influence the extent of oxidation and the resulting composite properties. The equation can represent the reaction:

$$a\mathrm{Ti_2CT}_x + \mathrm{H_2O_2} \rightarrow b\mathrm{TiO_2} + (1-b)\mathrm{Ti_2CT}_x + \mathrm{CO}_y$$

In the above equation, $Ti_2C$ is partially oxidized to form $TiO_2$ while retaining a portion of the original MXene structure. Using $H_2O_2$ is advantageous due to its highly effective oxygen content and environmentally safe properties, producing only water as a byproduct during oxidation (Hulea, Dumitriu, and Fajula 2021). The mild oxidation not only facilitates the formation of $TiO_2$ nanocrystals on the MXene surface but also enhances the structural stability and electrochemical performance of the composite material (Pogorielov et al. 2021). By optimizing the reaction conditions, they tailored the morphology and composition of the $TiO_2/Ti_2C$ composites, which are beneficial for applications in energy storage, such as LIBs (Ahmed et al. 2016). Compared to pristine $Ti_2C$ MXene, the $TiO_2/Ti_2C$ hybrid material exhibited a five-fold increase in specific surface area (Katarzyna and Teofil 2017; Alli et al. 2022). This significant enhancement was attributed to the opened layer spacing resulting from the growth of MXene-derived $TiO_2$ nanoparticles on the surface (Katarzyna and Teofil 2017; Pogorielov et al. 2021). The presence of $TiO_2$ nanoparticles helped maintain the separation between individual MXene sheets, preventing their restacking and preserving the high surface area (Alli et al. 2022). The improved $Li^+$ accessibility to the MXene surface, facilitated by the porous structure of the $TiO_2/Ti_2C$ hybrid, leads to significantly enhanced lithium-ion storage performance (Kumar et al. 2022; Alli et al. 2022). Notably, the hybrid material demonstrated outstanding power handling at high current densities, outperforming the pristine MXene. This improvement was attributed to the synergistic effect of the MXene and its derived $TiO_2$ nanoparticles, which provided efficient charge transport pathways and increased electrochemically active sites for $Li^+$ intercalation. In partial oxidation of $Ti_3C_2$ MXenes at room temperature, $Ti_3C_2$ MXenes could undergo partial oxidation at room temperature without the need for an oxidant. By stirring $Ti_3C_2$ in deionized water and alkaline solutions, different morphologies of $TiO_2$ were synthesized, specifically $Ti_3C_2/TiO_2$-nanoparticles and

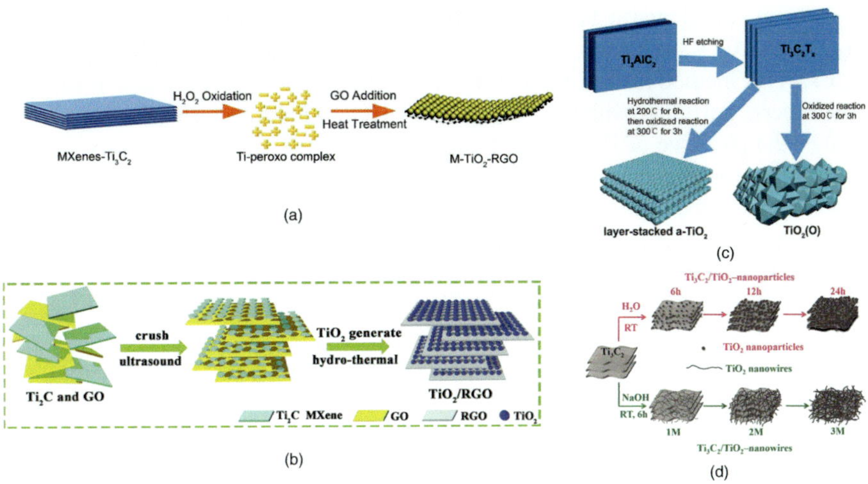


**FIGURE 2.4** (A) Oxidation of $Ti_3C_2$ MXenes and preparation of an MXene-derived $TiO_2$–graphene (M-$TiO_2$–RGO) composite. Adapted with permission from Wang et al. (2018). (B) Preparation of a $Ti_2C$-derived $TiO_2$/RGO composite. Adapted with permission from Fang et al. (2019). (C) Two different oxidation processes of $TiO_2$ from $Ti_3C_2T_x$ MXenes. Adapted with permission from Tang et al. (2016). (D) Schematic illustration of the preparation processes of $TiO_2/Ti_3C_2$ composites at room temperature. This figure is reproduced with permission from Cao et al. (2017).

$Ti_3C_2/TiO_2$-nanowires as depicted in Figure 2.4(D). The alkalinity of the reaction environment plays a crucial role in determining the morphology of the resulting $TiO_2$ (Cao et al. 2017).

Controlling the oxidation of MXenes through heat treatment can significantly enhance their properties, particularly via methods such as direct annealing and hydrothermal reactions (Wang et al. 2021; Zaharin et al. 2023; Rahman et al. 2022). The degree of oxidation can be precisely managed by carefully adjusting hydrothermal conditions, temperature, and duration. For instance, Yang et al. developed hierarchical accordion-like $TiO_2/Ti_3C_2$ nanohybrids for use as anodes in lithium-ion (LIBs) and sodium-ion batteries (SIBs) by partially oxidizing HF-etched $Ti_3C_2$ MXenes through a hydrothermal process. This method resulted in uniformly distributed nanoscale $TiO_2$ particles on the $Ti_3C_2$ surface, increasing interlayer spacing and more active sites for enhanced reversible capacity. The high electronic conductivity of $Ti_3C_2$ further facilitated rapid electron transfer (Rahman et al. 2022). The in situ derivation strategy using hydrothermal reactions can also be expanded to incorporate other active components, such as metal sulfides, metal oxides, and conducting polymers (Wang et al. 2020; He et al. 2020; Rahman et al. 2022). For example, by mixing ammonium heptamolybdate and nickel nitrate with MXenes, $MoS_2$ and NiO can be synthesized directly on the MXene surface while simultaneously undergoing partial oxidation during the hydrothermal process. Additionally, the oxidative polymerization of aniline under hydrothermal conditions has been utilized to create a hierarchical $TiO_2/Ti_3C_2T_x$ ternary composite, which is beneficial for high-performance electrochemical capacitors (Rahman et al. 2022) (He et al. 2020).

Li et al. developed a novel annealing strategy for synthesizing sandwich-like $Ti_3C_2/TiO_2$/rGO composites, which served as anodes in lithium-ion hybrid capacitors (LICs). This method involved calcining $Ti_3C_2T_x$/GO at 550°C for 2 hours in an argon atmosphere, partially oxidizing $Ti_3C_2T_x$ into $TiO_2$. This transformation was attributed to the presence of $H_2O$ in the interlayer of the MXene and oxygen-containing functional groups on the MXenes themselves (Li et al. 2024). The annealing process facilitated the in situ formation of both reduced graphene oxide (rGO) and $TiO_2$ nanoparticles between the $Ti_3C_2$ layers. This structural arrangement helped to prevent the restacking of MXene flakes, which is crucial as it increases the interlayer distance, thereby providing more active sites for lithium-ion diffusion and enhancing electron transfer kinetics (Kannan et al. 2020; Guo et al. 2023).

MXene composites, particularly $TiO_2$–$Ti_3C_2$ heterostructures, have shown significant potential as sulfur hosts in lithium–sulfur (Li–S) batteries. The synthesis of these heterostructures involved a fast oxidation treatment of accordion-like $Ti_3C_2$ MXenes by exposing them to air at controlled temperatures (200°C, 500°C, and 800°C). This process facilitates the growth of the new oxide phase $TiO_2$ on both the edges and surfaces of the MXene nanosheets, enhancing their polysulfide adsorption capabilities and increasing the interlayer spacing of MXenes. The remaining $Ti_3C_2$ nanosheets contribute to improved electronic conductivity within the composite materials (Rao et al. 2024; Lim and Kay 2022). In addition to the oxidation treatment, another $TiO_2$–MXene heterostructure was synthesized through a controlled hydrothermal method. Here, a few-layered $Ti_3C_2T_x$ MXenes were etched using LiF and HCl and treated at 180°C for varying durations (0, 2, 4, 6, and 8 hours). This hydrothermal oxidation allowed for controlled partial oxidation of $Ti_3C_2T_x$ nanosheets. After an 8-hour treatment, the (0002) peak of $Ti_3C_2T_x$ almost disappeared, indicating complete oxidation. During this process, $TiO_2$ nanoparticles were in situ formed on the surface of the MXenes. At the same time, most groups on the MXenes were replaced by –O terminations, enhancing their catalytic activity for lithium polysulfides (LiPSs) (Jiang et al. 2020). The resulting $TiO_2$–MXene heterostructures exhibit several advantages, such as a large specific surface area, strong polysulfide adsorption ability, excellent polysulfide conversion activity, and high electronic conductivity. These properties enable their combination with graphene to create an interlayer on lithium–sulfur battery separators.

### 2.3.5 Titanate/MXene Hybrid Structures Derived from Titanium-Based MXenes

MXene-derived titanates were synthesized by combining simultaneous oxidation and alkalization, allowing for the fabrication of titanate/MXene composites with controlled properties. Huang et al. developed sandwich-like $Na_{0.23}TiO_2/Ti_3C_2$ composites specifically for LIBs and SIBs via a one-step in situ transformation method (Sobolev et al. 2024). In this process, 2D $Ti_3C_2$nanosheets were partially oxidized using dissolved $O_2$ in a NaOH solution at room temperature. This reaction led to the growth of 1D ultrathin $Na_{0.23}TiO_2$ nanobelts on the $Ti_3C_2$ sheets, resulting in a three-dimensional sandwich-like structure. The transformation reaction can be represented as follows:

$$Ti_3C_2 + 4.69NaOH + 4.8275O_2 \rightarrow 3Na_{0.23}TiO_2 + 2Na_2CO_3 + 2.345H_2O$$

This $Na_{0.23}TiO_2/Ti_3C_2$ electrode exhibited remarkable cycling stability, maintaining nearly 100% capacity retention after 4000 cycles at a current density of 2 A/g for SIBs. The excellent performance is attributed to the well-defined sandwich architecture, where the amorphous nanobelts increase the interlayer spacing of $Ti_3C_2$ MXenes and shorten the ion diffusion pathway. Additionally, the $Ti_3C_2$nanosheets enhance electronic conductivity and facilitate electron transfer processes. Wang et al. further advanced this field by developing a layer-stacked $Li_4Ti_5O_{12}$–MXene composite using an in situ treatment with $H_2O_2$ and LiOH, followed by a thermal process (Zhang et al. 2022). In this method, $Li_4Ti_5O_{12}$ nanocrystals were derived from MXene layers, improving electrolyte accessibility and shortening diffusion pathways for $Li^+$ ions and electrons. The resultant $Li_4Ti_5O_{12}$–MXene composite achieved a 178 mA h/g capacity at a current density of 5 A/g after 500 cycles for LIBs. These studies highlight the potential of MXene-derived titanate composites in enhancing the performance of energy storage devices through tailored structural designs and optimized synthesis methods.

### 2.3.6 Other Partially Oxidized MXenes

Ti-based MXene derivatives have garnered significant attention in research. However, the derivatives from other transition metal-based MXenes are still emerging as promising candidates for energy storage applications. For example, Zhang et al. achieved the synthesis of orthorhombic $Nb_2O_5@Nb_4C_3$ hierarchical layered composites through one-step oxidation of $Nb_4C_3T_x$ MXenes in a $CO_2$ environment at 850°C (Estrada-Ruiz et al. 2024). This oxidation process led to the formation of amorphous carbon in conjunction with $Nb_2O_5$, represented by the following reaction:

$$Nb_4C_3T_x + 5CO_2 \rightarrow 2Nb_2O_5 + 8C$$

As an anode material in LIBs, the $Nb_2O_5@Nb_4C_3$ electrode demonstrated a capacity of 208 mA h/g at a current density of 50 mA/g, along with commendable rate performance, retaining 60% capacity at 1000 mA/g. This impressive performance is attributed to the synergistic effects arising from its hierarchical layered structure (Jin et al. 2020). The $Nb_2O_5$/amorphous carbon/ $Nb_4C_3$ architecture was formed following the $CO_2$ oxidation of the MXene layers. The $Nb_2O_5$ nanoparticles, which possess high capacity, are interconnected with the highly conductive $Nb_4C_3$ through disordered carbon that serves as an electron "bridge." Additionally, similar hierarchical structures have been obtained by oxidizing $Nb_2CT_x$ and $Ti_3C_2T_x$ using the same method (Zhang et al. 2016).

### 2.3.7 Characterization of MXene Derivatives

To understand the properties of MXenes derivatives, i.e., completely oxidized, deeply oxidized, and surface-oxidized MXene, a variety of characterization techniques were employed, such as XRD, which is crucial for confirming the transformation from MAX phase to MXene by analyzing the crystalline phase evolution during oxidation (Natu et al. 2021). XPS is also essential for analyzing surface functional groups and chemical bonding states, providing insights into the elemental composition and their oxidation states that influence material properties (Zhang et al. 2023). Raman spectroscopy was exploited to understand the information about surface species, and regarding the number of layers while also detecting the formation of anatase $TiO_2$ in distinct applications (Adomaviciute-Grabusove et al. 2024). Raman spectroscopy also helped to analyze the structural changes and defects caused by the oxidation of MXene (Wu et al. 2022). Electron microscopy techniques like SEM–EDX and HR-TEM examine the morphology, atomic structure, and crystallinity of the MXene derivatives. Thermogravimetric analysis (TGA) and UV–Vis spectroscopy were used to assess thermal stability and optical properties, respectively (He et al. 2020). The characterization of MXenes and their derivatives gave in-depth knowledge regarding the oxidation level and changes in the structure and electronic properties, which are essential for designing their applications.

## 2.4 CONCLUSION

MXenes are a rapidly growing family of 2DMs that have emerged as an excellent material for advanced applications due to their unique structural and chemical properties like hydrophilicity and metallic conductivity. The synthesis and functionalization of MXenes define their properties, enabling them for various applications, including energy storage, electronics, and biomedicine. In this chapter, we mainly focus on different methods of synthesis and characterization of MXenes. MXenes were typically synthesized using two approaches, namely the top-down approach and the bottom-up approach. The commonly used top-down approaches are HF etching, electrochemical etching, water-free etching, and alkali etching. In this approach, generally, the bulk MAX phases are etched and thin 2D MXenes are produced by introducing defects in the bulk material. However, in the bottom-up approach, the MXenes are produced from smaller molecules. This approach mainly consists of CVD, PEPLD, and the template methods which are ideal for preparing ultrathin films and crystals of MXenes. Moreover, this chapter also summarizes synthetic procedures for some of the MXene derivatives like completely, deeply, and surface-oxidized MXenes. It is essential to understand the synthesis and characterization of MXenes and their derivatives, which are used for different electronic, energy, and biomedical applications.

## REFERENCES

Adomaviciute-Grabusove Sonata ., Anton Popov, Simonas Ramanavicius, Valdas Sablinskas, Kateryna Shevchuk, Oleksiy Gogotsi, Ivan Baginskiy, Yury Gogotsi, and Arunas Ramanavicius. 2024. "Monitoring Ti(3)C(2)T(x) MXene degradation pathways using Raman spectroscopy." *ACS Nano* 18 (20):13184–13195. doi: 10.1021/acsnano.4c02150.

Ahmed, Bilal, Dalaver H. Anjum, Mohamed N. Hedhili, Yury Gogotsi, and Husam N. Alshareef. 2016. "$H_2O_2$ assisted room temperature oxidation of $Ti_2C$ MXene for Li-ion battery anodes." *Nanoscale* 8 (14):7580–7587. doi: 10.1039/C6NR00002A.

Alhabeb, Mohamed, Kathleen Maleski, Babak Anasori, Pavel Lelyukh, Leah Clark, Saleesha Sin, and Yury Gogotsi. 2017. "Guidelines for synthesis and processing of two-dimensional titanium carbide ($Ti_3C_2T_x$ MXene)." *Chemistry of Materials* 29 (18):7633–7644. doi: 10.1021/acs.chemmater.7b02847.

Alli, Uthman, Kieran McCarthy, Ioan-Alexandru Baragau, Nicholas P. Power, David J. Morgan, Steven Dunn, Seamus Killian, Tadhg Kennedy, and Suela Kellici. 2022. "In-situ continuous hydrothermal synthesis of $TiO_2$ nanoparticles on conductive N-doped MXene nanosheets for binder-free Li-ion battery anodes." *Chemical Engineering Journal* 430: 132976. doi: 10.1016/j.cej.2021.132976.

Byeon, Ayeong, Christine B. Hatter, Jae H. Park, Chi W. Ahn, Yury Gogotsi, and Jae W. Lee. 2017. "Molybdenum oxide/carbon composites derived from the $CO_2$ oxidation of $Mo_2CT_x$ (MXene) for lithium ion battery anodes." *Electrochimica Acta* 258:979–987. doi: 10.1016/j.electacta.2017.11.149.

Cao, Minjuan, Fen Wang, Lei Wang, Wenling Wu, Wenjing Lv, and Jianfeng Zhu. 2017. "Room temperature oxidation of $Ti_3C_2$ MXene for supercapacitor electrodes." *Journal of the Electrochemical Society* 164 (14):A3933. doi: 10.1149/2.1541714jes.

Chen Jizhang , Minfeng Chen, Weijun Zhou, Xinwu Xu, Bo Liu, Wenqing Zhang, and Chingping Wong. 2022a. "Simplified synthesis of fluoride-free Ti(3)C(2)T(x) via electrochemical etching toward high-performance electrochemical capacitors." *ACS Nano* 16 (2):2461–2470. doi: 10.1021/acsnano.1c09004.

Chen, Liang, Xinyue Dai, Wei Feng, and Yu Chen. 2022b. "Biomedical applications of MXenes: From nanomedicine to biomaterials." *Accounts of Materials Research* 3 (8):785–798. doi: 10.1021/accountsmr.2c00025.

Estrada-Ruiz, Marta, Daniel Cosano, Dolores Esquivel, Francisco J. Romero-Salguero, and José R. Ruiz. 2024. "3D hierarchical composites of hydrotalcite-coated carbon microspheres as catalysts in Baeyer–Villiger oxidation reactions." *Crystals* 14 (10):878. doi:10.3390/cryst14100878

Fang, Yongzheng, Rong Hu, Boya Liu, Yingying Zhang, Kai Zhu, Jun Yan, Ke Ye, Kui Cheng, Guiling Wang, and Dianxue Cao. 2019. "MXene-derived $TiO_2$/reduced graphene oxide composite with an enhanced capacitive capacity for Li-ion and K-ion batteries." *Journal of Materials Chemistry A* 7 (10):5363–5372. doi: 10.1039/C8TA12069B.

Gao, Yupeng, Hao Chen, Aiguo Zhou, Zhengyang Li, Fanfan Liu, Qianku Hu, and Libo Wang. 2015. "Novel hierarchical $TiO_2$/C nanocomposite with enhanced photocatalytic performance." *Nano* 10:150226001444006. doi: 10.1142/S1793292015500642.

Geng, Dechao, Xiaoxu Zhao, Zhongxin Chen, Weiwei Sun, Wei Fu, Jianyi Chen, Wei Liu, Wu Zhou, and Kian P. Loh. 2017. "Direct synthesis of large-area 2D $Mo_2C$ on in situ grown graphene." *Advanced Materials* 29 (35):1700072. doi: 10.1002/adma.201700072.

Gogotsi, Yury, and Babak Anasori. 2019. "The rise of MXenes." *ACS Nano* 13 (8):8491–8494. doi: 10.1021/acsnano.9b06394.

Gogotsi, Yury, and Qing Huang. 2021. "MXenes: Two-dimensional building blocks for future materials and devices." *ACS Nano* 15 (4):5775–5780. doi: 10.1021/acsnano.1c03161.

Guo, Zhang, Zhien Liu, Wan Chen, Xianzhong Sun, Xiong Zhang, Kai Wang, and Yanwei Ma. 2023. "Battery-type lithium-ion hybrid capacitors: Current status and future perspectives." *Batteries* 9 (2):74. doi: 10.3390/batteries9020074.

He, Shiyu, Qizhen Zhu, Razium A. Soomro, and Bin Xu. 2020. "MXene derivatives for energy storage applications." *Sustainable Energy & Fuels* 4 (10):4988–5004. doi: 10.1039/D0SE00927J.

He, Zhiquan, Tengda Rong, Yan Li, Junjie Ma, Quanshui Li, Furong Wu, Yuhang Wang, and Fengping Wang. 2022. "Two-dimensional TiVC solid-solution MXene as surface-enhanced Raman scattering substrate." *ACS Nano* 16 (3):4072–4083. doi: 10.1021/acsnano.1c09736.

Hulea, Vasile, Emil Dumitriu, and François Fajula. 2021. "Mild oxidation of organosulfur compounds with $H_2O_2$ over metal-containing microporous and mesoporous catalysts." *Catalysts* 11 (7):867. doi: 10.3390/catal11070867.

Jiang, Daixun, Xun Sun, Xilu Wu, Shuai Zhang, Xiaofei Qu, Liang Shi, Yelong Zhang, and Fanglin Du. 2020. "MXene-$Ti_3C_2$ assisted one-step synthesis of carbon-supported $TiO_2/Bi_4NbO_8Cl$ heterostructures for enhanced photocatalytic water decontamination." *Nanophotonics* 9 (7):2077–2088. doi: 10.1515/nanoph-2020–0088.

Jiang, Longbo, Ding Zhou, Jinjuan Yang, Shaoyu Zhou, Hou Wang, Xingzhong Yuan, Jie Liang, Xiaodong Li, Yaoning Chen, and Hui Li. 2022. "2D single- and few-layered MXenes: Synthesis, applications and perspectives." *Journal of Materials Chemistry A* 10 (26): 13651–13672. doi: 10.1039/D2TA01572B.

Jin, Fangzhou, Xingxing He, Jinlong Jiang, Weijun Zhu, Jianfeng Dai, and Hua Yang. 2020. "Synthesis of hierarchical porous $Ni_{1.5}Co_{1.5}S_4$/g-$C_3N_4$ composite for supercapacitor with excellent cycle stability." *Nanomaterials* 10 (9):1631.

Joshi, Siddharth, Qi Wang, Ajinkya Puntambekar, and Vidhya Chakrapani. 2017. "Facile synthesis of large area two-dimensional layers of transition-metal nitride and their use as insertion electrodes." *ACS Energy Letters* 2 (6):1257–1262. doi: 10.1021/acsenergylett.7b00240.

Kannan, Karthik, Kishor K. Sadasivuni, Aboubakr M. Abdullah, and Bijandra Kumar. 2020. "Current trends in MXene-based nanomaterials for energy storage and conversion system: A mini review." *Catalysts* 10 (5):495.

Katarzyna, Siwińska-Stefańska, and Jesionowski Teofil. 2017. "Advanced hybrid materials based on titanium dioxide for environmental and electrochemical applications." In *Titanium Dioxide*, edited by Janus Magdalena, Ch. 7, 69357. Rijeka: IntechOpen.

Kumar, Ravi, Azaj Ansari, Peter Comba, and Gopalan Rajaraman. 2024. "Rebound or cage escape? The role of the rebound barrier for the reactivity of non-heme high-valent FeIV=O species." *Chemistry – A European Journal* 30 (6):e202303300. doi: 10.1002/chem.202303300.

Kumar, Vinod, Sudheesh K. Shukla, Meenakshi Choudhary, Jalaj Gupta, Priyanka Chaudhary, Saurabh Srivastava, Mukesh Kumar, Manoj Kumar, Devojit K. Sarma, Bal C. Yadav, and Vinod Verma. 2022. "$Ti_2C$-$TiO_2$ MXene nanocomposite-based high-efficiency non-enzymatic glucose sensing platform for diabetes monitoring." *Sensors* 22 (15):5589.

Kurapati, Rajendra, Kostas Kostarelos, Maurizio Prato, and Alberto Bianco. 2016. "Biomedical uses for 2D materials beyond graphene: Current advances and challenges ahead." *Advanced Materials* 28 (29):6052–6074. doi: 10.1002/adma.201506306.

Late, Dattatray J., Parvez A. Shaikh, Ruchita Khare, Ranjit V. Kashid, Minakshi Chaudhary, Mahendra A. More, and Satishchandra B. Ogale. 2014. "Pulsed laser-deposited $MoS_2$ thin films on W and Si: Field emission and photoresponse studies." *ACS Applied Materials & Interfaces* 6 (18):15881–15888. doi: 10.1021/am503464h.

Li, Jingxiao, Shun Wang, Yulei Du, and Wenhe Liao. 2018a. "Enhanced photocatalytic performance of $TiO_2$@C nanosheets derived from two-dimensional $Ti_2CT_x$." *Ceramics International* 44 (6):7042–7046. doi: 10.1016/j.ceramint.2018.01.139.

Li, Pinda, Xueling Zhao, Yaxin Ding, Lifei Chen, Xin Wang, and Huaqing Xie. 2024. "Application studies on MXene-based flexible composites." *Frontiers in Thermal Engineering* 4: 1440165. doi: 10.3389/fther.2024.1440165.

Li, Tengfei, Lulu Yao, Qinglei Liu, Jiajun Gu, Ruichun Luo, Jinghan Li, Xudong Yan, Weiqiang Wang, Pan Liu, Bin Chen, Wang Zhang, Waseem Abbas, Raheela Naz, and Di Zhang. 2018b. "Fluorine-free synthesis of high-purity $Ti_3C_2T$ (T = OH, O) via alkali treatment." *Angewandte Chemie International Edition* 57 (21):6115–6119. doi: 10.1002/anie.201800887.

Lim, Jocelyn J. Y., and Andrew N. K. Lup. 2022. "Heterostructural $TiO_2$/$Ti_3C_2$ MXene aerogel composite for photocatalytic degradation of palm oil mill effluent." *Environmental Science: Advances* 1 (4):570–583. doi: 10.1039/D2VA00108J.

Liu, Fanfan, Aiguo Zhou, Jinfeng Chen, Jin Jia, Weijia Zhou, Libo Wang, and Qianku Hu. 2017. "Preparation of $Ti_3C_2$ and $Ti_2C$ MXenes by fluoride salts etching and methane adsorptive properties." *Applied Surface Science* 416:781–789. doi: 10.1016/j.apsusc.2017.04.239.

Liu, Liyuan, Hannes Zschiesche, Markus Antonietti, Mathieu Gibilaro, Pierre Chamelot, Laurent Massot, Patrick Rozier, Pierre-Louis Taberna, and Patrice Simon. 2023. "In situ synthesis of MXene with tunable morphology by electrochemical etching of MAX phase prepared in molten salt." *Advanced Energy Materials* 13 (7):2203805. doi: 10.1002/aenm.202203805.

Liu, Z. 2010. "4.09-Laser applied coatings." In *Shreir's Corrosion*, edited by Bob Cottis, Michael Graham, Robert Lindsay, Stuart Lyon, Tony Richardson, David Scantlebury, and Howard Stott, 2622–2635. Oxford: Elsevier.

Lukatskaya, Maria R., Olha Mashtalir, Chang E. Ren, Yohan Dall'Agnese, Patrick Rozier, Pierre Louis Taberna, Michael Naguib, Patrice Simon, Michel W. Barsoum, and Yury Gogotsi. 2013. "Cation intercalation and high volumetric capacitance of two-dimensional titanium carbide." *Science* 341 (6153):1502–1505. doi: 10.1126/science.1241488.

Mashtalir, Olha, Michael Naguib, Vadym N. Mochalin, Yohan Dall'Agnese, Min Heon, Michel W. Barsoum, and Yury Gogotsi. 2013. "Intercalation and delamination of layered carbides and carbonitrides." *Nature Communications* 4 (1):1716. doi: 10.1038/ncomms2664.

Naguib, Michael, Murat Kurtoglu, Volker Presser, Jun Lu, Junjie Niu, Min Heon, Lars Hultman, Yury Gogotsi, and Michel W. Barsoum. 2011. "Two-dimensional nanocrystals produced by exfoliation of $Ti_3AlC_2$." *Advanced Materials* 23 (37):4248–4253. doi: 10.1002/adma.201102306.

Naguib, Michael, Olha Mashtalir, Joshua Carle, Volker Presser, Jun Lu, Lars Hultman, Yury Gogotsi, and Michel W. Barsoum. 2012. "Two-dimensional transition metal carbides." *ACS Nano* 6 (2):1322–1331. doi: 10.1021/nn204153h.

Naguib, Michael, Olha Mashtalir, Maria R. Lukatskaya, Boris Dyatkin, Chuanfang Zhang, Volker Presser, Yury Gogotsi, and Michel W. Barsoum. 2014. "One-step synthesis of nanocrystalline transition metal oxides on thin sheets of disordered graphitic carbon by oxidation of MXenes." *Chemical Communications* 50 (56):7420–7423. doi: 10.1039/C4CC01646G.

Natu, Varun, Mohamed Benchakar, Christine Canaff, Aurélien Habrioux, Stéphane Célérier, and Michel W. Barsoum. 2021. "A critical analysis of the X-ray photoelectron spectra of $Ti_3C_2T_z$ MXenes." *Matter* 4 (4):1224–1251. doi: 10.1016/j.matt.2021.01.015.

Natu, Varun, Rahul Pai, Maxim Sokol, Michael Carey, Vibha Kalra, and Michel W. Barsoum. 2020. "2D $Ti_3C_2T_z$ MXene synthesized by water-free etching of $Ti_3AlC_2$ in polar organic solvents." *Chem* 6 (3):616–630. doi: 10.1016/j.chempr.2020.01.019.

Pogorielov Maksym, Kateryna Smyrnova, Sergiy Kyrylenko, Oleksiy Gogotsi, Veronika Zahorodna, and Alexander Pogrebnjak. 2021. "MXenes-A new class of two-dimensional materials: Structure, properties and potential applications." *Nanomaterials (Basel)* 11 (12): 3412. doi: 10.3390/nano11123412.

Rahman, Ubaid U., Muhammad Humayun, Usman Ghani, Muhammad Usman, Habib Ullah, Adil Khan, Nashwa M. El-Metwaly, and Abbas Khan. 2022. "MXenes as emerging materials: Synthesis, properties, and applications." *Molecules* 27 (15):4909.

Rao, K. S., Dipa D. Pathak, Balaji P. Mandal, and A. K. Tyagi. 2024. "MXene/rGO/sulfur loaded separator based current collector-free lithium sulfur batteries with high sulfur loading and cycle life." *Journal of Materials Science: Materials in Electronics* 35 (16):1060. doi: 10.1007/s10854-024-12818-8.

Salim, O., Khaled A. Mahmoud, Kamal Kishore Pant, and Rakesh Kumar Joshi. 2019. "Introduction to MXenes: Synthesis and characteristics." *Materials Today Chemistry* 14:100191. doi: 10.1016/j.mtchem.2019.08.010.

Sang, Xiahan, Yu Xie, Ming-Wei Lin, Mohamed Alhabeb, Katherine L. Van Aken, Yury Gogotsi, Paul R. C. Kent, Kai Xiao, and Raymond R. Unocic. 2016. "Atomic defects in monolayer titanium carbide ($Ti_3C_2T_x$) MXene." *ACS Nano* 10 (10):9193–9200. doi: 10.1021/acsnano.6b05240.

Sobolev, Kirill, Alexander Omelyanchik, Nikolai Shilov, Mikhail Gorshenkov, Nikolai Andreev, Antonio Comite, Sawssen Slimani, Davide Peddis, Yevgeniy Ovchenkov, Alexander Vasiliev, Kurban E. Magomedov, and Valeria Rodionova. 2024. "Iron oxide nanoparticle-assisted delamination of $Ti_3C_2T_x$ MXenes: A new approach to produce magnetic MXene-based composites." *Nanomaterials* 14 (1):97.

Sun Wanmei, Smit A. Shah, Yexiao Chen, Zeyi Tan, Huili Gao, Touseef Habib, Miladin Radovic, and Micah J. Green. 2017. "Electrochemical etching of $Ti_2AlC$ to $Ti_2CT_x$ (MXene) in low-concentration hydrochloric acid solution." *Journal of Materials Chemistry A* 5 (41):21663–21668. doi: 10.1039/C7TA05574A.

Tang, Huang, Shihao Zhuang, Zhihao Bao, Changshi Lao, and Yongfeng Mei. 2016. "Two-step oxidation of MXene in the synthesis of layer-stacked anatase titania with enhanced lithium-storage performance." *ChemElectroChem* 3 (6):871–876. doi: 10.1002/celc.201600078.

Tao, Mengli, Guangyuan Du, Tingting Yang, Wei Gao, Longcheng Zhang, Wenyan Du, Jian Jiang, Shujuan Bao, and Maowen Xu. 2020. "MXene-derived three-dimensional carbon nanotube network encapsulate $CoS_2$ nanoparticles as an anode material for solid-state sodium-ion batteries." *Journal of Materials Chemistry A* 8 (6):3018–3026. doi: 10.1039/C9TA12834D.

Tao, Mengli, Guangyuan Du, Youquan Zhang, Wei Gao, Dingyu Liu, Yushan Luo, Jian Jiang, Shujuan Bao, and Maowen Xu. 2019. "$TiO_xN_y$ nanoparticles/C composites derived from MXene as anode material for potassium-ion batteries." *Chemical Engineering Journal* 369:828–833. doi: 10.1016/j.cej.2019.03.144.

Tian, Zhengnan, Chaohui Wei, and Jingyu Sun. 2020. "Recent advances in the template-confined synthesis of two-dimensional materials for aqueous energy storage devices." *Nanoscale Advances* 2 (6):2220–2233. doi: 10.1039/D0NA00257G.

Wang, Changda, Hongwei Shou, Shuangming Chen, Shiqiang Wei, Yunxiang Lin, Pengjun Zhang, Zhanfeng Liu, Kefu Zhu, Xin Guo, Xiaojun Wu, Pulickel M. Ajayan, and Li Song. 2021. "Mxene synthesis: HCl-based hydrothermal etching strategy toward fluoride-Free MXenes." *Advanced Materials* 33 (27):2170209. doi: 10.1002/adma.202170209.

Wang, Di, Chenkun Zhou, Alexander S. Filatov, Wooje Cho, Francisco Lagunas, Mingzhan Wang, Suriyanarayanan Vaikuntanathan, Chong Liu, Robert F. Klie, and Dmitri V. Talapin. 2023. "Direct synthesis and chemical vapor deposition of 2D carbide and nitride MXenes." *Science* 379 (6638):1242–1247. doi: 10.1126/science.add9204.

Wang, Qingtao, Zhonghao Zhang, Lipeng Wang, Yongxia Zhang, Xiaozhong Zhou, and Guofu Ma. 2018a. "Ultrafast fabrication of $TiO_2$-Inlaid multilayered carbon for supercapacitor." *Chemistry Select* 3 (38):10811–10814. doi: 10.1002/slct.201802576.

Wang, Qinhua, Xiaofeng Pan, Changmei Lin, Haili Gao, Shilin Cao, Yonghao Ni, and Xiaojuan Ma. 2020. "Modified $Ti_3C_2T_X$ (MXene) nanosheet-catalyzed self-assembled, anti-aggregated, ultra-stretchable, conductive hydrogels for wearable bioelectronics." *Chemical Engineering Journal* 401:126129. doi: 10.1016/j.cej.2020.126129.

Wang, Rutao, Shijie Wang, Yabin Zhang, Dongdong Jin, Xinyong Tao, and Li Zhang. 2018b. "Graphene-coupled $Ti_3C_2$ MXenes-derived $TiO_2$ mesostructure: promising sodium-ion capacitor anode with fast ion storage and long-term cycling." *Journal of Materials Chemistry A* 6 (3):1017–1027. doi: 10.1039/C7TA09153B.

Wu, Zhihui, Da-Wen Sun, Hongbin Pu, Qingyi Wei, and Xuanran Lin. 2022. "$Ti_3C_2T_x$ MXenes loaded with Au nanoparticle dimers as a surface-enhanced Raman scattering aptasensor for AFB1 detection." *Food Chemistry* 372:131293. doi: 10.1016/j.foodchem.2021.131293.

Xu, Chuan, Libin Wang, Zhibo Liu, Long Chen, Jingkun Guo, Ning Kang, Xiu-Liang Ma, Hui-Ming Cheng, and Wencai Ren. 2015. "Large-area high-quality 2D ultrathin $Mo_2C$ superconducting crystals." *Nature Materials* 14 (11):1135–1141. doi: 10.1038/nmat4374.

Xu, Wenxin, Xin Zhao, Jiali Tang, Chao Zhang, Yu Gao, Shin-ichi Sasaki, Hitoshi Tamiaki, Aijun Li, and Xiao-Feng Wang. 2021. "Synthesis of Chl@$Ti_3C_2$ composites as an anode material for lithium storage." *Frontiers of Chemical Science and Engineering* 15 (3):709–716. doi: 10.1007/s11705-020-1984-z.

Xuan, Jinnan, Zhiqiang Wang, Yuyan Chen, Dujuan Liang, Liang Cheng, Xiaojing Yang, Zhuang Liu, Renzhi Ma, Takayoshi Sasaki, and Fengxia Geng. 2016. "Organic-base-driven intercalation and delamination for the production of functionalized titanium carbide nanosheets with superior photothermal therapeutic performance." *Angewandte Chemie International Edition* 55 (47):14569–14574. doi: 10.1002/anie.201606643.

Yuan, Wenyu, Laifei Cheng, Yani Zhang, Heng Wu, Shilin Lv, Liying Chai, Xiaohui Guo, and Lianxi Zheng. 2017. "2D-layered carbon/$TiO_2$ hybrids derived from $Ti_3C_2$MXenes for photocatalytic hydrogen evolution under visible light irradiation." *Advanced Materials Interfaces* 4 (20):1700577. doi: 10.1002/admi.201700577.

Zaharin, Haizum A., Mariyam J. Ghazali, Mohammad Khalid, Thachnatharen Nagarajan, Wong W. Pin, Farah Ezzah, Ong Gerard, Rashmi Walvekar, and Abdul K. Rasheed. 2023. "Tribological, oxidation and thermal analysis of advanced microwave–hydrothermal synthesised $Ti_3C_2T_x$ MXene as additives in outboard engine oil." *Lubricants* 11 (6):264.

Zhang, Chuanfang, Seon J. Kim, Michael Ghidiu, Meng-Qiang Zhao, Michel W. Barsoum, Valeria Nicolosi, and Yury Gogotsi. 2016. "Layered orthorhombic $Nb_2O_5$@$Nb_4C_3T$ and $TiO_2$@$Ti_3C_2T$ hierarchical composites for high performance Li-ion batteries." *Advanced Functional Materials* 26 (23):4143–4151. doi: 10.1002/adfm.201600682.

Zhang, Fan, Zhi Zhang, Huichao Wang, Cheuk H. Chan, Ngai Y. Chan, Xin X. Chen, and Ji-Yan Dai. 2017. "Plasma-enhanced pulsed-laser deposition of single-crystalline $Mo_2C$ ultrathin superconducting films." *Physical Review Materials* 1 (3):034002. doi: 10.1103/PhysRevMaterials.1.034002.

Zhang, Jizhen, Ken A. S. Usman, Mia A. N. Judicpa, Dylan Hegh, Peter A. Lynch, and Joselito M. Razal. 2023. "Applications of X-Ray-based characterization in MXene research." *Small Methods* 7 (8):2201527. doi: 10.1002/smtd.202201527.

Zhang, Peng, Razium A. Soomro, Zhaoruxin Guan, Ning Sun, and Bin Xu. 2020. "3D carbon-coated MXene architectures with high and ultrafast lithium/sodium-ion storage." *Energy Storage Materials* 29:163–171. doi: 10.1016/j.ensm.2020.04.016.

Zhang, Shunlong, Hangjun Ying, Pengfei Huang, Tiantian Yang, and Wei-Qiang Han. 2022. "Hierarchical utilization of raw $Ti_3C_2T_x$ MXene for fast preparation of various $Ti_3C_2T_x$ MXene derivatives." *Nano Research* 15 (3):2746–2755. doi: 10.1007/s12274-021-3847-4.

Zheng, Wei, Peigen Zhang, Li Yang, Jian Chen, Wubian Tian, Yamei Zhang, and ZhengMing Sun. 2018. "A fast approach to the synthesis of MO/CNT/Fe hybrid nanostructures built on MXene for enhanced Li-ion uptake." *Ceramics International* 44 (18):22456–22461. doi: 10.1016/j.ceramint.2018.09.013.

Zou, Guodong, Jianxin Guo, Qiuming Peng, Aiguo Zhou, Qingrui Zhang, and Baozhong Liu. 2016. "Synthesis of urchin-like rutile titania carbon nanocomposites by iron-facilitated phase transformation of MXene for environmental remediation." *Journal of Materials Chemistry A* 4 (2):489–499. doi: 10.1039/C5TA07343J.

Zou, Guodong, Baozhong Liu, Jianxin Guo, Qingrui Zhang, Carlos Fernandez, and Qiuming Peng. 2017. "Synthesis of nanoflower-shaped MXene derivative with unexpected catalytic activity for dehydrogenation of sodium alanates." *ACS Applied Materials & Interfaces* 9 (8):7611–7618. doi: 10.1021/acsami.6b13973.

# 3 Expansion of MXenes over Other 2D Materials for Electrochemical Sensing Applications

*Sudip Das, Arka Prava Roy, Arghya Chakravorty, Aarcha Appu Mini, Andrews Nirmala Grace, and Vimala Raghavan*

## 3.1 INTRODUCTION

### 3.1.1 MXenes – Revolutionizing Electrochemical Sensing in the Realm of 2D Materials

Electrochemical sensors play a pivotal role in various fields, including environmental monitoring, disease diagnosis, and food safety assessment. The relentless pursuit of miniaturization and enhanced sensitivity in analytical tools has propelled the field of electrochemical sensing toward a new era dominated by two-dimensional (2D) materials. Among these groundbreaking materials, MXenes, a unique class of transition metal carbides, nitrides, or carbonitrides, have emerged as a frontrunner with exceptional potential. This chapter delves into the exciting world of MXenes and explores their distinct advantages over other prominent 2D materials, ultimately highlighting their transformative role in the realm of electrochemical sensing applications.

The story of MXenes began in 2011 when a groundbreaking discovery took place at Drexel University by a team led by Professor Yury Gogotsi. Their research focused on exploring novel materials for energy storage applications (Anasori, Lukatskaya, and Gogotsi 2017). During their investigation, a seemingly simple experiment involving etching the MAX phase – a layered ternary carbide or nitride of a transition metal (M), a group 13 or 14 element (A), and carbon or nitrogen (X) – with hydrofluoric acid (HF) yielded an unexpected outcome. The etching process selectively removed the A-element layers, leaving behind a fascinating new material: a single sheet of transition metal carbide or nitride, christened MXene. This groundbreaking discovery marked the dawn of a new era in materials science, paving the way for a plethora of exciting applications.

MXenes, the name does not suggest a single, homogenous material; instead, they represent a vast and diverse family of 2D materials with a general formula of $M_{n+1}X_nT_x$, where M represents the transition metal (e.g., Ti, Zr, Nb, Mo), while $n$ is the number of C or N layers sandwiched between the M layers which can vary from


DOI: 10.1201/9781003736318-3

1 to 4, and $T_x$ signifies the surface terminations (e.g., –OH, –O, –F) introduced during the etching process. This inherent versatility in their chemical composition translates to a remarkable range of physical and chemical properties, making MXenes highly tunable and adaptable for various applications (Gogotsi and Anasori 2019).

One of the most captivating characteristics of MXenes is their exceptional surface chemistry. Unlike other 2D materials such as graphene, MXenes possess a naturally terminated surface with functional groups like hydroxyl (–OH), oxygen (–O), and fluorine (–F) atoms (Schultz et al. 2019). These functional groups readily participate in various chemical reactions, allowing for facile modification of the MXene surface (Naguib et al. 2012). This ability to tailor the surface chemistry enables the targeted immobilization of specific biomolecules or analytes, a crucial aspect for designing highly selective and sensitive electrochemical sensors. Another hallmark feature of MXenes is their impressive electrical conductivity. Depending on the specific composition, MXenes can exhibit metallic or near-metallic conductivity. This exceptional electrical property allows for efficient electron transfer between the electrode and the analyte molecule, a vital requirement for robust electrochemical sensing. The efficient electron transfer translates to enhanced sensor response, leading to sharper and more easily detectable signals.

The atomically thin structure of MXenes is another key advantage for electrochemical sensing applications. This unique morphology bestows upon MXenes a remarkably large surface area. Imagine a single sheet of paper unfolded to reveal its true surface area – that's the power of the 2D structure (Hantanasirisakul et al. 2020). This vast surface area provides a multitude of binding sites for analyte molecules, significantly increasing the sensor's sensitivity. In simpler terms, the larger the surface area, the more analyte molecules can interact with the electrode, leading to a stronger overall sensor response. Furthermore, MXenes possess inherent hydrophilicity, meaning they readily interact with water. This characteristic is particularly advantageous for electrochemical sensing applications that often involve aqueous solutions as the testing medium. The hydrophilicity facilitates the dispersion of MXenes in aqueous environments, promoting their interaction with water-soluble analytes, which are prevalent in biomolecule detection and environmental monitoring applications. Beyond their exceptional physicochemical properties, MXenes exhibit promising biocompatibility (Diedkova et al. 2023). Certain compositions of MXenes have demonstrated minimal cytotoxicity, making them potentially suitable for designing biosensors for medical diagnostics. This biocompatible nature opens a gateway toward the development of implantable or wearable sensors for real-time monitoring of biomolecules in the human body, a revolutionary advancement in healthcare technology. With this impressive repertoire of properties, MXenes have carved a niche for themselves in a wide range of applications beyond the realm of electrochemical sensing.

## 3.2 ADVANTAGES OF MXENES OVER OTHER 2D MATERIALS: GAME CHANGERS IN ELECTROCHEMICAL SENSING APPLICATIONS

The realm of electrochemical sensing is witnessing a paradigm shift with the emergence of 2D materials. Compared to other prominent 2D materials like graphene, transition metal dichalcogenides (TMDs), and phosphorene, MXenes offer a compelling set of advantages that elevate their potential for constructing high-performance electrochemical sensors. The key factors that distinguish MXenes and propel them to the forefront of electrochemical sensing have been articulated in the course of this chapter.

### 3.2.1 Tunable Surface Chemistry: The MXene Advantage in Electrochemical Sensing

The realm of electrochemical sensing thrives on the development of materials with exceptional properties, and in this arena, MXenes stand out due to their unique tunability, particularly in surface chemistry. Compared to other prominent 2D materials, MXenes offer a distinct advantage in their ability to be tailored for specific sensing applications. This section will delve deeper into how this tunable surface chemistry empowers MXenes for superior electrochemical sensing.

Electrochemical sensors rely on the interaction between the electrode surface and the target analyte. This interaction, often involving electron transfer, dictates the sensor's sensitivity and selectivity. Here's where the tunable surface chemistry of MXenes shines (Seredych et al. 2019). Unlike other 2D materials like graphene, which have inert surfaces, MXenes boast a naturally terminated surface rich in functional groups like hydroxyl (–OH), oxygen (–O), and fluorine (–F) atoms. These functional groups readily participate in various chemical reactions, allowing for the strategic modification of the MXene surface. This ability to modify the surface chemistry of MXenes opens a world of possibilities for sensor design. By introducing specific functional groups, researchers can tailor the MXene surface to selectively bind with target analytes. For instance, attaching amine ($–NH_2$) groups can enhance the binding of biomolecules with negatively charged functionalities (Dong et al. 2020). Similarly, incorporating thiol (–SH) groups can facilitate the capture of heavy metal ions through strong chelation. This targeted modification allows for the creation of highly selective sensors that can distinguish between different analytes in a complex sample (Bagheri et al. 2022). The tunable surface chemistry of MXenes not only improves selectivity but also contributes to enhanced sensitivity. Functional groups can act as anchoring sites for specific biorecognition molecules like antibodies or enzymes. These biorecognition molecules can selectively bind with their target analytes, amplifying the signal associated with the analyte's presence (Wu et al. 2020). Additionally, the functional groups can influence the local environment at the electrode surface, promoting efficient electron transfer between the analyte and the electrode (Su et al. 2022). This improves the overall sensitivity of the sensor, allowing for the detection of even minute quantities of the target analyte. The tunable surface chemistry of MXenes offers a multifaceted advantage for electrochemical sensing. Certain functional groups can be introduced to enhance the biocompatibility of MXenes, making them suitable for designing biosensors for applications like medical diagnostics. The surface chemistry can be manipulated to tune the electrocatalytic activity of MXenes, enabling them to participate in specific electrochemical reactions relevant to sensing applications. Modifying the surface chemistry can influence the hydrophilicity of MXenes, allowing for optimal dispersion in aqueous solutions – a crucial aspect for many sensing applications conducted in water-based environments.

The ability to tailor the surface chemistry of MXenes is a powerful tool that continues to be explored and refined by researchers. With ongoing advancements in surface modification techniques and a deeper understanding of the structure–function relationship in MXenes, the potential for even more sophisticated and sensitive electrochemical sensors is constantly expanding. This tunability of MXenes positions them as a frontrunner in the development of next-generation electrochemical sensing platforms, paving the way for advancements in healthcare, environmental monitoring, and beyond.

### 3.2.2 The Conductive Edge: MXenes Power Up Electrochemical Sensing

While other 2D materials hold promise, MXenes offer a distinct advantage – their remarkable electrical conductivity, ranging from metallic to near-metallic depending on their composition (Qiao et al. 2021). This exceptional property empowers MXenes to revolutionize the field of electrochemical sensing, leading to the development of highly sensitive and reliable sensors. In this part, it is discussed how MXenes leverage their high electrical conductivity and explore the potential of their semi-conducting counterparts.

Electrochemical sensing relies on the interaction between the electrode surface and the target analyte, often involving the transfer of electrons. This electron transfer plays a pivotal role in generating a measurable signal that indicates the presence and concentration of the analyte. MXenes, with their exceptional electrical conductivity, facilitate this process with remarkable efficiency. Electrons can effortlessly travel between the analyte molecule and the electrode surface, minimizing resistance and leading to a stronger overall signal. This translates to enhanced sensitivity of the sensor, allowing for the detection of even minute quantities of the target analyte. MXenes offer a compelling combination of high electrical conductivity, tunable surface chemistry, and relative ease of synthesis, making them a superior choice for many sensing applications (Wu et al. 2021b). While the metallic or near-metallic conductivity of certain MXenes is highly advantageous, their semi-conducting counterparts also hold promise for specific sensing applications. Semi-conducting MXenes exhibit a band gap, meaning a specific energy range where electron conduction is not possible (Wu et al. 2021b). By strategically tailoring the composition and surface chemistry of these semi-conducting MXenes, researchers can create sensors with unique properties. For instance, the band gap can be tuned to selectively respond to specific analytes or modulate the sensitivity of the sensor. This tunability opens new avenues for developing highly specialized electrochemical sensors for niche applications.

The band gap, the energy difference between the top of the valence band and the bottom of the conduction band, is a crucial property that determines a material's electrical conductivity. MXenes generally exhibit metallic or semimetallic behavior with minimal or no band gap. This is attributed to their strong metallic bonding and delocalized electrons. However, the band gap can be tuned by varying the composition, surface termination, and layer thickness of MXenes (Naguib et al. 2011). Graphene, on the other hand, is a zero-band-gap material, meaning it has no energy gap between its valence and conduction bands. This allows for high electron mobility and makes it an excellent conductor (Novoselov et al. 2004). TMDs, such as $MoS_2$ and $WS_2$, are typically semiconductors with a direct band gap. The band gap of TMDs can be tuned by controlling their thickness and the number of layers (Shi et al. 2018). Black phosphorus is an anisotropic semiconductor with a direct band gap that can be tuned by varying its thickness (Liu et al. 2014). The Density of States (DOS) represents the number of available energy states per unit energy interval. It provides insights into the electronic structure of a material and its conductivity. MXenes typically exhibit a high DOS near the Fermi level, indicating a large number of available states for electron conduction. This contributes to their high conductivity (Yu et al. 2018). Graphene has a linear DOS near the Fermi level, which is characteristic of a Dirac material. This linear dispersion leads to its high carrier mobility (Castro Neto et al. 2009). TMDs generally have a parabolic DOS near the band edges, similar to conventional semiconductors. The shape of the DOS can influence its electrical properties (He, Hummer, and Franchini 2014).

Black phosphorus has an anisotropic DOS due to its puckered structure. The DOS can vary depending on the direction of the applied electric field. MXenes, graphene, TMDs, and black phosphorus exhibit distinct band gaps, DOS, and conductivity properties, making them suitable for various applications. MXenes offer high conductivity and tunable properties, while graphene excels in its high carrier mobility. TMDs and black phosphorus exhibit semiconductor behavior with varying band gaps and conductivities. The choice of material depends on the specific requirements of the application, such as electronic devices, energy storage, or catalysis.

Research efforts are continuously exploring strategies to optimize the electrical conductivity of MXenes. This includes the development of novel synthesis techniques and the exploration of composite materials that combine MXenes with other conductive materials. As the understanding of these factors deepens, researchers can tailor the conductivity of MXenes to specific sensing applications, further pushing the boundaries of sensitivity and selectivity (Shi et al. 2023). In conclusion, the exceptional electrical conductivity of MXenes, along with the potential of their semi-conducting counterparts, provides a significant advantage over other 2D materials for electrochemical sensing. This remarkable property, coupled with other advantageous characteristics of MXenes, empowers researchers to create highly sensitive and versatile sensors. As research delves deeper into optimizing conductivity and exploring the potential of semi-conducting MXenes, their impact on the field of electrochemical sensing is poised to be transformative.

### 3.2.3 A Feast for Binding Sites: The Advantage of Large Surface Area in MXene

In this arena, MXenes hold a trump card – their exceptionally large surface area. Compared to other prominent 2D materials, MXenes boast an atomically thin structure that translates into a multitude of binding sites, significantly amplifying the potential for high-performance electrochemical sensors.

Electrochemical sensors rely on the interaction between the electrode surface and the target analyte. This interaction dictates the sensitivity and selectivity of the sensor. The larger the surface area of the electrode material, the more binding sites it possesses for the target analyte molecules to interact with. MXenes, with their 2D structure, offer a remarkable advantage in this regard (Zhao et al. 2023). Imagine a single sheet of paper unfolded to reveal its true area – that's the power of the 2D morphology. This vast surface area translates into a multitude of binding sites for analyte molecules, significantly increasing the sensor's sensitivity. In simpler terms, the more analyte molecules can interact with the electrode surface, the stronger the overall signal generated during the sensing process. While other 2D materials hold promise for electrochemical sensing applications, their surface area pales in comparison to MXenes. The large surface area of MXenes directly translates into enhanced sensitivity for electrochemical sensors. By providing a multitude of binding sites, MXenes allow for the capture of a larger number of analyte molecules, even at very low concentrations. This leads to a stronger overall signal, making it easier to detect the presence of the target analyte. This enhanced sensitivity is crucial for applications like environmental monitoring, where identifying trace amounts of pollutants is essential, or in medical diagnostics, where early detection of disease biomarkers is vital.

The large surface area of MXenes synergistically amplifies the impact of their other advantageous properties. The exceptional electrical conductivity of MXenes, when coupled with their vast surface area, allows for efficient electron transfer from a multitude of binding sites. This translates to a stronger overall signal and further enhances the sensor's sensitivity. Additionally, the tunable surface chemistry of MXenes can be leveraged to modify the electrode surface and introduce specific functional groups that promote even stronger binding of the target analyte molecules to the available binding sites (Ranjbar et al. 2022). This synergistic interplay between surface area, conductivity, and surface chemistry positions MXenes at the forefront of electrochemical sensing technology. Research efforts are continuously exploring strategies to further optimize the surface area of MXenes (Arshadi et al. 2021). This includes the development of novel synthesis techniques that promote the exfoliation of MXene layers, leading to a larger surface area per unit mass. Additionally, researchers are investigating methods for creating hierarchical structures with MXenes, incorporating features like pores or channels that can further increase the effective surface area available for analyte interaction. As the understanding of these factors deepens, researchers can tailor the surface area of MXenes to specific sensing applications, pushing the boundaries of sensitivity and detection limits.

In conclusion, the large surface area of MXenes stands out as a significant advantage over other 2D materials for electrochemical sensing. This remarkable characteristic allows for the creation of highly sensitive sensors capable of detecting even minute quantities of target analytes. Combined with their other advantageous properties, MXenes hold immense potential for revolutionizing the field of electrochemical sensing and paving the way for advancements in environmental monitoring, healthcare diagnostics, and beyond.

### 3.2.4 Embracing the Aqueous World: The Allure of Hydrophilicity in MXene

MXenes possess a distinct advantage in terms of their inherent hydrophilicity. While other 2D materials might require complex modifications for compatibility with water, MXenes readily interact and disperse in aqueous solutions, making them ideal for sensing applications involving water-soluble analytes.

Electrochemical sensing frequently utilizes aqueous solutions as the medium for analyzing samples. This translates to a need for electrode materials that readily interact with water. MXenes boast an inherent hydrophilicity, meaning they possess a natural affinity for water (Guan et al. 2020). This characteristic stems from the presence of functional groups like hydroxyl (–OH) and oxygen (–O) on their surface. These groups form hydrogen bonds with water molecules, promoting favorable interactions and facilitating the dispersion of MXenes within aqueous environments. This inherent hydrophilicity eliminates the need for complex surface modifications often required for other 2D materials, simplifying sensor fabrication processes. Several prominent 2D materials used for electrochemical sensing can exhibit limitations when it comes to hydrophilicity. MXenes, on the other hand, offer a compelling combination of inherent hydrophilicity, tunable surface chemistry, and relative ease of synthesis, making them a superior choice for electrochemical sensing applications that involve water-soluble analytes (Liu et al. 2022). The hydrophilicity of MXenes directly benefits the efficiency of electrochemical sensing. By readily interacting with water, MXenes ensure a uniform dispersion of the analyte molecules throughout the solution. The hydrophilicity of MXenes synergistically amplifies the impact of their other advantageous properties. The large surface area of MXenes, when coupled with their inherent affinity for water, maximizes the available area for interaction with water-soluble analytes. Additionally, the tunable surface chemistry of MXenes can be leveraged to further enhance the capture of specific analytes by introducing functional groups that promote even stronger interactions with water molecules. This interplay between hydrophilicity, surface area, and surface chemistry positions MXenes as a frontrunner in the development of next-generation electrochemical sensors for aqueous environments (Mini et al. 2024).

Researchers are continuously exploring strategies to tailor the hydrophilicity of MXenes for specific sensing applications. This might involve studying the influence of different surface functional groups or exploring methods to create hierarchical structures with MXenes that optimize the interaction with water molecules. As the understanding of these factors deepens, scientists can engineer MXenes with precisely controlled hydrophilicity, maximizing their effectiveness in diverse aqueous sensing applications. In conclusion, the hydrophilicity of MXenes offers a significant advantage over other 2D materials for electrochemical sensing in aqueous environments. This inherent property simplifies sensor fabrication, promotes efficient interaction with water-soluble analytes, and paves the way for advancements in crucial fields like healthcare diagnostics and environmental monitoring. By leveraging their unique combination of properties, MXenes stand poised to revolutionize the future of aqueous-based electrochemical sensing.

### 3.2.5 A Bridge between Technology and Biology: The Biocompatibility Advantage of MXenes

MXenes offer a distinct advantage – the potential for safe and reliable interaction with biological systems, opening doors to the development of innovative bioelectrochemical sensors.

Electrochemical biosensors play a vital role in healthcare by detecting and monitoring biomolecules like proteins, enzymes, and DNA (Amara et al. 2023). However, for these sensors to be effective for in vivo applications or implantable devices, they must be biocompatible. This translates to minimal cytotoxicity and the absence of adverse immune responses when in contact with living tissues. Certain compositions of MXenes have demonstrated promising biocompatibility in initial studies, suggesting their potential for safe integration within the human body (Wang et al. 2020). While other 2D materials hold promise for sensing applications, concerns regarding their biocompatibility often limit their potential for biosensing. MXenes offer a unique combination of high conductivity, large surface area, tunable surface chemistry, and promising biocompatibility, making them a compelling choice for the development of biocompatible electrochemical sensors. The tunable surface chemistry of MXenes plays a crucial role in their biocompatibility. By carefully selecting and introducing specific functional groups onto the MXene surface, researchers can create sensors that interact favorably with biomolecules. For instance, attaching biocompatible polymers or functional groups that mimic natural biomolecules can minimize cytotoxicity and promote better integration with biological systems. This tailored interface design allows for the development of biocompatible sensors that can selectively capture and detect target biomolecules with minimal disruption to the surrounding biological environment (Lim et al. 2021). The promising biocompatibility of MXenes opens a gateway toward the development of groundbreaking bioelectrochemical sensors. MXenes have been used to create implantable biosensors for real-time monitoring of biomolecules within the body, facilitating early diagnosis and treatment of diseases. Biocompatible MXene-based sensors could be integrated into wearable devices for continuous monitoring of vital biomolecular markers, promoting personalized healthcare management. MXenes, with their biocompatible nature and large surface area, could be explored for the development of targeted drug delivery systems, enhancing treatment efficacy and reducing side effects. 2D MXenes are also incorporated in making high-resolution neural interfaces that are essential tools for studying and modulating neural circuits underlying brain function and disease (Driscoll et al. 2018). $V_4C_3$ MXene has been used in immune profiling and modulation of T-cell and dendritic cell function and interaction (Fusco et al. 2023). Cytotoxicity assessment of $Ti_2AlC$, and $Ti_3AlC_2$ MAX Phases and $Ti_3C_2T_x$ MXenes has been conducted on human fibroblasts and cervical cancer cells to ensure the safety concentration and cytotoxicity (Scheibe et al. 2019). Fusco et al. propose a novel, label-free method to detect and analyze 2D materials (like MXenes) within single cells and tissues (Fusco et al. 2022). This method uses high-dimensional single-cell mass cytometry by time-of-flight (CyTOF) to provide detailed information about both the materials and the cells they interact with. The key findings included successful detection and quantification of MXenes in various human immune cells, assessment of MXenes' biocompatibility and influence on cellular functions, demonstration of MXene biodistribution in different organs of mice, and potential for broad applications in biomedical research. Figure 3.1 shows the environmental study of MXene films.

Research efforts are continuously striving to optimize the biocompatibility of MXenes. This includes exploring the influence of different surface modifications, studying the long-term biocompatibility of MXenes in vivo, and developing strategies

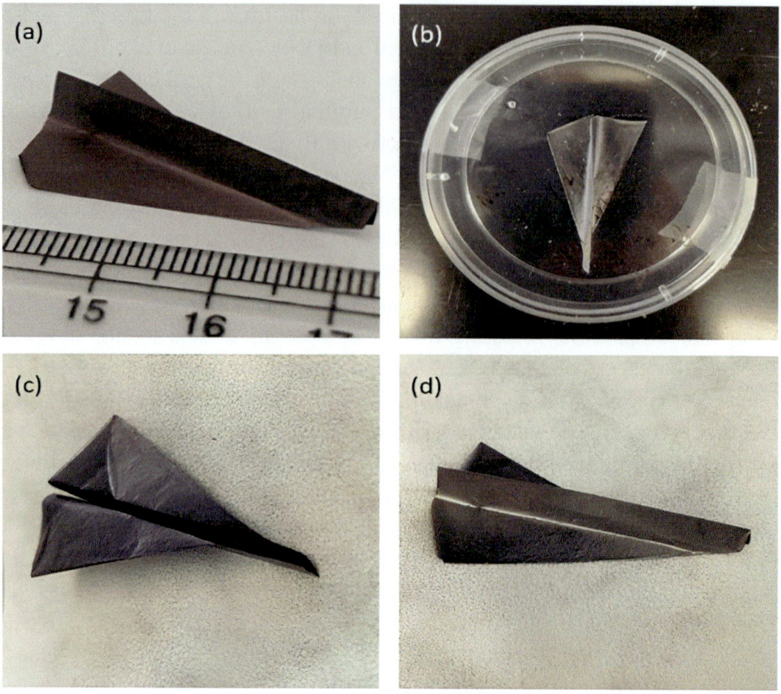


**FIGURE 3.1** Photographs of a $Ti_3C_2T_x$ paper plane from 2014 to 2024: (a) at the date of synthesis, (b) stored in a petri dish since 2014, (c) before drying, and (d) after drying at 200°C for 24 h to remove water. This study is to show the environmental study of MXene films (A. Lee et al. 2024) [CC BY 4.0].

to minimize any potential inflammatory responses. As research delves deeper into understanding the biocompatibility of MXenes and their interaction with biological systems, their impact on biosensing technologies is poised to be transformative. In conclusion, the promising biocompatibility of MXenes stands out as a significant advantage over other 2D materials for bioelectrochemical sensing. This unique property opens doors to the development of safe and reliable sensors for healthcare applications. Coupled with their other advantageous properties, MXenes offer immense potential for revolutionizing the field of biosensing and paving the way for advancements in early disease detection, personalized medicine, and beyond.

## 3.3 DIFFERENT TYPES OF MXENES USED FOR ELECTROCHEMICAL SENSORS

Since their discovery in 2011, researchers have explored various MXene-based sensors for a wide range of analytes, including heavy metals, organic pollutants, and biomolecules. To start with, titanium-based MXenes, such as $Ti_2CT_x$ and $Ti_3C_2T_x$, are among the most widely studied due to their excellent electrical conductivity and stability. These materials have been used for the detection of heavy metals like lead,

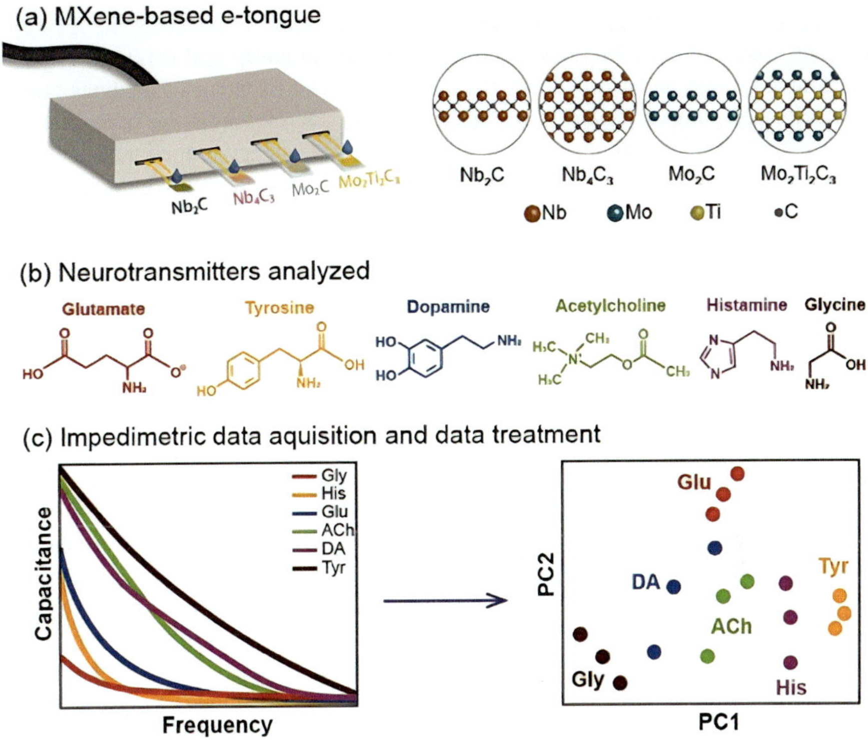


FIGURE 3.2 Schematic illustration of (a) the MXene-based impedimetric e-tongue (surface terminations of MXenes are not shown), (b) the structures of the nanotubes analyzed, and (c) the representation of the impedimetric analysis performed and the data treatment using principal component analysis (Facure et al. 2024).

cadmium, and mercury, as well as organic pollutants such as nitroaromatic compounds and phenols. Molybdenum-based MXenes, such as $Mo_2CT_x$ and $Mo_3C_2T_x$, have shown potential for sensing biomolecules. Their unique surface chemistry can be modified to selectively bind to specific biomolecules, enabling sensitive and selective detection. For example, $Mo_2CT_x$-based sensors have been developed for the detection of glucose and dopamine. Niobium-based MXenes, such as $Nb_2CT_x$ and $Nb_3C_2T_x$, have been investigated for sensing applications due to their high surface area and stability. They have been used for the detection of heavy metals and organic pollutants. Other MXenes, such as $V_2CT_x$, $Cr_2CT_x$, and $Ta_2CT_x$, have also been explored for electrochemical sensing applications, but their use is less common compared to Ti-, Mo-, and Nb-based MXenes (Facure et al. 2024). MXene-based impedimetric e-tongue for neurotransmitter monitoring using different MXenes is given in Figure 3.2.

MXene-based sensors can be fabricated using various electrochemical techniques, including amperometric sensing, which measures the current generated when the analyte is oxidized or reduced at the electrode surface. MXene-based electrodes can be used for amperometric detection of a wide range of analytes. Voltammetric sensing involves scanning the potential of the electrode and measuring the resulting current. Voltammetric methods, such as cyclic voltammetry and differential pulse voltammetry, can be used to study the electrochemical behavior of analytes and to quantify their concentration. Impedance spectroscopy measures the impedance of the electrode–electrolyte interface as a function of frequency. Impedance spectroscopy can be used to study the kinetics of electron transfer and to characterize the properties of the electrode–electrolyte interface.

## 3.4 APPLICATIONS OF MXENE-BASED ELECTROCHEMICAL SENSORS

MXene boasts a captivating combination of properties – high electrical conductivity, large surface area, tunable surface chemistry, and, in certain compositions, promising biocompatibility. This synergy of advantages positions MXenes at the forefront of developing next-generation electrochemical sensors with diverse applications across various fields. MXenes hold immense potential for environmental monitoring applications due to their ability to detect a wide range of pollutants. The large surface area and tunable surface chemistry of MXenes allow for the creation of sensors selective for specific heavy metals like lead, mercury, and cadmium. These sensors can be deployed in water bodies or soil to monitor for pollution and ensure environmental safety. For instance, research has shown that $Ti_3C_2T_x$ MXenes functionalized with specific ligands can effectively detect lead ions in water samples. MXenes can be tailored to detect organic pollutants like pesticides and herbicides. By incorporating specific functional groups onto the MXene surface, researchers can create sensors that selectively bind to these pollutants, facilitating their detection in environmental samples. For example, studies have demonstrated the use of functionalized MXenes for the detection of bisphenol A (BPA), a common environmental contaminant. The role of MXene in electrochemical sensing applications is summarized in Figure 3.3.

The food industry can benefit significantly from MXene-based sensors for ensuring food safety and quality control (Appu Mini and Raghavan 2024). MXenes can be functionalized with antibodies or aptamers that specifically bind to foodborne pathogens like E. coli or Salmonella. This allows for the rapid and sensitive detection of these pathogens in food samples, preventing outbreaks of foodborne illnesses. Researchers are actively exploring the use of MXene–antibody bioconjugates for the detection of various foodborne pathogens. The freshness and quality of food products can be monitored using MXene-based sensors. These sensors can detect specific biomarkers associated with spoilage, such as the presence of volatile organic compounds or changes in pH levels. Studies have shown promising results in using MXenes for the detection of volatile amines, a common indicator of fish spoilage (Ashiq et al. 2024; Wanniarachchi, Upul Kumarasinghe, and Jayathilake 2024).

The biocompatible nature of certain MXene compositions opens doors for their application in biosensing for healthcare diagnostics. MXenes can be designed to detect specific disease biomarkers present in blood, saliva, or other bodily fluids. This early detection of diseases like cancer or diabetes allows for timely intervention and improved treatment outcomes. For example, the electrochemical properties of MXene-coated fiber electrodes were utilized in hydrogen peroxide ($H_2O_2$)

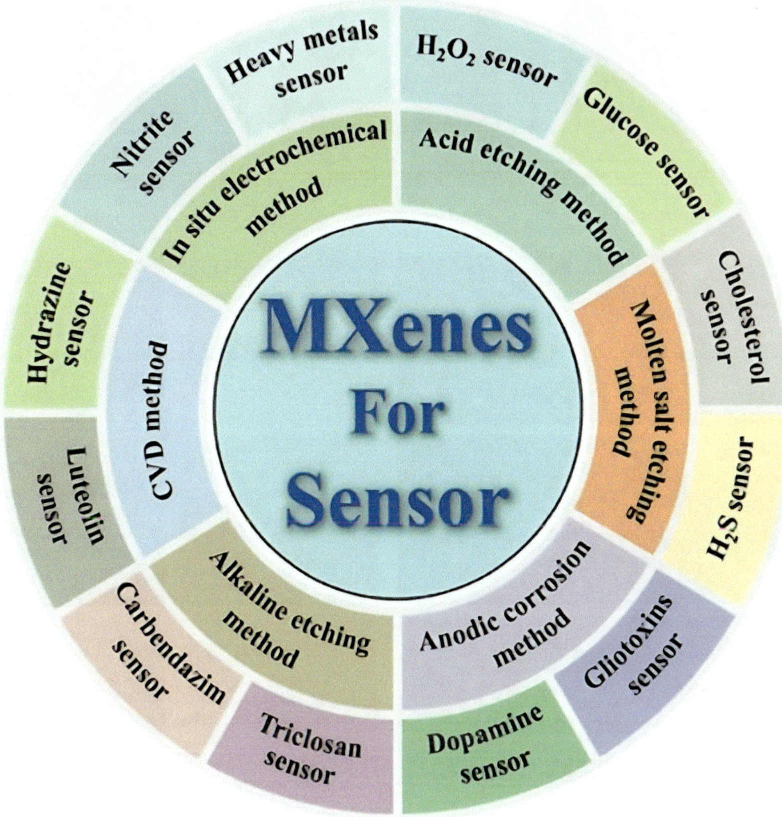


**FIGURE 3.3** Role of MXene in electrochemical sensing applications (Q. Wang et al. 2023) [CC BY 4.0].

sensing capability and showcased their applications *in vivo* (Figure 3.4) (Bi et al. 2024). MXene-based sensors can be used to monitor drug levels in the body, ensuring therapeutic efficacy and minimizing the risk of overdoses or underdoses. This personalized approach to medication management can significantly improve patient care. Studies are underway to explore the use of MXenes for monitoring therapeutic drugs used for various health conditions.

The versatility of MXenes extends beyond these prominent applications. MXenes can be used to create sensors for monitoring the health and performance of batteries and supercapacitors. This information is crucial for optimizing the efficiency and lifespan of energy storage devices. The unique properties of MXenes make them suitable for the development of sensors that can detect explosives with high sensitivity and selectivity. This technology has the potential to enhance security measures in airports and other sensitive locations. While the potential of MXene-based sensors is undeniable, certain challenges need to be addressed to realize their full potential. Developing efficient and cost-effective methods for large-scale MXene synthesis is crucial for widespread adoption in sensor fabrication. Standardized

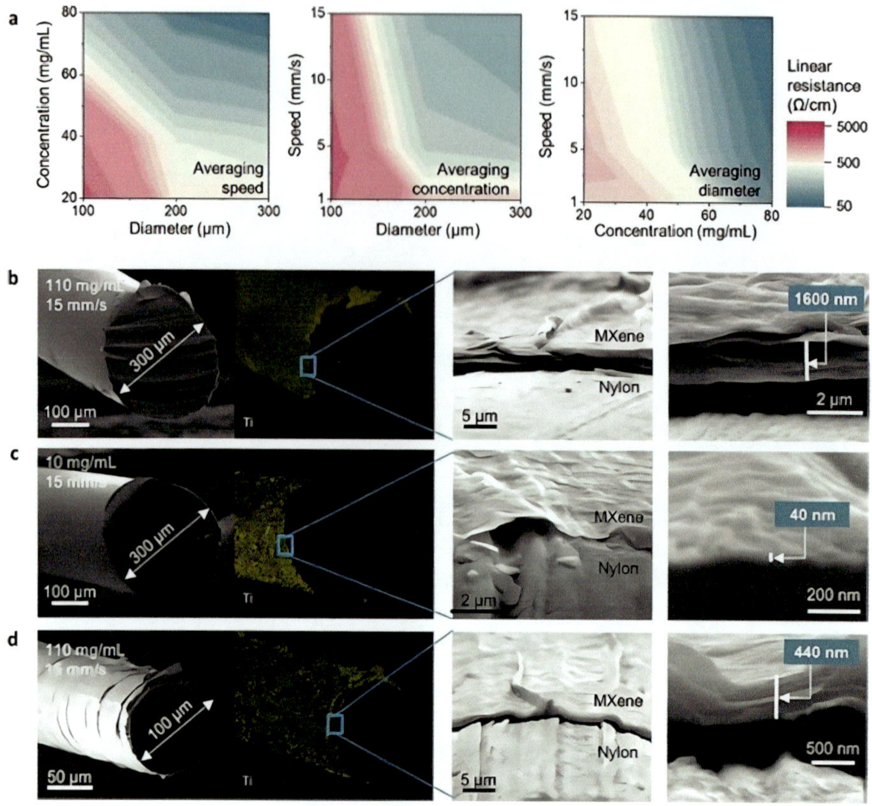


**FIGURE 3.4** Schematics of the fiber electrode fabrication and potential applications. (a) Schematic of the dip coating process consisting of a nylon filament running through a dispersion of single-layer MXene flakes. During this process, MXene flakes are aligned in the axial direction by the shear force present in the meniscus. (b) Schematics and digital photographs of the nylon filaments, MXene-coated nylon filaments, and Parylene-coated electrodes. (c) Schematic of cut electrodes with the cross-section of the MXene coating exposed and their demonstrated capabilities in this work. (d) Potential applications of the MXene electrodes in various biological systems, including cells, organoids, brains, and muscles [CC BY 4.0] (Bi et al. 2024).

protocols for MXene fabrication and surface modification are essential for ensuring reliable sensor performance and reproducibility. Further research is needed to ensure the long-term stability and biocompatibility of MXenes for in vivo applications. MXenes represent a transformative force in the field of electrochemical sensing. Their remarkable properties and diverse potential applications make them an exciting material for the future. As research continues to address existing challenges and explore new avenues, MXene-based sensors promise to revolutionize how we monitor our environment, ensure food safety, diagnose diseases, and develop next-generation technologies. The future of sensing is undoubtedly bright with MXenes at the forefront.

## 3.5 CHALLENGES AND FUTURE PROSPECTS

While MXenes offer a compelling set of advantages, they are not without their challenges. This chapter delves into both sides of the coin, exploring the current hurdles faced by MXene-based electrochemical sensors and their exciting future prospects, illustrated with specific examples. Despite their immense potential, several hurdles hinder the widespread adoption of MXene-based electrochemical sensors. MXenes, due to their inherent hydrophilicity, tend to restack or aggregate in solution. This reduces the available surface area, a crucial factor for capturing target analytes and achieving high sensitivity. Researchers at the University of Science and Technology of China explored the use of $Ti_3C_2T_x$ MXenes for dopamine detection (Han et al. 2018). However, restacking significantly reduced the sensor's sensitivity, hindering its practical application. The conventional synthesis of MXenes often involves harsh chemicals like hydrofluoric acid (HF). This not only poses environmental and safety concerns but can also leave residual fluoride ions on the MXene surface, impacting sensor performance. A study at Fudan University demonstrated the use of $Ti_3C_2T_x$ MXenes synthesized using HF for glucose detection. However, the presence of residual fluoride ions interfered with the sensor's response, limiting its accuracy (Wu et al. 2021a). While MXenes offer some tunability of their surface chemistry, achieving precise control over functional group placement remains a challenge. This limits the ability to tailor MXenes for specific sensing applications. The long-term stability of MXenes in aqueous environments, particularly under biological conditions, needs further investigation. Understanding their degradation processes is crucial for reliable sensor performance over time. A team of scientists at the National University of Singapore explored the use of functionalized MXenes for the detection of dopamine, a crucial neurotransmitter. The study revealed that the MXenes exhibited minimal cytotoxicity toward neural cells, making them promising candidates for implantable brain sensors that can monitor dopamine levels in real time, aiding in the diagnosis and treatment of neurological disorders like Parkinson's disease. However, their study revealed that the MXenes degraded over time in the culture medium, raising concerns about their stability for long-term biosensing applications.

Despite these challenges, researchers are actively exploring strategies to overcome them and unlock the full potential of MXenes in electrochemical sensing. Researchers are investigating various strategies to prevent restacking, including the use of intercalating agents, polymers, and the creation of hierarchical MXene structures with increased porosity. Scientists have explored using conductive polymers to coat MXenes, preventing restacking and improving the sensitivity of MXene-based sensors for detecting hydrogen peroxide (Lorencova et al. 2018). Alternative synthesis methods for MXenes that minimize the use of harsh chemicals and environmental impact are being actively explored. These include using milder acids, molten salts, or even electrochemical methods. A research team in Germany has successfully synthesized $Ti_3C_2T_x$ MXenes using a fluoride-free molten salt technique (Yang et al. 2018). This approach offers a more environmentally friendly alternative for MXene production. Researchers are developing novel techniques for precise surface modification of MXenes, allowing for the controlled introduction of functional groups tailored for specific target analytes. Strategies to improve the long-term stability of

MXenes in aqueous environments are being investigated. This includes surface passivation techniques and the development of composite materials with enhanced stability. MXenes offer a revolutionary approach to electrochemical sensing with their unique combination of properties like high conductivity, large surface area, and tunable surface chemistry. However, overcoming challenges like restacking and limited control over surface chemistry is crucial for their widespread adoption. Continuous research efforts exploring alternative synthesis techniques, advanced surface engineering, and stability enhancement strategies hold immense promise for unleashing the full potential.

## 3.6 ARTIFICIAL INTELLIGENCE AND MACHINE LEARNING: UNVEILING MXENE'S POTENTIAL FOR SUPERIOR ELECTROCHEMICAL SENSORS

Electrochemical sensors are revolutionizing various fields, from environmental monitoring to medical diagnostics. In this quest for better sensors, 2D materials like MXenes are gaining traction. Artificial Intelligence (AI) and Machine Learning (ML) are emerging as powerful tools to unlock MXene's true potential and establish it as a superior choice for electrochemical sensors. Developing high-performance electrochemical sensors involves a complex interplay between material properties and operational parameters. Traditional methods often rely on trial-and-error experimentation, which can be time-consuming and resource-intensive. For instance, factors like MXene surface chemistry, electrode design, and target analyte interactions significantly influence sensor sensitivity and selectivity. AI and ML offer a data-driven approach to optimize MXene-based sensors. ML algorithms can analyze vast datasets on MXene properties (surface area, pore size, conductivity) and their impact on sensor performance. This allows for the prediction of optimal MXene compositions for specific target analytes. For example, a study using ML predicted a specific MXene ($Ti_3C_2T_x$) with tailored surface functionalization for a tenfold increase in dopamine detection. AI can analyze sensor response data under various electrode configurations (size, shape, material) and environmental conditions (temperature, pH). This helps identify the ideal electrode setup for maximizing sensor sensitivity and minimizing noise. Electrochemical sensors generate complex data streams. ML algorithms can be trained to analyze these signals in real time, differentiating between target analytes and background noise. This improves sensor accuracy and reduces false positives, crucial for applications like disease diagnostics.

By leveraging AI and ML, researchers can unlock the full potential of these exceptional properties. A study used ML to design MXene/polymer composites for heavy metal ion detection was done by Yao et al. (2023). The ML-optimized sensor exhibited a remarkable selectivity for lead ions, highlighting the power of AI-driven material design. Researchers are exploring AI-powered sensor systems that combine MXene-based sensors with microfluidic platforms. These integrated systems offer real-time, on-site analysis with improved sensitivity and minimal sample preparation, making them ideal for environmental monitoring applications. The marriage of AI and MXene technology holds immense promise for the future of electrochemical sensors.

As research progresses, we can expect even more sophisticated ML algorithms and advanced MXene compositions, leading to sensors with ultra-high sensitivity and selectivity, enabling early-stage disease detection and trace molecule identification, enhanced stability and reusability, reducing costs and environmental impact, and miniaturization and integration into portable devices, paving the way for personalized and point-of-care diagnostics. In conclusion, AI and ML are acting as powerful catalysts in the development of MXene-based electrochemical sensors. By leveraging these tools, researchers are unlocking the true potential of MXenes, paving the way for a new generation of highly sensitive, selective, and versatile sensors that will revolutionize various sectors.

## 3.7 CONCLUSION

This chapter has explored the remarkable potential of MXenes, a unique class of 2D transition metal carbides, nitrides, or carbonitrides, in the realm of electrochemical sensing. Compared to other prominent 2D materials, MXenes stand out due to their exceptional combination of properties. Their tunable surface chemistry allows for tailored interactions with target analytes, significantly enhancing sensor selectivity. The exceptional electrical conductivity of MXenes facilitates efficient electron transfer, leading to stronger overall signals and improved sensitivity. The vast surface area of these 2D materials provides a multitude of binding sites, further amplifying the sensitivity of the sensors. Moreover, the inherent hydrophilicity of MXenes simplifies fabrication processes and optimizes their interaction with water-soluble analytes, crucial for many sensing applications. Furthermore, certain MXene compositions exhibit promising biocompatibility, paving the way for advancements in biosensing technologies. The potential to develop implantable sensors for real-time monitoring of biomolecules or biocompatible wearable devices opens exciting avenues for personalized healthcare management. MXenes also hold promise for targeted drug delivery systems, potentially revolutionizing treatment strategies. In conclusion, MXenes offer a compelling alternative to other 2D materials for electrochemical sensing applications. Their unique synergy of properties translates to highly sensitive, selective, and versatile sensors. As research delves deeper into optimizing the synthesis, surface modification, and biocompatibility of MXenes, their impact on the field of electrochemical sensing is poised to be transformative. We can expect to see MXene-based sensors playing a pivotal role in environmental monitoring, healthcare diagnostics, and beyond, shaping the future of this vital technology. However, challenges remain. Further research is needed to fully understand the long-term biocompatibility of MXenes and develop scalable and cost-effective synthesis methods. Nevertheless, the potential of MXenes is undeniable. This exciting class of 2D materials stands at the forefront of a revolution in electrochemical sensing, offering a glimpse into a future where sensors are not only highly functional but also biocompatible and seamlessly integrated into our lives.

## REFERENCES

Amara, Umay, Iftikhar Hussain, Muhmmad Ahmad, Khalid Mahmood, and Kaili Zhang. 2023. "2D MXene-based biosensing: A review." *Small* 19 (2):2205249. doi: 10.1002/smll.202205249.

Anasori, Babak, Maria R. Lukatskaya, and Yury Gogotsi. 2017. "2D metal carbides and nitrides (MXenes) for energy storage." *Nature Reviews Materials* 2 (2):16098. doi: 10.1038/natrevmats.2016.98.

Appu Mini, Aarcha, and Vimala Raghavan. 2024. "Mxene-derived $Na_2O_7Ti_3$ nanoribbon as a promising electrode material for the detection of ethyl paraoxon in complex matrices." *Microchemical Journal* 201:110674. doi: 10.1016/j.microc.2024.110674.

Arshadi, Faezeh, Munirah Mohammad, Ehsan Hosseini, Hadi Ahmadi, Mohsen Asadnia, Yasin Orooji, Asghar H. Korayem, Abdollah Noorbakhsh, and Amir Razmjou. 2021. "The effect of D-spacing on the ion selectivity performance of MXene membrane." *Journal of Membrane Science* 639:119752. doi: 10.1016/j.memsci.2021.119752.

Ashiq, Javaria, Unzila Saeed, Zheng Li, and Mian H. Nawaz. 2024. "Advances in meat spoilage detection: A review of methods involving 2D-based nanomaterials for detection of spoiled meat." *Journal of Food Composition and Analysis* 132:106295. doi: 10.1016/j.jfca.2024.106295.

Bagheri, Saman, Rylan Chilcott, Shengyuan Luo, and Alexander Sinitskii. 2022. "Bifunctional amine- and thiol-modified $Ti_3C_2T_x$ MXene for trace detection of heavy metals." *Langmuir* 38 (42):12924–12934. doi: 10.1021/acs.langmuir.2c02058.

Bi, Lingyi, Raghav Garg, Natalia Noriega, Ruocun J. Wang, Hyunho Kim, Kseniia Vorotilo, Justin C. Burrell, Christopher E. Shuck, Flavia Vitale, Bhavik A. Patel, and Yury Gogotsi. 2024. "Soft, multifunctional MXene-coated fiber microelectrodes for biointerfacing." *ACS Nano* 18 (34):23217–23231. doi: 10.1021/acsnano.4c05797.

Castro Neto, Antonio H., Francisco Guinea, Nuno M. R. Peres, Sir Kostya Novoselov, and Sir Andre Geim. 2009. "The electronic properties of graphene." *Reviews of Modern Physics* 81 (1):109–162. doi: 10.1103/RevModPhys.81.109.

Diedkova, Kateryna, Alexander D. Pogrebnjak, Sergiy Kyrylenko, Kateryna Smyrnova, Vladimir V. Buranich, Pawel Horodek, Pawel Zukowski, Tomasz N. Koltunowicz, Piotr Galaszkiewicz, Kristina Makashina, Vitaly Bondariev, Martin Sahul, Maria Čaplovičová, Yevheniia Husak, Wojciech Simka, Viktoriia Korniienko, Agnieszka Stolarczyk, Agata Blacha-Grzechnik, Vitalii Balitskyi, Veronika Zahorodna, Ivan Baginskiy, Una Riekstina, Oleksiy Gogotsi, Yury Gogotsi, and Maksym Pogorielov. 2023. "Polycaprolactone–MXene nanofibrous scaffolds for tissue engineering." *ACS Applied Materials & Interfaces* 15 (11):14033–14047. doi: 10.1021/acsami.2c22780.

Dong, Hui, Linlin Cao, Zhaoling Tan, Qing Liu, Jin Zhou, Pingping Zhao, Ping Wang, Yueyun Li, Wanshan Ma, and Yunhui Dong. 2020. "A signal amplification strategy of CuPtRh CNB-embedded ammoniated $Ti_3C_2$ MXene for detecting cardiac troponin I by a sandwich-type electrochemical immunosensor." *ACS Applied Bio Materials* 3 (1):377–384. doi: 10.1021/acsabm.9b00863.

Driscoll, Nicolette, Andrew G. Richardson, Kathleen Maleski, Babak Anasori, Oladayo Adewole, Pavel Lelyukh, Lilia Escobedo, D. Kacy Cullen, Timothy H. Lucas, Yury Gogotsi, and Flavia Vitale. 2018. "Two-dimensional $Ti_3C_2$ MXene for high-resolution neural interfaces." *ACS Nano* 12 (10):10419–10429. doi: 10.1021/acsnano.8b06014.

Facure, Murilo H. M., Gulnaz Gahramanova, Danzhen Zhang, Teng Zhang, Christopher E. Shuck, Luiza A. Mercante, Daniel S. Correa, and Yury Gogotsi. 2024. "All-MXene electronic tongue for neurotransmitters detection." *Biosensors and Bioelectronics* 262:116526. doi: 10.1016/j.bios.2024.116526.

Fusco, Laura, Arianna Gazzi, Christopher E. Shuck, Marco Orecchioni, Dafne Alberti, Sènan M. D'Almeida, Darawan Rinchai, Eiman Ahmed, Ofer Elhanani, Martina Rauner, Barbara Zavan, Jean-Charles Grivel, Leeat Keren, Giulia Pasqual, Davide Bedognetti, Klaus Ley, Yury Gogotsi, and Lucia G. Delogu. 2022. "Immune profiling and multiplexed label-free detection of 2D MXenes by mass cytometry and high-dimensional imaging." *Advanced Materials* 34 (45):2205154. doi: 10.1002/adma.202205154.

Fusco, Laura, Arianna Gazzi, Christopher E. Shuck, Marco Orecchioni, Eiman I. Ahmed, Linda Giro, Barbara Zavan, Açelya Yilmazer, Klaus Ley, Davide Bedognetti, Yury Gogotsi, and Lucia G. Delogu. 2023. "V4C3 MXene immune profiling and modulation of T cell-dendritic cell function and interaction." *Small Methods* 7 (8):2300197. doi: 10.1002/smtd.202300197.

Gogotsi, Yury, and Babak Anasori. 2019. "The rise of MXenes." *ACS Nano* 13 (8):8491–8494. doi: 10.1021/acsnano.9b06394.

Guan, Yanxue, Miaomiao Zhang, Juan Qin, Xingxing Ma, Chen Li, and Jilin Tang. 2020. "Hydrophilicity-dependent distinct frictional behaviors of different modified MXene nanosheets." *The Journal of Physical Chemistry C* 124 (25):13664–13671. doi: 10.1021/acs.jpcc.0c01551.

Han, Xiaoxia, Ju Huang, Han Lin, Zhigang Wang, Pan Li, and Yu Chen. 2018. "2D ultrathin MXene-based drug-delivery nanoplatform for synergistic photothermal ablation and chemotherapy of cancer." *Advanced Healthcare Materials* 7 (9):1701394. doi: 10.1002/adhm.201701394.

Hantanasirisakul, Kanit, Babak Anasori, Slavomir Nemsak, James L. Hart, Jiabin Wu, Yizhou Yang, Rajesh V. Chopdekar, Padraic Shafer, Andrew F. May, Eun J. Moon, Jun Zhou, Qinghua Zhang, Mitra L. Taheri, Steven J. May, and Yury Gogotsi. 2020. "Evidence of a magnetic transition in atomically thin $Cr_2TiC_2T_x$ MXene." *Nanoscale Horizons* 5 (12):1557–1565. doi: 10.1039/D0NH00343C.

He, Jiangang, Kerstin Hummer, and Cesare Franchini. 2014. "Stacking effects on the electronic and optical properties of bilayer transition metal dichalcogenides $MoS_2$, $MoSe_2$, $WS_2$, and $WSe_2$." *Physical Review B* 89 (7):075409. doi: 10.1103/PhysRevB.89.075409.

Lim, Gim P., Chin F. Soon, Nyuk L. Ma, Marlia Morsin, Nafarizal Nayan, Mohd K. Ahmad, and Kian S. Tee. 2021. "Cytotoxicity of MXene-based nanomaterials for biomedical applications: A mini review." *Environmental Research* 201:111592. doi: 10.1016/j.envres.2021.111592.

Liu, Han, Adam T. Neal, Zhen Zhu, Zhe Luo, Xianfan Xu, David Tománek, and Peide D. Ye. 2014. "Phosphorene: An unexplored 2D semiconductor with a high hole mobility." *ACS Nano* 8 (4):4033–4041. doi: 10.1021/nn501226z.

Liu, Jingkai, Xichen Jia, Yuan Liu, Yishun Wuliu, Jinyue Dai, Xinbei Zhu, and Xiaoqing Liu. 2022. "A synergetic strategy of well dispersing hydrophilic $Ti_3C_2T_x$ MXene into hydrophobic polybenzoxazine composites for improved comprehensive performances." *Composites Science and Technology* 219:109248. doi: 10.1016/j.compscitech.2021.109248.

Lorencova, Lenka, Tomas Bertok, Jaroslav Filip, Monika Jerigova, Dusan Velic, Peter Kasak, Khaled A. Mahmoud, and Jan Tkac. 2018. "Highly stable $Ti_3C_2T_x$ (MXene)/Pt nanoparticles-modified glassy carbon electrode for $H_2O_2$ and small molecules sensing applications." *Sensors and Actuators B: Chemical* 263:360–368. doi: 10.1016/j.snb.2018.02.124.

Mini, Aarcha A., Arghya Chakravorty, Sudip Das, Shikha Awasthi, Andrews N. Grace, Sarvesh K. Pandey, and Vimala Raghavan. 2024. "CuO nanoparticles passivated 2D MXene-based voltammetric sensor for detecting environmental hazardous pollutant." *Microchemical Journal* 201:110648. doi: 10.1016/j.microc.2024.110648.

Naguib, Michael, Murat Kurtoglu, Volker Presser, Jun Lu, Junjie Niu, Min Heon, Lars Hultman, Yury Gogotsi, and Michel W. Barsoum. 2011. "Two-dimensional nanocrystals produced by exfoliation of $Ti_3AlC_2$." *Advanced Materials* 23 (37):4248–4253. doi: 10.1002/adma.201102306.

Naguib, Michael, Olha Mashtalir, Joshua Carle, Volker Presser, Jun Lu, Lars Hultman, Yury Gogotsi, and Michel W. Barsoum. 2012. "Two-dimensional transition metal carbides." *ACS Nano* 6 (2):1322–1331. doi: 10.1021/nn204153h.

Novoselov, Kostya S., Sir Andre Geim, S. V. Morozov, Da Jiang, Y. Zhang, S. V. Dubonos, Irina V. Grigorieva, and Anatoly A. Firsov. 2004. "Electric field effect in atomically thin carbon films." *Science* 306 (5696):666–669. doi: doi:10.1126/science.1102896.

Qiao, Chunyang, Han Wu, Xiyan Xu, Zhengxin Guan, and Wei Ou-Yang. 2021. "Electrical conductivity enhancement and electronic applications of 2D $Ti_3C_2T_x$ MXene materials." *Advanced Materials Interfaces* 8 (24):2100903. doi: 10.1002/admi.202100903.

Ranjbar, Saba, Negar A. Astani, Maryam Atabay, Naimeh Naseri, Ali Esfandiar, and Mohammad R. Ejtehadi. 2022. "Electrochemical and computational studies of bio-mimicked $Ti_3C_2T_x$ MXene-based sensor with multivalent interface." *Journal of Colloid and Interface Science* 623:1063–1074. doi: 10.1016/j.jcis.2022.05.033.

Scheibe, Błażej, Jacek K. Wychowaniec, Magdalena Scheibe, Barbara Peplińska, Marcin Jarek, Grzegorz Nowaczyk, and Łucja Przysiecka. 2019. "Cytotoxicity assessment of Ti–Al–C based MAX phases and $Ti_3C_2T_x$ MXenes on human fibroblasts and cervical cancer cells." *ACS Biomaterials Science & Engineering* 5 (12):6557–6569. doi: 10.1021/acsbiomaterials.9b01476.

Schultz, Thorsten, Nathan C. Frey, Kanit Hantanasirisakul, Soohyung Park, Steven J. May, Vivek B. Shenoy, Yury Gogotsi, and Norbert Koch. 2019. "Surface termination dependent work function and electronic properties of $Ti_3C_2T_x$ MXene." *Chemistry of Materials* 31 (17):6590–6597. doi: 10.1021/acs.chemmater.9b00414.

Seredych, Mykola, Christopher E. Shuck, David Pinto, Mohamed Alhabeb, Eliot Precetti, Grayson Deysher, Babak Anasori, Narendra Kurra, and Yury Gogotsi. 2019. "High-temperature behavior and surface chemistry of carbide MXenes studied by thermal analysis." *Chemistry of Materials* 31 (9):3324–3332. doi: 10.1021/acs.chemmater.9b00397.

Shi, Jianping, Min Hong, Zhepeng Zhang, Qingqing Ji, and Yanfeng Zhang. 2018. "Physical properties and potential applications of two-dimensional metallic transition metal dichalcogenides." *Coordination Chemistry Reviews* 376:1–19. doi: 10.1016/j.ccr.2018.07.019.

Shi, Yantian, Tiantian Li, Lin Zhao, Yaojia Liu, Kexin Ding, Dongxiang Li, Peng He, Degang Jiang, Jing Liu, and Hong Zhou. 2023. "Ultrathin MXene nanosheet-based $TiO_2$/CdS heterostructure as a photoelectrochemical sensor for detection of CEA in human serum samples." *Biosensors and Bioelectronics* 230:115287. doi: 10.1016/j.bios.2023.115287.

Su, Tongming, Xiaohan Ma, Jianhua Tong, Hongbing Ji, Zuzeng Qin, and Zili Wu. 2022. "Surface engineering of MXenes for energy and environmental applications." *Journal of Materials Chemistry A* 10 (19):10265–10296. doi: 10.1039/D2TA01140A.

Wang, Dongyi, Lili Wang, Zheng Lou, Yiqiang Zheng, Kang Wang, Lianjia Zhao, Wei Han, Kai Jiang, and Guozhen Shen. 2020. "Biomimetic, biocompatible and robust silk Fibroin-MXene film with stable 3D cross-link structure for flexible pressure sensors." *Nano Energy* 78:105252. doi: 10.1016/j.nanoen.2020.105252.

Wanniarachchi, Piyumi C., K. G. Upul Kumarasinghe, and Chathuni Jayathilake. 2024. "Recent advancements in chemosensors for the detection of food spoilage." *Food Chemistry* 436:137733. doi: 10.1016/j.foodchem.2023.137733.

Wu, Minying, Qian Zhang, Yuanyuan Fang, Chao Deng, Fangzhou Zhou, Yi Zhang, Xingdong Wang, Yi Tang, and Yajun Wang. 2021a. "Polylysine-modified MXene nanosheets with highly loaded glucose oxidase as cascade nanoreactor for glucose decomposition and electrochemical sensing." *Journal of Colloid and Interface Science* 586:20–29. doi: 10.1016/j.jcis.2020.10.065.

Wu, Qiong, Ningbo Li, Ying Wang, Yanchao Xu, Jiandong Wu, Guangri Jia, Fujian Ji, Xuedong Fang, Fangfang Chen, and Xiaoqiang Cui. 2020. "Ultrasensitive and selective determination of carcinoembryonic antigen using multifunctional ultrathin amino-functionalized $Ti_3C_2$-MXene nanosheets." *Analytical Chemistry* 92 (4):3354–3360. doi: 10.1021/acs.analchem.9b05372.

Wu, Xinzhao, Pinyi Ma, Ying Sun, Fangxin Du, Daqian Song, and Guobao Xu. 2021b. "Application of MXene in electrochemical sensors: A review." *Electroanalysis* 33 (8): 1827–1851. doi: 10.1002/elan.202100192.

Yang, Sheng, Panpan Zhang, Faxing Wang, Antonio G. Ricciardulli, Martin R. Lohe, Paul W. M. Blom, and Xinliang Feng. 2018. "Fluoride-free synthesis of two-dimensional titanium carbide (MXene) using a binary aqueous system." *Angewandte Chemie International Edition* 57 (47):15491–15495. doi: 10.1002/anie.201809662.

Yao, Hang, Ruimei Wu, Jin Zou, Jiawei Liu, Guanwei Peng, Xu Wang, Weiqi Zhou, Shirong Ai, and Limin Lu. 2023. "A machine learning strategy-incorporated $BiFeO_3/Ti_3C_2$ MXene electrochemical platform for simple, rapid detection of $Pb^{2+}$ with high sensitivity." *Chemosphere* 340:139728. doi: 10.1016/j.chemosphere.2023.139728.

Yu, Peng, Gejin Cao, Sha Yi, Xiong Zhang, Chen Li, Xianzhong Sun, Kai Wang, and Yanwei Ma. 2018. "Binder-free 2D titanium carbide (MXene)/carbon nanotube composites for high-performance lithium-ion capacitors." *Nanoscale* 10 (13):5906–5913. doi: 10.1039/C8NR00380G.

Zhao, Yufei, Jinqiang Zhang, Xin Guo, Xianjun Cao, Shijian Wang, Hao Liu, and Guoxiu Wang. 2023. "Engineering strategies and active site identification of MXene-based catalysts for electrochemical conversion reactions." *Chemical Society Reviews* 52 (9):3215–3264. doi: 10.1039/D2CS00698G.

# 4 MXene-Modified Electrochemical Sensors for Biosensing

*Anson T. Mathew and Jeethu Raveendran*

## 4.1 INTRODUCTION

Modern biosensors are analytical instruments that comprise two fundamental components: a biorecognition element and a transducer (Perumal and Hashim 2014). Currently, these sensors constitute the foundation of point-of-care devices and wearable health monitors, with market analyses revealing that electrochemical biosensors represented nearly 35% of a global biosensor market valued at approximately USD 27 billion in 2021, and are anticipated to exceed USD 40 billion by 2027. This significant expansion is propelled by their portability, simplicity of use, and adaptability for commercial production (Research and Markets 2023, 2022; SkyQuest Technology Consulting Pvt 2024; MarketsandMarkets 2025).

The bioreceptor binds to particular target analytes such as antibody molecules, enzymes, or nucleic acids, while the transducer translates these biochemical interactions into measurable outputs (electrical, optical, etc.), influencing the total sensor efficacy (Altalhi 2023). As the field advances toward better sensitivity, quicker response times, and improved compactness, conventional transducer materials are gradually being augmented or even substituted by advanced nanomaterials (Vikesland and Wigginton 2010).

Researchers have investigated several advanced nanomaterials to improve sensor performance as transducer elements (Jayathilaka et al. 2019). Graphene, an established two-dimensional carbon nanomaterial, is renowned for its exceptional electrical conductivity and extensive surface area (Chowdury et al. 2024). However, its inherent hydrophobicity often requires complicated surface modifications to enhance biocompatibility and enzyme-loading capability.

MXenes have stood out as particularly appealing transducers among these novel materials due to their excellent electrical conductivity, extensive surface area, rich functional groups, and superior biocompatibility (Purbayanto et al. 2025). These two-dimensional transition metal-based carbides, nitrides, and carbonitrides increase electron transport and allow variable surface functionalization, while significantly improving sensor performance.

MXenes, transition metal carbides, nitrides, and carbonitrides, initially identified in 2011, are widely recognized for their hydrophilicity, extensive surface functional groups (–O, –OH, –F), and exceptional biocompatibility. In contrast to graphene, which requires chemical treatment to increase hydrophilicity, MXenes


DOI: 10.1201/9781003736318-4

naturally disperse in aqueous solutions, facilitating sensor development and improving interactions with biomolecules such as enzymes or antibodies. Their layered architecture offers an extensive surface area for the immobilization of biomolecules, which is essential for attaining improved sensitivity (Gautam, Marriwala, and Devi 2023). MXenes exhibit enzyme-loading capacities up to three times more than graphene oxide, because of their adjustable interlayer gap and surface chemical composition (Bilal et al. 2023; Yoon et al. 2020).

The unique trait of the MXene is its compositional adaptability. By modifying transition metals (e.g., Ti, Mo, Nb) and surface terminations, its electrical and catalytic characteristics can be precisely tuned. This versatility allows MXene to surpass not just graphene but also other nanomaterials such as carbon nanotubes (CNTs) and metal-organic frameworks (MOFs) (Wu et al. 2021b). For instance, $Ti_3C_2T_x$ MXene demonstrates metallic conductivity (~10,000 S/cm) (Berkinbayeva et al. 2025) comparable to graphene, together with inherent electrocatalytic properties that exceed those of CNTs in the detection of biomarkers. Moreover, in comparison to gold nanoparticles or quantum dots, which frequently experience aggregation or toxicity, the biocompatibility of MXene ensures consistent performance in physiological conditions, an important advantage for in vivo sensing (Libralato et al. 2017; Mathew and Rout 2021).

Although other nanomaterials possess certain characteristics, such as the elevated surface area of graphene or the catalytic efficacy of metal nanoparticles (Zhu and Xu 2016), MXenes combine these properties effectively. Graphene oxide (GO) lacks the hydrophilicity characteristic of MXenes and requires extensive functionalization to immobilize biomolecules, whereas metal nanoparticles frequently encounter difficulties in electron transfer efficiency (Ammar et al. 2016). MXenes provide a "ready-to-use" platform characterized by fundamentally active surfaces, rapid electron transfer kinetics, and mechanical flexibility, making them suitable for wearable or implantable biosensors (Jothiramalingama et al. 2024).

In this chapter, we will be discussing the MXene-based electrochemical sensors for the detection of biomolecules. The first and second sections provide an overview of MXene-based direct biomolecule sensing and its combination with biorecognition elements for detection. Three primary biorecognition elements, such as enzymes, antibodies, and DNA, are selected, and relevant studies on their applications in biosensing are thoroughly analyzed in the subsequent subsections.

## 4.2 MXENE-BASED ELECTROCHEMICAL SENSORS FOR THE DIRECT DETECTION OF BIOMOLECULES

Biosensors are crucial for the detection and analysis of biomolecules, which advances biotechnology, food safety, environmental monitoring, and healthcare. They facilitate drug discovery, aid in genetic and metabolic research, and allow for the quick and precise identification of disease markers. Biosensors are used in medicine to help with infection diagnosis, different biomarker monitoring, and treatment customization. Biomolecules with clinical significance are important markers of health and illness. Their identification plays an important role in modern medicine by aiding in the diagnosis of conditions like diabetes, cardiovascular diseases, kidney disorders, and hormonal imbalances, and many others (Hemdan et al. 2024). The use of MXenes in biological sensors to detect clinically significant molecules is discussed in this section. Direct electrochemistry of MXenes for the detection of these molecules is explained here. MXene biosensors for the detection of $H_2O_2$, glucose, uric acid, bilirubin, ascorbic acid, gliotoxin, dopamine, and $H_2S$ are explained in the subsequent paragraphs.

The levels of $H_2O_2$, a reactive oxygen species involved in cellular signaling, can reveal oxidative stress, which has been connected to a number of illnesses, including diabetes, cancer, Alzheimer's, and Parkinson's. Sensitive $H_2O_2$ detection in the nano- to micromolar range is essential for disease diagnosis and health monitoring. A report by Lenka Lorencova of pristine and oxidized $Ti_3C_2T_x$ MXene revealed that exposing the material to an anodic potential in an aqueous solution leads to its oxidation, resulting in the formation of a $TiO_2$ layer or $TiO_2$ domains. This is followed by the dissolution of $TiO_2$ due to $F^-$ ions, reducing the electrochemical activity of the material compared to pristine $Ti_3C_2T_x$. Notably, $Ti_3C_2T_x$ has demonstrated exceptional catalytic performance in $H_2O_2$ reduction, with an $H_2O_2$ sensor based on this material achieving the highest sensitivity reported to date, featuring a detection limit of 0.7 nM (Lorencova et al. 2017).

Few studies have examined the direct electrooxidation of glucose on MXene, and composites are used in most of the literature on MXene-based glucose biosensors. K.A. Saraswathi et al. developed a disposable $Ti_3C_2T_x$-modified screen-printed gold electrode for non-enzymatic glucose detection in sweat without incorporating additional doping or composite materials (Saraswathi et al. 2024). While $Ti_3C_2T_x$ is often combined with other substances to enhance sensitivity, selectivity, and stability, individual nanomaterials are more commercially viable since composites can lose their nanoscale properties during large-scale applications. In this study, different etching agents (HF, LiF + HCl, and $NH_4F$) were used to synthesize $Ti_3C_2T_x$. Among them, Li–$Ti_3C_2T_x$ exhibited the best performance due to the role of lithium ions in facilitating delamination and activating a higher number of active sites. This activation led to improved ion diffusion and faster electron transfer, surpassing the performance of HF- and $NH_4F$–HCl-etched MXene. The use of biocompatible disposable electrodes further enhanced the enzyme-free glucose detection process. The sensor demonstrated exceptional sensitivity of 569.40 μA/mM/cm² within a linear range of 1–350 μM and achieved an outstanding limit of detection (LOD) of 5 nM.

Measuring uric acid levels is essential for identifying and managing health conditions such as gout, kidney disease, and heart-related issues. MXene was used as the catalytic material for the simultaneous detection of crucial biomolecules, including ascorbic acid (AA), dopamine (DA), and uric acid (UA) (Murugan et al. 2021). To investigate its electrochemical performance, a glassy carbon electrode (GCE) was modified with $Ti_3C_2T_x$ and analyzed using cyclic voltammetry and differential pulse voltammetry. The $Ti_3C_2T_x$-modified GCE ($Ti_3C_2T_x$/GCE) exhibited excellent electrocatalytic properties, effectively separating the oxidation peaks of AA (0.01 V), DA (0.21 V), and UA (0.33 V). Moreover, the sensor enabled simultaneous detection of these biomolecules in a physiological pH range, with detection limits of 4.6 μM for AA, 0.06 μM for DA, and 0.075 μM for UA. The linear detection ranges were found to be 100–1000 μM for AA, 0.5–50 μM for DA, and 0.5–4 μM and 100–1500 μM for UA. In order to detect UA, a novel structure consisting of nitrogen and sulfur co-doped on the surface of $Ti_3C_2$ nanosheets was employed as an electrochemical sensor and peroxidase mimic (Chen et al. 2022). It is discovered that NS-$Ti_3C_2$ NSs exhibit significantly greater electrochemical and peroxidase-like activity than $Ti_3C_2$ NSs.

Monitoring bilirubin levels is essential for assessing liver health, detecting hemolytic disorders, and managing jaundice, especially in infants. In this area, a molecularly imprinted electrochemical sensor based on an Indium tin oxide (ITO) electrode modified with $Ti_3C_2T_x$ MXene was used to detect bilirubin. First, chemical etching is used to create $Ti_3C_2T_x$ MXene, which is then drop-cast onto the ITO electrode surface. A bilirubin-imprinted electrode is then created by electropolymerizing ortho-phenylenediamine as a monomer in the presence of a sodium acetate buffer solution that contains the bilirubin template (Manoj and Ghrera 2024). In a related study, a bilirubin-imprinted electrode was created by electropolymerizing pyrrole, a monomer, in the presence of a sodium acetate buffer solution that contained the bilirubin template (Ghrera 2024). Due to its ability to increase sensor sensitivity, $Ti_3C_2T_x$ MXene was incorporated on the sensor surface before electropolymerization.

Measuring ascorbic acid (vitamin C) is important because it functions as a powerful antioxidant, supports collagen formation, boosts immune health, and aids in disease prevention. Accurate detection is essential for various medical and scientific purposes. Zaidi et al. proposed a simple controlled oxidation method that increases the surface area of MXene nanosheets, thereby enhancing its effectiveness in detecting AA (Zaidi et al. 2023). The MAX phase was initially converted into MXene by a moderately intensive layer delamination method. A flash oxidation in an open-air environment lasting 1.5 hours ensued, resulting in the flocculation of the oxidized MXene in the presence of $H_2SO_4$, ultimately yielding crumpled nanosheets. The modified nanosheets were subsequently incorporated into a GCE, and their capacity to detect AA was validated using cyclic voltammetry, electrochemical impedance spectroscopy, and differential pulse voltammetry, demonstrating a linear range for detection from 300 to 0.005 μM and a threshold for detection of 2 nM (with a 2.8% RSD and S/N = 3) (Figure 4.1(i)).

Next is Gliotoxin, which was shown to inhibit actin dynamics and integrin activation, two processes essential for phagocytosis; comparable outcomes were observed in primary and immortalized phagocytes (Schlam et al. 2016). A gliotoxin electrochemical biosensor was developed by integrating $Ti_3C_2$ MXene nanosheets with deoxyribonucleic acid nanostructures (Wang et al. 2019). This approach utilizes phosphoryl moieties on the DNA for enhanced interaction with the MXene, hence facilitating electron transport between redox entities and the surface of the electrode. Upon binding with gliotoxin, DNA undergoes a conformational alteration, producing a quantifiable electrochemical signal. The biosensor achieved a detection range from 5 pM to 10 nM, with an LOD at 5 pM for gliotoxins. This platform also has potential for adaptation to detect other mycotoxins (Figure 4.1(ii)).

Dopamine detection is vital for analyzing its role in brain activity and diagnosing neurological disorders like Parkinson's and schizophrenia, facilitating early intervention and treatment. For dopamine detection, a group Zhang et al. worked to create

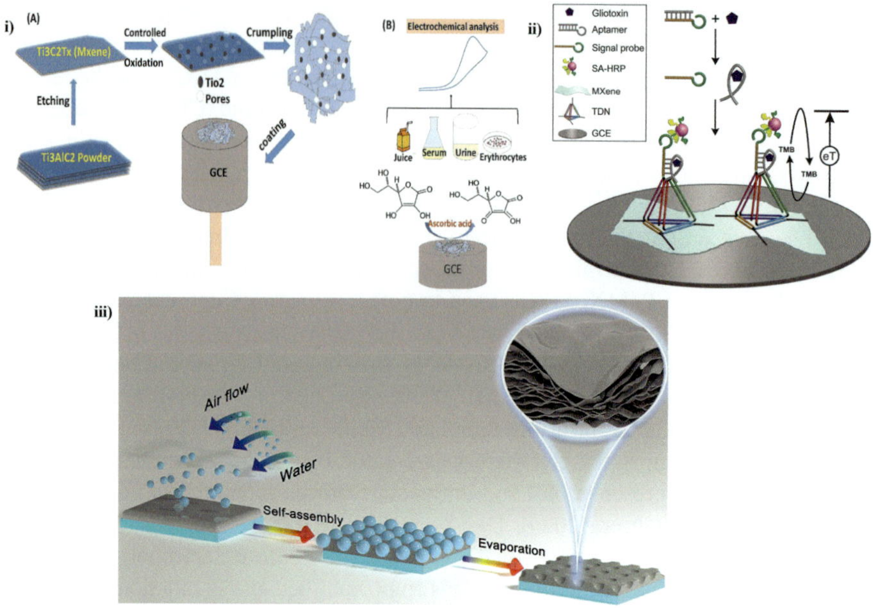


**FIGURE 4.1** (i) Schematic of (A) MXene formation, crumpled-MXene fabrication, and electrode design; (B) mechanism of AA detection and real sample evaluation using a cyclic voltammetry approach. Reproduced with permission from Zaidi et al. (2023) Copyright (2023) Elsevier. (ii) Schematic of tetrahedral DNA nanostructures/MXene-modified electrode and its principle. Reproduced with permission from Wang et al. (2019) Copyright (2019) Elsevier. (iii) Schematic for the three-dimensional porous MXene preparation process. Reproduced with permission from Zhang et al. (2023) Copyright (2023) Elsevier.

3D-porous MXene films by modifying them with a cationic surfactant functioning as an interlayer spacer to prevent the aggregation or restacking of nanosheets made of MXene and to improve their integration into the organic phase (Zhang et al. 2023) (Figure 4.1(iii)). MXene ($Ti_3C_2T_x$) and dimethyldioctadecylammonium bromide combination was agitating their solution (0.3 mg/mL) with just one-layer MXene dispersion (1 mg/mL) until the aqueous phase became translucent and the organic phase darkened. The compound was subsequently isolated and dehydrated at 60 °C for 12 hours. A honeycomb-patterned porous film was subsequently created by casting 50 μL of a 2 mg/mL MXene/DODA solution onto ITO glass at 75% humidity using the breath figure technique. Control films fabricated at varying humidity levels were additionally evaluated for electrochemical performance.

Ankitha et al. worked on. $Ti_3C_2T_x$ MXene, which was etched using 48% HF to remove the Al layer (Naguib et al. 2023), this MXene solution in dimethyl sulfoxide (DMSO) was dip-coated over carbon cloth yarn and dried to form a working electrode, which was further used in a bandage sensor setup with carbon thread and AgCl wire as counter and reference electrode, respectively. This sensor exhibited DA sensing capability within the range from 1 nM to 1 μM and also had an LOD of 316 pM (Ankitha, Shamsheera, and Rasheed 2024) (Figure 4.2).

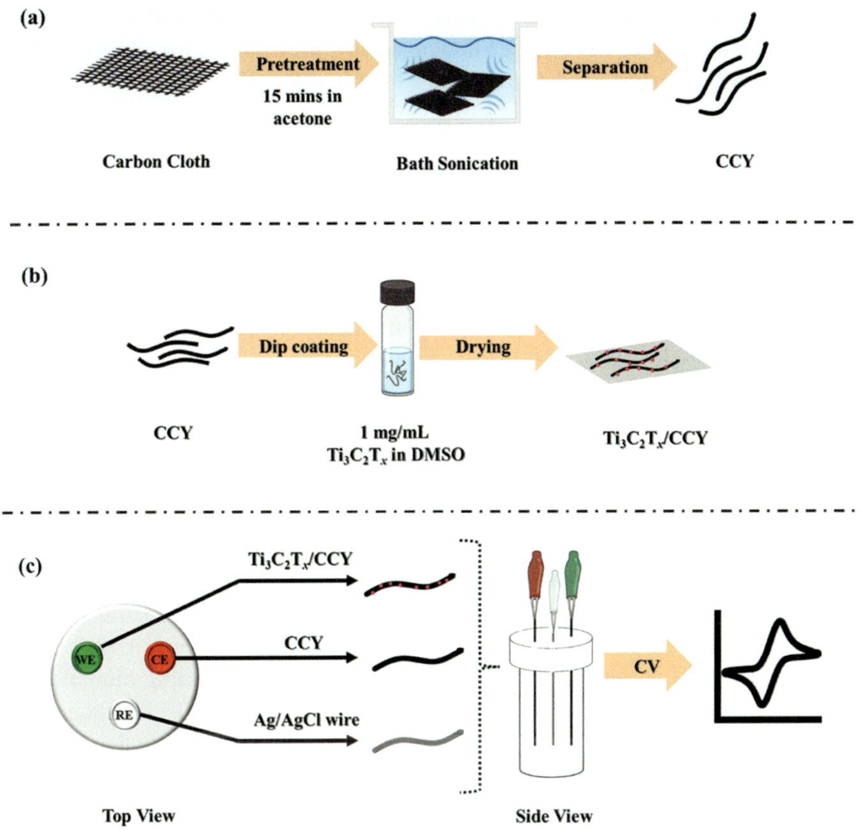


**FIGURE 4.2** (a–c) Schematic representing the process of making the MXene-modified working electrode and then using it for DA detection. Reproduced with permission from Ankitha, Shamsheera, and Rasheed (2024) Copyright (2024) ACS.

A study done by Zheng et al. showed the development of a 3D-printed electrochemical bioanalyte sensor by making graphene/poly-lactic acid electrodes with MXene quantum dots (QDs). The electrodes are first activated by exposing conductive graphene fibers, followed by the utilization of MXene QDs, enhancing electron transfer and increasing the active surface area. This change enables the sensor to accurately detect dopamine at very low concentrations (as low as 3 nM) across an extensive range (0.01–20 μM), ensuring great selectivity and stability (Wan et al. 2023).

Hydrogen sulfide ($H_2S$) is abundant in organisms, particularly within the brain, heart, gastrointestinal tract, vascular smooth muscle, and endothelial cells of animals (Nagy et al. 2014). In this portion of this chapter, we look on to the works that have included MXene-based sensors, which detect different levels of $H_2S$. Liu et al. reported a graphene-like structure consisting of multilayer $Ti_3C_2T_x$ MXene, effectively made using an easy hydrogen fluoride etching process, and it was employed for non-enzymatic $H_2S$ electrochemical sensing. The electrochemical test findings imply that the modified $Ti_3C_2T_x$ MXene/GCE has a high sensitivity of 0.587 μA/μM/$cm^2$ for the detection of trace $H_2S$. Simultaneously, it exhibits a low detection limit of 16.0 nM and an extensive detection range of 100 nM to 300 μM (Liu et al. 2023).

## 4.3 BIORECOGNITION MOLECULE-INCORPORATED MXENE-BASED ELECTROCHEMICAL SENSORS FOR THE DETECTION OF BIOMOLECULES

MXenes offer an exceptional platform for the incorporation of aptamers, enzymes, and DNA due to their unique properties, including a high surface area, excellent electrical conductivity, and adjustable surface chemistry (Ullah et al. 2022). These characteristics make MXenes highly suitable for functionalizing with various biorecognition elements, thereby improving the efficiency of electrochemical sensors. The expansive surface area allows for the attachment of a large number of aptamers, enzymes, or DNA, promoting enhanced interactions with target biomolecules. Additionally, MXenes' superior conductivity facilitates quick electron transfer, which is essential for generating sensitive and rapid electrochemical signals. Their surface can also be easily modified, enabling precise incorporation of aptamers for specific binding, enzymes for catalysis, and DNA for nucleic acid detection. This makes MXenes particularly effective for designing highly sensitive, selective, and durable electrochemical biosensors for detecting biomolecules (Khan and Andreescu 2020). Reports based on MXene-based electrochemical sensors for biomolecule detection using enzymes, antibodies, and nucleic acids as receptors are covered in this subsection.

### 4.3.1 Enzyme-based MXene Electrochemical Sensors

The high density of electroactive functional groups on few- and multi-layered MXenes enhances their capability to immobilize biomolecules, particularly enzymes, making them highly efficient for catalytic reactions and biosensing (Gürbüz and Ciftci 2024). The integration of enzymes with MXenes combines the electrocatalytic properties of both, creating an advanced platform for biosensing technologies. Biosensors based on enzyme-functionalized MXenes provide a viable solution to the limitations of traditional detection techniques, improving analytical performance across various real-world applications. Given their immense potential, this section explores enzyme-functionalized MXenes as an innovative interface for designing and developing next-generation bioanalytical monitoring systems.

MXenes boost the performance of immobilized enzymes by providing an optimized surface architecture, superior electrical conductivity, and a broad surface area, which support efficient electron exchange between the enzyme and the electrode. Their adjustable characteristics enhance detection accuracy, target specificity, and reaction speed in biosensor applications. Moreover, MXenes serve as electron transfer mediators and catalytic agents, further improving the electrochemical interactions essential for enzymatic function (Bilal et al. 2023). Enzymatic MXene-modified sensors for the detection of $H_2O_2$, glucose, creatinine, and aflatoxin B1 are explained here.

Horseradish peroxidase (HRP) enzymes were successfully immobilized onto two-dimensional MXene nanosheets with a vertically aligned junction structure to construct an electrochemical hydrogen peroxide biosensor. Incorporating MXene/chitosan onto GCE significantly amplified the redox peak signals compared to chitosan alone. After HRP immobilization, both HRP@chitosan/GCE and HRP@MXene/chitosan/GCE exhibited an increased reduction peak, with the latter showing the highest current enhancement (Ma et al. 2020). In metabolic processes, flavin adenine dinucleotide (FAD) functions as a redox cofactor and coenzyme. Using $Ti_3C_2T_x$ MXene, a novel immobilization method was created that improved the electrocatalytic reduction of hydrogen peroxide in a neutral medium and increased the redox activity of FAD (Nagarajan et al. 2021). However, the lack of catalytic and biocompatible sites made FAD immobilization on a bare GCE less successful. The catalyst material and charge transfer dynamics affected FAD's electron transfer efficiency. Higher FAD loading resulted from beneficial molecular interactions when the isoalloxazine ring remained exposed on the MXene surface, while the adenine ring of FAD promoted adsorption. The absence of scalable and reasonably priced biosensing technologies currently limits the effective real-time detection of $H_2O_2$ in cellular environments, which is crucial for early cancer diagnosis and treatment. The highly conductive MXene and biocompatible carboxymethyl cellulose (CMCS) are combined to create a highly precise water-based ink electrode (MXene/CMCS) with tunable viscosity (Figure 4.3(i)). Additional improvement via HRP nanoflower modification increased the stability, electrocatalytic activity, and electron transfer efficiency of the sensor. The electrochemical biosensor exhibited excellent biocompatibility and biosensing capabilities by efficiently detecting $H_2O_2$ released by HeLa cells (Wang et al. 2024).

By effectively electrostatically adhering glucose oxidase (GOx) polynanogel (PGOx) to MXene nanosheets, a nanocomposite was created. PGOx may improve the stability of enzymes, but MXene's large specific surface area may allow for the effective loading of PGOx nanocapsules and catalyze the breakdown of hazardous intermediate $H_2O_2$, which would lessen its impact on enzyme stability (Tong et al. 2024) (Figure 4.3(ii)). Yongjiu Lei et al. have designed a flexible, stretchable, and modular biosensor for analyzing perspiration in vitro. This advanced sensor incorporates a $Ti_3C_2T_x$ MXene/Prussian Blue (PB) composite combined with glucose oxidase (GOx) enzyme, allowing for reliable and highly sensitive detection of essential sweat biomarkers like lactate and glucose (Lei et al. 2019). With its durable construction and enhanced performance, this biosensor holds significant potential for real-time health monitoring applications. Minying Wu and colleagues developed a two-dimensional nanoreactor by immobilizing GOx onto $Ti_3C_2$ MXene nanosheets, enabling efficient glucose oxidation and hydrogen peroxide decomposition (Wu et al. 2021a). To enhance GOx attachment, the ultra-thin $Ti_3C_2$ nanosheets were modified with poly-L-lysine (PLL), a polypeptide rich in amino groups. This modification created a positively charged surface and allowed for a high enzyme-loading capacity of up to 50% by weight. The resulting $Ti_3C_2$-PLL-GOx nanosheets demonstrated significantly improved enzymatic activity compared to GOx immobilized on an inert porous silica substrate. This enhancement was largely due to the $Ti_3C_2$ nanosheets' ability to catalyze the breakdown of hydrogen peroxide, a potentially harmful byproduct of glucose oxidation.

(i)

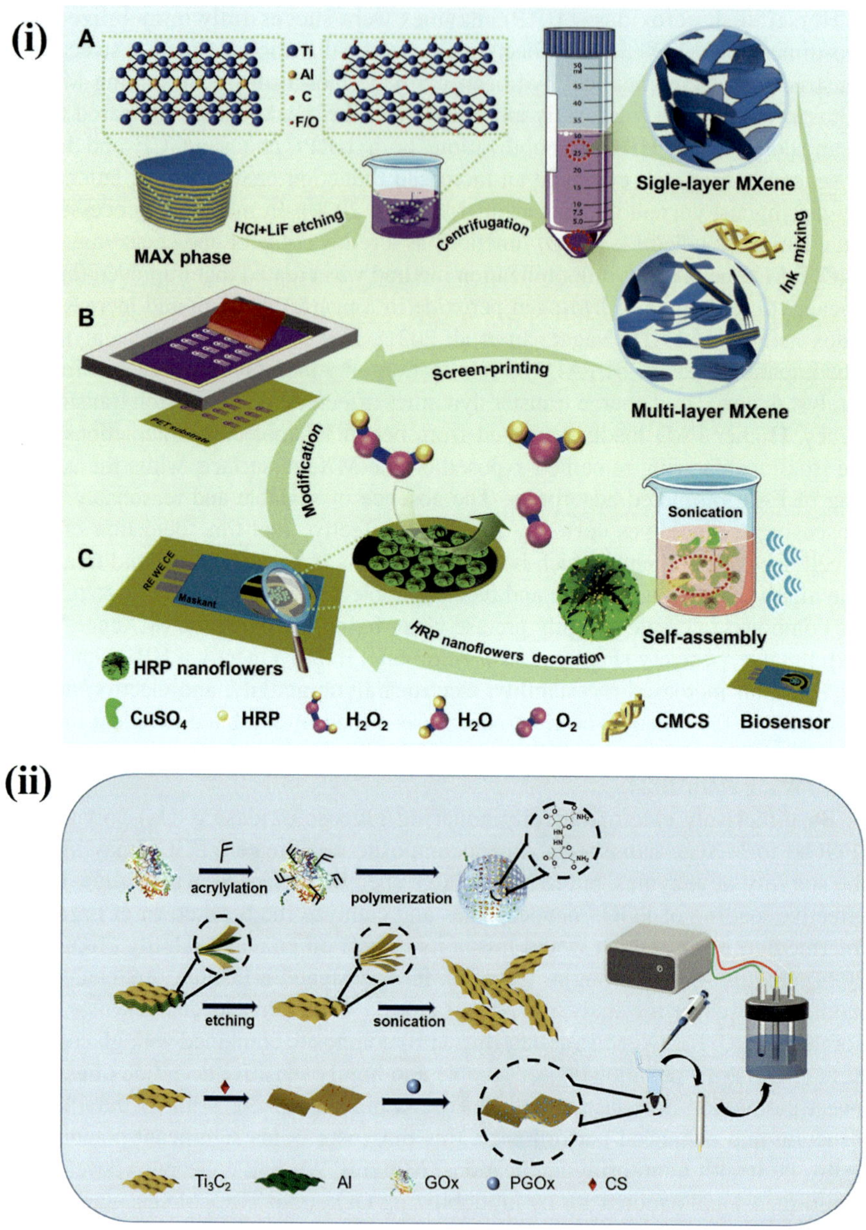


**FIGURE 4.3** (i) (A) The preparation method of MXene using the minimally intensive layer delamination (MILD) etching technique, (B) the screen-printing procedure employed for the rapid fabrication of the MXene-based three-electrode system, and (C) an illustration showing the modification process of the electrochemical biosensor designed for the detection of hydrogen peroxide. Reproduced with permission from Wang et al. (2024) Copyright (2024) RSC. (ii) Schematic representation of the synthesis pathway for glucose oxidase polynanogel integrated with MXene and chitosan. Reproduced with permission from Tong et al. (2024) Copyright (2024) Elsevier.

A rapid creatinine detection method was developed using screen-printed carbon electrodes (SPCEs) enhanced with MXene-based nanomaterials (Li et al. 2024). $Ti_3C_2T_x$ MXene, integrated with gold nanoparticles (MXene@AuNP) through in situ reduction, acted as a co-reactant accelerator, facilitating effective enzyme immobilization and improving sensor efficiency.

Xinyu Sun et al. reported the development of Au@MXene composites with high electrical conductivity and biocompatibility for biosensing applications (Sun et al. 2023). Liver microsomes were immobilized onto these composites and integrated with a GCE to detect aflatoxin B1 (AFB1) in phosphate-buffered solutions and maize samples. The layered structure and excellent biocompatibility of Au@MXene nanocomposites facilitated greater liver microsome adsorption while maintaining enzymatic activity. In this setup, CYP450 enzymes within the microsomes received electrons from the electrode under a negative potential, interacted with oxygen to form CYP-heme oxidants, and transferred reactive oxygen species to oxidize AFB1.

### 4.3.2 Aptamer-based MXene Electrochemical Sensors

Aptamers have drawn interest because of their potential to create affordable sensing devices due to their high production scalability, selectivity, and specificity. MXenes are a promising platform for integrating aptamers that target cancer-specific biomarkers because of their exceptional electroconductive qualities (Parihar et al. 2022). This combination addresses a vital healthcare need by providing substantial potential for early and accurate cancer diagnosis. A large number of MXene-based literature on cancer detection based on DNA biorecognition elements were found (Zhang et al. 2021a; Zhou et al. 2019; Yang et al. 2020; Wang et al. 2020). Similarly, Monireh Bolourinezhad et al. designed a DNA sensor for detecting COVID-19, utilizing GCE enhanced with an MXene/Pt/C nanocomposite (Bolourinezhad et al. 2023). The sensor employs a DNA/RNA hybridization approach to achieve high sensitivity and specificity in identifying SARS-CoV-2. In another study, a SPCE was utilized, which was first coated with $Ti_3C_2T_x$-MXene nanosheets and then functionalized with amino-modified probe DNA. This modification created a stable and efficient platform for detecting SARS-CoV-2 (Bharti et al. 2023). Using a lead-specific DNA oligonucleotide as the molecular recognition element, an electrochemical aptasensor with high selectivity and sensitivity was created to detect $Pb^{2+}$ (Rasheed et al. 2022). The sensor was built on an electrode modified with gold nanoparticles and $Nb_4C_3T_x$ MXene (Au@$Nb_4C_3T_x$), where an Au–S bond immobilized thiol-modified lead-binding DNA.

### 4.3.3 MXene-based Electrochemical Immunosensors

Electrochemical immunosensors make excellent diagnostic instruments that merge the great specificity of antigen–antibody interactions with the sensitivity of electrochemical movement. MXenes, a category of two-dimensional transition metal carbides, nitrides, and carbonitrides, are distinguished among sensing materials for their superior electrical conductivity, extensive surface area, and adaptable surface chemistry (Huang et al. 2022; Gogotsi and Huang 2021; Zou et al. 2022). MXene has distinctive characteristics that render it suitable for use as an electrode modifier and labels for detecting antibodies, offering potential advancements in early illness diagnosis, environmental monitoring, and food safety.

MXenes can be included in immunosensor systems in two principal capabilities. In one method, they function as electrode-modified nanomaterials (EMNs) that improve electron transport and enable the immobilization of capture antibodies. Salama's group created a label-free immunosensor for cancer-embryonic antigen (CEA) recognition by modifying $Ti_3C_2$ nanosheets of MXene with (3-Aminopropyl) triethoxysilane (APTES) to covalently bind anti-CEA antibodies. This sensor attained an exceptionally low detection limit, demonstrating the advantages of MXenes' strong conductivity and many surface sites for biomolecule attachment (Kumar et al. 2018). In prostate-specific antigen (PSA) detection, Chen et al. utilized $Ti_3C_2$ MXene-modified electrodes for immobilizing capture antibodies via $\pi$-stacking interactions. The subsequent application of horseradish peroxidase-labeled AuNPs for signal amplification enhanced the sensor's sensitivity, indicating that MXene-based platforms can compete with traditional enzyme-linked immunosorbent tests (ELISAs) in efficacy (Chen et al. 2019). Medetalibeyoglu et al. presented $Ti_3C_2$ MXene adorned with AuNPs as an electrocatalytic label for PSA detection, attaining an LOD of 3.0 fg/mL (Medetalibeyoglu et al. 2020). Zhang et al. incorporated MXene into a CuAu-LDH composite for dual-mode CEA detection (Zhang et al. 2021b). Liu et al. produced PdPtBP MNPs/MXene nanocomposites utilizing polydopamine-functionalized MXene as a substrate. These nanocomposites functioned as labels in a sandwich immunosensor, attaining remarkable sensitivity for KIM-1 detection (Liu et al. 2021).

## 4.4 CONCLUSION

The large surface area, high conductivity, and versatile functionalization properties of MXene materials, in particular, $Ti_3C_2T_x$, have made them attractive options for electrochemical sensing. By promoting electron transfer and enhancing surface interactions, these two-dimensional transition metal carbides and nitrides improve sensor performance. MXene-based electrochemical sensors have been successfully used to detect heavy metals, drugs, and biomolecules. These benefits have led to the growing use of MXene-based sensors in environmental analysis, food safety, and medical diagnostics, making them essential components for upcoming electrochemical sensing technologies.

## REFERENCES

Altalhi, Tariq. 2023. *Biosensors Nanotechnology*. John Wiley & Sons.

Ammar, Ali, Abdullah M. Al-Enizi, Mariam AlAli AlMaadeed, and Alamgir Karim. 2016. "Influence of graphene oxide on mechanical, morphological, barrier, and electrical properties of polymer membranes." *Arabian Journal of Chemistry* 9 (2):274–286.

Ankitha, Menon, Fathima Shamsheera, and Pathath Abdul Rasheed. 2024. "MXene-integrated single-stranded carbon yarn-based wearable sensor patch for on-site monitoring of dopamine." *ACS Applied Electronic Materials* 6 (1):599–610. doi: 10.1021/acsaelm.3c01668.

Berkinbayeva, Aknur S., Meruyert Nazhipkyzy, Muhammad Hashami, Zhanar B. Kudyarova, Togzhan Mashan, Aigul A. Beisebayeva, Kenes K. Kudaibergenov, and Madina Auyelkhankyzy. 2025. "The significant role of Mxene's derivatives and composites in supercapacitors: A review." *ES Energy & Environment* 27:1422.

Bharti, Anu, Sonam Singh, Dhanunjaya Munthala, Souradeep Roy, Soodkhet Pojprapai, Sanong Suksaweang, Sourav Sain, Susanta S. Roy, Julie J. Mohamed, and Devesh K. Avasthi. 2023. "Development of a nucleic acid-based screen printed electrochemical biosensor using $Ti_3C_2T_x$-MXene for the detection of SARS-CoV-2." *Microchemical Journal* 195:109521.

Bilal, Muhammad, Anil K. Singh, Hafiz M. N. Iqbal, and Grzegorz Boczkaj. 2023. "Enzyme-conjugated MXene nanocomposites for biocatalysis and biosensing." *Chemical Engineering Journal* 474:145020.

Bolourinezhad, Monireh, Majid Rezayi, Zahra Meshkat, Saman Soleimanpour, Majid Mojarrad, Seyed Hamid Aghaee-Bakhtiari, and Seyed M. Taghdisi. 2023. "Design of a rapid electrochemical biosensor based on MXene/Pt/C nanocomposite and DNA/RNA hybridization for the detection of COVID-19." *Talanta* 265:124804.

Chen, Da, Shuaibin Shao, Wei Zhang, Jingbo Zhao, and Meiling Lian. 2022. "Nitrogen and sulfur co-doping strategy to trigger the peroxidase-like and electrochemical activity of $Ti_3C_2$ nanosheets for sensitive uric acid detection." *Analytica Chimica Acta* 1197:339520.

Chen, Jialun, Ping Tong, Lingting Huang, Zhonghua Yu, and Dianping Tang. 2019. "$Ti_3C_2$ MXene nanosheet-based capacitance immunoassay with tyramine-enzyme repeats to detect prostate-specific antigen on interdigitated micro-comb electrode." *Electrochimica Acta* 319:375–381. doi: 10.1016/j.electacta.2019.07.010.

Chowdury, Md Shahjahan K., Ye J. Park, Sung B. Park, and Yong-il Park. 2024. "Two-dimensional nanostructured pristine graphene and heteroatom-doped graphene-based materials for energy conversion and storage devices." *Sustainable Materials and Technologies* 42: e01124.

Gautam, Rama, Nikhil Marriwala, and Reeta Devi. 2023. "A review: Study of Mxene and graphene together." *Measurement: Sensors* 25:100592.

Ghrera, Aditya S. 2024. "$Ti_3C_2T_x$MXene modified indium tin oxide (ITO) electrode for electrochemical sensing of bilirubin based on a molecularly imprinted pyrrole polymer." *Physica Scripta* 99 (5):055936.

Gogotsi, Yury, and Qing Huang. 2021. *MXenes: Two-dimensional Building Blocks for Future Materials and Devices*. ACS Publications.

Gürbüz, Berfin, and Fatih Ciftci. 2024. "Bio-electric-electronics and tissue engineering applications of MXenes wearable materials: A review." *Chemical Engineering Journal* 489:151230.

Hemdan, Mohamed, Mohamed A. Ali, Ahmed S. Doghish, Sherif S. A. Mageed, Ibrahim M. Elazab, Magdy M. Khalil, Mostafa Mabrouk, Diganta B. Das, and Alaa S. Amin. 2024. "Innovations in biosensor technologies for healthcare diagnostics and therapeutic drug monitoring: Applications, recent progress, and future research challenges." *Sensors* 24 (16):5143.

Huang, Hui, Caihong Dong, Wei Feng, Ying Wang, Bingcang Huang, and Yu Chen. 2022. "Biomedical engineering of two-dimensional MXenes." *Advanced Drug Delivery Reviews* 184:114178.

Jayathilaka, Wanasinghe A. D. M., Kun Qi, Yanli Qin, Amutha Chinnappan, William Serrano-García, Chinnappan Baskar, Hongbo Wang, Jianxin He, Shizhong Cui, and Sylvia W. Thomas. 2019. "Significance of nanomaterials in wearables: A review on wearable actuators and sensors." *Advanced Materials* 31 (7):1805921.

Jothiramalingama, R., T. Radhikab, N. P. Kavithac, H. Al-Lohedana, D. M. Aldhayan, and M. Karnand. 2024. "Synthesis, morphology and electrical property characteristics of MXene based titanium carbide ($Ti_3C_2T_x$) coating on non-woven cotton paper." *Digest Journal of Nanomaterials & Biostructures (DJNB)* 19 (1):15–24.

Khan, Reem, and Silvana Andreescu. 2020. "MXenes-based bioanalytical sensors: Design, characterization, and applications." *Sensors* 20 (18):5434.

Kumar, Saurabh, Yongjiu Lei, Niman H. Alshareef, M. A. Quevedo-Lopez, and Khaled N. Salama. 2018. "Biofunctionalized two-dimensional $Ti_3C_2$ MXenes for ultrasensitive detection of cancer biomarker." *Biosensors and Bioelectronics* 121:243–249.

Lei, Yongjiu, Wenli Zhao, Yizhou Zhang, Qiu Jiang, Jr-Hau He, Antje J. Baeumner, Otto S. Wolfbeis, Zhong L. Wang, Khaled N. Salama, and Husam N. Alshareef. 2019. "A MXene-based wearable biosensor system for high-performance in vitro perspiration analysis." *Small* 15 (19):1901190.

Li, Yilong, Yuteng Hang, Rusha Gopali, Xinxin Xu, Guanhua Chen, Xiaorong Guan, Ning Bao, and Yang Liu. 2024. "Point-of-care testing device platform for the determination of creatinine on an enzyme@CS/PB/MXene@AuNP-based screen-printed carbon electrode." *Microchimica Acta* 191 (9):534.

Libralato, Giovanni, Emilia Galdiero, Annarita Falanga, Rosa Carotenuto, Elisabetta De Alteriis, and Marco Guida. 2017. "Toxicity effects of functionalized quantum dots, gold and polystyrene nanoparticles on target aquatic biological models: a review." *Molecules* 22 (9):1439.

Liu, Changjin, Wei Yang, Xun Min, Decai Zhang, Xuhuai Fu, Shijia Ding, and Wenchun Xu. 2021. "An enzyme-free electrochemical immunosensor based on quaternary metallic/nonmetallic PdPtBP alloy mesoporous nanoparticles/MXene and conductive $CuCl_2$ nanowires for ultrasensitive assay of kidney injury molecule-1." *Sensors and Actuators B: Chemical* 334:129585.

Liu, Xinran, Liang He, Yuan Zhang, and Feng Li. 2023. "Improving the electrochemical properties of $Ti_3C_2T_x$ MXene for $H_2S$ electrochemical sensor by calcination." *Ceramics International* 49 (23):38575–38584.

Lorencova, Lenka, Tomas Bertok, Erika Dosekova, Alena Holazova, Darina Paprckova, Alica Vikartovska, Vlasta Sasinkova, Jaroslav Filip, Peter Kasak, and Monika Jerigova. 2017. "Electrochemical performance of $Ti_3C_2T_x$ MXene in aqueous media: Towards ultrasensitive $H_2O_2$ sensing." *Electrochimica acta* 235:471–479.

Ma, Bao-Kai, Mian Li, Ling-Zhi Cheong, Xin-Chu Weng, Cai Shen, and Qing Huang. 2020. "Enzyme-MXene nanosheets: Fabrication and application in electrochemical detection of $H_2O_2$." *无机材料学报* 35 (1) .

Manoj, and Aditya S. Ghrera. 2024. "Molecularly imprinted orthophenylenediamine with $Ti_3C_2T_x$MXene-based biosensor for sensitive detection of bilirubin." *Polymers for Advanced Technologies* 35 (3):e6346.

Markets and Markets. 2025. *Biosensors Market Size, Share, Trends & Industry Growth Analysis, by Type (Diagnosis, Monitoring, and Others), by Application (Point of Care Testing, Research Laboratories, Diagnostics Centre, and Others), Regional Insights, and Forecast from 2024 To 2032*. MarketsandMarkets.

Mathew, Minu, and Chandra S. Rout. 2021. "Electrochemical biosensors based on $Ti_3C_2T_x$ MXene: Future perspectives for on-site analysis." *Current Opinion in Electrochemistry* 30:100782.

Medetalibeyoglu, Hilal, Gül Kotan, Necip Atar, and Mehmet L. Yola. 2020. "A novel and ultrasensitive sandwich-type electrochemical immunosensor based on delaminated MXene@AuNPs as signal amplification for prostate specific antigen (PSA) detection and immunosensor validation." *Talanta* 220:121403.

Murugan, Nagaraj, Rajendran Jerome, Murugan Preethika, Anandhakumar Sundaramurthy, and Ashok K. Sundramoorthy. 2021. "2D-titanium carbide (MXene) based selective electrochemical sensor for simultaneous detection of ascorbic acid, dopamine and uric acid." *Journal of Materials Science & Technology* 72:122–131.

Nagarajan, Ramila D., Preethika Murugan, Kanagaraj Palaniyandi, Raji Atchudan, and Ashok K. Sundramoorthy. 2021. "Biocompatible MXene ($Ti_3C_2T_x$) immobilized with flavin adenine dinucleotide as an electrochemical transducer for hydrogen peroxide detection in ovarian cancer cell lines." *Micromachines* 12 (8):862.

Naguib, Michael, Murat Kurtoglu, Volker Presser, Jun Lu, Junjie Niu, Min Heon, Lars Hultman, Yury Gogotsi, and Michel W. Barsoum. 2023. "Two-dimensional nanocrystals produced by exfoliation of $Ti_3AlC_2$." In *MXenes*, edited ByYury Gogotsi, 15–29. Jenny Stanford Publishing.

Nagy, Péter, Zoltán Pálinkás, Attila Nagy, Barna Budai, Imre Tóth, and Anita Vasas. 2014. "Chemical aspects of hydrogen sulfide measurements in physiological samples." *Biochimica et Biophysica Acta (BBA)-General Subjects* 1840 (2):876–891.

Parihar, Arpana, Ayushi Singhal, Neeraj Kumar, Raju Khan, Mohd A. Khan, and Avanish K. Srivastava. 2022. "Next-generation intelligent MXene-based electrochemical aptasensors for point-of-care cancer diagnostics." *Nano-Micro Letters* 14 (1):100.

Perumal, Veeradasan, and Uda Hashim. 2014. "Advances in biosensors: Principle, architecture and applications." *Journal of Applied Biomedicine* 12 (1):1–15.

Purbayanto, Muhammad A. K., Volker Presser, Kacper Skarżyński, Marcin Słoma, Michael Naguib, and Agnieszka M. Jastrzębska. 2025. "MXenes: Multifunctional materials for the smart cities of tomorrow." *Advanced Functional Materials* 35 (10):2409953.

Rasheed, Pathath Abdul, Ravi P. Pandey, Khadeeja A. Jabbar, and Khaled A. Mahmoud. 2022. "$Nb_4C_3T_x$ (MXene)/Au/DNA aptasensor for the ultraselective electrochemical detection of lead in water samples." *Electroanalysis* 34 (10):1540–1546.

Research, and Markets. 2022. "Global biosensors market intelligence report to 2027 - Featuring ACON Laboratories, Bayer, Bio-Rad Laboratories and Biosensors International among others." *Business Wire*, 2022/09/09. https://www.businesswire.com/news/home/20220909005261/en/Global-Biosensors-Market-Intelligence-Report-to-2027---Featuring-ACON-Laboratories-Bayer-Bio-Rad-Laboratories-and-Biosensors-International-Among-Others---ResearchAndMarkets.com.

Research, and Markets. 2023. "Global biosensors market (2022 to 2027) - Industry trends, share, size, growth, opportunity and forecasts." *PR Newswire*, 2023/01/04. https://www.prnewswire.com/news-releases/global-biosensors-market-2022-to-2027---industry-trends-share-size-growth-opportunity-and-forecasts-301712431.html.

Saraswathi, K. A., M. S. B. Reddy, N. Jayarambabu, K. V. Rao, Shampa Aich, and T. V. Rao. 2024. "Non-invasive disposable 2D $Ti_3C_2T_x$ based enzyme free electrochemical sweat glucose biosensor." *Microchemical Journal* 205:111302.

Schlam, Daniel, Johnathan Canton, Marvin Carreño, Hannah Kopinski, Spencer A. Freeman, Sergio Grinstein, and Gregory D. Fairn. 2016. "Gliotoxin suppresses macrophage immune function by subverting phosphatidylinositol 3,4,5-trisphosphate homeostasis." *MBio* 7 (2), 10–1128. doi: 10.1128/mbio. 02242-15.

SkyQuest Technology Consulting Pvt, Ltd. 2024. *Biosensors Market Size, Share, Trends & Forecast | 2032*. SkyQuest Technology Consulting Pvt. Ltd.

Sun, Xinyu, Jiadi Sun, Yongli Ye, Jian Ji, Lina Sheng, Diaodiao Yang, and Xiulan Sun. 2023. "Metabolic pathway-based self-assembled Au@ MXene liver microsome electrochemical biosensor for rapid screening of aflatoxin B1." *Bioelectrochemistry* 151:108378.

Tong, Xinglai, Lin Jiang, Qi Ao, Xiaoxiao Lv, Ying Song, and Jun Tang. 2024. "Highly stable glucose oxidase polynanogel@MXene/chitosan electrochemical biosensor based on a multi-stable interface structure for glucose detection." *Biosensors and Bioelectronics* 248:115942.

Vikesland, Peter J., and Krista R. Wigginton. 2010. "Nanomaterial enabled biosensors for pathogen monitoring-A review." *Environmental Science & Technology* 44 (10):3656–3669.

Wan, Meijun, Asha Jimu, Honglin Yang, Jingqiu Zhou, Xuemei Dai, Yunchao Zheng, Jing Ou, Yaxin Yang, Jing Liu, and Lujun Wang. 2023. "MXene quantum dots enhanced 3D-printed electrochemical sensor for the highly sensitive detection of dopamine." *Microchemical Journal* 184:108180.

Wang, Haiyan, Jingjing Sun, Lin Lu, Xiao Yang, Jianfei Xia, Feifei Zhang, and Zonghua Wang. 2020. "Competitive electrochemical aptasensor based on a cDNA-ferrocene/MXene probe for detection of breast cancer marker Mucin1." *Analytica chimica acta* 1094:18–25.

Wang, Hui, Hui Li, Yue Huang, Menghua Xiong, Feng Wang, and Chao Li. 2019. "A label-free electrochemical biosensor for highly sensitive detection of gliotoxin based on DNA nanostructure/MXene nanocomplexes." *Biosensors and Bioelectronics* 142:111531. doi: 10.1016/j.bios.2019.111531.

Wang, Wenwu, Zeyu Ma, Qi Shao, Jiangwang Wang, Leixin Wu, Xiyao Huang, Zilu Hu, Nan Jiang, Jun Dai, and Liang He. 2024. "Multi-MXene assisted large-scale manufacturing of electrochemical biosensors based on enzyme-nanoflower enhanced electrodes for the detection of $H_2O_2$ secreted from live cancer cells." *Nanoscale* 16 (26):12586–12598.

Wu, Minying, Qian Zhang, Yuanyuan Fang, Chao Deng, Fangzhou Zhou, Yi Zhang, Xingdong Wang, Yi Tang, and Yajun Wang. 2021a. "Polylysine-modified MXene nanosheets with highly loaded glucose oxidase as cascade nanoreactor for glucose decomposition and electrochemical sensing." *Journal of Colloid and Interface Science* 586:20–29.

Wu, You, Xiaoming Li, Hui Zhao, Fubing Yao, Jiao Cao, Zhuo Chen, Xiaoding Huang, Dongbo Wang, and Qi Yang. 2021b. "Recent advances in transition metal carbides and nitrides (MXenes): Characteristics, environmental remediation and challenges." *Chemical Engineering Journal* 418:129296.

Yang, Xiao, Minghui Feng, Jianfei Xia, Feifei Zhang, and Zonghua Wang. 2020. "An electrochemical biosensor based on AuNPs/$Ti_3C_2$ MXene three-dimensional nanocomposite for microRNA-155 detection by exonuclease III-aided cascade target recycling." *Journal of Electroanalytical Chemistry* 878:114669.

Yoon, Jinho, Minkyu Shin, Joungpyo Lim, Ji-Young Lee, and Jeong-Woo Choi. 2020. "Recent advances in MXene nanocomposite-based biosensors." *Biosensors* 10 (11):185.

Zaidi, Shabi A., Hanan Sheikh, Muna Al-Mahasna, and Fathiya Elsin. 2023. "Crumpled MXene nanosheets for sensing of ascorbic acid in food, biological fluids, and erythrocytes in-vitro microenvironment." *International Journal of Biological Macromolecules* 249:126024. doi: 10.1016/j.ijbiomac.2023.126024.

Zhang, Huixin, Zonghua Wang, Feng Wang, Yimeng Zhang, Hongye Wang, and Yang Liu. 2021a. "$Ti_3C_2$ MXene mediated Prussian blue in situ hybridization and electrochemical signal amplification for the detection of exosomes." *Talanta* 224:121879.

Zhang, Junming, Yingyi Ma, Yuhang Han, Kaizheng Xu, Shun Yao, Lei Shi, and Min Zhu. 2023. "3D porous structure assembled from MXene via breath figure method for electrochemical detection of dopamine." *Chemical Engineering Journal* 452:139414. doi: 10.1016/j.cej.2022.139414.

Zhang, Mengmeng, Lisha Mei, Li Zhang, Xiao Wang, Xiaochen Liao, Xiuwen Qiao, and Chenglin Hong. 2021b. "$Ti_3C_2$ MXene anchors CuAu-LDH multifunctional two-dimensional nanomaterials for dual-mode detection of CEA in electrochemical immunosensors." *Bioelectrochemistry* 142:107943.

Zhou, Sijie, Chenxi Gu, Zhenzhen Li, Longyu Yang, Linghao He, Minghua Wang, Xiaoyu Huang, Nan Zhou, and Zhihong Zhang. 2019. "$Ti_3C_2T_x$ MXene and polyoxometalate nanohybrid embedded with polypyrrole: Ultra-sensitive platform for the detection of osteopontin." *Applied Surface Science* 498:143889.

Zhu, Qi-Long, and Qiang Xu. 2016. "Immobilization of ultrafine metal nanoparticles to high-surface-area materials and their catalytic applications." *Chem* 1 (2):220–245.

Zou, Jing, Jing Wu, Yizhou Wang, Fengxia Deng, Jizhou Jiang, Yizhou Zhang, Song Liu, Neng Li, Han Zhang, and Jiaguo Yu. 2022. "Additive-mediated intercalation and surface modification of MXenes." *Chemical Society Reviews* 51 (8):2972–2990.

# 5 Functionalised MXene-Modified Electrochemical Sensors

*Nishant Arya, Shivmuni Sarup, Niranjani Ramakrishnan, Mamta Kumari, and Sonu Gandhi*

## 5.1 INTRODUCTION TO FUNCTIONALISED MXENE-MODIFIED ELECTROCHEMICAL SENSORS

Electroanalysis has advanced significantly over time, particularly in the field of electrochemical sensors. Electrochemical sensors, which can be classified as either biosensors or non-biosensors, are detecting instruments that convert biological signals into electrical signals (Mahari & Gandhi, 2022; Roberts et al., 2023; Shrikrishna et al., 2024; Roberts, Mahari, and Gandhi 2022). They are classified based on the electrocatalytic and affinity characteristics of the modified electrodes for the detection of the target analytes. The distinction between biocatalytic and affinity sensors lies in the type of components incorporated into the detection device (Byakodi et al. 2022; Roberts and Gandhi 2022). For biocatalytic sensors, these components typically include enzymes or whole cells, whereas affinity-based sensors utilise entities such as antibodies or membrane receptors, enabling the generation of measurable electrical signals (Shrikrishna et al., 2025; Shrikrishna & Gandhi, 2025; Chouhan et al. 2023; Narlawar et al., 2022; Shahdeo & Gandhi, 2022). In addition to conventional electrochemical sensors, electrochemiluminescence and photochemical sensors have received substantial attention as innovative electroanalysis methods, owing to their high sensitivity, simple instrumentation, and low cost (Sagar Shrikrishna et al. 2024; Sahoo et al. 2024). With new achievements in sensors, nanomaterials with mouldable characteristics have attracted significant attention (Prakashan et al., 2023; Shahdeo et al., 2024). MXenes are among the several inorganic two-dimensional (2D) nanomaterials that have garnered significant attention from researchers due to their diverse applications, encompassing drug delivery, biosensing, and bioimaging, in addition to electromagnetic shielding, optics, and energy systems (Chouhan et al., 2023; Halder et al., 2024; Chouhan et al. 2023).

MXenes, which are early transition carbides or nitride-derived materials with highly tuneable physicochemical properties, are represented by the formula $M_{n+1}X_nT_x$, where M denotes a transition metal, $n$ ranges from 1 to 4, X represents carbon and/or nitrogen, and $T_x$ represents surface-terminating groups, such as –F, –O, –Cl, and –OH, as shown in Figure 5.1. Their chemical structures and single- or multi-layered microstructures are distinct, displaying interleaved M-layered sheets with

DOI: 10.1201/9781003736318-5 

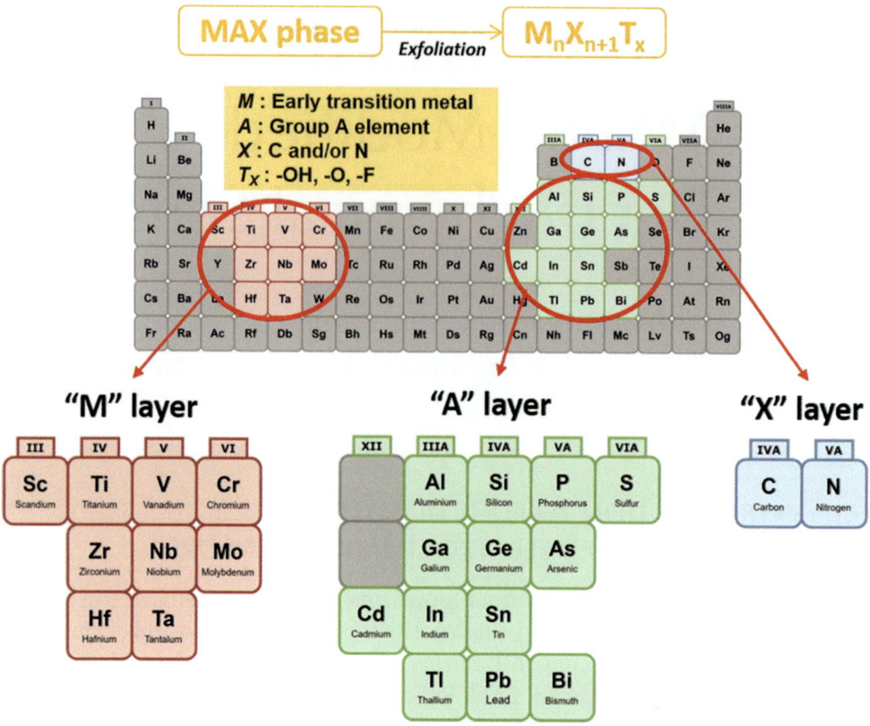


**FIGURE 5.1** The general element makeup of MXene and the MAX phase is as follows: M, metal of the early transition; $T_x$, surface functional group; A, Group A element; X, carbon (C) and/or nitrogen (N). Adapted with permission from Zamhuri et al. (2021). Copyright 2021. Copyright The Authors, some rights reserved; exclusive licensee Springer Nature. Distributed under Creative Commons Attribution Licence 4.0 (CC-BY).

exposed surface-terminating groups (X atoms) filling the octahedral positions between the M layers and $T_x$. In contrast to the precursor material, the MAX phase was denoted as $M_{n+1}AX_n$. Layer A was positioned between the M layers during the MAX phase. With the addition of surface-terminating groups and the removal of the A layer, MXenes are endowed with remarkable qualities, such as high electrical conductivity, high hydrophilicity, large specific surface area, and thermal stability. These characteristics make MXenes particularly suitable for electrochemical sensing applications (Liu et al. 2024; Sharma, Bhat, et al., 2025).

Surface modification of these nanomaterials is a promising method for controlling their properties and improving their stability (Choi et al. 2022). Altering surface characteristics can also introduce novel optical, chemical, and electronic properties. To achieve broader functionality, modifying the surface of MXenes with foreign molecules that possess distinctive attributes is highly desirable (Ji et al. 2019). Some examples of covalent modifications involve the integration of small molecules, such as organic amines, alkali metal hydroxides, epoxy compounds, and acid halides; the addition of single heteroatoms; or even polymerisation with macromolecules. These modifications not only improve the stability and performance but also widen their potential applications, making functionalised MXenes promising for study in materials science and engineering (Mozafari and Soroush 2021).

With ongoing advancements in MXene development, various functionalisation methods have been developed, leading to new applications. In this chapter, we comprehensively review different types of modification and functionalisation techniques for superior MXenes and their application in electrochemical sensing.

## 5.2 FROM MXENES TO FUNCTIONALISED MXENES: TECHNIQUES FOR FUNCTIONALISATION

Despite having numerous advantages, MXenes lack certain properties, such as stability and mechanical properties, for applications in various fields. The functionality of the surface is enhanced either by covalent or noncovalent modifications of the surface or by adjusting the modifications of the $T_x$ surface groups. These modifications are linked to the ability of the material to interact with the surroundings, and enhancing these functional groups has been linked to improving or tailoring the surface characteristics (Mozafari and Soroush 2021). The functionalisation of MXenes with enhanced surface properties can be achieved via chemical treatments, including hydrothermal, electrochemical anodisation, chemical vapour deposition, and solvothermal treatment approaches.

### 5.2.1 Hydrothermal Method

The hydrothermal method is a one-pot synthesis technique for the production of composite MXenes with enhanced surface functionalisation. This involves the introduction of different types of nanoparticles (NPs) onto the surface of MXenes under elevated temperatures and pressures. NPs were prepared under aqueous conditions and synthesised from metal oxides via dehydration and hydrolysis reactions. These reactions are carried out in a sealed environment above room temperature and at pressures higher than 1 atmosphere; this process produces composite MXenes with extraordinary tunability and predictable compositional variation. Li and coworkers used a hydrothermal method for the synthesis of $Bi_3TaO_7/Ti_3C_2$ for its application as a water contaminant remover, which has enhanced photocatalytic properties, as shown in Figure 5.2(a) (Li et al. 2020). Similarly, Cui et al. reported the synthesis of $Ti_3C_2$-loaded $Fe_3O_4$, which resulted in increased catalytic degradation of organic dyes, as shown in Figure 5.2(b) (Cui et al. 2020).

### 5.2.2 Solvothermal Method

The solvothermal process is the most widely used and efficient method for the functionalisation of MXenes. They are known for their simple operation, material crystallinity, distribution, and size and can be regulated more efficiently than hydrothermal processes. The only difference between solvothermal and hydrothermal processes is the replacement of aqueous conditions with nonaqueous organic solvents (Chen et al. 2018). This method was successfully used to fabricate nitrogen-doped MXenes for the detection of copper (Feng et al. 2020). Similarly, a Bi/BiOCl fabricated on

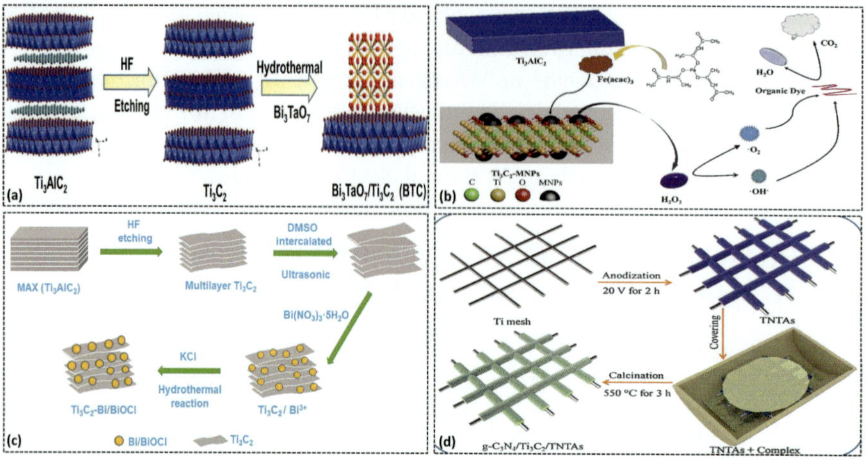


**FIGURE 5.2** (a) Hydrothermal synthesis of $Bi_3TaO_7/Ti_3C_2$ MXene. Adapted with permission from Li et al. (2020). Copyright 2020 Elsevier. (b) A one-step method for producing $Fe_3O_4$ nanoparticles on $Ti_3C_2$ MXene. Adapted with permission from Cui et al. (2020). Copyright 2020 Elsevier. (c) Solvothermal method for the preparation of $Ti_3C_2$-Bi/BiOCl composite. Adapted with permission from S. Wu et al. (2020). Copyright 2020 Elsevier. (d) Synthesis process involving electrochemical anodisation to produce g-$C_3N_4/Ti_3C_2$ MXene/TNTAs. Adapted with permission from Diao et al. (2020). Copyright 2020 Elsevier.

$Ti_3C_2$ material was developed, which exhibited enhanced light absorption and charge separation capabilities in the visible light spectrum, as shown in Figure 5.2(c) (Wu et al. 2020a).

### 5.2.3 Electrochemical Anodisation

An established technique for surface modification that is highly valued for its consistency and ease of application is electrochemical anodisation. This approach is especially attractive because it allows for fine control of the dimensions and properties of changed surfaces to fulfil specific application requirements by means of regulated anodic oxidation of metal substrates (Brownson and Banks 2012). Anodisation involves simultaneous redox reactions and ion diffusion propelled by the electric field, with a constant voltage or current provided between the anode and cathode. While anodisation has been employed for many years, it was not until the 1990s that researchers recognised the potential to create highly ordered nanoporous structures by meticulously adjusting parameters such as the electrolyte concentration and composition, anodisation voltage, and temperature (Beedri et al. 2014).

### 5.2.4 Chemical Vapour Deposition

Chemical vapour deposition (CVD) is a surface modification process that uses vapour-phase precursors in a chemical reaction to produce thin, frequently multilayered, solid coatings on substrate material. When applied to powders and nanomaterials, additional processes, such as efficient gas–solid mixing, individual particle treatment and submicron powder confinement at subatmospheric pressure, are needed (Verger et al. 2019). Powders can be treated with great efficacy using fluidised bed CVD (FBCVD), which involves forcing gas through a bed of solid particles at a critical flow rate to fluidise the bed. This makes it possible to polish flat surfaces—especially wafers—better than with other deposition techniques. FBCVD guarantees even and consistent deposition across and between wafers, particularly in low-pressure environments. Xu et al. fabricated high-quality, ultrathin $Mo_2C$ films exhibiting strong anisotropy with magnetic field orientation, consistent with Berezinskii–Kosterlitz–Thouless behaviour (Xu et al. 2015). Diao et al. synthesised ternary photocatalyst g-$C_3N_4$/$Ti_3C_2$/$TiO_2$ nanotube arrays (TNTAs) on titanium meshes using CVD, following the electrochemical anodisation of titanium nanotube arrays, as shown in Figure 5.2(d) (Diao et al. 2020).

## 5.3 IMPACT OF FUNCTIONALISATION ON PROPERTIES OF MXENES

MXenes have high conductivity and mechanical strength and are good materials for building electrodes. However, as compared to bare MXenes, functionalised MXenes exhibit improved sensing abilities. The main reason for this improvement can be attributed to the modification with compounds which might give the electrode electrocatalytic properties, increased specific surface area, or mass transfer. It has been observed that functionalising MXenes with certain metal NPs can enhance their electrocatalytic abilities. For instance, platinum (Pt)-, gold (Au)-, and palladium (Pd)-modified MXenes exhibit improved current signal, lower limit of detection (LOD), higher sensitivity, and selectivity (Rasheed et al. 2021; Prakashan et al., 2025; Ramya et al., 2025). For example, 10% Pt functionalisation of MXenes led to a rise in both cathodic and anodic current peaks with a high electron transfer rate, as shown in Figure 5.3(a), and a decrease in the $\Delta E_P$ (peak to peak separation) value.

In addition to enhancing the electrocatalytic properties, NPs functionalised MXenes also increase the surface area of the sensor. By modifying the MXenes with AuNPs for example, increased the active sites by increasing the surface area as in the case of AuNPs functionalised MXenes. This area can be utilised for the attachment of immunoglobulins, aptamers, or other detection molecules. In a study by Medetalibeyoglu, the specific surface area of bare MXene was calculated to be 42.13 $m^2/g$, whereas that of sulphur-doped MXene was calculated to be 269.70 $m^2/g$, which is more than a 6-time increment (Figure 5.3(b)) (Medetalibeyoglu et al. 2020b). Additionally, molybdenum disulfide ($MoS_2$) functionalised MXene exhibited a porous structure, increasing surface area, active areas and catalytic properties (Zhang et al. 2023). Wu et al. created 3D MXene aerogels and then applied a thin coating of polydimethylsiloxane (PDMS) to create compressible and electrically conductive MXene foams. The resultant porous structures demonstrated enhanced structural stability and durability (Figure 5.3(c)). The hybrid aerogel's unidirectional porous channels allowed it to exhibit an extraordinary conductivity of 2211 S/m (Wu et al. 2020b). Moreover, the functionalisation also resulted in the termination of various functional groups which help in the binding of analytes that enhances the detection limit. Carbon nanohorns (CNHs)-modified MXene (CNH/MXene) was used to build

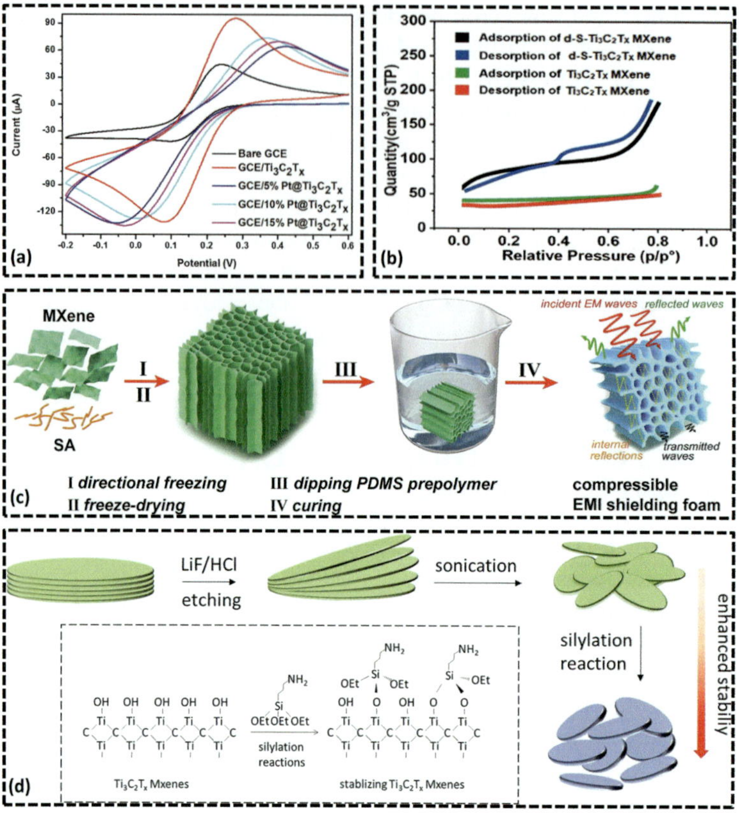

**FIGURE 5.3** (a) Cyclic voltammograms of Pt-modified MXene electrodes with different concentrations. The $\Delta E_P$ values were 103, 198, 463, 334, and 414 mV for bare GCE, $Ti_3C_2T_x$/GCE, 5%Pt@$Ti_3C_2T_X$/GCE, 10%Pt@$Ti_3C_2T_X$/GCE, and 15%Pt@$Ti_3C_2T_X$/GCE, in that order. The 10%Pt@$Ti_3C_2T_X$ nanocomposite demonstrated a higher rate for transfer of electrons than other Pt@$Ti_3C_2T_X$ nanocomposites, as evidenced by its lesser $\Delta E_P$ value and noticeable rise in both cathodic and anodic current peaks. Adapted with permission from Rasheed et al. (2021). Copyright 2021 Elsevier. (b) The specific surface areas of d-S-$Ti_3C_2T_x$ MXene exhibited a more than six-fold increase as compared to $Ti_3C_2T_x$ MXene, determined using the Brunauer–Emmett–Teller (BET) protocol, according to the nitrogen adsorption–desorption isotherms. These findings verified that the delamination process inhibited the flakes from stacking again, resulting from the wide pores accessible to sulphur. The result was a rise in d-S-$Ti_3C_2T_x$ MXene's specific surface area. Furthermore, d-S-$Ti_3C_2T_x$ MXene offered a surface area and electron transfer rate that were effective enough to capture numerous detection probes. Adapted with permission from Medetalibeyoglu, Beytur et al. (2020). Copyright 2020 Elsevier. (c) Polydimethylsiloxane (PDMS)-coated MXene foams were made by forming a three-dimensional (3D) MXene aerogel structure with the help of sodium alginate, and a thin coating of PDMS was applied for improvement in the porous architectures' structural stability and durability. These functionalised MXenes were compressible and electrically conductive. The 3D conductance network achieved good durability and compressibility due to the PDMS covering. Adapted with permission from X. Wu et al. (2020). Copyright 2020 Elsevier. (d) Modifying MXenes with silylation by functionalising with APTES enhanced the stability of MXenes. Adapted with permission from Ji et al. (2019). Copyright 2019 Elsevier.

a carbendazim (CBZ) sensor. CNH/MXene served as an anchor for the attachment of β-cyclodextrin–metal-organic frameworks, and this resulted in efficient electrocatalysis of CBZ (Tu et al. 2020).

One of the main issues with MXenes is their quick oxidation and degradation in aqueous environments and after atmospheric exposure, complicating the storage and maintenance of sensors. Ensuring sensor functionality over time is crucial. MXenes are particularly vulnerable to oxidative stress and temperature changes, mainly due to defects and vacancies created during synthesis. These defects act as nucleation centres for $TiO_2$ formation (Xia et al. 2021). It was observed that functionalisation of MXenes often resulted in significantly enhanced stability. Functionalisation of $Ti_3C_2T_x$ with hydrophobic groups in general led to protection from moisture and increased oxidative stability, e.g. silane coupling agents (Ji et al. 2019).

The intrinsic lack of oxidative stability of these materials was reduced by functionalising superhydrophobic self-assembled monolayers, such as fluoroalkylsilane, on MXenes as a molecular protective layer. By simply treating the surface of the $Ti_3C_2T_x$ nanosheets with a strongly fluorinated solution, a protective layer is created. Sufficient processing turns the hydrophilic $Ti_3C_2T_x$ into a superhydrophobic material. Fluoroalkylsilane-coated $Ti_3C_2T_x$ nanosheets demonstrated outstanding durability for environmental tests and stability in oxidising and humid settings in a study by Chen et al., exhibiting stability over a long period (Chen et al. 2020).

Additionally, the agglomeration of MXenes also poses a disruption in the electrochemical performance of MXenes. Effective prevention of MXene aggregation is essential for optimising its sensing performance (Prasanna et al. 2024). Functionalisation with 3-aminopropyltriethoxysilane (APTES) has been shown to prevent agglomeration (Zhang et al. 2020) and also improve the stability of MXenes (Figure 5.3(d)) (Ji et al. 2019). CNHs are also shown to prevent restacking of MXenes and thus result in increased specific surface area, charge transfer, and number of available active sites, leading to sensors which show improved current signal response (Chen et al. 2021).

## 5.4 APPLICATIONS OF FUNCTIONALISED MXENE-MODIFIED ELECTROCHEMICAL SENSORS

### 5.4.1 Functionalised MXene-Modified Electrochemical Sensors in Biomedical Sensing

Functionalised MXene-modified electrochemical sensors are serving as emerging advanced tools for the healthcare industry. These sensors can play a significant role in personalised medicine as well as real-time monitoring of analytes. For example, procalcitonin (PCT), a septicaemia biomarker, was detected using an electrochemical immunosensor formed of AuNPs functionalised with sulphur-doped MXene as a sensor platform and carboxylated graphitic carbon nitride (c-g-$C_3N_4$) as signal amplification, with an LOD of 2.0 fg/ml. $Ti_3AlC_2$ and sulphur were mixed and sintered at 1700°C for 3h to produce polycrystalline sulphur-doped MAX powder from which aluminium (Al) layer was etched out using HCl/LiF solution terminating the MXene with –F, –OH, –H functionalities (Medetalibeyoglu et al. 2020b). This technique was based on functionalisation by etching during synthesis. Additionally, the AuNPs functionalised on sulphur-doped MXene-modified glassy carbon electrode (GCE) enhanced the binding sites and acted as carrier for primary Ab against PCT. Signal amplification post binding of secondary Ab complex (c-g-$C_3N_4$) showed the presence of the analyte.

Furthermore, an electrochemical sensor was also developed for the detection of ifosfamide (IFO) (Kalambate et al. 2020), used in the treatment of several cancers but has certain deleterious effects on the body. Therefore, the detection of its concentration in body fluids may provide information to optimise the treatment. The MXene development in this sensor also involved etching out Al from MAX $Ti_3AlC_2$, although the etchant used was hydrofluoric acid (HF) to produce surface functional groups as in a previous study. The MXene was functionalised with embedded multi-walled carbon nanotubes (MWCNTs) and served as a spacer and bridged the gap between multilayers of MXenes and provided a conductive network, and chitosan by electrostatic forces was also functionalised on MXene for its film forming, water permitting, and mechanical properties. This functionalised MXene was coated on the GCE to make the working electrode. By functionalisation of MXene with these substances, the researchers enhanced the diffusion of analytes onto the electrode and highlighted the possibility of utilising functionalised MXene for drug analysis. The electrochemical sensing using this detection technique for acetaminophen, domperidone, and sumatriptan has also been done.

In addition to these, there has also been a development of a functionalised MXene-modified electrochemical sensor for the detection of urinary ammonia, an indicator for diseases such as renal tubular acidosis, Reye's syndrome, and lupus nephritis. Liu et al. developed a Pt-on-carbon nanotubes (CNTs)-modified MXene-based sensor for monitoring urinary ammonia. CNTs were created on MXene using the CVD method by pyrolysing zeolitic imidazolate framework-67 (ZIF-67) and melamine followed by electrodeposition of Pt on the CNTs. The urinary ammonia got adsorbed on the sensor surface and upon catalysis by Pt underwent dehydrogenation and subsequently led to the production of $N_2$. These reactions produced a change in current output which led to the accurate detection of urinary ammonia levels (Liu et al. 2024).

Research by Medetalibeyoglu et al. demonstrated the detection of prostate-specific antigen (PSA) electrochemically using an immunosensor (Figure 5.4(a)). This innovative approach utilised MXene-conjugated secondary antibody (Ab) as a detection molecule in the sensor rather than using MXene as the principal detection substrate. The sensor was made of p-aminothiophenol (ATP) conjugated AuNPs. This AuNPs–ATP duo was functionalised on graphene oxide (GO) and coated on GCE to make AuNPs–ATPGO/GCE. Primary Ab (1°Ab) against PSA was attached to the AuNPs of this modified electrode (1°Ab/AuNPs–ATPGO/GCE). To amplify the signal, a secondary Ab against PSA, conjugated with AuNPs in association with self-assembled delaminated MXene, was used. The Al was etched out from $Ti_3AlC_2$ MAX phase in the same way by using HCL/LiF solution to make $Ti_3C_2T_x$ MXene, leaving surface terminations ($T_X$ = –OH, –H, –F). $Ti_3C_2T_x$ MXene, upon treatment with Tetrabutylammonium hydroxide (TBAOH), yielded delaminated d-$Ti_3C_2T_x$ MXene, which underwent AuNP conjugation by $HAuCl_4$ to produce d-$Ti_3C_2T_x$ MXene@AuNPs, which further allowed the binding of secondary Ab (2°Ab to

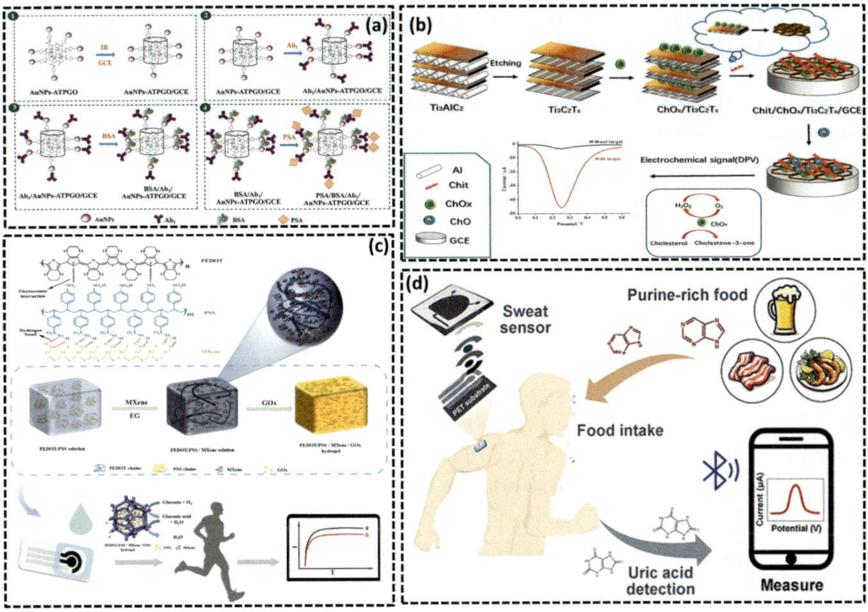


**FIGURE 5.4** (a) Prostate-specific antigen (PSA) was detected by using AuNPs-conjugated secondary antibody-functionalised MXene. Adapted with permission from Medetalibeyoglu, Kotan et al. (2020). Copyright 2020 Elsevier. (b) Cholesterol oxidase ($ChO_X$) was immobilised on $Ti_3C_2T_x$ MXene. Chitosan (Chit) and $Ti_3C_2T_x$ helped in the immobilisation of $ChO_X$ and played a role in conducting media for electron transfer. After the addition of a cholesterol-containing sample, the enzyme produced Cholestene-3-one and released $H_2O_2$, which resulted in an electrochemical signal that was detected by the sensor. Adapted with permission from Xia et al. (2021). Copyright 2021 Elsevier. (c) Glucose monitoring in sweat by a wearable sensor made of pol y(3,4-ethylenedioxythiophene):poly(styrene sulfonate) (PEDOT:PSS)-functionalised MXenes utilising an oxidase enzyme resulting in current changes which got detected by the sensor. Adapted with permission from Pan et al. (2024). Copyright 2024 Elsevier. (d) Detection of uric acid levels in sweat using 1,3,6,8-pyrene tetrasulfonic acid sodium salt (PyTS)-functionalised MXene. Uric acid (UA) oxidation and reduction resulted in the generation of current outputs, which aided in the detection of UA. Adapted with permission from F. Chen et al. (2024). Copyright 2024 American Chemical Society.

AuNPs. The binding of d-$Ti_3C_2T_x$ MXene@AuNPs/2°Ab to PSA bound to the 1°Ab/AuNPs–ATPGO/GCE provided a sensitive and specific detection of PSA with an LOD of 3.0 fg/ml (Medetalibeyoglu et al. 2020a).

Cholesterol imbalance has also garnered attention and therefore sensors have been made to detect cholesterol. To build the sensor detecting cholesterol levels, cholesterol oxidase ($ChO_X$) was immobilised on $Ti_3C_2T_x$ MXene. Chitosan (Chit) functionalisation of $Ti_3C_2T_x$ aided in the immobilisation of $ChO_X$ onto the MXene surface and played a role in conducting media for electron transfer. After the addition of the cholesterol-containing sample, the enzyme-catalysed reaction produced cholestene-3-one and released $H_2O_2$ which resulted in an electrochemical signal that was detected by the sensor (Xia et al. 2021). Figure 5.4(b) depicts a chitosan-modified MXene for the detection of cholesterol by $ChO_X$.

### 5.4.2 Flexible Sensing with Functionalised MXene-Modified Electrochemical Sensors

Significant progress in healthcare and analytical industries has focused on creating compact, reusable, and efficient point-of-care (POC) solutions. POC testing aims to develop diagnostic techniques that offer quick, reliable results, easy operation, cost-effectiveness, and minimal need for specialised equipment. Biosensors, which detect target molecules with high sensitivity and specificity, are central to these advancements (Chouhan et al. 2023). Flexible sensing technologies also known as wearable sensors (a part of POC testing) are crucial for real-time monitoring of analytes, addressing physiological perturbances as they occur (Sahoo et al., 2025). For instance, continuous tracking of glucose levels is essential to manage diabetes effectively, preventing adverse outcomes from glucose fluctuations. To address this, Pan et al. developed a wearable and non-invasive approach for monitoring glucose in sweat (Figure 5.4(c)). The monitoring of glucose was achieved via MXene-functionalised poly(3,4-ethylenedioxythiophene):poly(styrene sulfonate) (PEDOT:PSS) conducting composite hydrogels-based electrochemical biosensor incorporated on screen-printed carbon electrode (SPCE). MXene interacted with the hydrogel by hydrogen bonds (–OH, –H terminations of MXene), electrostatic bonds (negatively charged hydrophilic surfaces of MXene interacted with the PEDOT:PSS polymer chains, improving the mechanical properties and conductivity), and cross-linking (induced by ethylene glycol making a flexible hydrogel network incorporating MXene). However, the readings were taken after removing the SPCE patch and connecting it to an electrochemical workstation. The sensor displayed an LOD of 1.9 μM (Pan et al. 2024).

Similarly, sweat pH was also monitored using an electrochemical sensor utilising a flexible electrode synthesised on poly(ethylene terephthalate) substrate. $Ti_3C_2T_x$ MXene was functionalised with fluoroalkylsilane (FAS) to make FAS/$Ti_3C_2T_x$ and integrated it with polyaniline (PANI) to improve responsiveness, sensitivity, and reversibility. The resulting F/$Ti_3C_2T_x$/PANI pH sensor was used for real-time monitoring of sweat pH (Chen et al. 2022). In a similar approach, uric acid (UA) levels were monitored in sweat using an electrochemical sensor utilising 1,3,6,8-pyrene tetrasulfonic acid sodium salt (PyTS) functionalised MXene by pi–pi bonds (PyTS@$Ti_3C_2T_x$). This functionalised MXene showed improved electrocatalytic ability for UA oxidation and reduction by providing multiple oxidation–reduction active groups. The LOD for this sensor was 0.48 μM. The level of UA was observed by wireless electronics post sampling via flexible microfluidic sweat sampling integrated with PyTS@$Ti_3C_2T_x$-based flexible sensor (Figure 5.4(d)) (Chen et al. 2024a).

Ankitha et al.'s study created a novel method for the real-time detection of dopamine from synthetic human sweat by modifying a single-stranded carbon yarn-based sensor patch with MXene and employing a straightforward three-thread electrode design. The sensing electrode was a flexible conductive carbon yarn that had been treated with $Ti_3C_2T_x$ MXene using a dip-coating technique. With a detection limit of 316 pM, the MXene sensor's high surface area and superior conductivity allowed for increased sensitivity and selectivity in dopamine detection. The platform's potential for real-time physiological monitoring was further exhibited by testing using real samples, particularly in evaluating human performance under stress and making condition diagnoses using sweat analysis. Its comfortable, non-invasive on-body application and three-thread electrode arrangement make the patch a viable tool for continuous dopamine monitoring in patients with neurological diseases. This work demonstrated how wearable sensing technologies for POC diagnostics and personalised healthcare can be advanced by using MXene-integrated wearable sensor patches with simple electrode configurations (Ankitha, Shamsheera, and Rasheed 2024). The sensor patch could be easily sewn or attached to clothing, and positioned close to the skin, making it highly promising for wearable applications.

Human sweat consists of a lot of biomarkers and the detection of these biomarkers can reveal a lot about the physiological conditions. Most of the wearable biosensors detect the analytes present in sweat. Another such sensor detected the level of potassium in sweat. Potassium is an important biomarker for analysing the balance of body fluids. Abnormal potassium levels can cause complication which can even be potentially life threatening such as hypokalaemia. Zhang et al. developed an electrochemical sensor using MWCNTs-modified MXene for the detection of potassium. The sensor integrated a microfluidic sweat-collecting unit and utilised a hybrid multi-dimensional network of MWCNTs and MXene-$Ti_3C_2T_x$. Their design effectively immobilised the ionophore and improved ion diffusion, leading to heightened electrochemical stability and sensitivity. The microfluidic channel efficiently collected sweat and minimised contamination. Near Field Communication technology enabled wireless data transmission to a smartphone, making the system fully integrated and independent of external analysis equipment (Zhang et al. 2021).

These flexible sensors provide vast potential for real-time monitoring of physiological levels of analytes and aid in personalised intervention for diseased conditions as well as disease prevention.

### 5.4.3 Functionalised MXene-Modified Electrochemical Sensors in Environmental Sensing

The environment around us including soil, drinking water, air, etc. contains various contaminants hazardous to human health. Functionalised MXenes, play a major role in detecting contaminants in the environment. For example, Bisphenol A (BPA), a carcinogenic substance found in plastics abundant as pollutants in nature was detected by using platinum (Pt) functionalised MXenes ($Pt@Ti_3C_2T_x$) (Figure 5.5(a)). The nanocomposite was deposited on the surface of delaminated $Ti_3C_2T_x$ nanosheets by the self-reduction of Pt salt to Pt nanoparticles (PtNPs), where $Ti_3C_2T_x$ served as both a reducing agent and a conductive matrix at the same time. The presence of BPA in the sample resulted in a change in current output due to the two protons and two electrons that are transferred during BPA's electro-oxidation process, which was detected by the sensor (Rasheed et al. 2021).

Arsenic (As), another harmful substance present in the environment, poses a huge risk for drinking water contamination. Xiao et al. developed an amino-containing iron metal-organic framework (Fe-MOF)-modified MXene-based electrochemical sensor for the detection of As in water samples. The electrode was created in situ using a one-step hydrothermal process with MXene present. Based on the robust

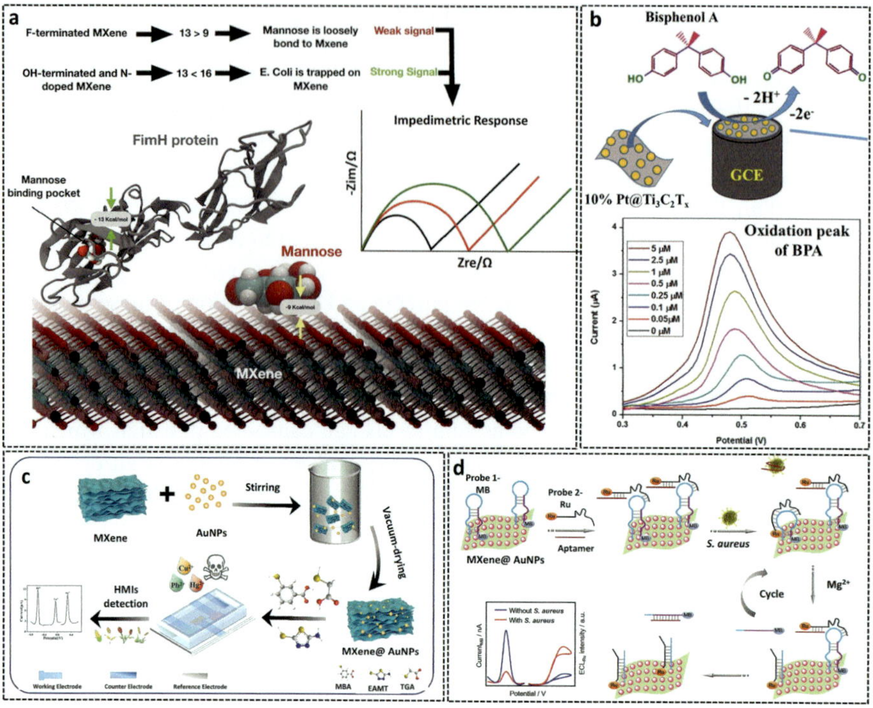


**FIGURE 5.5** (a) Detection of *E. coli* using mannose-functionalised MXene. Mannose binds with the fimH protein of *E. coli* by glucan multivalent interactions. This binding changed the current output which resulted in the detection of the bacteria. Adapted with permission from Ranjbar et al. (2022). Copyright 2022 Elsevier. (b) Detection of bisphenol A (BPA) by platinum (Pt)-functionalised MXenes ($Pt@Ti_3C_2T_x$). BPA in the sample resulted in a change in current output due to the two protons and two electrons that are transferred during BPA's electro-oxidation process. Adapted with permission from Rasheed et al. (2021). Copyright 2021 Elsevier. (c) Development of AuNP-functionalised MXene-based sensor for the detection of lead ($Pb^{2+}$), copper ($Cu^{2+}$), and mercury ($Hg^{2+}$). The sulfhydryl terminations for binding with these elements were provided by 3-mercaptobenzoic acid (MBA), 2-amino-5-mercapto-1,3,4-thiadiazole (EAMTD), and thioglycolic acid (TGA). Upon binding of the analyte with these sulfhydryl moieties, characteristic peaks were observed, which led to the detection of these elements. Adapted with permission from Y. Chen et al. (2024). Copyright 2024 Elsevier. (d) Aptasensor *S. aureus* detection used AuNPs–MXene and multiple probes, namely probe 1-MB (methylene blue) and probe 2-Ru. The former was a double-stem hairpin DNA attached with a methylene blue tag and also generated an electrochemical signal, apart from generating chemiluminescence; the latter probe was attached with a coordinate complex Ru(bpy)2(dcbpy)2+ and DNAzyme. Probe 2-Ru partially hybridised with an aptamer forming double-stranded DNA (dsDNA). This dsDNA was captured by the stem-loop probe 1-MB, and post addition of *S. aureus*, the aptamer dissociated from dsDNA and bound to the bacteria, which led to a conformational change and complete hybridisation of probe 1-MB with probe 2-Ru. The DNAzyme embedded in probe 2-Ru post release of aptamer got activated and cleaved probe 1-MB, which released a short oligonucleotide labelled with MB. The released oligonucleotide activated a competitive displacement reaction, further cleaving probe 1-MB and resulting in recycling amplification, which amplifies the signal by converting one target into multiple output signals, enhancing the sensitivity of the aptasensor. Adapted with permission from Shan et al. (2023). Copyright 2023 Elsevier.

As–hydroxyl bonding at the Fe-MOF/MXene electrode, a sensitive arsenite electrochemical sensor was created. The Fe-MOF/MXene-modified electrode showed an incredibly high current response to As(III), a super-low detection limit of 0.58 ng/l, and extraordinarily high sensitivity by employing square wave anodic-stripping voltammetry. The examination of actual water proved that the Fe-MOF/MXene-based sensor could detect As(III) in real-world applications (Xiao et al. 2022).

In a different approach, Feng et al. developed a sensor for detecting nitrite in tap water and vegetables. Nitrite, widely used in food and chemical industries, poses significant health risks as a carcinogenic contaminant, leading to conditions like methemoglobinemia. Its presence in water sources due to leakage or pollution has raised concerns, making the detection and monitoring of nitrite a critical focus for public health. A sensitive electrochemical sensor was developed to detect nitrite, which was based on carbon quantum dots (CQDs)-capped AuNPs-modified MXene-coated GCE. With a LOD of 0.078 μM, the sensor demonstrated excellent selectivity, high sensitivity, outstanding stability, and a wide linear range due to the robust catalytic activity of AuNPs and CQDs, the large specific surface area of MXene, and the exceptional electrical conductivity of both AuNPs and MXene (Feng et al. 2022).

In another study, the detection of metol, an organic compound and environmental contaminant was done using AuNPs on cerium stannate functionalised on MXene, synthesised via hydrothermal and ultrasonication methods. The LOD for this sensor was found to be 5.63 nM. Moreover, mercury ($Hg^{2+}$) which is a toxic element extensively found in the environment responsible for toxicity in living beings was detected using an amino-functionalised MXene. The amine group of the functionalised MXene made a coordinate effect with mercury and allowed for the accumulation of mercury on the MXene surface coated onto a GCE to be detected by the electrochemical sensor. The amine functionalisation was done by treating the MXene with APTES, which attaches to the surface of MXene through silane groups, while the amino groups remain available for interaction with mercury. The sensor showed an LOD of 8.27 nM (Liu et al. 2023b).

Other heavy metals such as lead ($Pb^{2+}$) and cadmium ($Cd^{2+}$) have also been detected using a functionalised MXene-modified electrochemical sensor. For instance, a bismuth NPs (BiNPs)-modified MXene electrochemical sensor was developed for simultaneous detection of $Pb^{2+}$ and $Cd^{2+}$. This sensor based on BiNPs@$Ti_3C_2T_x$ exhibited exceptional conductivity and a sizable effective surface area. When using anodic-stripping voltammetry to detect heavy metal ions, the BiNPs demonstrated exceptional efficacy. Following condition optimisation, $Pb^{2+}$ and $Cd^{2+}$ were detected concurrently by the BiNPs@$Ti_3C_2T_x$ nanosensor, with detection limits of 10.8 nM for $Pb^{2+}$ and 12.4 nM for $Cd^{2+}$. This sensor's large surface area, quick electron transfer, environmental friendliness, and ease of preparation made it a promising tool for detecting heavy metal ions (He et al. 2020).

The environmental sensing is not limited to the detection of toxic elements and substances such as mercury and BPA, but it also includes the detection of microorganisms such as *Escherichia coli* for the purpose of water, sanitation, and environmental protection. Ranjbar et al. developed a multivalent interface for detecting *E. coli* with MXene-based sensor (Figure 5.5(b)). MXene surface was functionalised with mannose, which bound to *E. coli*'s fimH protein by glucan multivalent interactions and resulted in detection. Al was etched out from bulk $Ti_3AlC_2$ with HF that produced more –OH groups, resulting in better contact with mannose by multivalent H-binding. The MXene in this case was modified with nitrogen, utilising the hydrothermal method using urea. This electrochemical device had an LOD of 10 CFU/ml and served a dual purpose by providing a nano filter to eliminate *E. coli* from water (Ranjbar et al. 2022).

### 5.4.4 Functionalised MXene-Modified Electrochemical Sensors in Food Safety

The detection of harmful substances in food is essential for maintaining food safety. A functionalised MXene-based sensor can serve an important role to detect hazardous elements and compounds in food and food products. For instance, CBZ, a fungicide with broad spectrum, and its residues cause chromosomal aberrations in organisms when they are exposed. Zhong et al. developed a modified MXene-based electrode with MXene@Ag nanoclusters (AgNCs) by directly reducing $AgNO_3$ solution in the existence of MXene and further modified it with amino-modified multi-walled carbon nanotubes ($NH_2$-MWCNTs) for the detection of CBZ in vegetable samples. This electrode was utilised for ratiometric electrochemical detection of CBZ with an LOD of 0.1 nM. AgNCs and $NH_2$-MWCNTs enhanced the electrocatalytic ability of MXene, and AgNCs provided an internal reference signal (Zhong et al. 2021).

Chen et al. developed AuNP-functionalised MXene via in situ growth of AuNPs on MXene surface by chemical reduction method and further modified the AuNPs with sulfhydryl groups to detect lead ($Pb^{2+}$), copper ($Cu^{2+}$) and $Hg^{2+}$ in food grains (Chen et al. 2024b). The analyte molecule upon adsorption on the sulfhydryl moiety resulted in its detection (Figure 5.5(c)). In a different approach, MXenes functionalised with aptamers have also been used for the development of sensors to regulate food safety. These aptasensors are able to detect heavy metals, pesticides, mycotoxins, antibiotics, and even viruses. A functionalised MXene-based aptasensor was developed to detect *Staphylococcus aureus*, a foodborne pathogenic bacterium with an innovative approach, which utilised AuNPs–MXenes which hybridised with multiple DNA probes and produced dual-mode ratiometric signals in the presence of the bacterium. The mechanism of detection involved an electro-chemiluminescent tagged probe which partially bound to another dsDNA probe with quenched DNAzyme. In the presence of a bacterium, the aptamer from the dsDNA probe got released, relieving the DNAzyme which cleaved the former probe and resulted in a dual signal: electrochemical and luminescent. The sensor showed an LOD of $5–10^8$ CFU/ml. The innovative mode of working of this sensor is depicted in Figure 5.5(d) (Shan et al. 2023). In a comparatively simpler approach, the detection of antibiotics was done in milk samples by functionalising aptamers on the MXene surface targeting tetracycline and kanamycin. The MXene coated on a GCE was functionalised with aptamers via pi–pi stacking of the aptamer over the MXene surface. The detection was done by measuring the change in current post binding of the analyte molecule with the aptamer (Xiong et al. 2024).

**TABLE 5.1**
**Summary of surface-modified MXenes and their advantages**

| Type of MXene | Functionalities | Synthesis methods | Advantage | References |
|---|---|---|---|---|
| $Ti_3C_2$ | $NaAlH_4$ | Ball milling | Enhanced hydrogen absorption and desorption properties | Jiang et al. (2020) |
| $Ti_3C_2$ | FePc | Self-assembly | Enhanced oxygen adsorption and reduction activity | Li et al. (2018) |
| $Ti_3C_2$ | $C_{12}E_6$ | Sonication | High energy and power density capacity | Xia et al. (2018) |
| $Ti_3C_2$ | Sodium alginate | Vacuum-assisted filtration | High electromagnetic interference (EMI) shielding efficiency | Shahzad et al. (2016) |
| $Ti_3C_2$ | PPy | In situ polymerisation | High volumetric and gravimetric capacitances | Boota et al. (2016) |
| $Ti_3C_2$ | PS | Electrostatic assembly | High electromagnetic interference (EMI) shielding efficiency | Sun et al. (2017) |
| $Ti_3C_2$ | PVA | Vacuum-assisted filtration | Enhanced mechanical and electrochemical properties | Ling et al. (2014) |
| $Ti_3C_2$ | PDDA | Vacuum-assisted filtration | Enhanced mechanical and electrochemical properties | Ling et al. (2014) |
| $Ti_3C_2$ | V-doped | Hydrothermal approach | High capacitance with greater stability | Gao, Zheng, and Lee (2019) |
| $Ti_3C_2$ | S-doped | Solution blending | High conductivity | Liang, Garsuch, and Nazar (2015) |
| $Ti_3C_2$ | S-doped | Heat and milling treatment | Enhanced electrochemical performance | Bao et al. (2019) |
| $Nb_2C$ | Et-modified | Hydrothermal approach | Enhanced electromagnetic wave absorption capacity | Jin et al. (2019) |
| $Nb_2C$ | N-doped | Hydrothermal approach | Enhanced electrochemical performance | Liu et al. (2019) |
| $V_4C_3$ | N-doped | Heat treatment | High specific capacitance and cycling stability | Li et al. (2019) |
| $V_2C$ | P-doped | Heat treatment | Enhanced catalytic performance | Yoon et al. (2019) |

Additionally, viruses are also present in foods, posing a threat to food safety. A novel electrochemical biosensor was developed for the sensitive detection of Norovirus (NoV), a significant foodborne pathogen causing acute gastroenteritis. The biosensor utilised a sophisticated peptide–target–aptamer sandwich approach, where AuNPs-conjugated black phosphorus nanosheets functionalised on MXenes (Au@BP@$Ti_3C_2$) and magnetic AuNPs-conjugated $ZnFe_2O_4$ magnetic NPs composited with magnetic covalent organic framework (Au@$ZnFe_2O_4$@COF) nanocomposites played key roles in enhancing the sensor's performance. The GCE was modified with these nanocomposites, and the specific binding interactions between the peptide, aptamer, and NoV facilitated the formation of a sandwich-like structure in the presence of the virus. The unique design of the biosensor allows it to detect NoV with unprecedented sensitivity, achieving a detection limit as low as 0.003 copies/ml. This remarkable sensitivity is attributed to the specific recognition capabilities of the peptide and aptamer, combined with the exceptional catalytic activity of the nanomaterials used (Liu et al. 2023a).

Moreover, the heightened use of pesticides such as organophosphates has led to their leaching in food chains via water contamination and food contamination. Accurate detection of organophosphorus pesticides (OPs) is crucial for food safety and environmental protection. This was achieved using an advanced electrochemical acetylcholinesterase (AChE) sensing platform. This platform, built on GCE decorated with Pt/$MoS_2$/$Ti_3C_2$ MXene (Pt/$MoS_2$/TM) and chitosan, excelled in detecting chlorpyrifos. The Pt/$MoS_2$/TM composite offered excellent biocompatibility, conductivity, stability, and a large surface area, enhancing the sensor's electrochemical performance. The sensor demonstrated high sensitivity and a wide detection range, with a low detection limit of 0.471 pM. It effectively detected OPs in fruits and vegetables, showing great potential for environmental monitoring and agricultural product safety (Wang et al. 2024).

## 5.5 FUTURE PERSPECTIVES AND CHALLENGES

Functionalised MXenes in electrochemical sensor technology have potential for remarkable advancements, driven by ongoing innovations in material science, nanotechnology, and interdisciplinary research. With the advent of newer functionalisation techniques, the performance of MXenes in electrochemical sensors is enhanced substantially (Banga et al., 2019a; Bhat et al., 2025). Future research is expected to focus on developing more efficient, precise, and eco-friendly methods for functionalising MXenes. The ability to modify the surface properties of MXenes with precision has enabled the creation of sensors with enhanced sensitivity, selectivity, and stability, built for specific applications in fields such as healthcare, environmental monitoring, and food safety.

The integration of an MXene-based sensors with emerging technologies such as nanotechnology, wearable devices, and the Internet of Things has the potential to broaden their applications. In particular, the miniaturisation of sensors and their incorporation into wearable devices is revolutionising personalised medicine, enabling continuous and real-time health monitoring ("Aptasensors for Point-of-Care Diagnostics of Cancer," 2023; Banga et al., 2019b; Nandi et al., 2020). Furthermore, the potential of MXenes in multi-analyte detection could lead to the development of sophisticated sensing platforms capable of providing comprehensive diagnostic information from a single device. Innovations in the design and functionalisation of MXenes are expected to lead to sensors that are capable of detecting a wider range of analytes with higher accuracy, even in complex and challenging environments. The development of multi-analyte sensors, in particular, holds great promise for applications that require the simultaneous detection of multiple substances, such as in healthcare diagnostics and environmental analysis.

One of the significant challenges for MXene-based sensors is the scalability of their production. Future efforts should focus on overcoming the barriers to mass production, ensuring the manufacturing of these material cost-effectively while maintaining good performance. Additionally, addressing regulatory and standardisation issues is crucial for the widespread commercialisation of MXene-based sensors, particularly in sensitive applications such as healthcare and environmental monitoring.

As the world increasingly prioritises sustainability, the development of eco-friendly functionalised MXene-based sensors is becoming a significant area of focus (Sharma, Sahoo, et al., 2025; Shrikrishna et al., 2021). This includes not only the use of green chemistry in the functionalisation processes but also the design of sensors that are sustainable throughout their life cycle, from production to disposal. Furthermore, the application of functionalised MXene-based sensors in environmental monitoring should contribute to global sustainability efforts, providing new tools for detecting and mitigating environmental hazards.

Despite the promising future of functionalised MXene-based sensors, several challenges remain, such as the use of eco-friendly safer solvents or etchants, minimising waste generation, and designing for degradation after use. Addressing these challenges requires concerted research efforts to improve the stability and reproducibility of functionalised MXene-based electrochemical biosensors, as well as to better understand their long-term behaviour in various environments.

In conclusion, the future of functionalised MXenes in electrochemical sensing is bright, with vast potential for innovation and impact across multiple domains. By continuing to advance the science and technology of MXenes, we can look forward to a new era of highly effective, sustainable, and versatile sensors that will play a crucial role in addressing the challenges of the 21st century.

## 5.6 CONCLUSION

Functionalised MXenes, with their distinctive structure and customisable surface properties, have become a powerful class of materials in the field of electrochemical sensing, providing numerous advantages. The various functionalisation approaches, including surface modifications, hybrid structures, and biocompatible coatings, have significantly enhanced the performance of MXene-based sensors. These advancements have enabled the precise detection of a wide range of analytes, from biomolecules in healthcare to pollutants in environmental monitoring. Despite the progress, challenges remain in ensuring the long-term stability of these sensors and in scaling up their production for commercial use. Future research will need to address these issues while also exploring the integration of MXene-based sensors with flexible and wearable devices. As the field continues to evolve, functionalised MXenes are poised to play a crucial role in advancing electrochemical sensing technologies, with broad applications across various industries.

## AUTHOR CONTRIBUTION

Conceptualisation, S.G. and N.A.; writing—original draft preparation, N.A., S.S., N.R., and M.K.; writing—review and editing, N.A.; supervision, S.G.

## CONFLICT OF INTEREST

The authors declare no conflict of interest.

## ACKNOWLEDGEMENTS

The authors are grateful to the Department of Biotechnology (DBT), New Delhi (Grant Number BT/PR52345/BSA/33/37/2024(CN19592)) and Indian Council of Medical Research (ICMR), New Delhi (Grant Number NIMR/Proj/Intra/HMM/134/24). Nishant Arya would like to acknowledge DBT Fellowship (DBT/2023–24/NIAB/2354), New Delhi.

## REFERENCES

Ankitha, Menon, Fathima Shamsheera, and P Abdul Rasheed. 2024. "MXene-Integrated Single-Stranded Carbon Yarn-Based Wearable Sensor Patch for On-Site Monitoring of Dopamine." *ACS Applied Electronic Materials* 6 (1): 599–610. doi:10.1021/acsaelm.3c01668.

Bao, Weizhai, Christopher E. Shuck, Wenxue Zhang, Xin Guo, Yury Gogotsi, and Guoxiu Wang. 2019. "Boosting Performance of Na–S Batteries Using Sulfur-Doped $Ti_3C_2T_x$ MXene Nanosheets with a Strong Affinity to Sodium Polysulfides." *ACS Nano* 13 (10): 11500–509. doi:10.1021/acsnano.9b04977.

Beedri, Niyamat I., Yusuf A. Inamdar, Suhail A. A. R. Sayyed, Arif V. Shaikh, Sandesh R. Jadkar, and Habib M. Pathan. 2014. "Electrochemical Anodization of Nano Crystalline ZnO Films in Acidic Bath and Their Properties." *Advanced Science Letters* 20 (5): 1147–50. doi:10.1166/asl.2014.5500.

Boota, Muhammad, Babak Anasori, Cooper Voigt, Meng-Qiang Zhao, Michel W. Barsoum, and Yury Gogotsi. 2016. "Pseudocapacitive Electrodes Produced by Oxidant-Free Polymerization of Pyrrole between the Layers of 2D Titanium Carbide (MXene)." *Advanced Materials* 28 (7): 1517–22. doi:10.1002/adma.201504705.

Brownson, Dale A. C., and Craig E. Banks. 2012. "The Electrochemistry of CVD Graphene: Progress and Prospects." *Physical Chemistry Chemical Physics* 14 (23): 8264. doi:10.1039/c2cp40225d.

Byakodi, Manisha, Narlawar S. Shrikrishna, Riya Sharma, Shekhar Bhansali, Yogendra Mishra, Ajeet Kaushik, and Sonu Gandhi. 2022. "Emerging 0D, 1D, 2D, and 3D Nanostructures for Efficient Point-of-Care Biosensing." *Biosensors and Bioelectronics: X* 12 (December): 100284. doi:10.1016/j.biosx.2022.100284.

Chen, Fan, Jinhao Wang, Lijuan Chen, Haoliang Lin, Dongxue Han, Yu Bao, Wei Wang, and Li Niu. 2024a. "A Wearable Electrochemical Biosensor Utilizing Functionalized $Ti_3C_2T_x$ MXene for the Real-Time Monitoring of Uric Acid Metabolite." *Analytical Chemistry* 96 (9): 3914–24. doi:10.1021/acs.analchem.3c05672.

Chen, Lijuan, Fan Chen, Gang Liu, Haoliang Lin, Yu Bao, Dongxue Han, Wei Wang, Yingming Ma, Baohua Zhang, and Li Niu. 2022. "Superhydrophobic Functionalized $Ti_3C_2T_x$ MXene-Based Skin-Attachable and Wearable Electrochemical PH Sensor for Real-Time Sweat Detection." *Analytical Chemistry* 94 (20): 7319–28. doi:10.1021/acs.analchem.2c00684.

Chen, Shuxian, Min Shi, Jing Yang, Yongfang Yu, Quan Xu, Jingkun Xu, Xuemin Duan, Yansha Gao, and Limin Lu. 2021. "MXene/Carbon Nanohorns Decorated with Conductive Molecularly Imprinted Poly(Hydroxymethyl-3,4-Ethylenedioxythiophene) for Voltammetric Detection of Adrenaline." *Microchimica Acta* 188 (12): 420. doi:10.1007/s00604-021-05079-3.

Chen, Winston Y., Sz-Nian Lai, Chao-Chun Yen, Xiaofan Jiang, Dimitrios Peroulis, and Lia A. Stanciu. 2020. "Surface Functionalization of $Ti_3C_2T_x$ MXene with Highly Reliable Superhydrophobic Protection for Volatile Organic Compounds Sensing." *ACS Nano* 14 (9): 11490–501. doi:10.1021/acsnano.0c03896.

Chen, Xu, Xueke Sun, Wen Xu, Gencai Pan, Donglei Zhou, Jinyang Zhu, He Wang, Xue Bai, Biao Dong, and Hongwei Song. 2018. "Ratiometric Photoluminescence Sensing Based on $Ti_3C_2$ MXene Quantum Dots as an Intracellular PH Sensor." *Nanoscale* 10 (3): 1111–18. doi:10.1039/C7NR06958H.

Chen, Yuanyuan, Yiyi Liu, Peng Zhao, Yi Liang, Yi Ma, Huan Liu, Jingzhou Hou, Changjun Hou, and Danqun Huo. 2024b. "Sulfhydryl-Functionalized 3D MXene-AuNPs Enabled Electrochemical Sensors for the Selective Determination of $Pb^{2+}$, $Cu^{2+}$ and $Hg^{2+}$ in Grain." *Food Chemistry* 446 (July): 138770. doi:10.1016/j.foodchem.2024.138770.

Cho, Il-Hoon, Dong H. Kim, and Sangsoo Park. 2020. "Electrochemical Biosensors: Perspective on Functional Nanomaterials for On-Site Analysis." *Biomaterials Research* 24 (1): 6. doi:10.1186/s40824-019-0181-y.

Chouhan, Raghuraj S., Maitri Shah, Drishya Prakashan, Ramya P. Rajan, Pratik Kolhe, and Sonu Gandhi. 2023. "Emerging Trends and Recent Progress of MXene as a Promising 2D Material for Point of Care (POC) Diagnostics." *Diagnostics* 13 (4): 697. doi:10.3390/diagnostics13040697.

Chouhan, Raghuraj Singh, Sonu Gandhi, Suresh K. Verma, Ivan Jerman, Syed Baker, and Marko Štrok. 2023. "Recent Advancements in the Development of Two-Dimensional Nanostructured Based Anode Materials for Stable Power Density in Microbial Fuel Cells." *Renewable and Sustainable Energy Reviews* 188: 113813. https://doi.org/10.1016/j.rser.2023.113813.

Cui, Yi, Meiying Liu, Hongye Huang, Dongao Zhang, Junyu Chen, Liucheng Mao, Naigen Zhou, Fengjie Deng, Xiaoyong Zhang, and Yen Wei. 2020. "A Novel One-Step Strategy for Preparation of $Fe_3O_4$-Loaded $Ti_3C_2$ MXenes with High Efficiency for Removal Organic Dyes." *Ceramics International* 46 (8): 11593–601. doi:10.1016/j.ceramint.2020.01.188.

Deepshikha Shahdeo, Sonu Gandhi. 2023. Aptasensors for point-of-care diagnostics of cancer, edited by Parihar, Arpana, Pushpesh Ranjan, and Raju Khan. IOP Publishing, 7–19. https://doi.org/10.1088/978-0-7503-5012-9.

Diao, Yabing, Ming Yan, Xiaoming Li, Chengyun Zhou, Bo Peng, Hui Chen, and Hao Zhang. 2020. "In-Situ Grown of $g\text{-}C_3N_4/Ti_3C_2/TiO_2$ Nanotube Arrays on Ti Meshes for Efficient Degradation of Organic Pollutants under Visible Light Irradiation." *Colloids and Surfaces A: Physicochemical and Engineering Aspects* 594 (June): 124511. doi:10.1016/j.colsurfa.2020.124511.

Feng, Xiwen, Guangda Han, Jihai Cai, and Xiaoying Wang. 2022. "Au@Carbon Quantum Dots-MXene Nanocomposite as an Electrochemical Sensor for Sensitive Detection of Nitrite." *Journal of Colloid and Interface Science* 607 (February): 1313–22. doi:10.1016/j.jcis.2021.09.036.

Feng, Yefeng, Furong Zhou, Qihuang Deng, and Cheng Peng. 2020. "Solvothermal Synthesis of in Situ Nitrogen-Doped $Ti_3C_2$ MXene Fluorescent Quantum Dots for Selective $Cu^{2+}$ Detection." *Ceramics International* 46 (6): 8320–27. doi:10.1016/j.ceramint.2019.12.063.

Gao, Zhi-Wen, Weiran Zheng, and Lawrence Y. S. Lee. 2019. "Highly Enhanced Pseudocapacitive Performance of Vanadium-Doped MXenes in Neutral Electrolytes." *Small* 15 (40): 1902649. doi:10.1002/smll.201902649.

G. Pratheeth Bhat, Narlawar Sagar Shrikrishna, Mamta Kumari, Sonu Gandhi. 2025. "Electrochemical Biosensing of Organophosphates." *Agricultural Electrochemistry*. ACS. 227–253, doi:10.1021/bk-2025-1496.ch010

Halder, Sayanti, Narlawar Sagar Shrikrishna, Riya Sharma, Pratheeth Bhat, and Sonu Gandhi. 2024. "Raising the Bar: Exploring Modern Technologies and Biomaterials for Enhancing Food Safety and Quality- a Comprehensive Review". *Food Control* 159 (May 2024): 110287. doi:10.1016/j.foodcont.2024.110287

He, Ying, Li Ma, Liya Zhou, Guanhua Liu, Yanjun Jiang, and Jing Gao. 2020. "Preparation and Application of Bismuth/MXene Nano-Composite as Electrochemical Sensor for Heavy Metal Ions Detection." *Nanomaterials* 10 (5): 866. doi:10.3390/nano10050866.

Ivneet Banga, Roshika Tyagi, Deepshikha Shahdeo, Sonu Gandhi. 2019. Biosensors and their application for the detection of Avian Influenza Virus. In *Nanotechnology in Modern Animal Biotechnology, Concepts and Applications*. Elsevier, 1–16. https://doi.org/10.1016/B978-0-12-818823-1.00001-6

Ji, Jingjing, Lufang Zhao, Yanfei Shen, Songqin Liu, and Yuanjian Zhang. 2019. "Covalent Stabilization and Functionalization of MXene via Silylation Reactions with Improved Surface Properties." *FlatChem* 17 (September): 100128. doi:10.1016/j.flatc.2019.100128.

Jiang, Ruicheng, Xuezhang Xiao, Jiaguang Zheng, Man Chen, and Lixin Chen. 2020. "Remarkable Hydrogen Absorption/Desorption Behaviors and Mechanism of Sodium Alanates in-Situ Doped with Ti-Based 2D MXene." *Materials Chemistry and Physics* 242 (February): 122529. doi:10.1016/j.matchemphys.2019.122529.

Jin, Zhaoyong, Yanfeng Fang, Xiaoxia Wang, Gengfang Xu, Mengli liu, Shuang Wei, Congli Zhou, Yanlin Zhang, and Yuanhong Xu. 2019. "Ultra-Efficient Electromagnetic Wave Absorption with Ethanol-Thermally Treated Two-Dimensional $Nb_2CT_x$ Nanosheets." *Journal of Colloid and Interface Science* 537 (March): 306–15. doi:10.1016/j.jcis.2018.11.034.

Jyotirmayee Sahoo, Riya Sharma, Sonu Gandhi. (2025). "Review on Biosensors for electroactive molecule detection." In *Recent Advancements in Voltammetry for Biosensing Applications*, edited by Manjunatha, J.G. Springer Nature. 17–44. doi: 10.1007/978-981-96-2595-6_2

Kalambate, Pramod K., Dhanjai, Ankita Sinha, Yankai Li, Yue Shen, and Yunhui Huang. 2020. "An Electrochemical Sensor for Ifosfamide, Acetaminophen, Domperidone, and Sumatriptan Based on Self-Assembled MXene/MWCNT/Chitosan Nanocomposite Thin Film." *Microchimica Acta* 187 (7): 402. doi:10.1007/s00604-020-04366-9.

Li, Hui, Xin Wang, Han Li, Shuai Lin, Bangchuan Zhao, Jianming Dai, Wenhai Song, Xuebin Zhu, and Yuping Sun. 2019. "Capacitance Improvements of $V_4C_3T$ by $NH_3$ Annealing." *Journal of Alloys and Compounds* 784 (May): 923–30. doi:10.1016/j.jallcom.2019.01.111.

Li, Kunshan, Xinyu Lu, You Zhang, Kuiliang Liu, Yongchao Huang, and Hong Liu. 2020. "$Bi_3TaO_7/Ti_3C_2$ Heterojunctions for Enhanced Photocatalytic Removal of Water-Borne Contaminants." *Environmental Research* 185 (June): 109409. doi:10.1016/j.envres.2020.109409.

Li, Zilan, Zechao Zhuang, Fan Lv, Han Zhu, Liang Zhou, Mingchuan Luo, Jiexin Zhu, Zhiquan Lang, Shihao Feng, Wei Chen, Liqiang Mai, Shaojun Guo. 2018. "The Marriage of the $FeN_4$ Moiety and MXene Boosts Oxygen Reduction Catalysis: Fe 3d Electron Delocalization Matters." *Advanced Materials* 30 (43): 1803220. doi:10.1002/adma.201803220.

Liang, Xiao, Arnd Garsuch, and Linda F. Nazar. 2015. "Sulfur Cathodes Based on Conductive MXene Nanosheets for High-Performance Lithium–Sulfur Batteries." *Angewandte Chemie* 127 (13): 3979–83. doi:10.1002/ange.201410174.

Ling, Zheng, Chang E. Ren, Meng-Qiang Zhao, Jian Yang, James M. Giammarco, Jieshan Qiu, Michel W. Barsoum, and Yury Gogotsi. 2014. "Flexible and Conductive MXene Films and Nanocomposites with High Capacitance." *Proceedings of the National Academy of Sciences* 111 (47): 16676–81. doi:10.1073/pnas.1414215111.

Liu, Huifang, Siying Ma, Guobao Ning, Runlin Zhang, Huan Liang, Feng Liu, Lin Xiao, Linjiang Guo, Yaping Zhang, Can-Peng Li, Hui Zhao. 2023a. "A 'Peptide-Target-Aptamer' Electrochemical Biosensor for Norovirus Detection Using a Black Phosphorous Nanosheet@$Ti_3C_2$-Mxene Nanohybrid and Magnetic Covalent Organic Framework." *Talanta* 258 (June): 124433. doi:10.1016/j.talanta.2023.124433.

Liu, Jinquan, Jiao Shi, Miao Zhong, Yating Wang, Xinxin Zhang, Wenyu Wang, Zhijun Chen, Yan Tan, Dongyun Xu, Shengyuan Yang, and Le Li. 2023b. "A Novel Electrochemical Sensing Method Based on an Amino-Functionalized MXene for the Rapid and Selective Detection of $Hg^{2+}$." *Analytical Methods* 16 (2): 244–52. doi:10.1039/D3AY01652H.

Liu, Rui, Wenkai Cao, Dongmei Han, Yudi Mo, Hui Zeng, Haocheng Yang, and Weihua Li. 2019. "Nitrogen-Doped $Nb_2CT_x$ MXene as Anode Materials for Lithium Ion Batteries." *Journal of Alloys and Compounds* 793 (July): 505–11. doi:10.1016/j.jallcom.2019.03.209.

Liu, Yu, Xu Zeng, Yulong Wu, Habib U. Khan, Fei Xiao, Bao Y. Xia, and Yimin Sun. 2024. "Functionalized MXene Fiber Electrode for the Electrochemical Sensing of Urinary Ammonia." *Advanced Sensor and Energy Materials* 3 (February): 100091. doi:10.1016/j.asems.2024.100091.

Mahari, Subhasis, and Sonu Gandhi. 2022. "Recent Advances in Electrochemical Biosensors for the Detection of Salmonellosis: Current Prospective and Challenges". *Biosensors* 6: 365. doi:10.3390/bios12060365.

Medetalibeyoglu, Hilal, Gül Kotan, Necip Atar, and Mehmet L. Yola. 2020a. "A Novel and Ultrasensitive Sandwich-Type Electrochemical Immunosensor Based on Delaminated MXene@AuNPs as Signal Amplification for Prostate Specific Antigen (PSA) Detection and Immunosensor Validation." *Talanta* 220 (December): 121403. doi:10.1016/j.talanta.2020.121403.

Medetalibeyoglu, Hilal, Murat Beytur, Onur Akyıldırım, Necip Atar, and Mehmet L. Yola. 2020b. "Validated Electrochemical Immunosensor for Ultra-Sensitive Procalcitonin Detection: Carbon Electrode Modified with Gold Nanoparticles Functionalized Sulfur Doped MXene as Sensor Platform and Carboxylated Graphitic Carbon Nitride as Signal Amplification." *Sensors and Actuators B: Chemical* 319 (September): 128195. doi:10.1016/j.snb.2020.128195.

Mozafari, Mohammad, and Masoud Soroush. 2021. "Surface Functionalization of MXenes." *Materials Advances* 2 (22): 7277–7307. doi:10.1039/D1MA00625H.

Sarthak Nandi, Ayusi Mondal, Akanksha Roberts, Sonu Gandhi. 2020. "Biosensor Platforms for Rapid HIV Detection". In *Advances in Clinical Chemistry*, edited by Gregory S. Makowski, Elsevier 98: 1–34. doi:10.1016/bs.acc.2020.02.001.

Narlawar, Sagar, Samraggi Coudhury, and Sonu Gandhi. "Magnetic Properties-Based Biosensors for Early Detection of Cancer". In *Biosensor Based Advanced Cancer Diagnostics*. Elsevier, 2022: 165–178. doi:10.1016/B978-0-12-823424-2.00010-7.

Pan, Yawen, Meiru He, Jing Wu, Hongbin Qi, and Yuanyuan Cheng. 2024. "One-Step Synthesis of MXene-Functionalized PEDOT:PSS Conductive Polymer Hydrogels for Wearable and Noninvasive Monitoring of Sweat Glucose." *Sensors and Actuators B: Chemical* 401 (February): 135055. doi:10.1016/j.snb.2023.135055.

Prakashan, Drishya, Ramya P. R., and Sonu Gandhi. 2023. "A Systematic Review on the Advanced Techniques of Wearable Point-of-Care Devices and Their Futuristic Applications". *Diagnostics* 13 (5): 916. doi:10.3390/diagnostics13050916.

Prakashan, Drishya, Maitri Shah, Pratik Kolhe, and Sonu Gandhi. 2025, 'Utility of Biosensors Fabricated from Smart and Intelligent Nanomaterials for Theragnostics'. In *Smart and Intelligent Nanostructured Materials for Next-Generation Biosensors*. Elsevier, 131–162. doi:10.1016/B978-0-443-19146-6.00002-2.

Prasanna, Sanjay B., Akash A. Jagtap, Gagankumar S. Kumar, Kumara S. Ningappa, Yu-Chien Lin, Yu-Chun Lu, Rajalakshmi Sakthivel, Ting-Yu Liu, and Ren-Jei Chung. 2024. "A Developed Highly Selective and Sensitive Electrochemical Sensor Using Gold Nanoparticles on Cerium Stannate/MXene as an Electrode Matrix for Detection of Metol in Various Environmental and Biological Samples." *Electrochimica Acta* 499 (September): 144703. doi:10.1016/j.electacta.2024.144703.

Ranjbar, Saba, Negar A. Astani, Maryam Atabay, Naimeh Naseri, Ali Esfandiar, and Mohammad R. Ejtehadi. 2022. "Electrochemical and Computational Studies of Bio-Mimicked $Ti_3C_2T_x$ MXene-Based Sensor with Multivalent Interface." *Journal of Colloid and Interface Science* 623 (October): 1063–74. doi:10.1016/j.jcis.2022.05.033.

Ramya, P. R., Sayanti Halder, and Sonu Gandhi. 2026. "Biosensing and imaging based on functionalized 2D nanomaterials." In *Two-Dimensional Nanomaterials for Biosensing and Imaging Applications*, edited by Ram Sevak Singh, Kalim Deshmukh, Chaudhery Mustansar Hussain, 415–448. doi:10.1002/9781394200375

Rasheed, P. A., Ravi P. Pandey, Khadeeja A. Jabbar, and Khaled A. Mahmoud. 2021. "Platinum Nanoparticles/$Ti_3C_2T_x$ (MXene) Composite for the Effectual Electrochemical Sensing of Bisphenol A in Aqueous Media." *Journal of Electroanalytical Chemistry* 880 (January): 114934. doi:10.1016/j.jelechem.2020.114934.

Roberts, Akanksha, and Sonu Gandhi. 2022. "A Concise Review on Potential Cancer Biomarkers and Advanced Manufacturing of Smart Platform-Based Biosensors for Early-Stage Cancer Diagnostics." *Biosensors and Bioelectronics: X* 11 (September): 100178. doi:10.1016/j.biosx.2022.100178.

Roberts, Akanksha, Subhasis Mahari, and Sonu Gandhi. 2022. "Biological/Synthetic Receptors (Antibody, Enzyme, and Aptamer) Used for Biosensors Development for Virus Detection." In *Advanced Biosensors for Virus Detection*, edited by Raju Khan, Arpana Parihar, Ajeet Kaushik, Ashok Kumar. 113–31. Elsevier. doi:10.1016/B978-0-12–824494-4.00007-2.

Roberts, Akanksha, Himani Dhanze, G. Taru Sharma, and Sonu Gandhi. 2023. "Point-of-care Detection of Japanese Encephalitis Virus Biomarker in Clinical Samples Using a Portable Smartphone-enabled Electrochemical "Sensit" Device." *Bioengineering & Translational Medicine* 8 (3): e10506. https://doi.org/10.1002/btm2.10506.

Sagar Shrikrishna, Narlawar, Riya Sharma, Jyotirmayee Sahoo, Ajeet Kaushik, and Sonu Gandhi. 2024. "Navigating the Landscape of Optical Biosensors." *Chemical Engineering Journal* 490 (June): 151661. doi:10.1016/j.cej.2024.151661.

Shrikrishna, Narlawar Sagar, and Sonu Gandhi. 2025. "Nanopore-Based Sensing for Biomarker Detection: From Fundamental Principles to Translational Diagnostics." *Journal of Nanobiotechnology* 23(1): 663. doi:10.1186/s12951-025-03762-8.

Sahoo, Jyotirmayee, Riya Sharma, Vivek Pachauri, and Sonu Gandhi. 2024. "Biomimetic/Bioderived Nanoengineered Interfaces for Biosensor Applications: A Review." *ACS Applied Nano Materials* 7 (17): 19854–75. doi:10.1021/acsanm.4c04554.

Shahdeo, Deepshikha, and Sonu Gandhi. 2022. "Next Generation Biosensors as a Cancer Diagnostic Tool". In *Biosensor Based Advanced Cancer Diagnostics*, edited by Editor(s): Raju Khan, Arpana Parihar, Sunil K. Sanghi, 179–196. Elsevier. doi:10.1016/B978-0-12-823424-2.00016-8

Shahdeo, Deepshikha, Drishya Prakashan, and Sonu Gandhi. 'Development of Advanced Diagnostic Platform for Detection of Infectious Diseases'. In *Molecular Pharmaceutics and Nano Drug Delivery*, edited by Umesh Gupta, Amit K. Goyal, 269–288. Elsevier. doi:10.1016/B978-0-323-91924-1.00008-3.

Shahzad, Faisal, Mohamed Alhabeb, Christine B. Hatter, Babak Anasori, Soon M. Hong, Chong M. Koo, and Yury Gogotsi. 2016. "Electromagnetic Interference Shielding with 2D Transition Metal Carbides (MXenes)." *Science* 353 (6304): 1137–40. doi:10.1126/science.aag2421.

Shan, Xia, Deqi Kuang, Qiumei Feng, Meisheng Wu, and Jie Yang. 2023. "A Dual-Mode Ratiometric Aptasensor for Accurate Detection of Pathogenic Bacteria Based on Recycling of DNAzyme Activation." *Food Chemistry* 423 (October): 136287. doi:10.1016/j.foodchem.2023.136287.

Sharma, Riya, G. Pratheeth Bhat, and Sonu Gandhi. 2025. "MXene-rGO Nanocomposite Based Electrochemical Immunosensor for Detection of Endosulfan – An Organochlorine Pesticide." *Chemosphere* 370: 143997. doi:10.1016/j.chemosphere.2024.143997.

Sharma, Riya, Jyotirmayee Sahoo, and Sonu Gandhi. 2025. 'Sustainable and Eco-Friendly 3D-Printed Electrochemical Sensors'. In *Additively Manufactured Electrochemical Sensors*, 1st edn, edited by J. G. Manjunatha and Chaudhery Mustansar Hussain, 369–400. Wiley. doi:10.1002/9781394303434.ch14.

Shrikrishna, Narlawar Sagar, Drishya Prakashan, Shivmuni Sarup, and Sonu Gandhi. 'Micro-Recycled Carbon Black from Waste Printer Cartridge: A Promising Magnetic Material for Single Stranded DNA Adsorption in SELEX'. *Small* 21 (30): 2500329. doi:10.1002/smll.202500329.

Shrikrishna, Narlawar Sagar, Sayanti Halder, Veerbhan Kesarwani, K. Nagamani, and Sonu Gandhi. 'Unveiling the Potential: High-Affinity Aptamers for Point of Care Detection of SARS-CoV-2 RBD Protein and It's Validation in Clinical Samples'. *Chemical Engineering Journal* 493 (August 2024): 152841. https://doi.org/10.1016/j.cej.2024.152841.

Shrikrishna, Narlawar Sagar, Subhasis Mahari, Naina Abbineni, S. A. Eremin, and Sonu Gandhi. 'New Trends in Biosensor Development for Pesticide Detection'. In *Biosensors in Agriculture: Recent Trends and Future Perspectives*, edited by Ramesh Namdeo Pudake, Utkarsh Jain, and Chittaranjan Kole. Springer International Publishing, 2021. https://doi.org/10.1007/978-3-030-66165-6_8.

Sun, Renhui, Hao-Bin Zhang, Ji Liu, Xi Xie, Rui Yang, Yue Li, Song Hong, and Zhong-Zhen Yu. 2017. "Highly Conductive Transition Metal Carbide/Carbonitride(MXene)@polystyrene Nanocomposites Fabricated by Electrostatic Assembly for Highly Efficient Electro magnetic Interference Shielding." *Advanced Functional Materials* 27 (45): 1702807. doi:10.1002/adfm.201702807.

Tu, Xiaolong, Feng Gao, Xue Ma, Jin Zou, Yongfang Yu, Minfang Li, Fengli Qu, Xigen Huang, and Limin Lu. 2020. "Mxene/Carbon Nanohorn/β-Cyclodextrin-Metal-Organic Frameworks as High-Performance Electrochemical Sensing Platform for Sensitive Detection of Carbendazim Pesticide." *Journal of Hazardous Materials* 396 (September): 122776. doi:10.1016/j.jhazmat.2020.122776.

Verger, Louisiane, Chuan Xu, Varun Natu, Hui-Ming Cheng, Wencai Ren, and Michel W. Barsoum. 2019. "Overview of the Synthesis of MXenes and Other Ultrathin 2D Transition Metal Carbides and Nitrides." *Current Opinion in Solid State and Materials Science* 23 (3): 149–63. doi:10.1016/j.cossms.2019.02.001.

Wang, Fei, Yao Zhu, Long Qian, Yuhao Yin, Ziyu Yuan, Yuting Dai, Tao Zhang, Dongya Yang, and Fengxian Qiu. 2024. "Lamellar $Ti_3C_2$ MXene Composite Decorated with Platinum-Doped $MoS_2$ Nanosheets as Electrochemical Sensing Functional Platform for Highly Sensitive Analysis of Organophosphorus Pesticides." *Food Chemistry* 459 (November): 140379. doi:10.1016/j.foodchem.2024.140379.

Wu, Shanshan, Yumei Su, Yi Zhu, Yuanming Zhang, and Mingshan Zhu. 2020a. "In-Situ Growing Bi/BiOCl Microspheres on $Ti_3C_2$ Nanosheets for Upgrading Visible-Light-Driven Photocatalytic Activity." *Applied Surface Science* 520 (August): 146339. doi:10.1016/j.apsusc.2020.146339.

Wu, Xinyu, Bingyong Han, Hao-Bin Zhang, Xi Xie, Tingxiang Tu, Yu Zhang, Yang Dai, Rui Yang, and Zhong-Zhen Yu. 2020b. "Compressible, Durable and Conductive Polydimethylsiloxane-Coated MXene Foams for High-Performance Electromagnetic Interference Shielding." *Chemical Engineering Journal* 381 (February): 122622. doi:10.1016/j.cej.2019.122622.

Xia, Tianzi, Guangyan Liu, Junjie Wang, Shili Hou, and Shifeng Hou. 2021. "MXene-Based Enzymatic Sensor for Highly Sensitive and Selective Detection of Cholesterol." *Biosensors and Bioelectronics* 183 (July): 113243. doi:10.1016/j.bios.2021.113243.

Xia, Yu, Tyler S. Mathis, Meng-Qiang Zhao, Babak Anasori, Alei Dang, Zehang Zhou, Hyesung Cho, Yury Gogotsi, and Shu Yang. 2018. "Thickness-Independent Capacitance of Vertically Aligned Liquid-Crystalline MXenes." *Nature* 557 (7705): 409–12. doi:10.1038/s41586-018-0109-z.

Xiao, Ping, Guodong Zhu, Xiaohong Shang, Bin Hu, Boshuang Zhang, Zhaoyu Tang, Jianmao Yang, and Jianyun Liu. 2022. "An Fe-MOF/MXene-Based Ultra-Sensitive Electrochemical Sensor for Arsenic(III) Measurement." *Journal of Electroanalytical Chemistry* 916 (July): 116382. doi:10.1016/j.jelechem.2022.116382.

Xiong, Shuyue, Zijian Wu, Fang Li, Pei Zhao, Haitao Li, Huanxin Bao, and Xu Yang. 2024. "Self-Assembled Sensing Interface Based on Specific Aptamer Functionalized $Ti_3C_2$ MXene for Ultrasensitive Antibiotic Detection in Milk." *LWT* 200 (May): 116186. doi:10.1016/j.lwt.2024.116186.

Xu, Chuan, Libin Wang, Zhibo Liu, Long Chen, Jingkun Guo, Ning Kang, Xiu-Liang Ma, Hui-Ming Cheng, and Wencai Ren. 2015. "Large-Area High-Quality 2D Ultrathin $Mo_2C$ Superconducting Crystals." *Nature Materials* 14 (11): 1135–41. doi:10.1038/nmat4374.

Yoon, Yeoheung, Anand P. Tiwari, Min Choi, Travis G. Novak, Wooseok Song, Hyunju Chang, Taehyoung Zyung, Sun S. Lee, Seokwoo Jeon, and Ki-Seok An. 2019. "Precious-Metal-Free Electrocatalysts for Activation of Hydrogen Evolution with Nonmetallic Electron Donor: Chemical Composition Controllable Phosphorous Doped Vanadium Carbide MXene." *Advanced Functional Materials* 29 (30): 1903443. doi:10.1002/adfm.201903443.

Zhang, Guoliang, Tiecheng Wang, Zehai Xu, Minmin Liu, Chong Shen, and Qin Meng. 2020. "Synthesis of Amino-Functionalized $Ti_3C_2T_x$ MXene by Alkalization-Grafting Modification for Efficient Lead Adsorption." *Chemical Communications* 56 (76): 11283–86. doi:10.1039/D0CC04265J.

Zhang, Shipeng, Md Abu Zahed, Md. Sharifuzzaman, Sanghyuk Yoon, Xue Hui, Sharat C. Barman, Sudeep Sharma, Hyo S. Yoon, Chani Park, and Jae Y. Park. 2021. "A Wearable Battery-Free Wireless and Skin-Interfaced Microfluidics Integrated Electrochemical Sensing Patch for On-Site Biomarkers Monitoring in Human Perspiration." *Biosensors and Bioelectronics* 175 (March): 112844. doi:10.1016/j.bios.2020.112844.

Zhang, Yue, Zheng Wang, Xiaohao Liu, Yating Liu, Yuxuan Cheng, Daxiang Cui, Feng Chen, and Wentao Cao. 2023. "A MXene/$MoS_2$ Heterostructure Based Biosensor for Accurate Sweat Ascorbic Acid Detection." *FlatChem* 39 (May): 100503. doi:10.1016/j.flatc.2023.100503.

Zhong, Wei, Feng Gao, Jin Zou, Shuwu Liu, Mingfang Li, Yansha Gao, Yongfang Yu, Xiaoqiang Wang, and Limin Lu. 2021. "MXene@Ag-Based Ratiometric Electrochemical Sensing Strategy for Effective Detection of Carbendazim in Vegetable Samples." *Food Chemistry* 360 (October): 130006. doi:10.1016/j.foodchem.2021.130006.

# 6 Doped MXenes Synthesis and Their Electrochemical Sensing Applications

*Mufeeda Mundummal*

## 6.1 INTRODUCTION

Over the past decade, two-dimensional transition metal carbides, nitrides, and carbonitrides, collectively referred to as MXenes, have garnered considerable attention due to their potential as fascinating materials. Their outstanding electrical properties, large surface area, exceptional biocompatibility, hydrophilic functional groups, and low toxicity render them ideal candidates for applications in electrochemical sensors (Zhu et al. 2021). MXenes are typically denoted by the formula $M_{n+1}X_nT_x$ ($n$ = 1–4), where M represents an early transition metal cluster (such as Mo, Ti, V, Sc, Nb, Cr, Zr, and Hf), X stands for nitrogen (N) and/or carbon (C), and $T_x$ refers to surface bound entities (such as –O–, –Cl, –OH, or –F)(Sang et al. 2018; Naguib et al. 2011). Research findings indicate that beyond the organized in-plane structures or atomic sandwiches made of transition metal planes, the field of MXene development has expanded from monometallic MXenes such as $Ti_3C_2$ and $Mo_2C$ to include bimetallic MXenes like $Mo_2TiC_2$ and $(Mo_{2/3}Y_{1/3})_2C$ (Wang et al. 2023; Shi et al. 2020). The adaptable nature of MXene structures, along with the functional moieties present on the outer layer of the material and their diverse compositions, allows MXenes to be highly versatile for use in numerous areas, including energy storage, nanogenerators, photocatalysis, environmental cleanup, sensing., as well as biomedical and therapeutic uses (Pei et al. 2021; Tee et al. 2021; Li et al. 2022a; Li et al. 2022b). Nevertheless, the drawbacks of MXenes, such as diminished flexibility, susceptibility to stacking, and inadequate stability in oxygen-rich environments, significantly constrain the continuous advancement of unmodified MXene (Amiri et al. 2020; Zhang et al. 2017). In the past few years, motivated by the doping technique in graphene and materials like transition metal dichalcogenides (Zhang et al. 2021a; Wen et al. 2017; Zhu et al. 2020), doping MXene sheets with foreign atoms or elements has become an effective approach for enhancing the intrinsic MXene properties like electrical, optical, magnetic, and so on. By introducing heteroatoms, MXenes undergo electron redistribution within their structure, thereby enhancing electron transfer and, consequently, augmenting their electronic properties (Wang et al. 2022). Generally, elemental doping offers the following benefits: It involves

DOI: 10.1201/9781003736318-6

converting inactive sites into electrochemically active species, which increases the material's conductivity (Larijani and Khorshidian 2019), it successfully causes the Fermi energy to fluctuate which results in enhanced electrical conductivity by modifying electronic band structure, and improved catalytic activity due to the surface reactivity (Lu et al. 2020). Doping is a method of deliberately modifying a material's electrical structure to produce defects, and it changes its electrical (Filatov et al. 2024), optical (Jana et al. 2013), thermal (Zhang and Li 2010), and magnetic characteristics (Chakraborty et al. 2017). It modifies the functional groups or chemical bonds on the surface, which hinders MXene rearrangement and promotes polysulfide transitions(Bao et al. 2021), by altering the grating range, and it increases the photometric range and the photoelectric effect of the material (Dillon et al. 2016). Doping also enhances a material's band gap and surface chemical properties and thus promotes the material's catalytic activity (Ran et al. 2017). Similarly, doping improves magnetic properties by altering dipoles (Xu et al. 2024). Generally, elemental doping alters the characteristics of the material by either modifying its surface or replacing elements in the lattice of the original MXene. This process substantially enhances the stability and performance of devices used in applications such as energy storage, catalytic sensors, and environmental remediation (An et al. 2021; Han et al. 2020; Huang et al. 2020).

Electrochemical sensors are composed of a set of chemical sensors wherein an electrode functions as a transducer component in the presence of an analyte (Simões and Xavier 2017). They offer an economical and convenient resolution for detecting diverse target molecules and are extensively applied in food industries, agriculture, and also in ecological and medical contexts (Baranwal et al. 2022). Two-dimensional MXenes and their composites are broadly used as materials for electrochemical sensors since their early times (Mufeeda, Ankitha, and Rasheed 2023; Vaishag et al. 2023; Arjun et al. 2023). Song et al. synthesized a composite of $Ti_3C_2$ MXene/Au NPs to analyze acetylcholinesterase (AChE) and developed an electrochemical biosensor for detecting organophosphorus pesticides (OPs), methamidophos, achieving a minimal detectable concentration (MDC) of 0.134 pM (Song et al. 2019). Rasheed et al. engineered a pliable and highly responsive sensor system by enhancing a glassy carbon electrode (GCE) with a lamellar $Ti_3C_2T_x$/Nafion composite to identify bromate ions ($BrO^{3-}$) in aqueous solution (water). The MXene material displayed remarkable electrocatalytic properties, facilitating the efficient reduction of $BrO^{3-}$. The constructed sensor demonstrated a proportional response to $BrO^{3-}$ concentrations ranging from 50 nM to 5 μM, achieving a detection threshold as low as 41 nanomolar (Rasheed et al. 2018). The researchers have explored additional methods to combine MXenes with other functional nanomaterials to enhance the performance of biosensors. Wang et al. modified MXenes with $TiO_2$ nanoparticles, thereby expanding the active surface area for the absorption of protein and the enzymatic activity of the sensor. The Hb-based biosensor they developed showed improved detection capabilities for $H_2O_2$, with a limit of detection (LOD) of 14 nM, surpassing the performance of the version lacking $TiO_2$NPs (Wang et al. 2015). Additionally, the practical applicability of sensors was boosted by elemental doping, which reduced the spacing between the layers of MXene sheets (Huang et al. 2020). Shuvo et al. discovered that when compared to undoped counterparts, S-doped $Ti_3C_2T_x$ MXenes exhibit high gas-sensing

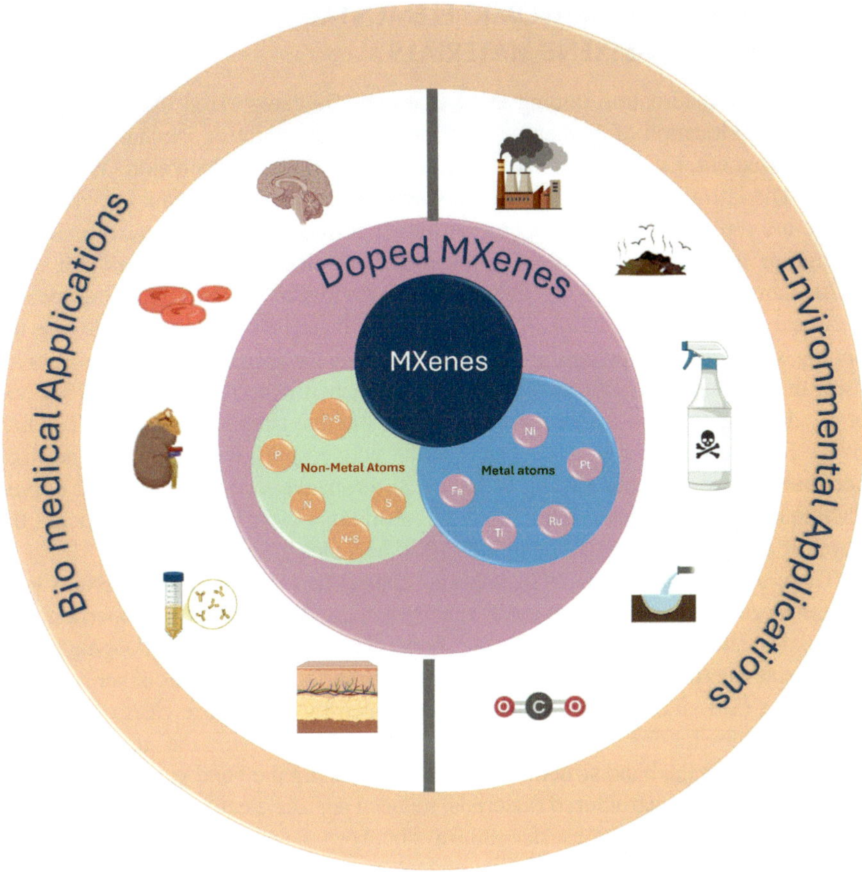


**FIGURE 6.1** Schematic diagram depicting elementally doped MXenes and their potential applications.

capabilities for VOCs (volatile organic compounds) and distinctive selectivity toward toluene. This change is attributed to the functionalization, which notably boosts the binding energy of benzene molecules to MXenes functionalized with sulfur, leading to altering the charge distribution of sulfur atoms around the gas molecules that are adsorbed and thus promoting an augmented gas response (Shuvo et al. 2020). Moreover, the layered N-doped MXene-derived materials demonstrate advantageous characteristics for both the co-adsorption and diffusion of formaldehyde (HCHO) and water molecules, culminating in a sensor endowed with commendable repeatability, prolonged stability, and discerning selectivity (Wang, Zhou, and Wang 2020). This chapter elaborates on the recent advancements in electrochemical sensing applications of doped MXenes. Figure 6.1 shows the schematic diagram depicting the elementally doped MXenes and their potential applications. We aim to investigate diverse doping strategies utilized across different materials, varied synthesis methods in their production, and their subsequent application as electrochemical sensors.

## 6.2 STRATEGIES FOR ATOMIC ELEMENTAL DOPING IN MXENE MATERIALS

Typically, the heteroatom doping in MXenes can be categorized into three groups based on the location of element atom doping and its operating mechanism: (1) lattice replacement, (2) functional substitution, and (3) surface adsorption (Wang et al. 2022). The lattice structure is modified when specific sites or vacancies within the lattice are replaced by dopant elements, either through ex-situ or in-situ methods, leading to substantial alterations to the material's electronic properties (Lu et al. 2020; Kamysbayev et al. 2020). The practical substitution involves replacing specific chemical linkages or functional clusters on the material's surface, employing the doping substitution approach to imbue MXenes with distinct surface functionalities, thereby significantly boosting their electrochemical efficacy (Li 2016; Zhao et al. 2021). It leads to an increased number of active sites and enhanced conductivity. Surface absorption takes place at the active spots within the 2D network configuration of MXenes (Wang et al. 2022). From the various compound doping techniques outlined above, multifaceted coordination contributes to enhancing the physical and chemical attributes of individual MXenes.

There are various forms of heteroatom doping that demonstrate a certain level of enhancement in electrochemical characteristics, ferromagnetic properties, and mechanical attributes of MXenes (Chakraborty et al. 2017). Heteroatom doping attains remarkable electronic conductivity through reducing electron densities of surface functional sites (Shi and Liu 2021). Lower electron density can reduce electron–electron scattering, allowing for higher carrier mobility. Also, modifying the surface electron density can alter the band structure, potentially reducing the band gap and increasing conductivity. In nanostructured materials, reduced surface electron density can facilitate more efficient electron transfer pathways between nanoparticles or layers. Surface adsorption allows for the modification of MXene properties without altering its core structure. It can introduce new functional groups or atoms that enhance specific properties, such as catalytic activity or sensing capabilities. This method is often simpler and less destructive compared to lattice substitution or functional group replacement (Gomaa et al. 2024). Additionally, doping enhances the surface area and pore volume as well as MXene's adsorption capacity, resulting in enhanced battery performance with respect to rate capability, reversible capacity, and cycling stability (Bao et al. 2018). X-ray photoelectron spectroscopy (XPS) can distinguish between the three modes of nitrogen doping in MXenes, as evidenced by the N 1s spectrum showing distinct peaks at 396.0 eV (lattice substitution), 399.7 eV (functional substitution), and 401.9 eV (surface adsorption) (Yang et al. 2021). Here, we will discuss the various synthesis strategies of doped MXenes using the above-mentioned mechanisms in detail, with their properties and advantages from the literature. To highlight exceptional characteristics of MXene in electrochemistry, magnetism, and mechanics, various complementary doping methods are also employed.

## 6.3 SYNTHESIS OF DOPED MXENES

Based on their constituent properties, MXene dopants can be subdivided into two divisions: nonmetals and metals. According to experiments, the ways of doping or substituting elements in MXenes are classified as ex-situ and in-situ processes. The in-situ route assumes that a doped MAX phase should be synthesized before MXene synthesis, while the ex-situ method implies the post-synthesis tailoring of MXenes (Figure 6.2). This section focuses on systematically reviewing experimental results obtained from doping and substitution of different elements within MXenes, especially nonmetals (N, S, P) and metals (Gao et al. 2021). Density Functional Theory (DFT)

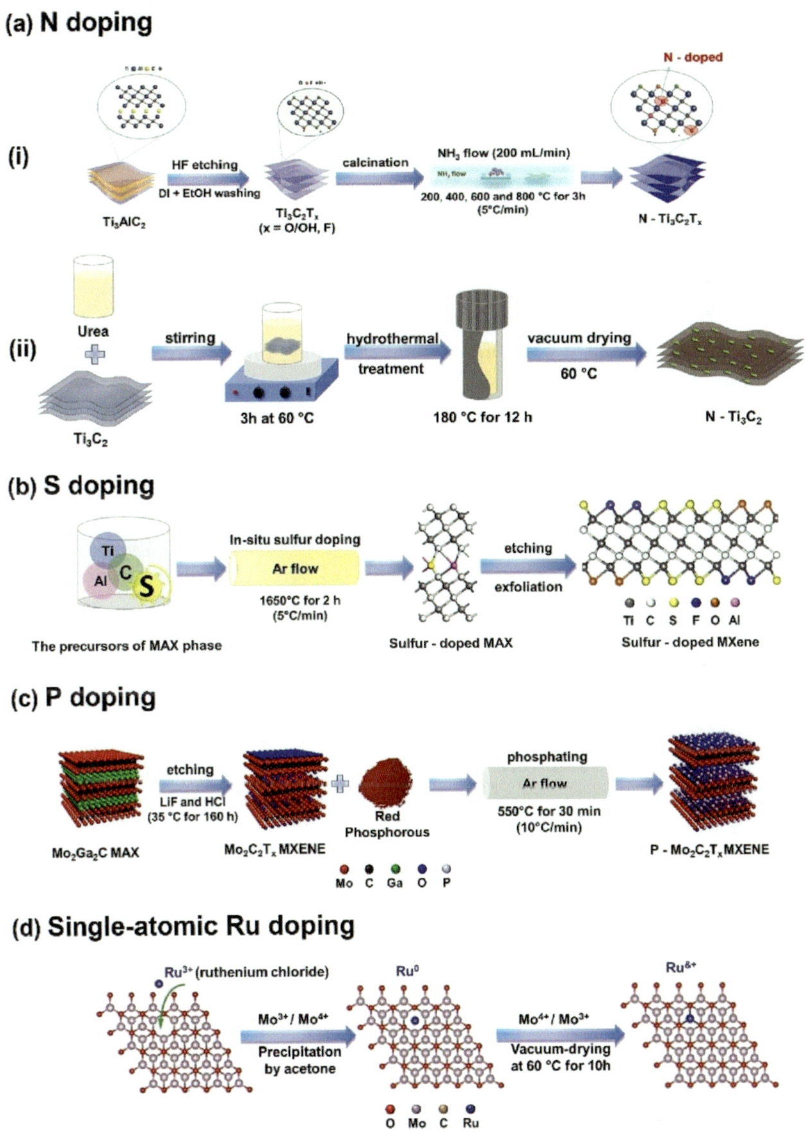


**FIGURE 6.2** Schematic diagram depicting the structures involved in in-situ (from parent molecules) and ex-situ (post-synthetic) doping techniques. Reprinted with permission from Bao et al. (2019), Le et al. (2019), Qu et al. (2018), and Tang et al. (2017).

offers significant potential for investigating the effects of doping in MXenes. By solving the Kohn–Sham equations, DFT provides an in-depth understanding of how dopants alter the electronic band structure, thereby influencing the material's conductivity, magnetic properties, and chemical reactivity. The electronic density of states, band gaps, and charge distributions of the product can be predicted using the DFT calculations, and the dopant type that is needed to depict essential properties can be identified. In addition to DFT, ab initio molecular dynamics is used to investigate the stability and diffusion of dopants in actual MXene structures, and finite-difference and tight-binding schemes were utilized to investigate large-scale dopant systems and for vaster property predictions. The studies using DFT on X substitutions (Si, S, B, N substituting for C) in MXenes show how these dopants affect the M–X interatomic distance of the materials, subsequently altering the layer thickness, unit cell dimensions, and electronic characteristics of the 2D nanomaterials. Besides, first-principles calculations have been applied to studies on the formation energies and electronic properties of MXenes with different functional groups (T = OH, F, O, S, Cl) for M = Nb, V, Cr, Ti, Zr, Ta, Sc MXene systems, allowing the investigation of the single and multilayer MXenes geometric structures and electronic features (Dey et al. 2023). This section will provide an overview of recent methods for the successful and targeted introduction of dopants into MXenes, exploring the latest literature on the topic.

### 6.3.1 Nonmetal Doping

Nonmetal doping in MXenes includes the process of adding nonmetal elements ,like nitrogen, sulfur, and phosphorus, for specific purposes that affect the ability or performance of the material in certain ways. Table 6.1 summarizes the nonmetal-doped MXene synthesis methods reported in the literature. For example, nitrogen can be incorporated into MXenes by introducing nitrogen-containing gases during the process of MXene preparation. In 2012, Naguib et al. pioneered the in-situ synthesis of nitrogen-doped MXenes by integrating nitrogen into the MAX phase precursor. Their approach involved heating a 3:1:1 mixture of titanium, aluminum nitride, and graphite at 1500°C for 2 h to form $Ti_3AlCN$, followed by selective etching of aluminum by hydrofluoric acid to produce the nitrogen-doped MXene $Ti_3CNT_x$ (Naguib et al. 2012). Lu et al. recently explored N-doping in $Ti_3C_2T_x$ (T = O, OH, and F) using DFT simulations to calculate the energetics of various nitrogen configurations. Their research pinpointed three potential sites for nitrogen doping: substitution within the carbon lattice (LS), replacement of the –OH functional group (FS), and surface adsorption at the –O termination (SA), with formation energies of –1.31, –4.71, and –2.87 eV, respectively. They explored three methods to incorporate nitrogen into $Ti_3C_2$ MXenes: (i) in-situ etching of $Ti_3AlCN$, (ii) hydrothermal treatment using urea, and (iii) cold plasma exposure with nitrogen gas. To develop an advanced electrochemiluminescence (ECL) aptasensor, Jiang et al. employed a technique involving the ultrasonication of a solution containing glycine as the nitrogen dopant, $Zn(NO_3)_2{\cdot}6H_2O$, and $Ti_3C_2$ MXene nanosheets for 1 h. The precursor solution was then subjected to thermal annealing in an argon atmosphere at 500°C with a heating rate of 5°C/min, which promoted the formation of Zinc oxide quantum dots on the

**TABLE 6.1**
**Nonmetal-doped MXenes and their synthesis methods**

| Doping element | Compound name | Synthesis process | Ref. |
|---|---|---|---|
| Nitrogen | $Ti_3CN_x$, $Ti_3C_{2-x}N_x$ | In-situ doping with AlN in MAX phase | Naguib et al. (2012), Liang et al. (2023) and Tang et al. (2020) |
| | N-$Ti_3C_2T_x$ | Ex-situ, hydrothermal doping precursor: urea | Cui et al. (2023) and Yang et al. (2017) |
| | N-$V_4C_3T_x$, N-$Ti_3C_2T_x$ | Ex-situ, hydrothermal doping precursor: ammonia | Zhang et al. (2022a), Wu et al. (2023) and Xia et al. (2021) |
| | N-$Ti_3C_2T_x$ | Ex-situ, hydrothermal doping precursor: chitosan | Pu et al. (2022) |
| | N-$Ti_3C_2T_x$ | Ex-situ, hydrothermal doping precursor: diethanolamine | Yang et al. (2019) |
| | N-$Ti_3C_2T_x$ QDs | Ex-situ, hydrothermal doping precursor: O-phenylenediamine | Yan et al. (2021) |
| | N-$Ti_3C_2T_x$ | Ex-situ, Hydrothermal doping precursor: diethylene tetramine | Feng et al. (2020) |
| | N-$Nb_2CT_x$ | $Nb_2CT_x$ MXene is mixed with $H_2PtCl_6$ (hexachloroplatinic acid) followed by overnight stirring | Shabana et al. (2023) |
| Sulfur | $Ti_3C_2S_x$ | In-situ, MAX phase precursor: S powder | Bao et al. (2019) |
| | S-$Nb_4C_3T_x$ | Ex-situ, hydrothermal precursor: sodium sulfide nonahydrate | Wu et al. (2022) |
| | S-$Ti_3C_2T_x$ | Ex-situ, MXene powder is mixed with sodium sulfide for 48 h at room temperature, then vacuum-dried for 12 h at 80°C | Sun et al. (2019) |
| | S-$Ti_3C_2T_x$ | Phosphoric acid, phytic acid, sodium dihydrogen phosphate | Shuvo et al. (2020) and Myagmarsereejid et al. (2022) |
| Tantalum | $Ti_xTa_{4-x}C_3$ | In-situ, MAX phase precursor: titanium–tantalum alloy | Maldonado-Lopez et al. (2022) |
| Phosphorous | P-$Ti_3C_2T_x$ | Ex-situ, hydrothermal precursor: phytic acid, phosphoric acid, sodium dihydrogen phosphate | Gupta et al. (2022) |
| Nitrogen and sulfur | N, S-$Ti_3C_2T_x$ | Ex-situ, MXene powder mixed with thiourea and annealed at 500°C for 3 h under $N_2$ atmosphere | Yang et al. (2017) |
| | S, N-$Nb_2CT_x$ | Ex-situ, N-$Nb_2CT_x$ prepared hydrothermally using thiourea and S, N-$Nb_2CT_x$ prepared by ball milling with benzyl disulfide | Zhang et al. (2021b) |
| | N, S-$Mo_2CT_x$ | Ex-situ, hydrothermal precursor: thioacetamide | Ang et al. (2015) |

nitrogen-doped titanium carbide ($Ti_3C_2$) nanosheet matrix (Jiang et al. 2022). The other method is post-synthesis modification, namely, a chemical modification of the synthesized MXene is performed, where it reacts with nonmetal agents or elements like sulfur or phosphoric acid. The following are the methods of gas-phase doping: In this method, MXenes are exposed to gaseous species different from metals, such as sulfur vapor or phosphorus pentachloride, that can penetrate the MXene lattice. Moreover, the approach of solution-based approach is based on using nonmetallic precursor solutions to incorporate dopants into the basal planes of MXene layers. These methods allow for very fine control of the doping process and are crucial for modifying the electronic, catalytic, and environmental performance of MXenes for various technologies.

Sulfur has garnered interest as a novel dopant capable of increasing the interlayer spacing through its significant atomic radius. This modification facilitates the formation of M–S bonds between adjacent MXene layers, enhancing ion diffusion and thereby improving the electrochemical performance of MXenes. Many studies have reported on sulfur-doped MXenes via ex-situ or in-situ methods. Weizhai et al. proposed an in-situ S doping strategy via the vacuum freeze drying method, and the fabricated electrodes were applied in Na–S batteries. The MXene/sulfur electrode demonstrates both high areal sulfur loading, reaching up to 4.5 $mg/cm^2$, and excellent electrochemical performance (Bao et al. 2019). Methods like thermal annealing treatment and hydrothermal are used for ex-situ development of S-doped MXenes. For example, Wu et al. synthesized S-doped $Ti_3C_2$ MXene, in which the sulfur content was 6.16% and was successfully treated using a basic solution soaking technique. In their work, etched $Ti_3C_2$ MXene was dispersed in $Na_2S$ solution, which made S-doped $Ti_3C_2$ MXene via electrostatic attraction and/or oxygen bridge conjunction (Sun et al. 2019). Moreover, recent reports have shown the synthesis of 3D S-doped MXenes via a sacrificial templating route. Xu et al synthesized a sulfur template using sodium thiosulfate with polyvinyl pyrrolidone and excess HCl. The obtained sulfur suspension was mixed with the etched MXene and further heated at 300°C for 3 h to eliminate the sulfur template. With the help of PVP (polyvinyl pyrrolidone) as an additive, the MXene/S structure was formed uniformly, excluding the formation of large sulfur particles and their clumping (Zhao et al. 2019).

Apart from sulfur, phosphorus can also increase the interlayer distance and distort the structure of MXenes to create more defects and reactive sites which in turn enhances the electrochemical efficiency of the MXene-related materials. Dual doping by two different heteroatoms, like N and S or N and P, introduces more defects and active sites during material formation. Modulation of the interlayer spacing within 2D materials through this process facilitates coordination with single metal atoms, hence enhancing their catalytic properties for extended applications. For example, nitrogen- and phosphorus-doped MXene quantum dots (MQDs) were prepared by Bai et al. They fabricated a dual-modal colorimetric and fluorometric sensor for nitrite detection. They prepared a solution containing $Ti_3C_2T_x$ MXene, phosphoric acid, and formamide, pH adjusted to 9.0 by ammonia. Afterwards, 800 W microwave irradiation was carried out at 120°C for 30 min. It was centrifuged at 5000 rpm for 10 min, and later membrane-filtered at 220 nm. The products were vacuum-dried to obtain N, P-doped $Ti_3C_2T_x$ (Bai et al. 2022). In another study, Xia et al. fabricated

N- and P-doped $Ti_3C_2T_x$ MXene ribbons for a novel electrochemical stripping analysis sensor to detect various heavy metal ions. They prepared the ribbons by immersing pristine $Ti_3C_2T_x$ MXene in a nitrogen atmosphere at 25°C for 108 h. This was followed by washing and drying the obtained nanoribbons. The dried ribbons, as synthesized, were further treated with a diammonium phosphate solution, followed by hydrothermal processing at 120°C for 12 h. The final product was termed N, P-$Ti_3C_2T_x$. The ribbons were further characterized with XPS, which revealed peaks for the graphitic-N, pyrrolic-N, and pyridinic-N N 1s peaks at 401.6, 399.8, and 397.5 eV, respectively. The peaks of P 2p included peaks at 134.0, 133.3, 132.8, and 130.1 eV corresponding to the P–O, P–N, P–C, and P–Ti bonding (Gupta et al. 2022).

Besides the above elements, halide elements like F, Cl, Br, and I are also used for doping through in-situ as well as ex-situ techniques. From the very beginning, Naquib et al. incorporated fluoride groups into MXenes by etching with either HF or a LiF/HCl solution (Naguib et al. 2011). The process of utilizing molten salts for the etching of MAX phase enables the conversion of MXenes with termination groups, which can be readily tailored to a specific MXene composition type. It has facilitated the easy introduction of halides other than Fluorine, including Chlorine, Bromine, or Iodine, with up to 100% coverage as termination groups, effectively eliminating any unwanted oxidation or hydrolysis (Dey et al. 2023). Halide-doped MXenes have a wide range of applications in energy storage devices as they show high electrical conductivity, capacitance, and energy density compared to their undoped counterparts. The content of halide ions has an enhancing effect on the electrochemically active sites on the MXene surface and thus a higher specific capacitance of the supercapacitor. Also, it was shown that halide doping can enhance the stability of MXenes as electrodes in batteries and enhance cycling performance, thus increasing device lifetimes. Oxygen is another effective dopant, which leads to structural changes in MXene, including a change in interlayer spacing and overall morphology. Most of the reported methods for synthesizing oxygen-doped MXenes have relied on ex-situ techniques. In contrast, Tian et al. explored an in-situ approach for creating oxygen-doped MXenes for supercapacitor electrodes. They combined experimental methods with DFT calculations to investigate the introduction of aluminum oxide as the oxygen dopant during the preparation of $Ti_3AlC_{2-x}O_x$ MAX phases (Tian et al. 2022).

### 6.3.2 Metal Doping

The modification of the electronic structure by metal doping can provide substantial changes in the properties of MXenes and provide the opportunity to be used in various applications. Table 6.2 gives an outline of methods for the synthesis of metal-doped MXenes. This is an interesting possibility that suggests that metal doping creates new electronic states in the MXene band structure, which could in turn open a band gap in what otherwise would be metallic MXenes. They effect a change in the Fermi level or adjust the band structure to allow the use of MXenes in the semiconductor industry. The studies show that terminal functionalization with certain metal dopants can enhance the ability of MXenes in increasing global electrical conductivity. For instance, doping with metals such as Cu or Ag can increase concentrations of charge carriers in the structure – this is useful in applications that demand

**TABLE 6.2**
**Metal-doped MXenes and their synthesis methods**

| Dopant element | Compound name | Synthesis process | Ref. |
|---|---|---|---|
| Titanium | $Mo_2TiC_2T_x$, $Mo_2Ti_2C_3T_x$, $(V_{1-x}Ti_x)_2C$, $(V_xTi_{1-x})_2C$, $Ti_{0.2}V_{1.8}C$ | In-situ, MAX phase precursor: Ti powder | Anasori et al. (2015), Zhou et al. (2017), Wang et al. (2019), and Simonenko et al. (2022) |
| Iron | Fe-$Ti_3C_2T_x$ | In-situ, MAX phase precursor: Fe powder | Zhao et al. (2021) |
| Niobium | TiNbC, $Ti_2NbC_x$, $V_{1.8}Nb_{0.2}CT_x$ | In-situ, MAX phase precursor: Nb powder | Naguib et al. (2012), Liu et al. (2023), and Xu et al. (2023) |
| Cobalt and iron | $Mo_2CT_x$:Co(Fe) | In-situ, with MAX phase precursor: cobalt nitrate hexahydrate (iron(III) nitrate nonahydrate) | Kuznetsov et al. (2019) and Kuznetsov et al. (2021) |
| Aluminum | Al-$Ti_3C_2Cl_x$ | Ex-situ method by heating MXene with Al powder | Cheng et al. (2023) |
| Palladium | Pd/B-N-$Ti_3C_2T_x$ | B-$Ti_3C_2T_x$ prepared hydrothermally using boric acid precursor and further mixed with ammonia to form B-N-$Ti_3C_2T_x$ and dispersed in $PdCl_2$ solution | Chen et al. (2022b) |
| Nickel | Ni-$Nb_2C$ | Ex-situ, hydrothermal precursor: nickel nitrate hexahydrate | Chen et al. (2022b) |
| Cesium | $Cs^+$-$Ti_3C_2T_x$ | Ex-situ, MXene powder mixed with $CsNO_3$ solution | Jun et al. (2020) |
| Erbium | Er-$Nb_2C$ | Ex-situ, hydrothermal doping precursor: erbium nitrate hexahydrate | Gul et al. (2021) |

high conductivity. Wei Peng et al. presented single-atomic ruthenium-modified $Mo_2CT_x$ MXene nanosheets as an efficient electrocatalyst for the nitrogen reduction reaction. This catalyst exhibited a Faradaic efficiency of 25.77% and an ammonia yield rate of 40. 57 μg/h/mg at −0.3 V against the reversible hydrogen electrode in a 0.5 M $K_2SO_4$ solution. This single-atomic Ru located on the surface of MXene nanosheets acts as the center for electron back-donation for $N_2$ activation, as well as provides an improved site for nitrogen adsorption and causes a decrease in the thermodynamic energy barrier for the hydrogenation step (Peng et al. 2020). Sun et al. studied the electronic and magnetic tailoring of metal-doped $Cr_2C$ MXenes. Thus, they stated that doping with Fe greatly changed the electronic and magnetic properties of $Cr_8C_4X_8$ (where X = F or OH). While the introduction of doping of metal atoms decreases the energy band gap in the system, it keeps being a bipolar antiferromagnetic semiconductor (Sun, Fu, and Yang 2020). Similarly, optimal metal bonding is possible for Nb, also a transition metal with partially filled d orbitals. When Nb replaces Ti, the complexity of the density of the electronic states and the lattice defects enhance as well as the width of the band gap decreases, which are considerate of the capacitance performance (Fatima et al. 2020). The electronic, structural, and magnetic properties of hetero-M $M_2C$ have been investigated because of the relatively small atomic size of the compounds. As illustrated by Li et al., it is evident that the Sc-doping results in an enhancement of the lattice parameter and a decrease of the layer thickness. In addition, the increase in the content of V can increase the mechanical properties of $Ti_2C$ at the same time to obtain the maximum of elastic strength with 425 GPa in $(Ti_{0.25}V_{0.75})_2CO_2$, which according to data three times surpasses the density of mechanical properties of Sc-doped $(Ti_{0.125}Sc_{0.875})_2$, $CO_2$ (Feng et al. 2017). In 2016, Ling et al. found that while pristine $V_2CO_2$ is ineffective for the hydrogen evolution reaction (HER), doping $V_2CO_2$ with Fe/Co/Ni results in extremely high catalytic activity (Ling et al. 2016). Cheng et al. used DFT to investigate the structural stability and electrocatalytic properties of the M-doped $Cr_2CO_2$ with various TM atoms: Ag, Au, Co, Fe, Ir, Mn, Mo, Ni, Pd, and Ru. Thus, under the guidance of the ultrahigh formation pressure of Ni oxide and the electron transfer from Ni to O*, the Ni atoms are closely anchored to the $Cr_2CO_2$ surface uniformly with high coverage and low overpotential of 0.16 V for the hydrogen evolution reaction (HER) and showing good catalytic activity.

## 6.4 APPLICATIONS OF DOPED MXENES IN ELECTROCHEMICAL SENSORS

The distinctive attributes of MXene make it a favorable substance for electrochemical detectors. The regulation of chemical composition and structural arrangement of modified MXenes through targeted elemental doping plays a pivotal role in shaping the efficacy of MXene-modified sensors. Element doping facilitates the manipulation of surface terminal functional clusters and interlayer distances within MXenes, exerting direct influence on the configuration and utility of functionalized MXenes. Here, we take up details on different electrochemical sensors based on doped MXenes to two different categories: electrochemical sensors for biomedical applications and environmental remediation.

### 6.4.1 Electrochemical Sensors for Biomedical Applications

Electrochemical biosensors are sophisticated analytical devices that convert specific biological interactions or reactions into quantifiable electrical signals, enabling the detection and measurement of target analytes with high sensitivity and selectivity (Lowe 1984). Biosensors typically exhibit remarkable specificity, enabled by the ability to customize the interaction of substances through the interaction of biological recognition components onto the sensor base, which possess a distinct affinity for the target molecule. Commonly employed sensing elements in biosensors are enzymes, nucleic acids, antibodies, cells, and receptors.

Guangyan et al. developed an enzymatic biosensor based on Eu-doped $SnO_2$ embedded in MXene for the detection of lactate molecule (Liu et al. 2022). The sensor was constructed on GCE, which revealed a linear relationship in a range of $1 \times 10^{-9}$ to $1 \times 10^{-4}$ M of lactate concentration with a detection limit of $3.38 \times 10^{-10}$ M and an excellent sensitivity of 4.815 mA/nmol/L/cm$^2$. Further, Hila et al. developed a sandwich-type electrochemical immunosensor for the detection of procalcitonin. They have utilized a delaminated sulfur-doped MXene-coated GCE along with AuNPs as a platform for an immunosensor. The sensor has given a good performance toward procalcitonin detection with an LOD of 2.0 fg/mL in a linear range of 0.01–1.0 pg/mL with excellent stability, repeatability, and reusability. Dual doping can also improve the chemical stability of MXenes in the harsh conditions of experiments. For example, the incorporation of multiple dopants might prevent the surface of the MXene from degrading, thus enhancing the life of the sensor. Lian M. et al reported a N and S co-doped $Nb_2C$ sensor for electrochemical detection of dopamine in gastric juice. NS-$Nb_2C$ MXene was prepared via the one-step treatment with thiourea as the dopant, which effectively improved the interlayer distance, the surface area, and the electrical conductivity. Thus, using the concept of dual doping in the 2D MXene nanoparticle, it is possible to sense dopamine with very high sensitivity (S/N = 3), and a relatively low LOD of 0.12 μM (Lian et al. 2022) (Figure 6.3). Further, Yaru Hou et al. designed another sensor with N, S-co-doped C dots along with Ti MXene for dopamine detection. Incorporating N, S-CDs into $Ti_3C_2T_x$ MXene nanosheets prevents flake aggregation, increasing interlayer spacing to 1.578 nm and tripling the specific surface area. Using this N, S-CDs/$Ti_3C_2T_x$ composite on GCEs, a dopamine (DA) sensor was developed with a detection range of 1–1000 μM and an LOD of 0.91 μM (Hou et al. 2024). Table 6.3 shows a comparison of heteroatom-doped MXenes for biomedical sensing applications

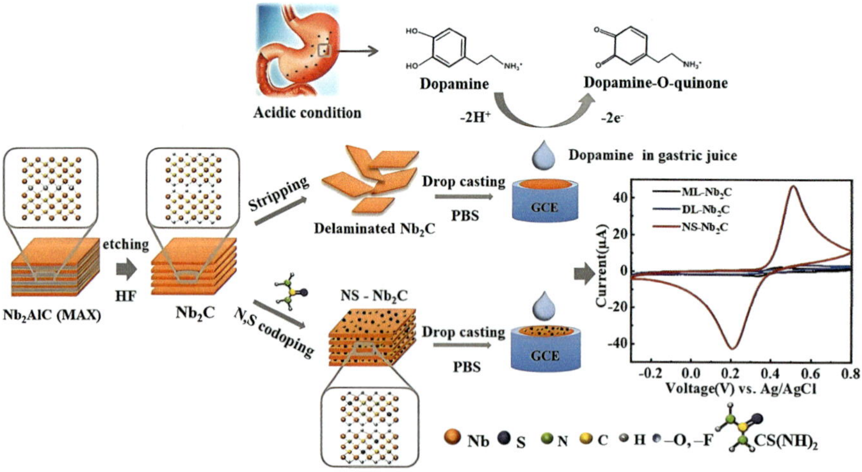


**FIGURE 6.3** Schematic diagram of electrochemical detection of dopamine using N, S-co-doped Nb2C MXene. Reproduced with permission from Lian et al. (2022).

**TABLE 6.3**
**Heteroatom-doped MXenes for biomedical sensing applications**

| Sensing material | Doped element | Sensing molecule | Technique used | Ref. |
|---|---|---|---|---|
| $Ti_3C_2T_x$/N-rGO | N | Adrenaline | CV, DPV | Chen et al. (2021) |
| d-S-$Ti_3C_2T_x$ | Delaminated S | Procalcitonin biomarker | CV, EIS | Medetalibeyoglu et al. (2020) |
| N, S-CDs/$Ti_3C_2T_x$ | N, S | Dopamine | CV, EIS | Hou et al. (2024) |
| NS-$Nb_2C$ | N, S | Dopamine | CV | Lian et al. (2022) |
| ZnO/N-$Ti_3C_2$ | N | Chloramphenicol | ECL | Jiang et al. (2022) |
| UiO-66-$NH_2$-$Ti_3C_2$ | UiO-66-$NH_2$ | Dopamine | CV, EIS | Wen et al. (2022) |
| $Ti_3C_2$-MXene@ Eu-$SnO_2$ | Eu | Lactate | CV | Liu et al. (2022) |
| NS-$Ti_3C_2$ NSs | N, S | Uric acid | CV | Chen et al. (2022a) |
| PEDOT:SCX/ MXene/GOx/GCE | PEDOT | Glucose | CV | Murugan et al. (2022) |

### 6.4.2 Electrochemical Sensors for Environmental Remediation

Besides biosensing applications of biomolecules, doped MXenes are also involved prominently in other electrochemical detection techniques apart from biosensors. For example, they can effectively adsorb and catalyze the extraction of dangerous materials, such as heavy metals and pesticides from water, or nitrogen dioxide ($NO_2$) and carbon monoxide (CO) in the air. The described influence of doping on MXenes' properties thus places doped MXenes at the forefront of all the applications related to environmental sensing and remediation, having potential for addressing key issues in the field of water and air quality. Yixuan et al. developed N and P co-doped $Ti_3C_2T_x$ MXenes nanoribbons (N, P-$Ti_3C_2T_x$R) that had been for the first time applied in an electrodeposition-free electrochemical sensing assay. This new material could spontaneously adsorb and reduce $Cu^{2+}$ and $Hg^{2+}$ ions and the deposition of $Cu^0$ and $Hg^0$ on the sensor surface. This stems from the novel strategy that leads to the development of a concise and rather sensitive platform for the detection of $Cu^{2+}$ and $Hg^{2+}$ ions (Xia et al. 2022) (Figure 6.4). Similarly, Xiao Zhang et al. fabricated a $Ti_3C_2$@N-C MXene heterostructure sensor for electrochemical detection of both $Cd^{2+}$ and $Pb^{2+}$ in seawater and tap water matrices via the square-wave anodic stripping voltammetry (SWASV) method. The proposed sensor has very low LODs of 2.55 nM for $Cd^{2+}$ and 1.10 nM for $Pb^{2+}$, high selectivity with respect to matrix interferences, and a higher signal-to-noise ratio due to the higher conductivity of the N-C and $Ti_3C_2T_x$MXene heterostructure (Zhang et al. 2022b). In addition to that, newer developments in electrochemical sensors, particularly those that use doped MXenes, provide enhanced sensitivity, selectivity, and efficiency in detecting toxic gases like $CO_2$ and CO, which is a critical aspect of environmental monitoring, industrial safety, and public health. Yong Zhou introduced a new sensor based on N-doped $Ti_3C_2T_x$MXene/PEI/rGO hybrid film for the detection of $CO_2$ gas in the concentration of 8–3000 ppm.

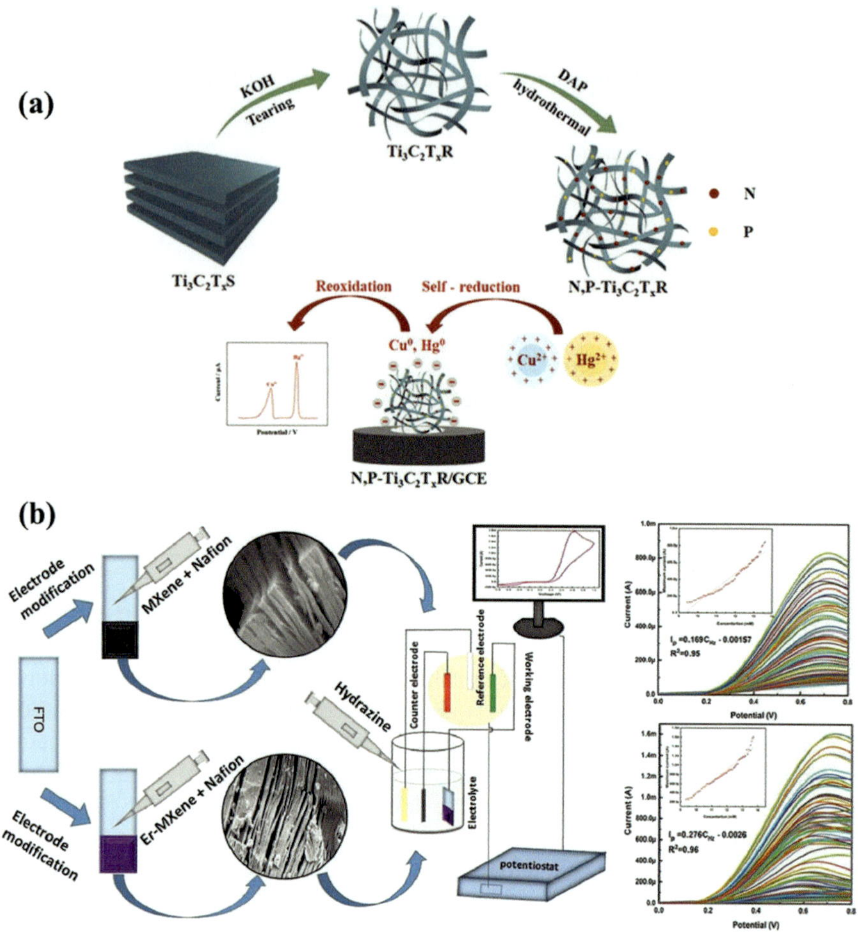


**FIGURE 6.4** (a) Schematic diagrams for preparing N,P-Ti3C2TxR and electrochemical sensing of $Cu^{2+}$ and $Hg^{2+}$. Reprinted with permission from Xia et al. (2022). (b) Er-doped Nb2C MXene for detection of hydrazine (Gul et al. 2021).

A ternary sensor with polyethyleneimine (PEI) with 0.01 mg/mL concentration under 62% relative humidity at 20°C offered the best performance. This sensor had a very good detection limit of 8 ppm; selectivity, repeatability, and stability over a long period of time were also observed. The rGO nanosheets offered a supply channel for rapid charge carrier transfer during the sensing process of $CO_2$ (Zhou et al. 2020). Table 6.4 shows a comparison of doped MXenes for environmental remediation applications.

## 6.5 CONCLUSION

As 2D nanostructured materials with highly stable surface chemistry, MXenes exhibit high electrical conductivity and well-developed surface area, which enable them to

**TABLE 6.4**
**Doped MXenes for environmental remediation applications**

| Sensing material | Doped element | Sensing molecule | Technique used | Ref. |
|---|---|---|---|---|
| $Ti_3C_2$@N-C | N | $Cd^{2+}$ and $Pb^{2+}$ | SWASV | Zhang et al. (2022b) |
| rGO/N-$Ti_3C_2T_x$/ $TiO_2$ | N | Formaldehyde | Time-resolved resistance | Wang, Zhou, and Wang (2020) |
| rGO/N-$Ti_3C_2T_x$/ PEI | N | Carbon dioxide | Time-resolved resistance | Zhou et al. (2020) |
| d-S-$Ti_3C_2T_x$ | Delaminated S | Chlorpyrifos insecticide | CV, EIS | Kadirsoy, Atar, and Yola (2020) |
| S-$Ti_3C_2T_x$ | S | VOCs | Dynamic impedimetric sensing | Shuvo et al. (2020) |
| 3D melamine-rGO/ $Ti_3C_2T_x$ | 3D melamine | $Zn^{2+}$, $Pb^{2+}$, $Cd^{2+}$ | CV, DPASV | Chen et al. (2023) |
| N, P-$Ti_3C_2T_xR$ | N, P | $Cu^{2+}$, $Hg^{2+}$ | CV, EIS | Xia et al. (2022) |
| Co-$Ti_3C_2$ nanoenzyme | Co | Organophosphate pesticides | DPV | Yu et al. (2022) |
| Un-Er-$Nb_2C$ | Un | Hydrazine | CV, LSV | Gul et al. (2021) |
| $MnMoO_4$-MXene | Mn | Hydroquinone, catechol | CV, EIS | Ranjith et al. (2022) |
| $Ti_3C_2T_x$/ $Pd(OH)_2$/C | $Pd(OH)_2$ | Nicotine | CV, EIS | Magesh et al. (2022) |
| MXene–titania–Nafion composite | Titania–Nafion | Capsaicin | CV, EIS | Lee, Lee, and Lee (2023) |
| ZnO/N-$Ti_3C_2$ | N | Chloramphenicol | ECL | Jiang et al. (2022) |

demonstrate high sensitivity and efficiency toward various applications including energy storage and sensing. Doping these materials with other elements enhances their performance by incorporating new functional sites. In this chapter, we discuss different synthesis techniques used to prepare doped MXenes incorporating non-metal and metal heteroatoms. In addition, we also provide a thorough overview of doped MXene nanosheets for their applications in electrochemical sensors, focusing on biomedical and environmental remediation. Although the versatility of MXene chemistry allows for the fine-tuning of their properties across a range of applications, there are still significant challenges and opportunities for future research in the realm of doped MXenes. While dopant atoms are believed to be the primary active centers in doped MXenes, there is much more to the surface chemistry originating from dopant atoms being involved in reaction mechanisms and adsorption processes. To enhance these materials, modeling and predictions are needed to predict their behavior, describe reaction mechanisms, and optimize new doped MXene structures and related architectures with better kinetics. In case of synthesis techniques more methods should be introduced that ensure reproducibility, controlled surface groups, and maintain the stability of doped MXenes in diverse environments. In terms of electrochemical sensors, there exist wide opportunities for wearable sensors using doped MXenes, utilizing their high sensitivity and selectivity toward various analytes. Similarly, doped MXenes can be designed to improve release patterns and targeting efficiency in drug delivery systems. The attachment of therapeutic agents and targeted ligands is made possible by their high surface area and adjustable surface chemistry, which can enhance the specificity and regulated release of medications. Overall, integration of doped MXenes into these applications will likely drive advancements in various fields from energy storage to medical diagnostics.

## REFERENCES

Amiri, Ahmad, Yijun Chen, Chew B. Teng, and Mohammad Naraghi. 2020. "Porous nitrogen-doped MXene-based electrodes for capacitive deionization." *Energy Storage Materials* 25:731–739.

An, Yongling, Yuan Tian, Chengkai Liu, Shenglin Xiong, Jinkui Feng, and Yitai Qian. 2021. "Rational design of sulfur-doped three-dimensional $Ti_3C_2T_x$ MXene/ZnS heterostructure as multifunctional protective layer for dendrite-free zinc-ion batteries." *Acs Nano* 15 (9):15259–15273.

Anasori, Babak, Yu Xie, Majid Beidaghi, Jun Lu, Brian C. Hosler, Lars Hultman, Paul R. C. Kent, Yury Gogotsi, and Michel W. Barsoum. 2015. "Two-dimensional, ordered, double transition metals carbides (MXenes)." *ACS Nano* 9 (10):9507–9516.

Ang, Huixiang, Hui T. Tan, Zhi M. Luo, Yu Zhang, Yuan Y. Guo, Guilue Guo, Hua Zhang, and Qingyu Yan. 2015. "Hydrophilic nitrogen and sulfur Co-doped molybdenum carbide nanosheets for electrochemical hydrogen evolution." *Small* 11 (47):6278–6284.

Arjun, Ajith Mohan, Menon Ankitha, Neermunda Shabana, Pushpalatha V. Vaishag, Fathima Shamsheera, Mundummal Mufeeda, and Pathath Abdul. Rasheed. 2023. "An overview on surface modification of niobium MXenes for diagnostic and prognostic applications." *FlatChem*, 41, 100538.

Bai, Yuxuan, Yu He, Manman Wang, and Gongwu Song. 2022. "Microwave-assisted synthesis of nitrogen, phosphorus-doped $Ti_3C_2$ MXene quantum dots for colorimetric/fluorometric dual-modal nitrite assay with a portable smartphone platform." *Sensors and Actuators B: Chemical* 357:131410.

Bao, Weizhai, Christopher E. Shuck, Wenxue Zhang, Xin Guo, Yury Gogotsi, and Guoxiu Wang. 2019. "Boosting performance of Na–S batteries using sulfur-doped $Ti_3C_2T_x$ MXene nanosheets with a strong affinity to sodium polysulfides." *ACS Nano* 13 (10): 11500–11509.

Bao, Weizhai, Lin Liu, Chengyin Wang, Sinho Choi, Dan Wang, and Guoxiu Wang. 2018. "Facile synthesis of crumpled nitrogen-doped MXene nanosheets as a new sulfur host for lithium–sulfur batteries." *Advanced Energy Materials* 8 (13):1702485.

Bao, Weizhai, Ronghao Wang, Chengfei Qian, Zherui Zhang, Ruijun Wu, Yuhao Zhang, Fangyang Liu, Jingfa Li, and Guoxiu Wang. 2021. "Porous heteroatom-doped $Ti_3C_2T_x$ MXene microspheres enable strong adsorption of sodium polysulfides for long-life room-temperature sodium–sulfur batteries." *ACS Nano* 15 (10):16207–16217.

Baranwal, Jaya, Brajesh Barse, Gianluca Gatto, Gabriela Broncova, and Amit Kumar. 2022. "Electrochemical sensors and their applications: A review." *Chemosensors* 10 (9):363.

Chakraborty, Poulami, Tilak Das, Dhani Nafday, Lilia Boeri, and Tanusri Saha-Dasgupta. 2017. "Manipulating the mechanical properties of $Ti_2C$ MXene: Effect of substitutional doping." *Physical Review B* 95 (18):184106.

Chen, Da, Shuaibin Shao, Wei Zhang, Jingbo Zhao, and Meiling Lian. 2022a. "Nitrogen and sulfur co-doping strategy to trigger the peroxidase-like and electrochemical activity of $Ti_3C_2$ nanosheets for sensitive uric acid detection." *Analytica Chimica Acta* 1197: 339520.

Chen, Shuxian, Min Shi, Quan Xu, Jingkun Xu, Xuemin Duan, Yansha Gao, Limin Lu, Feng Gao, Xiaoqiang Wang, and Yongfang Yu. 2021. "$Ti_3C_2T_x$ MXene/nitrogen-doped reduced graphene oxide composite: A high-performance electrochemical sensing platform for adrenaline detection." *Nanotechnology* 32 (26):265501.

Chen, Yuanyuan, Peng Zhao, Yi Liang, Yi Ma, Yiyi Liu, Jinsong Zhao, Jingzhou Hou, Changjun Hou, and Danqun Huo. 2023. "A sensitive electrochemical sensor based on 3D porous melamine-doped rGO/MXene composite aerogel for the detection of heavy metal ions in the environment." *Talanta* 256:124294.

Chen, Zhangxin, Jiajie Cao, Xiaohui Wu, Dongqin Cai, Minghui Luo, Shuyu Xing, Xiuli Wen, Yongyin Chen, Yanxian Jin, and Dan Chen. 2022b. "B, N Co-doping sequence: An efficient electronic modulation of the Pd/MXene interface with enhanced electrocatalytic properties for ethanol electrooxidation." *ACS Applied Materials & Interfaces* 14 (10):12223–12233.

Cheng, Zongju, Zhiguo Du, Hao Chen, Qi Zhao, Yu Shi, Haiyang Wang, Yuxuan Ye, and Shubin Yang. 2023. "Metal-bonded atomic layers of transition metal carbides (MXenes)." *Advanced Materials* 35 (39):2302141.

Cui, Lidan, Jianfeng Wen, Quanhao Deng, Xin Du, Tao Tang, Ming Li, Jianrong Xiao, Li Jiang, Guanghui Hu, and Xueli Cao. 2023. "Improving the photocatalytic activity of $Ti_3C_2$ MXene by surface modification of N doped." *Materials* 16 (7):2836.

Dey, Avishek, Silvia Varagnolo, Nicholas P. Power, Naresh Vangapally, Yuval Elias, Lois Damptey, Bright N. Jaato, Saianand Gopalan, Zahra Golrokhi, and Prashant Sonar. 2023. "Doped MXenes—A new paradigm in 2D systems: Synthesis, properties and applications." *Progress in Materials Science*, 139, 101166.

Dillon, Andrew D., Michael J. Ghidiu, Alex L. Krick, Justin Griggs, Steven J. May, Yury Gogotsi, Michel W. Barsoum, and Aaron T. Fafarman. 2016. "Highly conductive optical quality solution-processed films of 2D titanium carbide." *Advanced Functional Materials* 26 (23):4162–4168.

Fatima, Mahjabeen, Jameela Fatheema, Nasbah B. Monir, Ahmad H. Siddique, Bushra Khan, Amjad Islam, Deji Akinwande, and Syed Rizwan. 2020. "Nb-doped MXene with enhanced energy storage capacity and stability." *Frontiers in Chemistry* 8:168.

Feng, Li, Xian-Hu Zha, Kan Luo, Qing Huang, Jian He, Yijun Liu, Wei Deng, and Shiyu Du. 2017. "Structures and mechanical and electronic properties of the $Ti_2CO_2$ MXene incorporated with neighboring elements (Sc, V, B and N)." *Journal of Electronic Materials* 46:2460–2466.

Feng, Yefeng, Furong Zhou, Qihuang Deng, and Cheng Peng. 2020. "Solvothermal synthesis of in situ nitrogen-doped $Ti_3C_2$ MXene fluorescent quantum dots for selective $Cu^{2+}$ detection." *Ceramics International* 46 (6):8320–8327.

Filatov, Nikita M., Alexander A. Kolchugin, Alexander A. Pankratov, and Liliya A. Dunyushkina. 2024. "Impact of dopants on electrical conductivity of proton-conducting $SrHfO_3$ perovskite." *Ceramics International* 50 (20):40282–40291.

Gao, Lingfeng, Wenli Bao, Artem V. Kuklin, Shan Mei, Han Zhang, and Hans Ågren. 2021. "Hetero-MXenes: Theory, synthesis, and emerging applications." *Advanced Materials* 33 (10):2004129.

Gomaa, Islam, Nasser M. Hosny, Hanan Elhaes, Hend A. Ezzat, Maryam G. Elmahgary, and Medhat A. Ibrahim. 2024. "Two-dimensional MXene as a promising adsorbent for trihalomethanes removal: A density-functional theory study." *Nanomaterials* 14 (5):454.

Gul, Sundus, Martha I. Serna, Syedah A. Zahra, Nimrah Arif, Mudassir Iqbal, Deji Akinwande, and Syed Rizwan. 2021. "Un-doped and Er-adsorbed layered $Nb_2C$ MXene for efficient hydrazine sensing application." *Surfaces and Interfaces* 24:101074.

Gupta, Nisha, Ranjan K. Sahu, Trilochan Mishra, and Pallab Bhattacharya. 2022. "Microwave-assisted rapid synthesis of titanium phosphate free phosphorus doped $Ti_3C_2$ MXene with boosted pseudocapacitance." *Journal of Materials Chemistry A* 10 (29):15794–15810.

Han, Xin, Lin An, Yue Hu, Yaogang Li, Chengyi Hou, Hongzhi Wang, and Qinghong Zhang. 2020. "$Ti_3C_2$ MXene-derived carbon-doped $TiO_2$ coupled with g-$C_3N_4$ as the visible-light photocatalysts for photocatalytic $H_2$ generation." *Applied Catalysis B: Environmental* 265:118539.

Hou, Yaru, Fanyong Yan, Xinyi Bai, Xin Wei, Liang Chen, and Yang Fu. 2024. "Designing advanced 0D/2D heterojunctions of N, S-co-doped carbon dots/$Ti_3C_2T_x$ MXene nanosheets in electrochemical sensor applications." *Ceramics International* 50 (7):10735–10745.

Huang, Weichun, Lanping Hu, Yanfeng Tang, Zhongjian Xie, and Han Zhang. 2020. "Recent advances in functional 2D MXene-based nanostructures for next-generation devices." *Advanced Functional Materials* 30 (49):2005223.

Jana, Debnarayan, Chia-Liang Sun, Li-Chyong Chen, and Kuei-Hsien Chen. 2013. "Effect of chemical doping of boron and nitrogen on the electronic, optical, and electrochemical properties of carbon nanotubes." *Progress in Materials Science* 58 (5):565–635.

Jiang, Ding, Meng Wei, Xiaojiao Du, Ming Qin, Xueling Shan, and Zhidong Chen. 2022. "One-pot synthesis of ZnO quantum dots/N-doped $Ti_3C_2$ MXene: Tunable nitrogen-doping properties and efficient electrochemiluminescence sensing." *Chemical Engineering Journal* 430:132771.

Jun, Byung-Moon, Min Jang, Chang M. Park, Jonghun Han, and Yeomin Yoon. 2020. "Selective adsorption of $Cs^+$ by MXene ($Ti_3C_2T_x$) from model low-level radioactive wastewater." *Nuclear Engineering and Technology* 52 (6):1201–1207.

Kadirsoy, Selcen, Necip Atar, and Mehmet L. Yola. 2020. "Molecularly imprinted QCM sensor based on delaminated MXene for chlorpyrifos detection and QCM sensor validation." *New Journal of Chemistry* 44 (16):6524–6532.

Kamysbayev, Vladislav, Alexander S. Filatov, Huicheng Hu, Xue Rui, Francisco Lagunas, Di Wang, Robert F. Klie, and Dmitri V. Talapin. 2020. "Covalent surface modifications and superconductivity of two-dimensional metal carbide MXenes." *Science* 369 (6506):979–983.

Kuznetsov, Denis A., Zixuan Chen, Paula M. Abdala, Olga V. Safonova, Alexey Fedorov, and Christoph R. Muller. 2021. "Single-atom-substituted $Mo_2CT_x$: Fe-layered carbide for selective oxygen reduction to hydrogen peroxide: Tracking the evolution of the MXene phase." *Journal of the American Chemical Society* 143 (15):5771–5778.

Kuznetsov, Denis A., Zixuan Chen, Priyank V. Kumar, Athanasia Tsoukalou, Agnieszka Kierzkowska, Paula M. Abdala, Olga V. Safonova, Alexey Fedorov, and Christoph R. Müller. 2019. "Single site cobalt substitution in 2D molybdenum carbide (MXene) enhances catalytic activity in the hydrogen evolution reaction." *Journal of the American Chemical Society* 141 (44):17809–17816.

Larijani, H. Tavassoli, and Majid Khorshidian. 2019. "Theoretical insight into the role of pyridinic nitrogen on the catalytic activity of boron-doped graphene towards oxygen reduction reaction." *Applied Surface Science* 492:826–842.

Le, Thi A., Quoc V. Bui, Ngoc Q. Tran, Yunhee Cho, Yeseul Hong, Yoshiyuki Kawazoe, and Hyoyoung Lee. 2019. "Synergistic effects of nitrogen doping on MXene for enhancement of hydrogen evolution reaction." *ACS Sustainable Chemistry & Engineering* 7 (19): 16879–16888.

Lee, Sang J., Don H. Lee, and Won-Yong Lee. 2023. "An electrochemical sensor for capsaicin based on two-dimensional titanium carbide (MXene)-doped titania-Nafion composite film." *Microchemical Journal* 185:108216.

Li, Longhua. 2016. “Lattice dynamics and electronic structures of $Ti_3C_2O_2$ and $Mo_2TiC_2O_2$ (MXenes): The effect of Mo substitution.” *Computational Materials Science* 124: 8–14.

Li, Xingxing, Yanan Ma, Yang Yue, Guosheng Li, Chuankun Zhang, Minglei Cao, Yongchen Xiong, Jintao Zou, Yongheng Zhou, and Yihua Gao. 2022a. “A flexible Zn-ion hybrid micro-supercapacitor based on MXene anode and $V_2O_5$ cathode with high capacitance.” *Chemical Engineering Journal* 428:130965.

Li, Yuanlin, Yanxiang Liu, Tianfang Zheng, Shin-ichi Sasaki, Hitoshi Tamiaki, and Xiao-Feng Wang. 2022b. “Chlorophyll derivative sensitized monolayer $Ti_3C_2T_x$ MXene nanosheets for photocatalytic hydrogen evolution.” *Journal of Photochemistry and Photobiology A: Chemistry* 427:113792.

Lian, Meiling, Yuqing Shi, Wei Zhang, Jingbo Zhao, and Da Chen. 2022. “Nitrogen and sulfur co-doped $Nb_2C$-MXene nanosheets for the ultrasensitive electrochemical detection dopamine under acidic conditions in gastric juice.” *Journal of Electroanalytical Chemistry* 904:115849.

Liang, Licheng, Linyan Bian, Yanping Fan, Shuyan Guan, Xianyun Liu, Qunying Sun, and Baozhong Liu. 2023. “Nitrogen doping excited Ru and $Ti_3C_{2-x}N_x$ support for hydrogen generation from ammonia borane.” *Fuel* 339:127445.

Ling, Chongyi, Li Shi, Yixin Ouyang, and Jinlan Wang. 2016. “Searching for highly active catalysts for hydrogen evolution reaction based on O-terminated MXenes through a simple descriptor.” *Chemistry of Materials* 28 (24):9026–9032.

Liu, Guangyan, Tianzi Xia, Xilin Liang, Shili Hou, and Shifeng Hou. 2022. “Enzymatic electrochemical biosensor from Eu-doped $SnO_2$ embedded in MXene for high performance sensing lactate.” *ChemElectroChem* 9 (22):e202200848.

Liu, Wei, Jin Cao, Feng Song, Dong-Dong Zhang, Manunya Okhawilai, Jin Yi, Jia-Qian Qin, and Xin-Yu Zhang. 2023. “A double transition metal $Ti_2NbC_2T_x$ MXene for enhanced lithium-ion storage.” *Rare Metals* 42 (1):100–110.

Lowe, Christopher R. 1984. “Biosensors.” *Trends in Biotechnology* 2 (3):59–65.

Lu, Chengjie, Li Yang, Bingzhen Yan, Liangbo Sun, Peigen Zhang, Wei Zhang, and ZhengMing Sun. 2020. “Nitrogen-doped $Ti_3C_2$ MXene: Mechanism investigation and electrochemical analysis.” *Advanced Functional Materials* 30 (47):2000852.

Magesh, Vasanth, Ashok K. Sundramoorthy, Dhanraj Ganapathy, Raji Atchudan, Sandeep Arya, Razan A. Alshgari, and Ahmed M. Aljuwayid. 2022. “Palladium hydroxide (Pearlman’s catalyst) doped MXene ($Ti_3C_2T_x$) composite modified electrode for selective detection of nicotine in human sweat.” *Biosensors* 13 (1):54.

Maldonado-Lopez, Daniel, Jassiel R. Rodriguez, Vilas G. Pol, Ravuri Syamsai, Nirmala G. Andrews, Sandra Julieta Gutiérrez-Ojeda, Rodrigo Ponce-Perez, Ma G. Moreno-Armenta, and Jonathan Guerrero-Sanchez. 2022. “Atomic-scale understanding of Li storage processes in the $Ti_4C_3$ and chemically ordered $Ti_2Ta_2C_3$ MXenes: A theoretical and experimental assessment.” *ACS Applied Energy Materials* 5 (2):1801–1809.

Medetalibeyoglu, Hilal, Murat Beytur, Onur Akyıldırım, Necip Atar, and Mehmet L. Yola. 2020. “Validated electrochemical immunosensor for ultra-sensitive procalcitonin detection: Carbon electrode modified with gold nanoparticles functionalized sulfur doped MXene as sensor platform and carboxylated graphitic carbon nitride as signal amplification.” *Sensors and Actuators B: Chemical* 319:128195.

Mufeeda, Mundummal, Menon Ankitha, and Pathath Abdul Rasheed. 2023. “$Nb_2CT_x$/protonated carbon nitride nanocomposite for electrochemical detection of serotonin.” *ACS Applied Nano Materials* 6 (22):21152–21161.

Murugan, Preethika, Jayshree Annamalai, Raji Atchudan, Mani Govindasamy, Deepak Nallaswamy, Dhanraj Ganapathy, Anatoly Reshetilov, and Ashok K. Sundramoorthy. 2022. “Electrochemical sensing of glucose using glucose oxidase/PEDOT: 4-sulfocalix [4] arene/MXene composite modified electrode.” *Micromachines* 13 (2):304.

Myagmarsereejid, Purevlkham, Munkhjargal Bat-Erdene, Abdulaziz S. R. Bati, Batjargal Sainbileg, Michitoshi Hayashi, Joseph G. Shapter, Yu L. Zhong, and Munkhbayar Batmunkh. 2022. "Sulfur-functionalized titanium carbide $Ti_3C_2T_x$ (MXene) nanosheets modified light absorbers for ambient fabrication of $Sb_2S_3$ solar cells." *ACS Applied Nano Materials* 5 (9):12107–12116.

Naguib, Michael, Murat Kurtoglu, Volker Presser, Jun Lu, Junjie Niu, Min Heon, Lars Hultman, Yury Gogotsi, and Michel W. Barsoum. 2011. "Two-dimensional nanocrystals produced by exfoliation of $Ti_3AlC_2$." In *MXenes*, edited by Yuri Gogotsi, 15–29. Jenny Stanford Publishing.

Naguib, Michael, Olha Mashtalir, Joshua Carle, Volker Presser, Jun Lu, Lars Hultman, Yury Gogotsi, and Michel W. Barsoum. 2012. "Two-dimensional transition metal carbides." *ACS Nano* 6 (2):1322–1331.

Pei, Yangyang, Xiaoli Zhang, Zengyu Hui, Jinyuan Zhou, Xiao Huang, Gengzhi Sun, and Wei Huang. 2021. "$Ti_3C_2T_X$ MXene for sensing applications: recent progress, design principles, and future perspectives." *ACS Nano* 15 (3):3996–4017.

Peng, Wei, Min Luo, Xiandong Xu, Kang Jiang, Ming Peng, Dechao Chen, Ting-Shan Chan, and Yongwen Tan. 2020. "Spontaneous atomic ruthenium doping in $Mo_2CT_X$ MXene defects enhances electrocatalytic activity for the nitrogen reduction reaction." *Advanced Energy Materials* 10 (25):2001364.

Pu, Liuyue, Jiaoxia Zhang, Nguetsa K. L. Jiresse, Yifeng Gao, Haijun Zhou, Nithesh Naik, Peng Gao, and Zhanhu Guo. 2022. "N-doped MXene derived from chitosan for the highly effective electrochemical properties as supercapacitor." *Advanced Composites and Hybrid Materials* 5:1–14.

Qu, Guoxing, Yang Zhou, Tianli Wu, Guoliang Zhao, Fangfang Li, Yijin Kang, and Chen Xu. 2018. "Phosphorized MXene-phase molybdenum carbide as an earth-abundant hydrogen evolution electrocatalyst." *ACS Applied Energy Materials* 1 (12):7206–7212.

Ran, Jingrun, Guoping Gao, Fa-Tang Li, Tian-Yi Ma, Aijun Du, and Shi-Zhang Qiao. 2017. "$Ti_3C_2$ MXene co-catalyst on metal sulfide photo-absorbers for enhanced visible-light photocatalytic hydrogen production." *Nature Communications* 8 (1):13907.

Ranjith, Kugalur S., A. T. Ezhil Vilian, Seyed M. Ghoreishian, Reddicherla Umapathi, Seung-Kyu Hwang, Cheol W. Oh, Yun S. Huh, and Young-Kyu Han. 2022. "Hybridized 1D–2D $MnMoO_4$–MXene nanocomposites as high-performing electrochemical sensing platform for the sensitive detection of dihydroxybenzene isomers in wastewater samples." *Journal of Hazardous Materials* 421:126775.

Rasheed, Pathath Abdul, Ravi P. Pandey, Kashif Rasool, and Khaled A. Mahmoud. 2018. "Ultra-sensitive electrocatalytic detection of bromate in drinking water based on Nafion/$Ti_3C_2T_x$ (MXene) modified glassy carbon electrode." *Sensors and Actuators B: Chemical* 265:652–659.

Sang, Xiahan, Yu Xie, Dundar E. Yilmaz, Roghayyeh Lotfi, Mohamed Alhabeb, Alireza Ostadhossein, Babak Anasori, Weiwei Sun, Xufan Li, and Kai Xiao. 2018. "In situ atomistic insight into the growth mechanisms of single layer 2D transition metal carbides." *Nature Communications* 9 (1):2266.

Shabana, Neermunda, Ajith M. Arjun, Karuvatta Nubla, Menon Ankitha, and Pathath Abdul Rasheed. 2023. "Platinum nanoparticles decorated $Nb_2CT_x$ MXene as an efficient dual functional catalyst for hydrogen evolution and oxygen reduction reaction." *International Journal of Hydrogen Energy* 48 (21):7698–7707.

Shi, Junjie, Siliang Wang, Qiang Wang, Xing Chen, Xiaoyu Du, Min Wang, Yanjie Zhao, Chen Dong, Limin Ruan, and Wei Zeng. 2020. "A new flexible zinc-ion capacitor based on δ-$MnO_2$@carbon cloth battery-type cathode and MXene@cotton cloth capacitor-type anode." *Journal of Power Sources* 446:227345.

Shi, Yacheng, and Yang Liu. 2021. "Vacancy and N dopants facilitated $Ti^{3+}$ sites activity in 3D $Ti_{3-x}C_2T_y$ MXene for electrochemical nitrogen fixation." *Applied Catalysis B: Environmental* 297:120482.

Shuvo, Shoumya N., Ana M. U. Gomez, Avanish Mishra, Winston Y. Chen, Avinash M. Dongare, and Lia A. Stanciu. 2020. "Sulfur-doped titanium carbide MXenes for room-temperature gas sensing." *ACS Sensors* 5 (9):2915–2924.

Simões, Fabio R., and Miguel G. Xavier. 2017. "Electrochemical sensors." *Nanoscience and Its Applications* 1:155–178.

Simonenko, Nikolay P., Olga E. Glukhova, Ilya A. Plugin, Dmitry A. Kolosov, Ilya A. Nagornov, Tatiana L. Simonenko, Alexey S. Varezhnikov, Elizaveta P. Simonenko, Victor V. Sysoev, and Nikolay T. Kuznetsov. 2022. "The $Ti_{0.2}V_{1.8}C$ MXene ink-prepared chemiresistor: From theory to tests with humidity versus VOCs." *Chemosensors* 11 (1):7.

Song, Dandan, Xinyu Jiang, Yanshan Li, Xiong Lu, Sunrui Luan, Yuanzhe Wang, Yan Li, and Faming Gao. 2019. "Metal–organic frameworks-derived $MnO_2/Mn_3O_4$ microcuboids with hierarchically ordered nanosheets and $Ti_3C_2$ MXene/Au NPs composites for electrochemical pesticide detection." *Journal of Hazardous Materials* 373:367–376.

Sun, Qian, Zhaoming Fu, and Zongxian Yang. 2020. "Tunable magnetic and electronic properties of the Cr-based MXene ($Cr_2C$) with functional groups and doping." *Journal of Magnetism and Magnetic Materials* 514:167141.

Sun, Shuijing, Zhenlang Xie, Yurong Yan, and Songping Wu. 2019. "Hybrid energy storage mechanisms for sulfur-decorated $Ti_3C_2$ MXene anode material for high-rate and long-life sodium-ion batteries." *Chemical Engineering Journal* 366:460–467.

Tang, Yi, Chenhui Yang, Yapeng Tian, Yangyang Luo, Xingtian Yin, and Wenxiu Que. 2020. "The effect of in situ nitrogen doping on the oxygen evolution reaction of MXenes." *Nanoscale Advances* 2 (3):1187–1194.

Tang, Yi, Jianfeng Zhu, Wenling Wu, Chenhui Yang, Wenjing Lv, and Fen Wang. 2017. "Synthesis of nitrogen-doped two-dimensional $Ti_3C_2$ with enhanced electrochemical performance." *Journal of the Electrochemical Society* 164 (4):A923.

Tee, Si Y., Enyi Ye, Choon P. Teng, Yuki Tanaka, Karen Y. Tang, Khin Y. Win, and Ming-Yong Han. 2021. "Advances in photothermal nanomaterials for biomedical, environmental and energy applications." *Nanoscale* 13 (34):14268–14286.

Tian, Yapeng, Maomao Ju, Yijia Luo, Xiaoqing Bin, Xiaojie Lou, and Wenxiu Que. 2022. "In situ oxygen doped $Ti_3C_2T_x$ MXene flexible film as supercapacitor electrode." *Chemical Engineering Journal* 446:137451.

Vaishag, Pushpalatha V., Sabarija A. Mohandas, Mundummal Mufeeda, Praveena Gangadharan, and Pathath A. Rasheed. 2023. "Conversion of electronic waste to an electrochemical sensor for dopamine: Using MXene-modified liquid crystal display panels." *ACS Sustainable Chemistry & Engineering* 11 (34):12771–12779.

Wang, Fei, Shujuan Wang, Feng Tian, Fuqian Wang, Xuewen Xia, Qi Zhang, Zhongya Pang, Xing Yu, Guangshi Li, and Hsien-Yi Hsu. 2023. "Advances in molten-salt-assisted synthesis of 2D MXenes and their applications in electrochemical energy storage and conversion." *Chemical Engineering Journal*: 470, 144185.

Wang, Fen, ChenHui Yang, Max Duan, Yi Tang, and JianFeng Zhu. 2015. "$TiO_2$ nanoparticle modified organ-like $Ti_3C_2$ MXene nanocomposite encapsulating hemoglobin for a mediator-free biosensor with excellent performances." *Biosensors and Bioelectronics* 74:1022–1028.

Wang, Ronghao, Muhan Li, Kaiwen Sun, Yuhao Zhang, Jingfa Li, and Weizhai Bao. 2022. "Element-doped Mxenes: Mechanism, synthesis, and applications." *Small* 18 (25): 2201740.

Wang, Ying, Wei Zheng, Peigen Zhang, Wubian Tian, Jian Chen, and ZhengMing Sun. 2019. "Preparation of $(V_x, Ti_{1-x})_2C$ MXenes and their performance as anode materials for LIBs." *Journal of Materials Science* 54:11991–11999.

Wang, Yuhang, Yong Zhou, and Yanjie Wang. 2020. "Humidity activated ionic-conduction formaldehyde sensing of reduced graphene oxide decorated nitrogen-doped MXene/titanium dioxide composite film." *Sensors and Actuators B: Chemical* 323:128695.

Wen, Mingzhen, Ying Xing, Guangyan Liu, Shili Hou, and Shifeng Hou. 2022. "Electrochemical sensor based on $Ti_3C_2$ membrane doped with UIO-66-$NH_2$ for dopamine." *Microchimica Acta* 189 (4):141.

Wen, Yangyang, Thomas E. Rufford, Xingzhu Chen, Neng Li, Miaoqiang Lyu, Liming Dai, and Lianzhou Wang. 2017. "Nitrogen-doped $Ti_3C_2T_x$ MXene electrodes for high-performance supercapacitors." *Nano Energy* 38:368–376.

Wu, Fengxian, Chunyan Xu, Xu Yang, Liying Yang, and Shougen Yin. 2022. "S-doped multilayer niobium carbide ($Nb_4C_3T_x$) electrocatalyst for efficient hydrogen evolution in alkaline solutions." *International Journal of Hydrogen Energy* 47 (39):17233–17240.

Wu, Shanshan, Yefeng Feng, Kaidan Wu, Wenqin Jiang, Zhifeng Xue, Deping Xiong, Li Chen, Zuyong Feng, Kunhua Wen, and Zhaoying Li. 2023. "Mxene $Ti_3C_2$ generated $TiO_2$ nanoparticles in situ and uniformly embedded in rGO sheets as high stable anodes for potassium ion batteries." *Journal of Alloys and Compounds* 930:167414.

Xia, Yi, Sufang He, Junming Wang, Liexing Zhou, Jing Wang, and Sridhar Komarneni. 2021. "MXene/$WS_2$ hybrids for visible-light-activated $NO_2$ sensing at room temperature." *Chemical Communications* 57 (72):9136–9139.

Xia, Yixuan, Yong Zhao, Fengxiang Ai, Yinhui Yi, Tingting Liu, Huiyu Lin, and Gangbing Zhu. 2022. "N and P co-doped MXenes nanoribbons for electrodeposition-free stripping analysis of Cu(II) and Hg(II)." *Journal of Hazardous Materials* 425:127974.

Xu, Chengyuan, Lin Wen, Anjian Pan, Lizhong Zhao, Yuansen Liu, Xuefeng Liao, Yu Pan, and Xuefeng Zhang. 2024. "First-principles study of Ti-doping effects on hard magnetic properties of $RFe_{11}Ti$ magnets." *Crystals* 14 (6):507.

Xu, Jianguang, Zhiyong Liu, Qiang Wang, Junsheng Li, Yuxiang Huang, Mengnan Wang, Linyu Cao, Wei Yao, Haijiang Wu, and Chi Chen. 2023. "Facile tailoring of surface terminations of MXenes by doping Nb element: Toward extraordinary pseudocapacitance performance." *ACS Applied Materials & Interfaces* 15 (12):15367–15376.

Yan, Fanyong, Jingru Sun, Yueyan Zang, Zhonghui Sun, Hao Zhang, Jinxia Xu, and Xiang Wang. 2021. "Solvothermal synthesis of nitrogen-doped MXene quantum dots for the detection of alizarin red based on inner filter effect." *Dyes and Pigments* 195:109720.

Yang, Chenhui, Wenxiu Que, Xingtian Yin, Yapeng Tian, Yawei Yang, and Meidan Que. 2017. "Improved capacitance of nitrogen-doped delaminated two-dimensional titanium carbide by urea-assisted synthesis." *Electrochimica Acta* 225:416–424.

Yang, Chenhui, Yi Tang, Yapeng Tian, Yangyang Luo, Xingtian Yin, and Wenxiu Que. 2019. "Methanol and diethanolamine assisted synthesis of flexible nitrogen-doped $Ti_3C_2$ (MXene) film for ultrahigh volumetric performance supercapacitor electrodes." *ACS Applied Energy Materials* 3 (1):586–596.

Yang, Fangli, Dylan Hegh, Dongxing Song, Jizhen Zhang, Ken Aldren S. Usman, Zhiyu Wang, Peng Zhang, Weigang Ma, Wenrong Yang, and Si Qin. 2021. "A nitrogenous pre-intercalation strategy for the synthesis of nitrogen-doped $Ti_3C_2T_x$ MXene with enhanced electrochemical capacitance." *Journal of Materials Chemistry A* 9 (10):6393–6401.

Yu, Lei, Jiafu Chang, Xinyu Zhuang, Haiyin Li, Ting Hou, and Feng Li. 2022. "Two-dimensional cobalt-doped $Ti_3C_2$ MXene nanozyme-mediated homogeneous electrochemical strategy for pesticides assay based on in situ generation of electroactive substances." *Analytical Chemistry* 94 (8):3669–3676.

Zhang, Chuanfang J., Sergio Pinilla, Niall McEvoy, Conor P. Cullen, Babak Anasori, Edmund Long, Sang-Hoon Park, Andrés Seral-Ascaso, Aleksey Shmeliov, and Dileep Krishnan. 2017. "Oxidation stability of colloidal two-dimensional titanium carbides (MXenes)." *Chemistry of Materials* 29 (11):4848–4856.

Zhang, Gang, and Baowen Li. 2010. "Impacts of doping on thermal and thermoelectric properties of nanomaterials." *Nanoscale* 2 (7):1058–1068.

Zhang, Gujia, Luhua Wang, Rongjian Sa, Chao Xu, Zhaohui Li, and Lianzhou Wang. 2022a. "Interconnected N-doped MXene spherical shells for highly efficient capacitive deionization." *Environmental Science: Nano* 9 (1):204–213.

Zhang, Shunlong, Hangjun Ying, Pengfei Huang, Jianli Wang, Zhao Zhang, Zhihao Zhang, and Wei-Qiang Han. 2021a. "Ultrafine Sb pillared few-layered $Ti_3C_2T_x$ MXenes for advanced sodium storage." *ACS Applied Energy Materials* 4 (9):9806–9815.

Zhang, Wenyang, Huixin Jin, Yiqun Du, Guowen Chen, and Jianxin Zhang. 2021b. "Sulfur and nitrogen codoped $Nb_2C$ MXene for dendrite-free lithium metal battery." *Electrochimica Acta* 390:138812.

Zhang, Xiao, Dong An, Zhaoshun Bi, Wei Shan, Beibei Zhu, Long Zhou, Liangmin Yu, Han Zhang, Shuwei Xia, and Meng Qiu. 2022b. "$Ti_3C_2$-MXene@N-doped carbon heterostructure-based electrochemical sensor for simultaneous detection of heavy metals." *Journal of Electroanalytical Chemistry* 911:116239.

Zhao, Nana, Yijun Yang, Ding Yi, Yubao Xiao, Ke Wang, Weibin Cui, and Xi Wang. 2021. "Regulating Fe–O bond in $Ti_3C_2T_x$ MXene anode for high-capacity Li-ion batteries." *Chemical Engineering Journal* 422:130018.

Zhao, Qian, Qizhen Zhu, Jiawei Miao, Peng Zhang, Pengbo Wan, Lingzhang He, and Bin Xu. 2019. "Flexible 3D porous MXene foam for high-performance lithium-ion batteries." *Small* 15 (51):1904293.

Zhou, Jian, Shaohua Gao, Zhonglu Guo, and Zhimei Sun. 2017. "Ti-enhanced exfoliation of $V_2AlC$ into $V_2C$ MXene for lithium-ion battery anodes." *Ceramics International* 43 (14):11450–11454.

Zhou, Yong, Yuhang Wang, Yanjie Wang, and Xian Li. 2020. "Humidity-enabled ionic conductive trace carbon dioxide sensing of nitrogen-doped $Ti_3C_2T_x$ MXene/polyethyleneimine composite films decorated with reduced graphene oxide nanosheets." *Analytical Chemistry* 92 (24):16033–16042.

Zhu, Haoyue, Xin Gan, Amber McCreary, Ruitao Lv, Zhong Lin, and Mauricio Terrones. 2020. "Heteroatom doping of two-dimensional materials: From graphene to chalcogenides." *Nano Today* 30:100829.

Zhu, Xiaohua, Youyu Zhang, Meiling Liu, and Yang Liu. 2021. "2D titanium carbide MXenes as emerging optical biosensing platforms." *Biosensors and Bioelectronics* 171:112730.

# 7 MXene–Metal Composite–Based Electrochemical Sensors

*Jithin Thomas Chacko and P. Abdul Rasheed*

## 7.1 INTRODUCTION

Two-dimensional (2D) materials have increasingly gained prominence in material science due to their unique properties that distinguish them significantly from bulk materials. These properties, which include enhanced mechanical strength, remarkable electrical conductivity, and superior thermal properties, make 2D materials highly desirable for applications in a wide array of industries (Li et al. 2018a). The key distinguishing factor of 2D materials is their thickness, which typically ranges from a few atomic layers to a single atomic layer. This ultra-thin structure provides these materials with an enormous surface area relative to their volume, which leads to unique electronic, mechanical, and optical properties that are not seen in their three-dimensional counterparts (Ji et al. 2020). Graphene is a well-known 2D material that consists of a single layer of carbon atoms arranged in a hexagonal lattice. The exceptional properties of graphene, including its high electrical conductivity, strength, and flexibility, have prompted intense research into other 2D materials that could offer similar or even superior characteristics for a variety of applications (Wu et al. 2013).

Despite the tremendous potential of graphene, its widespread use has been limited by several practical challenges, such as difficulty in large-scale production, high production costs, and limited functionalization capabilities (Lawal 2015). As a result, the field of 2D materials has expanded to explore other promising alternatives. One of the most significant developments in this area has been the discovery and subsequent exploration of MXenes, a new family of 2D materials composed of transition metal carbides, nitrides, or carbonitrides. MXenes were first synthesized in 2011 by researchers at Drexel University, marking a breakthrough in the study of 2D materials (Zhan et al. 2020). MXenes are derived from MAX phases, which are a group of layered materials that include the general formula $M_{n+1}AX_n$, where M represents a transition metal, A is an element such as aluminum or silicon, and X is either carbon or nitrogen (Zhan et al. 2020).

The unique synthesis method used to prepare MXenes is one of the key features that sets them apart from other 2D materials. MXenes are synthesized by selectively etching away the Al-layer of MAX phases, a process that results in the formation of highly conductive and stable 2D sheets (Guo et al. 2015). The etching process also introduces surface functional groups, such as –OH, –O, and –F, which can be further


DOI: 10.1201/9781003736318-7

modified to enhance the properties of MXenes and enable for a wide range of potential applications (Jiang et al. 2024). The high surface area and tunable surface chemistry of MXenes make them ideal candidates for use in energy storage devices, such as supercapacitors and lithium-ion batteries, as well as in various catalytic processes, sensing technologies, and electromagnetic shielding applications (Ghidiu et al. 2014).

While MXenes possess outstanding intrinsic properties, there is still a need to further enhance their performance for many practical applications. One promising approach to improving the properties of MXenes is to combine them with other materials, such as metals, to create composite materials (Chen et al. 2020). MXene–metal composites combine the unique properties of both MXenes and metals to produce materials that outperform either of the individual components. Metals, particularly noble metals like Au (Liu et al. 2023), Ag (Xia et al. 2024), Pt (Rasheed et al. 2021), and Pd (Phuong et al. 2021; Rasheed et al. 2019), are known for their excellent electrical conductivity, high catalytic activity, and thermal stability. The incorporation of Au nanoclusters on MXene makes the composites a promising material for electrochemical sensing with high electrical conductivity, large surface area, outstanding catalytic activity, and good thermal stability (Liu et al. 2023). The Ag–MXene composites can provide a larger specific surface area and good conductivity, which enables more active sites to bind analytes toward improved sensor sensitivity in electrochemical sensing applications in addition to the antifouling properties (Xia et al. 2024). The presence of Pd and Pt on MXenes can enhance electron transfer owing to the electrical conductivity and sheet-like structure of MXene and the presence of Pt and Pd nanoparticles (NPs), which enhances the surface to volume ratio (Rasheed et al. 2021). Additionally, the introduction of Pt and Pd NPs can enhance the electrochemical stability of the composite material with respect to the unmodified MXenes (Rasheed et al. 2019). Therefore, by integrating these metals with MXenes, it is possible to improve the electrical conductivity, mechanical strength, catalytic activity, and thermal as well as electrochemical stability of the composite material.

The preparation of MXene–metal composites is an area of intense research, and several synthesis methods have been developed to create these hybrid materials. One of the most commonly used methods is chemical reduction, in which metal salts or metal precursors are reduced to their nanostructures in the presence of MXenes (Li et al. 2021a). Here, MXene is acting as the reducing agent, and the reduction process leads to the deposition of metal NPs onto the surface of the MXene sheets. The size and distribution of the metal particles can be controlled by adjusting various reaction conditions, such as the concentration of metal precursors, temperature, and reaction time. This approach has been successfully used to create MXene composites with metals like Au, Ag, and Pt which enhance the electrochemical properties of MXenes for applications in energy storage, sensing, and catalysis (Rajavel et al. 2020; Zhang et al. 2020). Another commonly employed synthesis method for MXene–metal composites is in situ chemical reduction in the presence of a reducing agent (Zuo et al. 2019). MXenes are mixed with metal precursors in an aqueous solution along with a reducing agent, and metal particles are deposited onto the MXene sheets through chemical deposition. This synthesis method offers greater flexibility than chemical reduction, as it allows the incorporation of a wide range of metals into the MXene structure. It also enables better control over the size, distribution, and morphology of the metal particles, which is particularly important for applications that require uniformity in particle dispersion, such as in catalysis and sensing. This technique has been widely used to create MXene composites with metals such as Pt (Thirumal et al. 2022) and Pd (Huang et al. 2023), which have demonstrated excellent performance in applications such as hydrogen evolution, oxygen reduction, $CO_2$ reduction, and electrochemical sensing applications (Lorencova et al. 2018). In addition, metal NPs can be electrodeposited on MXene nanosheets and can be utilized as a sensing platform in electrochemical sensors with enhanced sensitivity (Chen et al. 2025; Li et al. 2023). Atomic layer deposition and ball milling methods are also used for the synthesis of MXene–metal composites.

The applications of MXene–metal composites are vast and span a wide range of fields, from energy storage to catalysis and sensing. One of the most promising applications for MXene–metal composites is in energy storage devices, such as supercapacitors and lithium-ion batteries (Tian et al. 2019). MXene–metal composites have also been widely explored for use in electrochemical sensing applications, particularly in the detection of biomolecules (Feng et al. 2023) and environmental pollutants (Xiao et al. 2022). The high surface area and reactivity of MXenes, coupled with the catalytic properties of metal NPs, in which metal particles can serve as active sites for the detection, make these composites can be explored for developing highly sensitive electrochemical sensors. The integration of metals, such as Au and Pt, into MXenes further improves the selectivity and sensitivity of these sensors, making them suitable for a wide range of applications, including environmental monitoring, healthcare diagnostics, and industrial process control.

While the advantages of MXene–metal composites are compelling, several challenges remain to be addressed. Issues such as the agglomeration of metal particles (Zhou et al. 2021), potential leaching of metals, and the reproducibility of synthesis methods can hinder the effectiveness and reliability of these sensors. The stability of these hybrid composites under operational conditions is crucial for their practical applications. Factors such as the environmental stability of MXenes, potential oxidation of metallic components (Kumar et al. 2024), and the need for adequate protective coatings must be critically addressed (Zheng et al. 2024). Furthermore, optimizing synthesis processes and exploring different MXene compositions can lead to further improvements in sensor performance, paving the way for innovative applications in real-world scenarios (Mim et al. 2024).

## 7.2 PROPERTIES OF MXENE–METAL COMPOSITES

MXene–metal composites synergistically combine the exceptional characteristics of MXenes with the functional advantages of metallic components, and this integration gives rise to multifunctional materials that exhibit superior properties compared to either component alone. A comprehensive overview of the key properties of MXene–metal composites is given in the following section, highlighting their significance for advanced applications. MXenes are known for their metallic-level electrical conductivity (up to 20–24 kS/cm for a pure film), which is further enhanced when integrated with conductive metals such as Ag, Al, and Li (Park, Kim, and Kim 2024). This composite structure enables superior electron mobility and reduced internal resistance, which are vital for high-performance applications in energy storage and electrochemical devices. The synergistic effect between MXene nanosheets and metallic phases enhances charge transfer kinetics, making these materials highly effective as electrodes in electrochemical applications (Nguyen et al. 2021).

MXene–metal composites also show improved thermal properties, including elevated thermal conductivity and thermal stability (Wyatt, Nemani, and Anasori 2021). The metallic phase contributes to effective heat dissipation, which is essential in electronic packaging, power electronics, and thermal interface materials (Wang et al. 2021). Additionally, MXenes enhance the composite's thermal resistance, allowing them to maintain structural integrity and functional performance at elevated temperatures, which is essential for aerospace and high-temperature industrial applications (Ma et al. 2022).

One of the critical advantages of MXene–metal composites is their enhanced resistance to corrosion and environmental degradation. The 2D layered structure of MXenes acts as a physical barrier, reducing the diffusion of corrosive agents. Furthermore, surface functional groups (e.g., –OH, –F, –O) on MXenes interact with corrosive ions, providing chemical passivation and extending the material's service life. The presence of MXenes in metal matrices improves the overall chemical stability of the composite, especially under harsh conditions such as marine, acidic, or oxidative environments (Saharudin, Che Nasir, and Hasbi 2022). However, the specific chemical behavior can vary depending on the type of metal used and the synthesis conditions, necessitating optimized design strategies for long-term stability.

MXene–metal composites exhibit remarkable electrocatalytic performance due to the combination of large surface area and metallic conductivity of MXenes with the catalytic properties of incorporated metals such as Pd (Yao et al. 2023), Pt (Abdullah et al. 2020), or Ni (Zhao et al. 2024). These composites provide additional active sites and enhanced electron transfer, which accelerate redox reaction kinetics, making them ideal for electrochemical sensors. The high surface area and reactivity of MXenes, along with the catalytic properties of metal NPs enhances the electrochemical properties of the composite.

## 7.3 SYNTHESIS OF MXENE–METAL COMPOSITES

MXenes, with their high conductivity, large surface area, and abundant surface functional groups, provide an ideal platform for the nucleation, anchoring, and stabilization of metallic NPs. This section systematically presents the synthesis methods employed for various MXene–metal composites, emphasizing their preparation techniques, structural characteristics, and potential implications for advanced applications. Figure 7.1 shows the various synthetic approaches for the synthesis of MXene–metal nanocomposites, including in situ chemical reduction, in situ hydrothermal synthesis, atomic layer deposition, electrodeposition, and ball milling.

### 7.3.1 In Situ Chemical Reduction

The synthesis of MXene–metal composites by chemical reduction methods involves synthesis methods that include in situ reduction methods without using

**FIGURE 7.1** Various synthetic approaches for the synthesis of MXene–metal nanocomposites.

any reducing agents. In the in situ chemical reduction method, MXene acts as the reducing agent to convert metal salt into metal nanostructures. For example, the $Au/Ti_3C_2T_x$ nanocomposite was synthesized by simply mixing MXene and $HAuCl_4$ solution, followed by stirring for a few hours. Here, the self-reduction reaction is initiated by the addition of a metal salt, and Au-decorated MXene is formed (Yin et al. 2020). In another work, Ag–MXene ($Ag–Ti_3C_2T_x$ and $Ag–Nb_2CT_x$) composites were prepared by the self-chemical reduction method by using MXene as a reducing agent and silver nitrate as a silver source. The MXene is dispersed in distilled water by sonication, and silver nitrate is added to it under continuous magnetic stirring and sonication (Rajavel et al. 2020). Similarly, $Pd@Ti_3C_2T_x$ composite has been synthesized by reacting MXene and palladium salt, followed by sonication and overnight stirring (Rasheed et al. 2019). In another report, $Pt/Nb_2CT_x$ MXene composite was synthesized by reacting MXene and metal salt by Fan et al. (2021). In these composites, the metal NPs are formed and decorated on the surface of MXene sheets. An important feature of this method is that the MXenes serve as both the reducing agent and the substrate without the need for external reducing agents or high-energy inputs. This approach simplifies the synthesis process as well as reduces the environmental impact by eliminating toxic chemicals. However, this reduction method on MXene often results in random metal ion nucleation and non-uniform deposition of metal NPs on the MXene surface.

In order to have more control over the metal deposition into uniform patterns, a novel in situ reduction strategy for the synthesis of $Ti_3C_2T_x$ MXene–metal composites has been introduced by Zhang et al. (2025). They proposed that the redox process is driven by the reaction between the metal ions and the reduced Ti species ($Ti^{2+}$/$Ti^{3+}$) on the MXene surface. The increase in the $Ti^{4+}$ fraction after metal deposition indicates that the electron for the reduction of metal ions comes from Ti ions. Here, the metal atoms initially nucleate on vacancies in the MXene surface, and as metal loading increases, these atoms aggregate together to form larger NPs. In addition, Ti leaching was observed during the metal deposition, and this leads to the defect formation in the MXene structure, which enables additional active sites for further growth and overall deposition of metal NPs. They further investigated the deposition of Au, Pd, Ag, Pt, Rh, Ru, and Cu on the $Ti_3C_2T_x$ surface by this method and found that metals (Pt, Ru, Rh, and Cu) adopt a monoatomic state at low loadings and transform to a nanoparticle state with higher loadings. The formed Au, Ag, and Pd NPs on the MXene surface show different distributions, such as a worm-like structure of Pd and an uneven deposition of Ag. They also proposed that this growth mechanism of metal NPs can be applied to other MXene substrates. such as $Mo_2CT_x$, $V_2CT_x$, $Ti_3CNT_x$, $Nb_4C_3T_x$, and $Mo_2Ti_2C_3T_x$.

### 7.3.2 In Situ Chemical Reduction with Reducing Agents

In this method, a reducing agent is added to the metal salt, followed by adding the MXene or adding metal salt and reducing to MXene dispersion to get the MXene–metal composites. Here, metal NPs are formed by the reducing agent and decorated or conjugated on the MXene nanosheets. In this approach, Au/$Ti_3C_2T_x$ nanocomposite was synthesized by adding a reducing agent into a gold precursor ($HAuCl_4$) solution under magnetic stirring to facilitate in situ reduction of $Au^{3+}$ ions. Following this, exfoliated $Ti_3C_2T_x$ MXene nanosheets were introduced into the reaction medium and ultrasonicated as shown in Figure 7.2(a). This process facilitated the nucleation and uniform deposition of gold NPs onto the MXene surface (Rakhi et al. 2016). Figure 7.2(b) and (c) shows the X-ray diffraction pattern of and transmission electron microscopy (TEM) image of Au/MXene nanocomposite. The X-ray diffraction (XRD) pattern shows the presence of Au along with that of $Ti_3C_2T_x$ MXene, and the TEM image revealed almost uniform distribution of Au nanoparticles over the MXene surface. In another report, $Ti_3C_2T_x$/DNA/Pd/Pt nanocomposite was synthesized by modifying double-stranded DNA (dsDNA) with MXene, followed by adding Pd and Pt salts as well as a reducing agent. The DNA backbone played a critical role in nanoparticle dispersion and adhesion, enhancing catalytic performance, particularly for dopamine detection (Zheng et al. 2018). Similarly, a RhNi bimetallic nanoparticle-decorated MXene nanocomposite was fabricated using a one-step wet chemical reduction route by reacting the metal salts (Rh and Ni salts) and MXene, followed by the addition of the reducing agent. The microscopy analysis revealed monodispersed RhNi NPs (~2.8 nm) uniformly dispersed across the MXene basal planes (Liu et al. 2018).

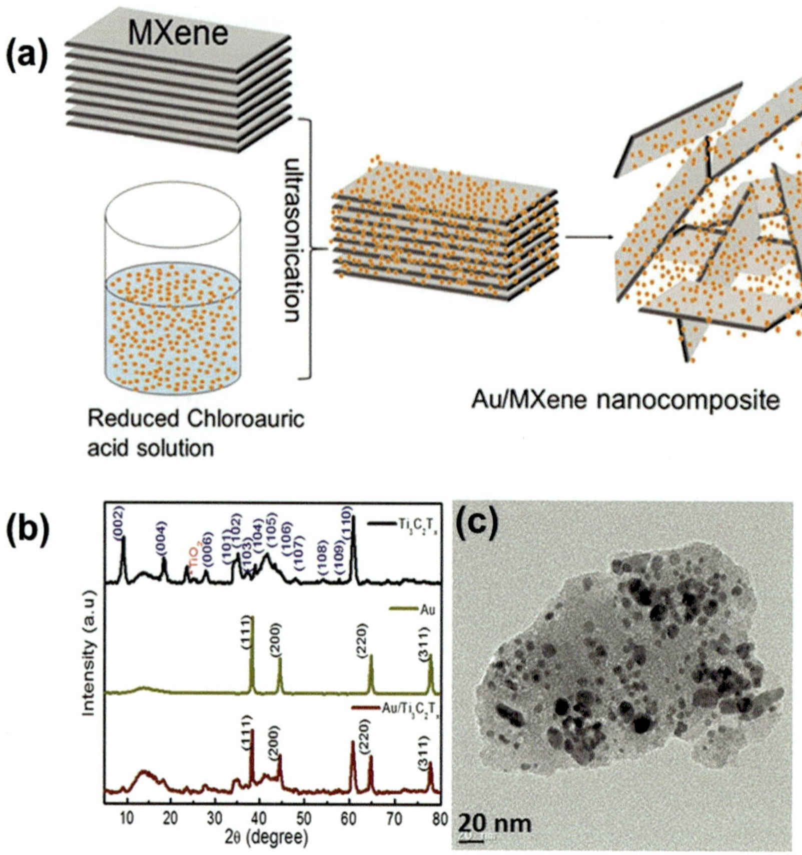


**FIGURE 7.2** (a) Schematic illustration of the synthesis of Au/$Ti_3C_2T_x$ nanocomposite. (b) Powder X-ray diffraction pattern of $Ti_3C_2T_x$ MXene, AuNPs, and Au/MXene nanocomposite. (c) TEM image of Au/MXene nanocomposite. Reproduced with permission from Rakhi et al. (2016). Copyright © 2016, The Author(s).

### 7.3.3 In Situ Hydrothermal Synthesis

MXene–metal composites can be synthesized by in situ hydrothermal assembly of metal particles on the MXene layer. Toward this approach, $Ti_3C_2T_x$/Ni composite was synthesized by reacting the metal salt $NiCl_2{\cdot}6H_2O$ with a reducing agent and exfoliated $Ti_3C_2T_x$, which were mixed using stirring. After getting the uniform dispersion, the reaction mixture is transferred into a Teflon-lined stainless-steel autoclave for hydrothermal treatment to get the composite (Li et al. 2019).

### 7.3.4 Atomic Layer Deposition

Atomic layer deposition (ALD) is a variant of chemical vapor deposition (CVD) that enables the deposition of materials in discrete, single-atom layers. This technique achieves precise control over film thickness by sequentially applying precursor chemicals, with each cycle resulting in the deposition of a single atomic layer. The ALD method is used for the fabrication of MXene–metal composites with relatively low metal contents. Toward this approach, Zhang et al. synthesized Pt NP-deposited $Ti_3C_2T_x$ MXene with uniform and small Pt NPs (~2 nm in diameter) with a low Pt content of 1.70 wt% by ALD toward using in the hydrogen evolution reaction (Zhang et al. 2020). They have controlled the Pt content in the composite by varying the ALD deposition cycles.

### 7.3.5 Electrodeposition

In this method, the metal NPs are electrodeposited on the MXene sheets by applying electrochemical potential and this is generally used for developing the electrochemical applications. The fabrication process began with the drop-coating of MXene onto the electrode surface, followed by the electrodeposition of metal NPs. Chen et al. fabricated Au/MXene on a screen-printed electrode by drop casting of MXene, followed by electrodeposition of An NPs through cyclic voltammetry in the electrolyte containing metal salt (Chen et al. 2025). Similarly, Liu et al. reported in situ electrodeposition of Au NPs and $Ti_3C_2T_x$ MXenes together on the screen-printed electrode (SPE) toward electrochemical sensing applications (Liu et al. 2024). In another work, Jian et al. synthesized platinum nanoparticle-electrodeposited $Ti_3C_2T_x$ MXene on carbon fiber through continuous cyclic voltammetry scan for 1000 cycles (Jian et al. 2022).

### 7.3.6 Ball Milling

Ball milling is yet another method that has been used to prepare MXene–metal composites. In this process, MXenes are mixed with metal powders and subjected to high-energy ball milling (Li et al. 2021b). This method causes mechanical alloying and leads to the formation of composite materials with a uniform distribution of metal particles within the MXene matrix. Ball milling has the advantage of scalability, making it suitable for large-scale production of MXene–metal composites. Additionally, the mechanical alloying process often results in the formation of new metal–MXene phases that exhibit enhanced properties compared to the individual components. This method has been used to create MXene composites with metals like copper (Cu) (Gao et al. 2024), nickel (Ni) (Li et al. 2021a), and iron (Fe) (Wang, Zhao, et al. 2025), which exhibit improved mechanical and electrochemical properties.

## 7.4 APPLICATION OF MXENE–METAL COMPOSITES IN ELECTROCHEMICAL SENSING

The incorporation of metal nanostructures on MXene makes composites with enhanced surface area and good stability in addition to the outstanding catalytic activity, particularly in the presence of noble metals (Khan et al. 2022). The large surface area of MXene–metal composites enables more active sites to bind analytes and enhances sensor sensitivity in electrochemical sensing applications (Yao et al. 2023). Different MXene–metal composites have been used in the electrochemical sensing of physiological biomarkers, food-related applications, and environmental applications.

### 7.4.1 Electrochemical Sensing of Physiological Biomarkers

MXene–metal nanocomposites with metal as Au, Ag, Pt, and Pd have been used for developing the electrochemical sensors for the detection of various biomolecules. Elumalai et al. reported a hybrid nanocomposite consisting of AuNPs anchored onto two-dimensional $Ti_3C_2T_x$ MXene, synthesized through an ultrasonication-assisted in situ reduction. This AuNP@$Ti_3C_2T_x$ composite featured a layered architecture and demonstrated remarkable electrocatalytic activity toward the voltammetric detection of uric acid (UA) and folic acid (FA) by using a glassy carbon electrode (GCE) modified with AuNP@$Ti_3C_2T_x$ composite. The modified electrode (AuNP@$Ti_3C_2T_x$@GCE) achieved a well-resolved peak separation of 340 mV between UA and FA (oxidation potential of +0.35 V and 0.70 V, respectively), facilitating their simultaneous detection. Under optimized analytical conditions, the sensor showed a wide linear range for UA (0.03–1520 μM) and FA (0.02–3580 μM) with detection limits of 11.5 nM for UA and 6.20 nM for FA. The developed sensor was successfully applied for simultaneous quantification of UA and FA in real human urine, reinforcing its potential for practical bioanalytical applications (Elumalai et al. 2019).

Yola and colleagues developed an advanced electrochemical immunosensor in which a self-assembled delaminated MXene–gold nanoparticles (d-$Ti_3C_2T_x$ MXene@AuNPs) was used as a signal amplifier after conjugating with a secondary antibody ($Ab_2$) to prostate-specific antigen (PSA). The $Ab_2$ was conjugated to $Ti_3C_2T_x$@AuNPs composite by interaction between the gold atom of the composite and $-NH_2$ of $Ab_2$, as shown in Figure 7.3(a). The sensor platform was made by incorporating AuNPs and p-aminothiophenol-functionalized graphene oxide (AuNPs-ATP-GO), followed by the addition of primary antibody ($Ab_1$) and bovine serum albumin (BSA). The PSA can bind to $Ab_1$ and form the complete sensing platform PSA/BSA/$Ab_1$/AuNPs-ATPGO/GCE. The antigen–antibody interaction reaction between d-$Ti_3C_2T_x$ MXene@AuNPs/$Ab_2$ and PSA/BSA/$Ab_1$/AuNPs-ATPGO/GCE occurs by the reaction between PSA and $Ab_2$, as shown in Figure 7.3(b). The developed immunosensor demonstrated a wide dynamic detection range from 0.01 to 1.0 pg/mL, with an impressively low limit of detection (LOD) of 3.0 fg/mL for PSA. The sensor exhibited excellent performance attributes, including high stability, selectivity, repeatability, and reproducibility. Moreover, its promising analytical features suggest strong potential for clinical diagnostic applications, contributing to improved healthcare outcomes (Medetalibeyoglu et al. 2020).

In a study by Zheng et al., a $Ti_3C_2$/DNA/Pd/Pt nanocomposite was developed through an in situ reduction method and used for electrochemical detection of dopamine. Here, the DNA template on MXene plays a major role in the formation of small and uniform deposition of Pd/Pt NPs and in forming $Ti_3C_2$/DNA/Pd/Pt nanocomposite. The sensor demonstrated a linear amperometric response in the dopamine concentration range of 0.2–1000 μM with a detection limit of 30 nM. The sensor exhibited excellent selectivity against common interferents such as ascorbic acid, UA, and glucose, making it a promising tool for in vivo and clinical dopamine detection (Zheng et al. 2018).

In another report, Rasheed et al. (2019) synthesized a Pd@$Ti_3C_2T_x$ nanocomposite by growing Pd NPs on $Ti_3C_2T_x$ nanosheets through a self-reduction approach. This composite was then used to fabricate a biosensor for the electrochemical detection

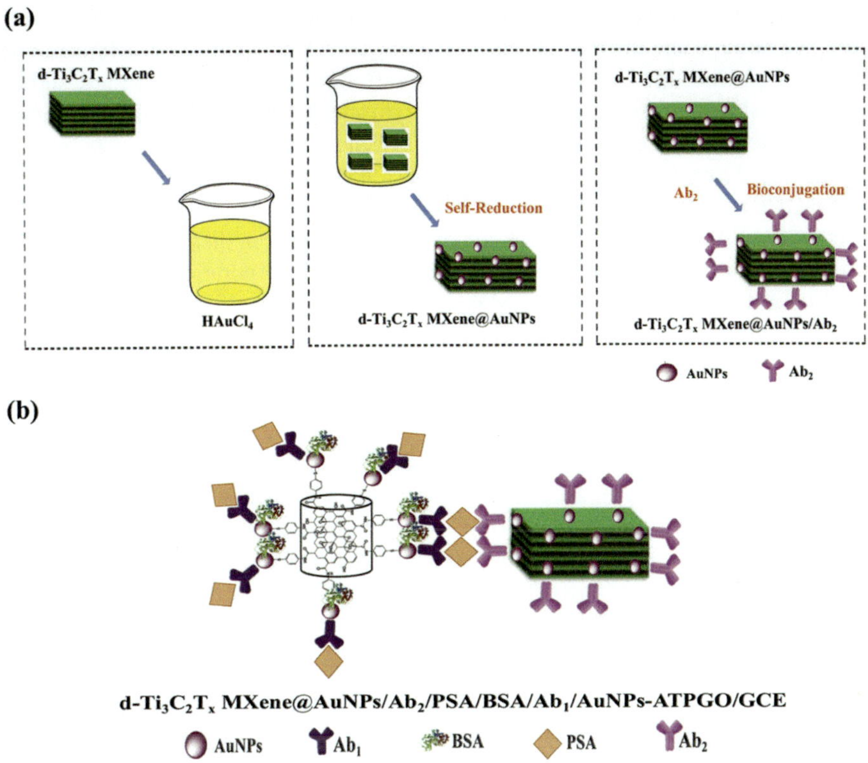


**FIGURE 7.3** (a) Schematics showing the synthesis of $Ab_2$-modified $Ti_3C_2T_X$@AuNPs ($Ti_3C_2T_x$ MXene@AuNPs/$Ab_2$); (b) the antigen–antibody interaction reaction between $Ti_3C_2T_x$ MXene@AuNPs/$Ab_2$ and PSA/BSA/$Ab_1$/AuNPs-ATPGO/GCE toward the sensing of PSA. Reprinted with permission from Medetalibeyoglu et al. (2020). 

of L-Cys by modifying it onto GCE. The sensor exhibited a wide detection range from 0.5 to 10 μM with a detection limit of 0.14 μM. The Pd@$Ti_3C_2T_x$-based sensor demonstrated excellent stability, making it suitable for long-term use in biological and clinical applications. Additionally, it showed high selectivity over common interfering ions such as ascorbic acid, dopamine, and glucose, which are often present in complex biological samples (Rasheed et al. 2019).

MXene–silver (Ag) composites have been evaluated for their electrochemical sensing capabilities. Recently, Xia et al. developed a novel and highly effective strategy for the sensitive and selective detection of programmed death ligand 1 (PD-L1) using a dual-functional sensing platform composed of hydrophilic Ag@MXene and a custom-designed capture peptide (CP) on a gold electrode. The large surface area of Ag@MXene can load more CP probes, while Ag@MXene can form a fouling barrier so that this strategy enhances the sensor response in addition to the resistance to contamination. The designed electrochemical sensor showed promising selectivity and a wide linear range of 0.1–50 ng/mL with a detection limit of 24.54 pg/mL for the detection of PD-L1 (Xia et al. 2024).

Composites of MXenes and Cu-based metal-organic framework (CuMOF) have been synthesized and utilized for electrochemical sensing applications. By utilizing this strategy, Chen et al. developed an enzyme-free electrochemical sensor designed for the sensitive and selective detection of tyrosine. In this work, they have made a composite material consisting of $Ti_3C_2T_x$ MXene, carbon nanotubes (CNTs), and copper-based metal-organic frameworks (CuMOFs) and used it to construct the sensor on GCE as shown in Figure 7.4. The integration of CNTs into MXene nanosheets effectively mitigated the common issue of sheet aggregation, thereby enhancing the structural stability of the composite. Furthermore, the incorporation of porous, octahedral CuMOF particles into the MXene/CNTs matrix significantly improved the overall porosity and catalytic activity of the sensing interface. Under optimized conditions, the fabricated sensor exhibited excellent linear response for tyrosine detection across a wide concentration range of 0.53 to 232.46 μM, with an impressive LOD of 0.19 μM. The sensor also demonstrated high stability, reproducibility, and selectivity, enabling its application in the reliable detection of tyrosine in human serum samples (Chen et al. 2022). In another study, Cheng et al. constructed a novel non-enzymatic hydrogen peroxide ($H_2O_2$) electrochemical sensor based on copper metal-organic frameworks (Cu-MOFs) hybridized with $Ti_3C_2T_x$ MXene nanosheets. The CuMOF/$Ti_3C_2T_x$ MXene composite was prepared via a self-assembly process, resulting in a 3D flower-like morphology, which offered a large active surface area

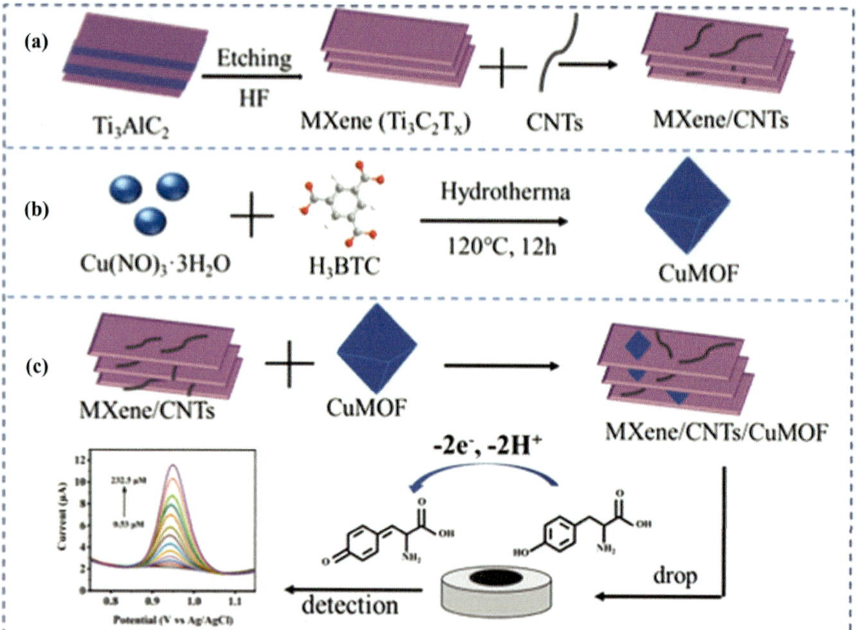


**FIGURE 7.4** Schematic diagram of the MXene/CNTs/CuMOF synthesis procedure and the sensing mechanism for detecting tyrosine. (a) Synthesis of MXene/CNT composite, (b) synthesis of CuMOF, and (c) synthesis of MXene/CNTs/CuMOF and sensing mechanism. Reprinted with permission from Chen et al. (2022). Copyright © 2022 Published by Elsevier Ltd.

and facilitated efficient electron transfer. The resulting CuMOF/MXene/GCE displayed a broad linear response range from 1 μM to 6.12 mM and a low detection limit of 0.35 μM for $H_2O_2$ under a detection potential of −0.35 V. The sensor demonstrated excellent analytical performance in real-sample matrices such as serum and milk, thus emphasizing its potential for practical biomedical and food safety monitoring applications (Cheng et al. 2021).

### 7.4.2 Electrochemical Sensing of Environmental Pollutants and Food Contaminants

Rasheed et al. used AuNPs/$Nb_4C_3T_x$ MXene composite for developing the electrochemical aptasensor for the detection of $Pb^{2+}$ toward environmental sensing applications. The AuNPs/$Nb_4C_3T_x$ was modified with Pb-specific binding DNA oligonucleotide through conjugation with thiol-modified DNA and AuNPs by Au–S bond. The DNA-Au@$Nb_4C_3T_x$-modified GCE showed a promising sensor response toward the detection of $Pb^{2+}$ in the range of 10 nM to 5 μM with a detection limit of 4 nM. From this work, they proposed that the $Nb_4C_3T_x$ can be used as a robust immobilization platform for DNA oligonucleotides toward various electrochemical sensing applications (Rasheed et al. 2022). Song et al. fabricated an electrochemical sensing platform for ultrasensitive determination of organophosphorus pesticides (OPs) using a composite made of $Ti_3C_2$ MXene/Au NPs and MOF-derived $MnO_2$/$Mn_3O_4$. The sensor platform was modified with chitosan and acetylcholinesterase (AChE), where AChE acted as the recognition element for methamidophos. Under the optimum conditions, the reported sensing platform AChE-Chit/MXene/Au NPs/$MnO_2$/$Mn_3O_4$/GCE can be utilized to detect methamidophos in a broad concentration range (1 pM to 1 μM) with an LOD of 0.134 pM. Additionally, the practical application of the developed sensor was validated by detecting methamidophos in real samples (Song et al. 2019). In another report, a biosensor electrode was fabricated with laccase (Lac) immobilized on Au/MXene on CE using the Nafion solution for the detection of catechol by Chandran et al. (2021). The developed sensor showed a linear amperometric response in the catechol concentration range of 0.05–0.15 μM with a detection limit of 0.05 μM. This Lac/Au/MXene/GCE biosensor could be used to detect catechol from industrial effluents and water bodies.

In another study by Zhao et al. (2020), a bimetallic MXene/Au–Pd nanocomposite was developed for the electrochemical detection of OPs using an acetylcholinesterase (AChE) biosensor. The sensor demonstrated excellent catalytic activity and sensitivity, with a detection limit as low as 1.75 ng/L for paraoxon, a model OP pesticide. The linear range of the sensor spanned from 0.1 to 1000 μg/L, and it was successfully applied to detect pesticides in real food samples such as pears and cucumbers (Zhao et al. 2020).

Rasheed et al. evaluated the electrochemical performance of PtNPs/$Ti_3C_2T_x$ nanocomposite (Pt@$Ti_3C_2T_x$) toward environmental sensing applications, especially sensing of bisphenol A (Rasheed et al. 2021). The composite was prepared by a self-reduction strategy in which $Ti_3C_2T_x$ simultaneously acted as a reducing agent as well as a conductive matrix, as shown in Figure 7.5(a). The Pt content was optimized in the composite, and it was found that 10%Pt@$Ti_3C_2T_x$-modified GCE has

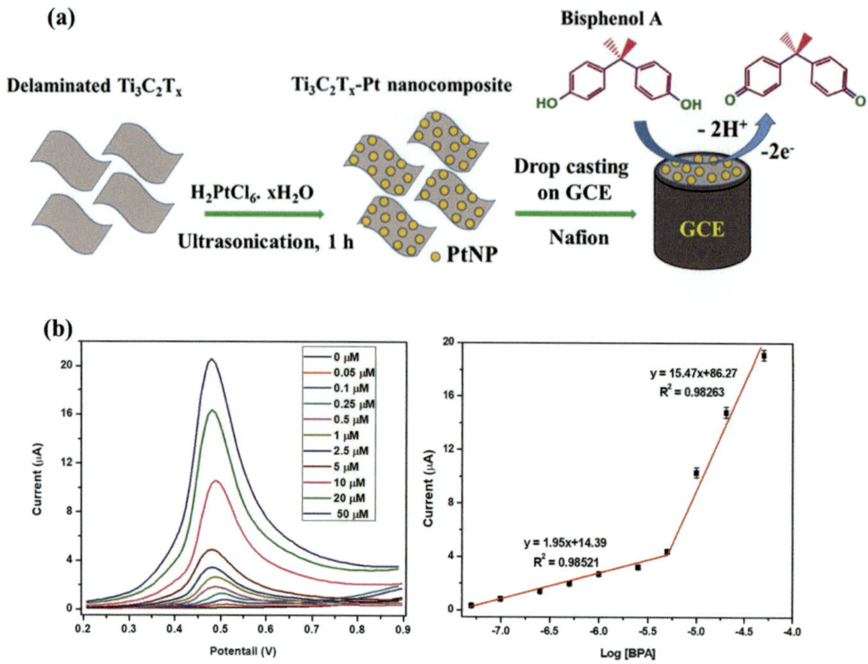


**FIGURE 7.5** (a) Schematics showing the synthesis of Pt@$Ti_3C_2T_x$ composite and the sensing of bisphenol A after modifying the composite on GCE. (b) Differential pulse voltammetry (DPV) response of 10%Pt@$Ti_3C_2T_x$/GCE electrode in the presence of Bisphenol A (BPA) at different concentrations and corresponding calibration plot between current and concentration (logarithmic). Reprinted with permission from Rasheed et al. (2021). Copyright © 2020 Elsevier B.V. All rights reserved.

the highest electrochemical activity in the anodic potential window. The developed sensor was able to detect bisphenol A in the concentration range from 50 nM to 5 μM with a detection limit of 32 nM (Figure 7.5(b)). They have evaluated the practical application of the sensor with drinking water and fresh milk samples.

Very recently, Wang et al. developed an ultrasensitive electrochemical biosensor designed to detect staphylococcal enterotoxin B (SEB), a toxin that commonly leads to food contamination and foodborne diseases (Wang, Huang, et al. 2025). This biosensor uses an innovative electrochemical nanoprobe consisting of MXene/AgNPs decorated with para-sulfonatocalix[4]arene (pSC4) and methylene blue (MB) as an electroactive signal molecule (MXene/AgNPs@pSC4/MB). The combination of MXene and AgNPs enhances the overall conductivity and signal molecule loading capacity, leading to an amplified biosensor response. Moreover, the pSC4 serves as a stable linker, facilitating the binding of the nanoprobe to SEB by interacting with its amino acid residues. The specific aptamer immobilized on the electrode effectively captures SEB, leading to the formation of a complex with the electrochemical nanoprobes, which results in an enhanced electrochemical signal. This approach enables the biosensor to detect SEB at a remarkably low concentration of $1.76 \times 10^{-7}$ ng/mL, with the additional benefits of easy operation and rapid response time.

**TABLE 7.1**
**Comparison table showing synthesis and electrochemical sensing applications of various MXene–metal composites**

| Composite | Synthesis technique | Analyte detected | Linear range | Limit of detection (LOD) | Ref. |
|---|---|---|---|---|---|
| AuNP@$Ti_3C_2T_x$ | In situ chemical reduction | Uric acid (UA), folic acid (FA) | 0.03–1520 µM (UA), 0.02–3580 µM (FA) | 11.5 nM (UA), 6.20 nM (FA) | Elumalai et al. (2019) |
| d-$Ti_3C_2T_x$@AuNPs | In situ chemical reduction | Prostate-specific antigen (PSA) | 0.01–1.0 pg/mL | 3.0 fg/mL | Medetalibeyoglu et al. (2020) |
| $Ti_3C_2$/DNA/Pd/Pt | In situ chemical reduction with a reducing agent | Dopamine | 0.2–1000 µM | 30 nM | Zheng et al. (2018) |
| Pd@$Ti_3C_2T_x$ | In situ chemical reduction | L-Cysteine | 0.5–10 µM | 0.14 µM | Rasheed et al. (2019) |
| Ag@$Ti_3C_2T_x$ | In situ chemical reduction | Programmed death ligand 1 (PD-L1) | 0.1–50 ng/mL | 24.54 pg/mL | Xia et al. (2024) |
| Lac/Au/$Ti_3C_2T_x$ | In situ chemical reduction with reducing agent | Catechol | 0.05–0.15 µM | 0.05 µM | Chen et al. (2022) |
| AChE-Chit/$Ti_3C_2T_x$/Au NPs/$MnO_2$/$Mn_3O_4$ | In situ chemical reduction | Methamidophos | 1 pM to 1 µM | 0.134 pM | Song et al. (2019) |
| $Ti_3C_2T_x$/Au-Pd/GA/AChE | In situ chemical reduction | Paraoxon | 0.1–1000 µg/L | 1.75 ng/L | Zhao et al. (2020) |
| Pt@$Ti_3C_2T_x$ | In situ chemical reduction | Bisphenol A | 0.05–5 µM | 32 nM | Rasheed et al. (2021) |
| $Ti_3C_2T_x$/AgNPs@pSC4/MB | In situ chemical reduction | Staphylococcal enterotoxin B (SEB) | $1 \times 10^{-4}$ to 100 ng/mL | $1.76 \times 10^{-7}$ ng/mL | Wang, Huang, et al. (2025) |

Furthermore, the biosensor's performance has been tested with real food samples, confirming its potential applicability for monitoring food safety. Table 7.1 shows the comparison table showing synthesis and electrochemical sensing applications of various MXene–metal composites.

## 7.5 CONCLUSION AND FUTURE PERSPECTIVES

In conclusion, MXene–metal composites are emerging as a highly promising material class with significant potential for electrochemical sensing. The exceptional blend of MXene's high conductivity, expansive surface area, and adaptable surface chemistry, paired with the catalytic capabilities of metals, has enabled the creation of sensors that are highly sensitive, selective, and durable for detecting a variety of analytes such as gases, pollutants, and biomolecules. The combination of MXenes with metals results in composites that display improved catalytic activity, electrical properties, and stability, making them suitable for advanced electrochemical sensing applications. This chapter has reviewed the key properties of MXene–metal composites and described the various synthesis techniques used to create these materials, underscoring their adaptability in enhancing sensor performance.

As MXene–metal composites represent a transformative advancement in electrochemical sensing, offering a powerful combination of high sensitivity, selectivity, and scalability. When MXenes are used in biosensors, the main issue is the surface termination of fluorine in MXene, which is not favorable under different conditions. Hence, it is suggested to remove the fluorine termination in biosensing applications, and this will be highly beneficial for specific biomedical applications as well. It should be noted that the electrochemical stability of MXene can be enhanced after compositing with metal nanoparticles, which can decrease the electrochemical performance. There should be an optimum balance between the stability and electrochemical activity while considering a specific MXene–metal composite for electrochemical sensing applications. Future work on MXene–metal composites should focus on the design and synthesis methods to optimize the sensing capability without losing other useful properties. More bimetallic structures with MXenes or ternary composites can be introduced to achieve the best results in properties as well as in applications. As research continues, further breakthroughs in material engineering and fabrication techniques will undoubtedly expand their role in next-generation sensor technologies, driving significant progress in various scientific and industrial domains.

## ACKNOWLEDGMENTS

PAR acknowledges the Ramalingaswami fellowship (BT/RLF/Re-entry/75/2020) from the Department of Biotechnology (DBT), Government of India. The authors acknowledge the Indian Institute of Technology Palakkad for providing the facilities.

## REFERENCES

Abdullah, Norulsamani, R. Saidur, Azran Mohd Zainoodin, and Navid Aslfattahi. 2020. "Optimization of electrocatalyst performance of platinum–ruthenium induced with MXene by response surface methodology for clean energy application." *Journal of Cleaner Production* 277:123395. doi: 10.1016/j.jclepro.2020.123395.

Chandran, M., Aswathy, E., Shamna, I., Vinoba, M., Kottappara, R., & Bhagiyalakshmi, M. 2021. "Laccase immobilized on Au confined MXene based electrode for electrochemical detection of catechol". *Materials Today: Proceedings*, 46, 3136–3143. doi: 10.1016/j.matpr.2021.02.697

Chen, Jing, Ying Chen, Shuying Li, Jiao Yang, Jianbin Dong, and Xiaoquan Lu. 2022. "MXene/CNTs/Cu-MOF electrochemical probe for detecting tyrosine." *Carbon* 199:110–118. doi: 10.1016/j.carbon.2022.07.021.

Chen, Junyu, Qiang Huang, Hongye Huang, Liucheng Mao, Meiying Liu, Xiaoyong Zhang, and Yen Wei. 2020. "Recent progress and advances in the environmental applications of MXene related materials." *Nanoscale* 12 (6):3574–3592. doi: 10.1039/C9NR08542D.

Chen, Yongmei, Xinyu Zhang, Yan Liu, Zhengkun Liu, Zijie Li, Zhiqi Li, Xiaman Chen, Jiayi Liu, Zihao Chen, Guangquan Mo, Yong Xia, and Lei Mou. 2025. "A Gold Nanoparticles and MXene Nanocomposite Based Electrochemical Sensor for Point-of-Care Monitoring of Serum Biomarkers." *ACS Nano*. doi: 10.1021/acsnano.5c03194.

Cheng, Dan, Peipei Li, Xiaohua Zhu, Meiling Liu, Youyu Zhang, and Yang Liu. 2021. "Enzyme-free Electrochemical Detection of Hydrogen Peroxide Based on the Three-Dimensional Flower-like Cu-based Metal Organic Frameworks and MXene Nanosheets†." *Chinese Journal of Chemistry* 39 (8):2181–2187. doi: 10.1002/cjoc.202100158.

Feng, Lei, Wenjing Qin, Yanli Wang, Changshun Gu, Xinxin Li, Junji Chen, Jianxing Chen, Hanying Qiao, Min Yang, Zhenhao Tian, and Shougen Yin. 2023. "Ti3C2Tx MXene/Graphene/AuNPs 3D porous composites for high sensitivity and fast response glucose biosensing." *Microchemical Journal* 184:108142. doi: 10.1016/j.microc.2022.108142.

Jiang, Y., Lao, J., Dai, G., & Ye, Z. (2024). "Advanced insights on MXENES: categories, properties, synthesis, and applications in alkali metal ion batteries." *ACS Nano*, 18(22), 14050–14084. doi: 10.1021/acsnano.3c12543

Elumalai, Satheeshkumar, Veerappan Mani, Nithiya Jeromiyas, Vinoth K. Ponnusamy, and Masahiro Yoshimura. 2019. "A composite film prepared from titanium carbide $Ti_3C_2T_x$ (MXene) and gold nanoparticles for voltammetric determination of uric acid and folic acid." *Microchimica Acta* 187 (1):33. doi: 10.1007/s00604-019-4018-0.

Fan, X., P. Du, X. Ma, R. Wang, J. Ma, Y. Wang, D. Fan, Y. Long, B. Deng, K. Huang, and H. Wu. 2021. "Mechanochemical Synthesis of Pt/Nb(2)CT(x) MXene Composites for Enhanced Electrocatalytic Hydrogen Evolution." *Materials (Basel)* 14 (9). doi: 10.3390/ma14092426

Gao, Chenyu, Jikun Deng, Jiaxing He, Zechuan Wang, Yuanyuan Sheng, Junwu Liu, Honghai Zhong, Guoqing Tong, and Yang Jiang. 2024. "Electrochemical deposition of superconductive Cu/MXene laminated composite foils with low thermal expansion coefficient." *Journal of Materials Research* 39 (14):2082–2094. doi: 10.1557/s43578-024-01367-9.

Ghidiu, Michael, Maria R. Lukatskaya, Meng-Qiang Zhao, Yury Gogotsi, and Michel W. Barsoum. 2014. "Conductive two-dimensional titanium carbide 'clay' with high volumetric capacitance." *Nature* 516 (7529):78–81. doi: 10.1038/nature13970.

Guo, Jianxin, Qiuming Peng, Hui Fu, Guodong Zou, and Qingrui Zhang. 2015. "Heavy-metal adsorption behavior of two-dimensional alkalization-intercalated MXene by first-principles calculations." *The Journal of Physical Chemistry C* 119 (36):20923–20930. doi: 10.1021/acs.jpcc.5b05426.

Huang, Huajie, Di Xiao, Zihan Zhu, Chi Zhang, Lu Yang, Haiyan He, Jungmok You, Quanguo Jiang, Xingtao Xu, and Yusuke Yamauchi. 2023. "A 2D/2D heterojunction of ultrathin Pd nanosheet/MXene towards highly efficient methanol oxidation reaction: The significance of 2D material nanoarchitectonics." *Chemical Science* 14 (36):9854–9862. doi: 10.1039/D3SC03735E.

Jian, Xuan, Tan Li, Shanshan Guo, Loujun Gao, Feng Fu, Yue Tian, and Yucheng Wu. 2022. "Platinum Nanoparticle-Electrodeposited Ti3C2Tx MXene as a Binder-Free Electrocatalyst for Improved Hydrogen Evolution." *ACS Applied Energy Materials* 5 (3):3092–3099. doi: 10.1021/acsaem.1c03708.

Ji, Zhengjia, Lin Zhang, Guoxin Xie, Wenhu Xu, Dan Guo, Jianbin Luo, and Braham Prakash. 2020. "Mechanical and tribological properties of nanocomposites incorporated with two-dimensional materials." *Friction* 8 (5):813–846. doi: 10.1007/s40544-020-0401-4.

Khan, Maaz Ullah, LiJing Du, Shuai Fu, Detian Wan, Yiwang Bao, Qingguo Feng, Salvatore Grasso, and Chunfeng Hu. 2022. "Preparations and Applications of MXene–Metal Composites: A Review." *Coatings* 12 (4):516.

Kumar, Sunil, Nitu Kumari, Tej Singh, and Yongho Seo. 2024. "Shielding 2D MXenes against oxidative degradation: Recent advances, factors and preventive measures." *Journal of Materials Chemistry C* 12 (23):8243–8281. doi: 10.1039/D4TC00884G.

Lawal, Abdulazeez T. 2015. "Synthesis and utilisation of graphene for fabrication of electrochemical sensors." *Talanta* 131:424–443. doi: 10.1016/j.talanta.2014.07.019.

Liu, Fuxin, Lingling Han, Yali Yang, Zhonghua Xue, Xiaoquan Lu, and Xiuhui Liu. 2023. "Designable synthesis of a novel layered MXene loaded gold nanocluster composite for efficient electrochemical sensing of homocysteine in biological samples." *Chemical Engineering Journal* 461:141928. doi: 10.1016/j.cej.2023.141928.

Li, Han, and Zhaoqun Du. 2019. "Preparation of a highly sensitive and stretchable strain sensor of MXene/silver nanocomposite-based yarn and wearable applications." *ACS Applied Materials & Interfaces* 11 (49):45930–45938. doi: 10.1021/acsami.9b19242.

Li, Mengqi, Shoukun Wang, Qinhuan Wang, Fuqiang Ren, and Yu Wang. 2021a. "Microstructure and tensile properties of Ni nano particles modified MXene reinforced copper matrix composites." *Materials Science and Engineering:* A 808:140932. doi: 10.1016/j.msea.2021.140932.

Li, Mengqi, Shoukun Wang, Qinhuan Wang, Fuqiang Ren, and Yu Wang. 2021b. "Preparation, microstructure and tensile properties of two dimensional MXene reinforced copper matrix composites." *Materials Science and Engineering:* A 803:140699. doi: 10.1016/j.msea.2020.140699.

Li, Quan-Fu, Xu Chen, Hai Wang, Ming Liu, and Hui-Ling Peng. 2023. "Pt/MXene-Based Flexible Wearable Non-Enzymatic Electrochemical Sensor for Continuous Glucose Detection in Sweat." *ACS Applied Materials & Interfaces* 15 (10):13290–13298. doi: 10.1021/acsami.2c20543.

Liu, Tong, Qingtao Wang, Jingzhi Yuan, Xue Zhao, and Guanhui Gao. 2018. "Highly dispersed bimetallic nanoparticles supported on titanium carbides for remarkable hydrogen release from hydrous hydrazine." *ChemCatChem* 10 (10):2200–2204. doi: 10.1002/cctc.201701633.

Liu, Yan, Yuetao Zhang, Zijie Li, Zhiqi Li, Shaoke Zhou, Yong Xia, and Lei Mou. 2024. "Fast and in-situ electrodeposition of MXene/AuNPs composite for multiplexed and sensitive detection of infectious biomarkers using an electrochemical biosensor." *Microchemical Journal* 207:112064. doi: 10.1016/j.microc.2024.112064.

Lorencova, Lenka, Tomas Bertok, Jaroslav Filip, Monika Jerigova, Dusan Velic, Peter Kasak, Khaled A. Mahmoud, and Jan Tkac. 2018. "Highly stable Ti3C2Tx (MXene)/Pt nanoparticles-modified glassy carbon electrode for H2O2 and small molecules sensing applications." *Sensors and Actuators B: Chemical* 263:360–368. doi: 10.1016/j.snb.2018.02.124.

Ma, Zhenping, Hui Feng, Yongbao Feng, Xin Ding, Xianzhen Wang, Wei Wang, Xiaojie Zhang, Shuo Kong, Xiong Lan, and Qiulong Li. 2022. "An ultralight and thermally conductive $Ti_3C_2T_x$ MXene–silver nanowire cellular composite film for high-performance electromagnetic interference shielding." *Journal of Materials Chemistry C* 10 (38):14169–14179. doi: 10.1039/D2TC02856E.

Medetalibeyoglu, Hilal, Gül Kotan, Necip Atar, and Mehmet L. Yola. 2020. "A novel and ultrasensitive sandwich-type electrochemical immunosensor based on delaminated MXene@AuNPs as signal amplification for prostate specific antigen (PSA) detection and immunosensor validation." *Talanta* 220:121403. doi: 10.1016/j.talanta.2020.121403.

Mim, Mumtahina, Khairul Habib, Sazratul N. Farabi, Syed A. Ali, Md Abu Zaed, Mohammad Younas, and Saidur Rahman. 2024. "MXene: A roadmap to sustainable energy management, synthesis routes, stabilization, and economic assessment." *ACS Omega* 9 (30): 32350–32393. doi: 10.1021/acsomega.4c04849.

Nguyen, Viet P., Mikyung Lim, Kyung-Shik Kim, Jae-Hyun Kim, Ji S. Park, Jong M. Yuk, and Seung-Mo Lee. 2021. "Drastically increased electrical and thermal conductivities of Pt-infiltrated MXenes." *Journal of Materials Chemistry A* 9 (17):10739–10746. doi: 10.1039/D1TA00331C.

Park, Dabin, Minsu Kim, and Jooheon Kim. 2024. "High-performance thermoelectric and electromagnetic interference shielding derived from MXene/$Ag_2Se$ nanowire composite film." *Advanced Electronic Materials* 10 (9):2400018. doi: 10.1002/aelm.202400018.

Phuong Doan, Thanh Hoang, Won G. Hong, and Jin-Seo Noh. 2021. "Palladium nanoparticle-decorated multi-layer $Ti_3C_2T_x$ dual-functioning as a highly sensitive hydrogen gas sensor and hydrogen storage." *RSC Advances* 11 (13):7492–7501. doi: 10.1039/D0RA10879K.

Rajavel, Krishnamoorthy, Yougen Hu, Pengli Zhu, Rong Sun, and Chingping Wong. 2020. "MXene/metal oxides-Ag ternary nanostructures for electromagnetic interference shielding." *Chemical Engineering Journal* 399:125791. doi: 10.1016/j.cej.2020.125791.

Rakhi, Raghavan Baby., Pranati Nayak, Chuan Xia, and Husam N. Alshareef. 2016. "Novel amperometric glucose biosensor based on MXene nanocomposite." *Scientific Reports* 6 (1):36422. doi: 10.1038/srep36422.

Rasheed, Pathath. Abdul., Ravi P. Pandey, Khadeeja A. Jabbar, Janarthanan Ponraj, and Khaled A. Mahmoud. 2019. "Sensitive electrochemical detection of l-cysteine based on a highly stable Pd@$Ti_3C_2T_x$ (MXene) nanocomposite modified glassy carbon electrode." *Analytical Methods* 11 (30):3851–3856. doi: 10.1039/C9AY00912D.

Rasheed, P. Abdul, Pandey, R. P., Jabbar, K. A., & Mahmoud, K. A. 2022. "$Nb_4C_3T_x$ (MXene)/Au/DNA aptaSensor for the ultraselective electrochemical detection of lead in water samples." *Electroanalysis*, 34(10), 1540–1546. doi: 10.1002/elan.202100685

Rasheed, P. Abdul, Ravi P. Pandey, Khadeeja A. Jabbar, and Khaled A. Mahmoud. 2021. "Platinum nanoparticles/Ti3C2Tx (MXene) composite for the effectual electrochemical sensing of Bisphenol A in aqueous media." *Journal of Electroanalytical Chemistry* 880:114934. doi: 10.1016/j.jelechem.2020.114934.

Saharudin, Mohd Shahneel, Nur Ahza Che Nasir, and Syafawati Hasbi. 2022. "Tensile and Corrosion Resistance Studies of MXenes/Nanocomposites: A Review." In *Design in Maritime Engineering*, edited by Azman Ismail, Wardiah Mohd Dahalan and Andreas Öchsner, 189–198. Cham: Springer International Publishing.

Song, D., Jiang, X., Li, Y., Lu, X., Luan, S., Wang, Y., Li, Y., & Gao, F. 2019. "Metal–organic frameworks-derived MnO2/Mn3O4 microcuboids with hierarchically ordered nanosheets and Ti3C2 MXene/Au NPs composites for electrochemical pesticide detection". *Journal of Hazardous Materials*, 373, 367–376. doi: 10.1016/j.jhazmat.2019.03.083

Thirumal, V., R. Yuvakkumar, P. Senthil Kumar, G. Ravi, and Dhayalan Velauthapillai. 2022. "Facile preparation and characterization of MXene@Platinum nanocomposite for energy conversion applications." *Fuel* 317:123493. doi: 10.1016/j.fuel.2022.123493

Tian, Yuan, Yongling An, Chuanliang Wei, Baojuan Xi, Shenglin Xiong, Jinkui Feng, and Yitai Qian. 2019. "Flexible and free-standing $Ti_3C_2T_x$ MXene@Zn paper for dendrite-free aqueous zinc metal batteries and nonaqueous lithium metal batteries." *ACS Nano* 13 (10):11676–11685. doi: 10.1021/acsnano.9b05599.

Wang, Dezhao, Han Wei, Ying Lin, Pingkai Jiang, Hua Bao, and Xingyi Huang. 2021. "Achieving ultrahigh thermal conductivity in Ag/MXene/epoxy nanocomposites via filler-filler interface engineering." *Composites Science and Technology* 213:108953. doi: 10.1016/j.compscitech.2021.108953.

Wang, Yuhui, Huang, Y., Zhang, Y., & Wang, H. 2025. "Ultrasensitive detection of staphylococcal enterotoxin B using electrochemical nanoprobe based on AgNPs@para-sulfonatocalix[4] arene-functionalized MXene". *Sensors and Actuators B: Chemical*, 426, 137126. doi: 10.1016/j.snb.2024.137126

Wang, Yuhui, Wenjing Zhao, Yaqian Pang, Shuhan Fu, Zhenyang Xu, Yixiao Zou, Yishuo Li, Shangyi Li, Yong Lu, and Tingting Zhang. 2025. "In-situ growth of Fe clusters on MXene for fluorinated fire-fighting foam wastewater purification in peroxone system." *Chemical Engineering Journal* 508:160883. doi: 10.1016/j.cej.2025.160883.

Wu, Shixin, Qiyuan He, Chaoliang Tan, Yadong Wang, and Hua Zhang. 2013. "Graphene-based electrochemical sensors." Small, 9(8), 1160–1172. doi: 10.1002/smll.201202896.

Wyatt, Brian C., Srinivasa K. Nemani, and Babak Anasori. 2021. "2D transition metal carbides (MXenes) in metal and ceramic matrix composites." *Nano Convergence* 8 (1):16. doi: 10.1186/s40580-021-00266-7.

Xia, Junjie, Yangyang Zhou, Yindian Wang, Yawen Liu, Qiang Chen, Kwangnak Koh, Xiaojun Hu, and Hongxia Chen. 2024. "Ultrasensitive electrochemical sensor based on synergistic effect of Ag@MXene and antifouling cyclic multifunctional peptide for PD-L1 detection in serum." *Microchimica Acta* 191 (7):380. doi: 10.1007/s00604-024-06470-6.

Xiao, Ping, Guodong Zhu, Xiaohong Shang, Bin Hu, Boshuang Zhang, Zhaoyu Tang, Jianmao Yang, and Jianyun Liu. 2022. "An Fe-MOF/MXene-based ultra-sensitive electrochemical sensor for arsenic(III) measurement." *Journal of Electroanalytical Chemistry* 916:116382. doi: 10.1016/j.jelechem.2022.116382.

Yao, Chenxue, Qiang Zhang, Bo Cheng, Qianqian Chen, Renjie Wang, Xuefei Zhou, Ying Ren, Fengchun Wei, and Meng Zhang. 2023. "Palladium nanoparticles anchored on MXene-based N-doped porous carbon nanosheets as an advanced electrocatalyst for ethanol oxidation." *Journal of Alloys and Compounds* 953:169983. doi: 10.1016/j.jallcom.2023.169983.

Yin, Juanjuan, Fangke Zhan, Tifeng Jiao, Wenhan Wang, Guangcong Zhang, Jinghua Jiao, Guiyuan Jiang, Qingrui Zhang, Jianmin Gu, and Qiuming Peng. 2020. "Facile preparation of self-assembled MXene@Au@CdS nanocomposite with enhanced photocatalytic hydrogen production activity." *Science China Materials* 63 (11):2228–2238. doi: 10.1007/s40843-020-1299-4.

Zhan, Xiaoxue, Chen Si, Jian Zhou, and Zhimei Sun. 2020. "MXene and MXene-based composites: Synthesis, properties and environment-related applications." *Nanoscale Horizons* 5 (2):235–258. doi: 10.1039/C9NH00571D.

Zhang, Qingxiao, Jia-ao Wang, Qinghua Yu, Qizhen Li, Runze Fan, Chong Li, Yiyi Fan, Cong Zhao, Weihua Cheng, Peiyi Ji, Jie Sheng, Chenhao Zhang, Songhai Xie, Graeme Henkelman, and Hui Li. 2025. "Metal/MXene composites via in situ reduction." *Nature Synthesis* 4 (2):252–261. doi: 10.1038/s44160-024-00660-z.

Zhang, Xiaobao, Baiyi Shao, Zemin Sun, Zhe Gao, Yong Qin, Ce Zhang, Fangming Cui, and Xiaojing Yang. 2020. "Platinum nanoparticle-deposited $Ti_3C_2T_x$ MXene for hydrogen evolution reaction." *Industrial & Engineering Chemistry Research* 59 (5):1822–1828. doi: 10.1021/acs.iecr.9b05046.

Zhao, Fengnian, Yao Yao, Chengmei Jiang, Yuzhou Shao, Damià Barceló, Yibin Ying, and Jianfeng Ping. 2020. "Self-reduction bimetallic nanoparticles on ultrathin MXene nanosheets as functional platform for pesticide sensing." *Journal of Hazardous Materials* 384:121358. doi: 10.1016/j.jhazmat.2019.121358.

Zhao, Xin, Wan-Peng Li, Yanhui Cao, Arsenii Portniagin, Bing Tang, Shixun Wang, Qi Liu, Denis Y. W. Yu, Xiaoyan Zhong, Xuerong Zheng, and Andrey L. Rogach. 2024. "Dual-atom Co/Ni electrocatalyst anchored at the surface-modified $Ti_3C_2T_x$ MXene enables efficient hydrogen and oxygen evolution reactions." *ACS Nano* 18 (5):4256–4268. doi: 10.1021/acsnano.3c09639.

Zheng, Jiushang, Bin Wang, Ailing Ding, Bo Weng, and Jiucun Chen. 2018. "Synthesis of MXene/DNA/Pd/Pt nanocomposite for sensitive detection of dopamine." *Journal of Electroanalytical Chemistry* 816:189–194. doi: 10.1016/j.jelechem.2018.03.056.

Zheng, Zhong, Qian Yang, Shuyi Song, Yifan Pan, Huan Xue, and Jing Li. 2024. "Anti-oxidized self-assembly of multilayered F-Mene/MXene/TPU composite with improved environmental stability and pressure sensing performances." *Polymers* 16 (10):1337.

Zhou, Weiwei, Zhou Zhenxing, Fan Yuchi, and Naoyuki Nomura. 2021. "Significant strengthening effect in few-layered MXene-reinforced Al matrix composites." *Materials Research Letters* 9 (3):148–154. doi: 10.1080/21663831.2020.1861120.

Zuo, Ding-chuan, Sheng-chao Song, Chang-sheng An, Lin-bo Tang, Zhen-jiang He, and Jun-chao Zheng. 2019. "Synthesis of sandwich-like structured Sn/SnOx@MXene composite through in-situ growth for highly reversible lithium storage." *Nano Energy* 62:401–409. doi: 10.1016/j.nanoen.2019.05.062.

# 8 MXene–Metal Oxide–Based Electrochemical Sensors

## *Advancing Sensitivity and Selectivity in Electrochemical Detection*

*Kannankuzhiyan Niyas*

## 8.1 INTRODUCTION

The biosensor field has experienced significant expansion in recent decades, propelled by rising needs in healthcare, environmental surveillance, and food safety. Electrochemical biosensors have emerged as promising tools among various other biosensing technologies, offering cost-effectiveness and potential for miniaturization with high sensitivity. These devices represent a unique intersection of biological recognition elements and electrochemical transducers that offer rapid, selective, and reliable detection of diverse analytes, making them invaluable tools for contemporary analytical challenges.

Two-dimensional materials have been used more and more as possible alternatives for advanced sensing systems in the future over the past few years. Researchers are currently more focused on MXenes, which belong to a class of two-dimensional nanomaterials consisting of transition metal carbides, nitrides, or carbonitrides, because of the special chemical and physical qualities they have. MXenes are very good at conducting electricity, have a lot of surface area, and can have their surface chemistry changed, which makes them very appealing for electrochemical sensing uses.

MXenes constitute an emerging category of two-dimensional materials that have rapidly gained significant interest among researchers and scholars since their discovery in 2011 (Anasori, Lukatskaya, and Gogotsi 2017; Ataca, Şahin, and Ciraci 2012). These materials offer a distinct set of characteristics that render them especially suited for electrochemical sensing applications. The MAX phases, having the general formula $M_{n+1}AX_n$, are the source of MXenes. M represents an early transition metal (example: Ti, V, Nb, Mo), A typically represents an element from groups 13 or 14 (example: Al, Si, Ga), X represents carbon and/or nitrogen, and n can be 1, 2, 3, or 4. Selectively etching the A layer from the MAX phase yields the $M_{n+1}X_nT_x$

 DOI: 10.1201/9781003736318-8

structure, where $T_x$ represents surface terminations like –F, –O, or –OH. This process creates a structure consisting of $n + 1$ layers of M atoms sandwiching n layers of X atoms, with a thickness of just a few atomic layers. The predominant technique for the synthesis of MXenes involves the selective etching of MAX phases, succeeded by the process of delamination (Thakur et al. 2023). High-temperature methods such as spark plasma sintering or hot pressing typically synthesize the MAX phase precursor. Hydrofluoric acid (HF) or a mixture of hydrochloric acid (HCl) and fluoride salts is employed to selectively etch the A layer from the MAX phase (Thakur et al. 2023). After etching, delamination processing is done to obtain single- or few-layer MXene flakes, often achieved through sonication or intercalation with large organic molecules like Tetramethylammonium hydroxide (TMAOH), Dimethyl sulfoxide (DMSO), and Tetrabutylammonium hydroxide (TBAOH) to get a multi-layered structure (Prabhakar et al. 2024). The metallic conductivity of MXenes is attributed to their layered architecture and the inclusion of transition metal elements. The d-orbitals of the transition metals overlap, creating a continuous electron network. This allows for rapid electron transfer, which is critical in electrochemical sensing because it enables fast signal transduction and improved sensitivity. The two-dimensional nanostructured MXenes result in a high surface-to-volume ratio. When exfoliated, MXene sheets can have thicknesses of just a few atomic layers, exposing a large proportion of their atoms to the environment (Natu et al. 2019). The enhanced surface area-to-volume ratio affords a multitude of active sites for the adsorption of the target analytes and facilitates specific interactions, potentially leading to lower detection limits and improved sensor performance. Various methods, such as thermal annealing, chemical treatments, and plasma processing, can further modify surface terminal functional groups of MXene, mainly –F, –O, and –OH. This tunability allows researchers to optimize MXene surfaces for specific analytes, enhancing selectivity and sensitivity in sensing applications. Because of their surface terminations, MXenes are inherently hydrophilic, particularly –OH groups. This property makes them highly compatible with the aqueous electrolytes commonly used in electrochemical sensing (Munteanu et al. 2021). The hydrophilicity facilitates better dispersion in water-based solutions and promotes efficient interaction with dissolved analytes, which is crucial for many biosensor applications. You can easily integrate MXenes into flexible substrates or form them into freestanding films. This flexibility allows for the development of bendable, stretchable, or wearable sensors (Munteanu et al. 2021). For instance, researchers have incorporated MXene-based sensors into textiles for sweat analysis or fabricated them as skin-attachable patches for ongoing physiological assessment (Li et al. 2022; Kumar and Singh 2022).

Metal oxides have been recognized as valuable materials in electrochemical applications because of their distinctive electronic properties and high surface area, which enhance reaction kinetics and enhance the efficiency in energy conversion, energy storage devices and electrocatalysis (Poizot et al. 2000). The metal oxide nanocomposites are extensively used and acknowledged as efficient electrocatalysts for detecting a range of elements such as various biomarkers, environmental contaminants, and pharmaceutical drugs. Metal oxide nanoparticles can be employed in the immobilization of biomolecules for the development of immunosensors, DNA aptamer sensors, and enzymatic sensors with a lower Michaelis–Menten constant values that favors adsorption of biomolecules on metal oxides fast and high affinity binding (Chauhan and Pundir 2011; Ezhil Vilian et al. 2015; Rahman et al. 2010). The nanoscale domain has been demonstrated to be a productive environment for progress in catalysis and sensing technologies, due to the improved catalytic efficiency and structural stability of nanomaterials in comparison to their larger counterparts. Metal oxide nanocomposites, which include structures in one, two, and three dimensions, have shown exceptional effectiveness in catalytic and sensing tasks. The synthesis methodology is crucial in accurately manipulating the shape and size of nanocomposites based on transition metal oxides.

Although there has been notable advancement, our comprehension of the intricate specifics of these nanocomposites is still lacking. We still lack a comprehensive understanding of the precise measurements, subtle variations in composition, crystalline structure, and alignment, concentrations of defects, electrical properties, and other essential features of metal oxide nanocomposites. An important obstacle in this field is understanding the complex connections between these structural characteristics and the resulting electrocatalytic properties (Ma et al. 2024). Even though metal oxide materials are widely used in electrochemical sensing, they are subject to a number of constraints that can have an effect on their performance and the applications to which they can be applied. The comparatively low electrical conductivity of these materials presents a significant challenge since it might potentially impede the movement of electrons and impair the sensitivity of sensors (Liao et al. 2015; Aydoğdu et al. 2013). In addition, a great number of metal oxides encounter difficulties in maintaining their stability in hostile chemical conditions, which results in degradation and a shorter lifespan for sensors. Ensuring precise control over the fabrication of nanostructured metal oxide with uniform morphology and composition can be quite difficult. This can lead to differences from batch to batch, which can have an impact on the repeatability of sensors (Aravind and Ramaprabhu 2012). Although metal oxide sensors have a surface area that is generally good, it is possible that it is not always appropriate for certain sensing applications. Furthermore, one major limitation with nanostructured metal oxide particles is their wide band gap, and the large band gap inherent to metal oxide nanoparticles renders them semiconductors or insulators (Xiao et al. 2012). As a result of these constraints, research has been conducted into alternate materials and composite architectures in order to address the deficiencies that are associated with conventional metal oxide sensors in electrochemical applications. These limitations have driven research into the development of composite materials by introducing highly conductive two-dimensional material support, like graphene and MXene (Wang et al. 2016; Dong et al. 2012), upon which transition metal oxide nanostructures are uniformly dispersed, constituting an efficacious approach to mitigating these issues. This facilitates the movement of ions and electrons across the interface, which aims to overcome these drawbacks while leveraging the beneficial properties of both components. Having a sophisticated comprehension is crucial for maximizing the design and efficiency of these materials in different applications. As research advances, understanding these connections between structure and function will be crucial for fully harnessing the capabilities of metal oxide nanocomposites in future catalytic and sensing technologies.

This chapter looks at the area of electrochemical sensors that use MXene–metal oxide nanocomposites. As we start this study, it is clear that MXene–metal oxide electrochemical sensors are a big step toward better and more accurate ways to find analytes like various organic molecules. The fact that they were able to totally change how sense works in many areas shows how important it is to keep studying and developing this interesting subject.

## 8.2 FUNDAMENTALS OF MXENE–METAL OXIDE COMPOSITES IN ELECTROCHEMICAL SENSING

The integration of the unique properties of two-dimensional MXenes with the versatile characteristics of metal oxides has led to the development of an advanced class of MXene–metal oxide composites. The synergistic integration has opened up new avenues for enhancing the efficacy of a range of electrochemical applications, which include electrocatalysis, energy storage devices, and sensor electrode fabrication (Ali 2016; Wee et al. 2024; Cai et al. 2018; Shirsat and Hianik 2023). The integration of MXenes' exceptional electrical conductivity and substantial surface volume ratio endowed with numerous surface functional groups, alongside the diverse physical and chemical characteristics inherent to metal oxides, has culminated in the development of nanocomposites that frequently surpass the performance of their constituent elements (Ouikhalfan et al. 2019).

The distinctive design maintains the optimal characteristics of both materials while introducing novel functionalities at their interface. The two-dimensional structure of MXenes disperses and anchors metal oxide nanoparticles, thereby preventing aggregation and maximizing the exposure of active sites (Wee et al. 2024; Hussain et al. 2023). The composites of MXene and metal oxides are critical in considerably improving the electron transport and movement of ions, and diffusion of electrons and ions over the electrode surface is facilitated by the presence of layered MXene sheets. This integration further enhances the electrical conductivity and charge transfer kinetics of the nanostructured composite materials. Metal oxide nanostructures have the ability to stabilize MXene nanostructures, which means they prevent the layers from clumping together and help maintain their large surface area and reactivity by preventing the oxidation of the MXene surface.

Several techniques can be utilized to create composites of MXene and metal oxides. In situ growth techniques, such as hydrothermal or solvothermal processes (Guo et al. 2023), promote the direct formation of metal oxide nanostructures above the MXene surface. This method leads to an even distribution of the metal oxide over MXene and strong interfacial interactions. Solution mixing methods by the support of sonication, which involve the combination of MXene and metal oxide precursors followed by reduction or precipitation, provide scalability and simplicity (Thukkaram et al. 2023; Ankitha et al. 2024). Atomic layer deposition (ALD) is distinguished by its capacity to generate ultrathin, conformal layers of metal oxides on MXene surfaces with atomic-level precision. The formation and properties of these composites are significantly influenced by the surface chemistry of MXenes (Ustad et al. 2023; Lu et al. 2022). The composite's overall electrochemical performance is also affected by the numerous surface terminal functional groups, including –F, –O, and –OH terminations, which not only function as anchorage sites for metal oxide nanoparticles. These functional groups also have the potential to enhance the catalytic activity and sensing capabilities of the composite by participating in redox reactions, facilitating ion adsorption, and modifying the electronic structure (Gopal et al. 2022).

In addition to the more common metal oxides, researchers are also investigating the incorporation of more complex oxide systems with MXenes. Such as perovskite oxides, spinel structures, and layered double hydroxides (Wu et al. 2023; Subramania et al. 2023; Ramadoss et al. 2024; Xue et al. 2023; Yang et al. 2022b). These advanced composites have the potential to provide even greater performance enhancements and novel functionalities. Furthermore, research is underway to investigate the integration of these composites with other nanomaterials, including carbon nanotubes, graphene, and conductive polymers, to develop even more sophisticated hybrid materials with synergistic properties. Understanding and controlling the interfacial interactions are crucial for improving the effectiveness of these composite materials. Sophisticated characterization methodologies, including transmission electron microscopy (TEM) and synchrotron-based X-ray spectroscopies, are employed to examine atomic and electronic configurations of the material interfaces, as well as their transformations during electrochemical processes.

However, the utilization of MXene–metal oxide composites for electrochemical sensing has also garnered considerable attention. The advancement in sensors characterized by heightened sensitivity and selectivity for various analytes is facilitated by the unique recognition capabilities of metal oxides, coupled with the extensive surface area and conductive properties inherent to MXenes. The application of these composites has proven to be effective in the detection of biomolecules, environmental pollutants, and heavy metals, and frequently exhibits wider linear ranges and lower detection limits than traditional sensing materials (Yang, Lu, and Liu 2022a; Rhouati et al. 2022).

## 8.3 ELECTROCHEMICAL SENSING PLATFORMS USING MXENE–METAL OXIDE COMPOSITE

Electrochemical sensing plays a vital role in real-time monitoring of neurotransmitters, detecting biomarkers, analyzing flavonoids, and other biomolecules. Its applications range from understanding neural processes and diagnosing neurological disorders to enabling rapid medical diagnostics and enhancing food quality assessment. The efficacy of these sensors has been considerably increased by the integration of nanotechnology and innovative designs, thereby expanding their utility in the medical, environmental, and agricultural sectors. Electrochemical sensors are essential instruments in research and practical applications for health, safety, and quality monitoring due to their ability to facilitate immediate analysis and intervention and their high temporal resolution and point-of-care testing capabilities (Arjun, Shinde, and Slaughter 2023; Ankitha et al. 2023).

### 8.3.1 Glucose Sensors

Glucose is regarded as the primary biomarker due to the fact that the majority of diseases are correlated with variations in blood glucose levels. Advanced biosensors are required for continuous glucose monitoring, facilitating the acquisition of real-time data on diurnal and nocturnal glucose level variations. At present, glucose biosensors encompass both enzymatic and non-enzymatic types for the detection of glucose. Development of extremely sensitive, non-enzymatic glucose sensors has shown promise using MXene–metal oxide composites. The MXene–copper oxide ($Cu_2O$) composite is extensively investigated for the non-enzymatic electrochemical sensing of glucose. Alanazi et al. fabricated ternary nanocomposites consisting of $Cu_2O$, MXene, and rGO as an electrode material for glucose detection (Alanazi et al. 2023). The integration of delaminated MXene and graphene oxide into a 3D structure called MGA was facilitated by ascorbic acid. Furthermore, several volume ratios of $Cu_2O$ to MGA variants, specifically MGA-25, MGA-50, MGA-75, and MGA-100, were synthesized with the objective of enhancing the electrochemical properties of the modified electrode. The inclusion of $Cu_2O$ on the sensor resulted in an enhancement in electrical current, with the most significant effect observed in the MGA-75 composite. Quantitative detection of glucose is done by using the chronoamperometry technique, within a solution containing 0.2 M NaOH and a known amount of glucose. Analysis of the experimental data revealed the presence of two distinct linear ranges that correspond to different glucose concentrations, say 0.1–14 mM, as well as 15–40 mM, and exhibited sensitivity of 264.52 and 137.95 μA/μM/cm², respectively, having the lowest limit detection limit (LOD) of 1.1 μM. Cui et al. built a laser-induced graphene (LIG) electrode modified with a novel Au@CuO/$V_2CT_x$ MXene combination for non-enzymatic electrochemical sensors for the urine glucose detection (Cui et al. 2023), with a sensitivity of $1.124 \times 10^6$ μA/mM/cm² and the lowest LOD of 1.8 μM within a detection range from 0.005 to 5.0 mM achieved by the electrochemical sensor with a small response time of 1.5 seconds. The tetrabutylammonium hydroxide (TBAOH) intercalated an MXene monolayer mixed with 2 wt% Poly diallyl dimethylammonium chloride (PDDA) solution was added to the Au@CuO dispersion and followed by the LIG electrode (Figure 8.1(a)). Similarly, Tamil et al. analyzed the function of graphene oxide in MXene($Ti_3C_2T_x$)–rGO (MG) composite as a detection tool for glucose sensing applications. Further, they were applied to synthesize a ternary composite, which possesses a higher defect density and a smaller crystalline rGO-supported MXene-based $Cu_2O$ composite (MG–$Cu_2O$), employed by the coprecipitation method. In comparison to MXene–$Cu_2O$, these MG–$Cu_2O$ composites may result in an increase in conductivity and an amplification of the signal for glucose oxidation. Consequently, the composite that has been prepared exhibits a sensitivity of around 125.6 μA/mM/cm and a wide linear detection range of 0.01–30 mM. Additionally, it demonstrates remarkable stability, selectivity, and reproducibility (Gopal et al. 2022). Myndrul et al. analyzed the possibility of ZnO tetrapods/MXene composite for an enzymatic electrochemical glucose sensor (Myndrul et al. 2022). The efficacy of a skin-wearable, flexible enzymatic electrochemical glucose sensor is enhanced through the application of ZnO tetrapod coatings. with MXene ($Ti_3C_2T_x$) nanoflakes. The sensors that were developed showed an excellent sensitivity of 29.88 μA/mM/cm² when tested in artificial sweat, with an LOD around 17 μM, and a wide linear detection window spanning from 0.05 to 0.7 mM. Vinod et al. synthesized $Ti_2C$–$TiO_2$ nanocomposites in situ through the controlled oxidation of $Ti_2C$-MXene nanosheets with $H_2O_2$ under ambient conditions (Kumar et al. 2022). The oxidation procedure facilitates the generation of $TiO_2$ nanocrystals and the subsequent

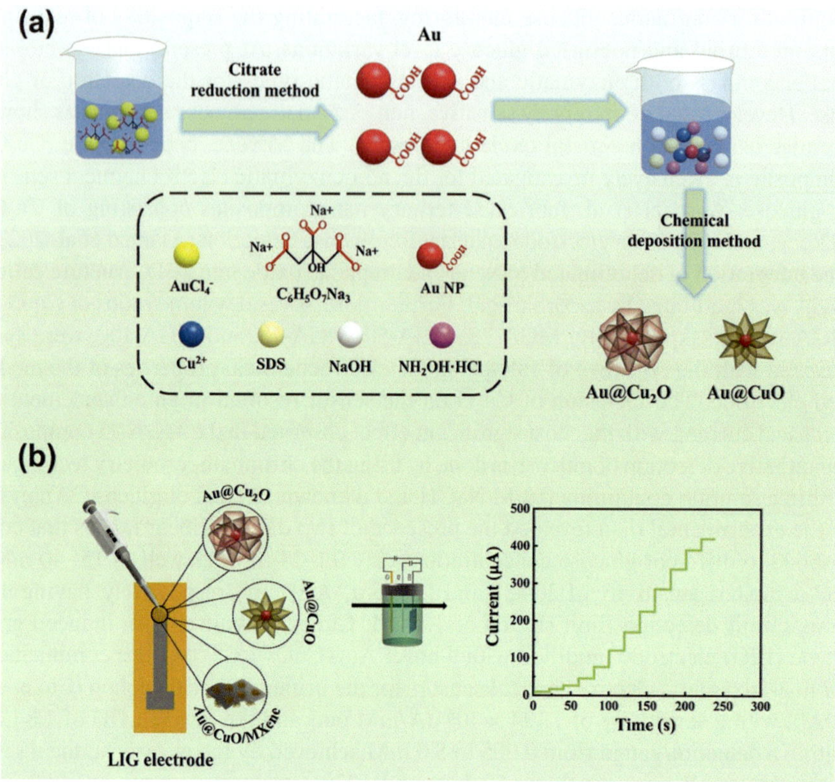


**FIGURE 8.1** Different types of MXene–metal oxide composite electrode material for electrochemical glucose sensing. (a) Au@CuO/$V_2CT_x$ MXene-modified flexible LIG electrode. Reprinted with permission from Cui et al. (2023). Copyright 2023, Elsevier. (b) $Co_3O_4$/MXene@CFE electrode. Reprinted with permission from Manoj et al. (2022). Copyright 2023, Elsevier.

exfoliation of $Ti_2C$-MXene nanosheets. This material is presently being employed as a novel electrode for the precise and sensitive detection of glucose without the use of enzymes. The nanocomposite-based non-enzymatic glucose sensor that has been developed exhibits a broad linear range from 0.1 to 200 μM and LOD of 0.12 μM with substantial sensitivity of 75.32 μA/mM/cm$^2$ and an exceptionally rapid response time of approximately 3 seconds. The enhanced sensing capabilities of the recently developed non-enzymatic glucose sensor are ascribed to the synergistic interaction between MXene and $TiO_2$ nanoparticles. Alternative strategies have been undertaken for the development of non-enzymatic electrochemical glucose sensors using diverse electrode configurations. Manoj et al. integrated $Co_3O_4$ nanocubes over a MXene ($Ti_3C_2T_x$) on a carbon fiber electrode (CFE) to develop a flexible electrode for glucose sensing (Manoj et al. 2022). The $Co_3O_4$–MXene dispersion was incorporated over the CFE by the hydrothermal method, as shown in Figure 8.1(b). The resultant $Co_3O_4$/MXene@CFE exhibited a remarkable sensitivity of 19.3 μA/mM/cm$^2$, a comprehensive linear detection range extending from 0.05 μM to 7.44 mM, and an LOD of 10 nM in solution containing 0.1 M KOH and known glucose concentration under the optimized conditions. Subramanian et al. developed a double MXene-$NiCo_2O_4$ hybrid composite with the support of (3-Aminopropyl)triethoxysilane (APTES) functionalization. The $NiCo_2O_4/Ti_2NbC_2$ nano hybrid composite exhibited glucose sensing within a linear range from 0.025 to 1 mM of glucose concentration in 0.1 M KOH solution and an LOD of 298 μM (Subramania et al. 2023).

An electrochemical biosensor for glucose detection, inspired by sponge biomimetics, was designed by Jiang et al. The MXene ($Ti_3C_2T_x$) was incorporated into a $CeO_2$ solution and agitated with chitosan hydrogel to form homogeneous MXene@$CeO_2$/chitosan hydrogels. MXene@$CeO_2$/chitosan glucose sensor achieved the detection of natural sweat from thermoregulation, having an LOD of 0.004 μM and sensitivity of 801.27 μA/μM/$cm^2$ within a linear detection range of 0.01–2.5 μM of glucose concentration (Jiang et al. 2023).

### 8.3.2 $H_2O_2$ Sensors

Ramila et al. developed an electrochemical sensor podium consisting of $Ti_3C_2T_x$-$Fe_2O_3$ electrode material for the detection of $H_2O_2$ (hydrogen peroxide). The highly selective $Ti_3C_2T_x$-$Fe_2O_3$/GCE was synthesized by simple ultrasonication of a dispersion including a 10:1 ratio of $Ti_3C_2T_x$ and α-$Fe_2O_3$, and drop-cast onto the glassy carbon electrode (GCE). The highly sensitive and reproducible sensor showed a detection range of 10–1000 nM and had an LOD of 7.46 nM for the detection of $H_2O_2$ (Nagarajan, Sundaramurthy, and Sundramoorthy 2022). Zhou et al. synthesized CuO–$CeO_2$ powders using the sol–gel method and surface tagged above $Ti_3C_2T_x$ MXene to get CuO–$CeO_2$/Ti3C2$T_x$ nanocomposite (Zhou et al. 2022). The GCE modified CuO–$CeO_2$/$Ti_3C_2T_x$/GCE showed LOD about 1.67 μM with linear detection ranging from 5 to 100 μM of $H_2O_2$.

Sima et al. designed another $H_2O_2$ sensor for detection in cancer cells by the support of MXene–$Co_3O_4$ nanocomposites (Singh et al. 2023). The MXene–$Co_3O_4$ composite was synthesized by the hydrothermal method of a solution containing cobalt acetate tetrahydrate and $Ti_3C_2$ MXene. MXene–$Co_3O_4$, further modified over a screen-printed electrode (SPE) disposable sensor, detects $H_2O_2$ within a dynamic linear range from 0 to 75 μm and an LOD of 0.5 μm. Also, they have evaluated the variation of $H_2O_2$ in cancer cells evaluated because of therapeutic treatment. Ramachandran et al. designed a hybrid composite comprising nickel oxide microsphere and $Ti_3C_2T_x$ as a bifunctional electrode material for a non-enzymatic $H_2O_2$ sensor and battery-type supercapacitor electrode (Ramachandran et al. 2019). The dual-function NiO/$Ti_3C_2T_x$ electrode was derived from the Ni-MOF/$Ti_3C_2T_x$ hybrid by a calcination process. The highly selective and reproducible NiO/$Ti_3C_2T_x$/GCE sensor detects $H_2O_2$ in a detection range of 10–4.5 mM and exhibited lower LOD of 0.34 μM.

### 8.3.3 Flavonoid Sensors

Gao et al. developed an electrochemical sensor designed for selective and ultrasensitive luteolin detection (Gao et al. 2024). A composite of $SnO_2$ quantum dots (QDs) and MXene is synthesized, leading to the creation of smaller-sized $SnO_2$ QDs that are arranged over the MXene sheet. The spacing between the MXene layers ($Ti_3C_2$) expanded as a result of $SnO_2$ QDs being integrated into the densely packed MXene layers. The $SnO_2$ QDs-MXene/GCE electrode demonstrates outstanding sensing efficiency toward luteolin detection, having an extremely low limit of detection of 0.14 nM and a high sensitivity of 12.4 μA/μM with a wide linear range spanning from 0.1 to 1200 nM.

Similarly, Wang et al. designed an exceptional composite material made of MXene ($Ti_3C_2T_x$) and ruthenium dioxide ($RuO_2$) incorporated onto a carbon cloth (CC) for the electrochemical kaempferol detection, which exhibited two linear detection ranges from 0.06 to 1 μM and 1 to 15 μM, having an LOD of about 0.039 μM (Wang et al. 2023). $RuO_2$ was deposited onto CC utilizing pulsed laser deposition under an oxygen partial pressure of $10^{-3}$ Torr and at a higher temperature. Further, MXene@ $RuO_2$/CC composite was synthesized utilizing an electrostatic self-assembly technique, by mixing MXene with cetyltrimethylammonium bromide (CTAB) assisted solution of $RuO_2$/CC. This new sensor exhibits exceptional sensitivity and selectivity toward detecting kaempferol due to the substantial active surface area of MXene and the synergistic interaction of the combined MXene and $RuO_2$.

A $TiO_2$–MXene/PEDOT:PSS (poly(3,4-ethylenedioxythiophene):poly(styrenesul fonate) composite was meticulously engineered and employed as an electrode for the archive precise electrochemical sensing of baicalein by Xue et al. (2023). PEDOT:PSS is a conductive polyelectrolyte complex with the structure of PEDOT-rich cores surrounded by a PSS-rich shell. The PEDOT:PSS films exhibited uniform and dense characteristics, resulting in enhanced conductivity and dispersion of $TiO_2$–MXene. Furthermore, the films further improved the electrocatalytic activity as well as stability of $TiO_2$–MXene by using the synergistic mechanism between PEDOT:PSS and $TiO_2$–MXene. The $TiO_2$–MXene/PEDOT:PSS composite exhibited a substantially increased electroactive surface area with excellent conductivity and exceptional electrocatalytic performance, which facilitates efficient electron transfer and baicalein adsorption. The composite material's characteristics allowed the electrochemical sensor to have a broad linear range of 0.007–10.0 μM and an ultralow LOD of 2.33 nM. Ramadoss et al. synthesized a bimetallic oxide-MXene composite for the selective determination of Ruthin. Hydrothermally developed $CuMoO_4$ nanoflakes were incorporated over $Ti_3C_2T_x$ layers by the sono-chemical method to form the final CuMoO4/$Ti_3C_2T_x$ nanocomposite. The highly selective and stable reproducible $CuMoO_4$/$Ti_3C_2T_x$-GCE nanocomposite detects Ruthin with a linearity ranging from 1 to 15 μM, having an LOD of around 42.9 nM (Ramadoss et al. 2024). Similarly, Ranjith et al. developed another bimetallic oxide-MXene-based electrochemical sensor for Ruthin (Ranjith et al. 2021). $FeWO_4$ nanofibers are synthesized by electrospinning and subsequently undergo thermal decomposition. The $FeWO_4$ nanofibers were further integrated above the surface of delaminated single-layered $Ti_3C_2T_x$ MXenes by the hydrothermal method to form a 2D-1D MXene-$FeWO_4$ heterostructure. MXene-$FeWO_4$/GCE exhibited an ultra-sensitivity, demonstrating a sensitivity of 0.3799 μA/nM/cm$^2$ toward Ruthin oxidation. In addition, it demonstrated exceptional linearity ranging from 4 to 147 nM, with an extremely low LOD of 0.42 nM.

### 8.3.4 Environmental Contaminant Sensors

The ability to perform real-time analyses is a significant advantage of electrochemical sensors. They enable automated chemical analyses in complex matrices, providing rapid, reliable, and cost-effective measurements of both inorganic and organic pollutants. This capability is especially advantageous for tracking contaminants of growing concern, such as pharmaceuticals, pesticides, and personal care products.

Chen et al. synthesized an MXene-yttrium oxide ($Y_2O_3$) composite electrode for sensitive and selective detection of catechol in water bodies. Polyphenol oxidase (PPO) was introduced into a composite of MXene-$Y_2O_3$ (Figure 8.2(a)). By the addition of chitosan-assisted PPO into the MXene $Y_2O_3$ surface, the MXene-$Y_2O_3$/CS-PPO electrode was developed via electrostatic interaction. This developed sensor results in a remarkable sensitivity for catechol detection, about 2351.4 mA/M in river water and 1506.7 mA/M in seawater samples within a linear detection range of 0.04–3.2 μM and an LOD of 7 nM (Chen et al. 2024). Yang et al. proposed an alternative way for the detection of imidacloprid in fruits and vegetables by using an electrochemical sensing platform (Yang et al. 2023). Oxygen vacancy-rich $TiO_2$@MXene heterostructure performed as a highly selective and stable electrode material for the detection of imidacloprid. $TiO_2$ nanoparticles, grown in situ over MXene, serve as a superstructure to restrict the stacking of MXene nanosheets. The concentration of surface oxygen vacancies (VO) over $TiO_2$@MXene was further increased by an electrochemical activation strategy. The VO-$TiO_2$@MXene/GCE electrode exhibited two distinct linear ranges from 0.07 to 10.0 μM and 10.0 to 70.0 μM, having a lowest LOD of around 23.3 nM for the analysis of imidacloprid.

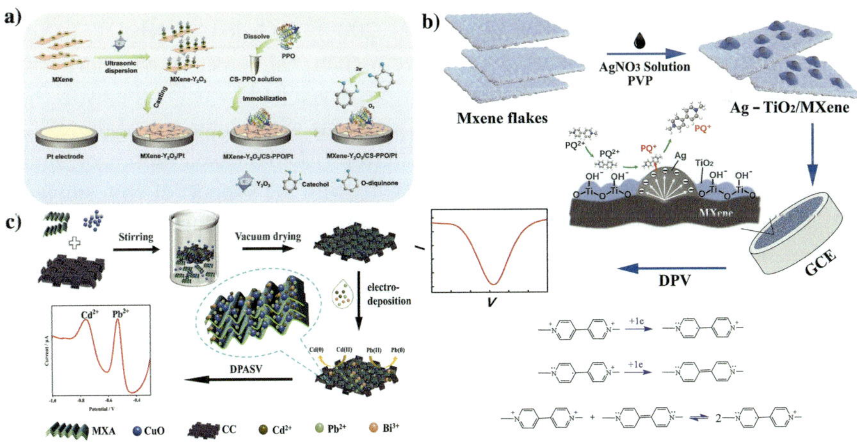


**FIGURE 8.2** Schematic representation of the fabrication and modification strategies to develop electrochemical sensors for the detection of environmental contaminants. (a) MXene-$Y_2O_3$/CS-PPO. Reprinted with permission from Chen et al. (2024). Copyright 2024, Elsevier. (b) Ag–$TiO_2$/MXene composite. Reprinted with permission from Hou et al. (2023). Copyright 2022, Springer Nature. (c) MXene aerogel-CuO-modified carbon cloth electrode. Reprinted with permission from Wen et al. (2022). Copyright 2022, Elsevier.

Hou et al. tested silver-incorporated $TiO_2$ and MXene-based composite material for the electrochemical sensing of paraquat (Hou et al. 2023). In situ prepared Ag–$TiO_2$/MXene nanocomposite was modified above GCE for making the sensor electrode (Figure 8.2(b)). Ag–$TiO_2$/MXene/GCE demonstrates exceptional electrochemical activity toward paraquat. The LOD of the electrochemical sensor is found to be approximately 0.01 μM, within its two linear detection ranges of 0.05–1 μM and 1–15 μM.

Wen et al. developed an electrochemical sensing podium of the simulation detection of two prevalent and highly detrimental heavy metal pollutants, such as cadmium (Cd) and lead (Pb) (Wen et al. 2022). An innovative MXene aerogel-CuO/CC electrode was developed through the incorporation of MXene ($Ti_3C_2T_x$) aerogel and CuO onto the top of a carbon cloth, as shown in Figure 8.2(c). Electrode made of MXene aerogel-CuO/CC exhibits outstanding detection capabilities, allowing for the detection of $Cd^{2+}$ and $Pb^{2+}$ simultaneously within a detection range of 4–800 μg/L and 4–1200 μg/L, respectively. The lower detection limits for $Cd^{2+}$ are 0.3 μg/L, and $Pb^{2+}$ is 0.2 μg/L, which provides viable options for developing efficient electrochemical sensors. Thukkaram et al. found the possibility of a titanium MXene ($Ti_3C_2T_x$) and vanadium pentoxide ($V_2O_5$) composite-based electrochemical sensor for the bisphenol A detection (Thukkaram et al. 2023). The $V_2O_5/Ti_3C_2T_x$ composite was synthesized by adding $V_2O_5$ with delaminated $Ti_3C_2T_x$ MXene into a bath sonication system at room temperature. It provided a remarkably ultra-low LOD of 87 nM, accompanied by a dynamic and extensive linear range from 411 nM to 31.2 μM for the detection of bisphenol A. Liu et al. (2015) successfully utilized MXene in hemoglobin-based biosensing, offering valuable insights into the application of enzymatic biosensors within the environmental domain. Enzymatic biosensors primarily consist of a protein known as acetylcholinesterase (AChE) and are specifically designed to detect pesticides (Zhao et al. 2009). $SnO_2$-loaded $Nb_2CT_x$ MXene composite analyzed by Guo et al. for the electrochemical detection of chlorpyrifos (Guo et al. 2023). $SnO_2/Nb_2CT_x$ composites were developed by using the hydrothermal method, which maintained a number of electrons at the interface of the 0D/2D heterojunction. To produce an acetylcholinesterase (AChE) based biosensor, $SnO_2/Nb_2CT_x$/GCE was further drop-casted with AChE with the support of chitosan solution. The electrochemical biosensor showed excellent performance. It had a linear range of detection from $5.1 \times 10^{-14}$ to $5.1 \times 10^{-7}$ M for chlorpyrifos, and a lowest LOD of 51 pM. Song et al. successfully developed a three-dimensional $MnO_2/Mn_3O_4$ hierarchical microcuboid, which is synthesized from Mn-MOF nanosheets and decorated with MXene ($Ti_3C_2$)/AuNPs composite (Song et al. 2019). An AChE (Acetylcholinesterase) biosensor was made with $MnO_2/Mn_3O_4$ and MXene/Au NPs by drop casting above GCE, followed by drop casting of AChE and Chitosan (Chit) mixture. The AChE-Chit/MXene/Au NPs/$MnO_2/Mn_3O_4$/GCE biosensor was utilized for detecting methamidophos using the differential pulse voltammetry (DPV) method. Under the optimized sensing conditions, the biosensor can detect methamidophos in a broad concentration range of $10^{-12}$ to $10^{-6}$ M, having an LOD of $1.34 \times 10^{-13}$ M.

Zhong et al. engineered a sensor for the specific detection of benomyl within fruit samples (Zhong et al. 2022). The hierarchical nano-$Cu_xO$ was modified MWCNTs-COOH/MXene (nano-$Cu_xO$-MWCNTs-COOH/MXene/GCE) electrode detects benomyl with a broad linear range from 10.0 nM to 10.0 μM with an LOD of 3.0 nM using the DPV technique. The recovery rates of benomyl vary from 99.7% to 102.2% with a relative standard deviation (RSD) of 3.4%. The satisfactory outcomes demonstrate the reliability of the proposed sensor in detecting benomyl in real-life samples.

Shen et al. designed a new type of sensor using nucleic acid aptamers to sense Aflatoxin B1, with high sensitivity and specificity (Shen et al. 2023). A MXene/MWCNTs/$NiCo_2O_4$ composite was synthesized by ultrasonication of MWCNTs/$NiCo_2O_4$ dispersion and layered MXene and drop-casted onto the surface-graphenized pencil graphite electrode (SGPGE). The electrode was further modified by dopamine through electro-polymerization in order to successfully attach the $NH_2$-cDNA-aptamer to the surface MXene/MWCNTs/$NiCo_2O_4$/SGPGE. The final MCH/cDNA/PDA/MXene/MWCNTs/$NiCo_2O_4$/SGPGE sensor detects Aflatoxin B1 (AfB) within a detection window from 2.5 to 200 ng/mL, along with an LOD of 1.890 ng/mL.

A voltammetric sensor based on two-dimensional MXene and CuO nanoparticles is synthesized by Mini et al. for the detection of hazardous diclofenac sodium (Mini et al. 2024). The MXene/CuO/GCE nanocomposite electrode demonstrates exceptional performance with an extensive linear range of 0.34 μM to 31.67 μM, attaining a minimal detection limit of 8.3 nM in the Square Wave Voltammetry (SWV) and 8.5 nM in the DPV method, with a quantification limit of 0.0252 μM and 0.0256 μM for SWV and DPV, respectively.

### 8.3.5 Drug Molecule Sensors

Ankitha et al. introduced $SnO_2$ nanoparticle/$Nb_2CT_x$ MXene nanosheets composite-based flexible biosensor as shown in Figure 8.3(a), for the ultrasensitive detection of propofol ranges in a detection window 1 μM to 300 μM along with an LOD of 0.24 μM (Ankitha et al. 2024). Incorporation of $SnO_2$ nanoparticle on MXene was done by a new approach using ascorbic acid by in situ reaction. $SnO_2$/$Nb_2CT_x$ composite modified CC electrode possessed a high sensitivity of about 0.26 μA/μM/$cm^2$ toward

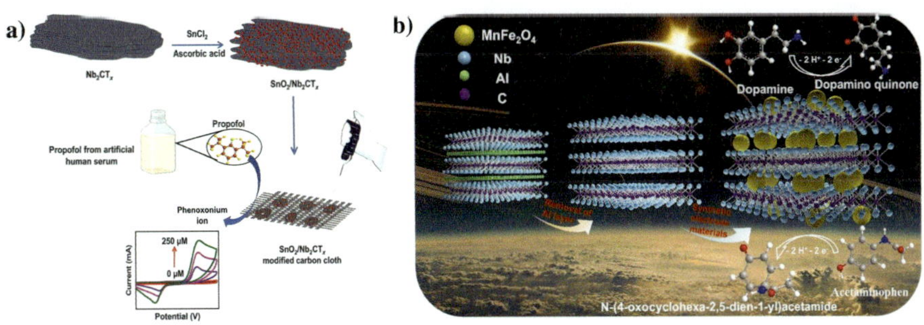


**FIGURE 8.3** Schematic representation of the fabrication strategies for pharmaceutical sensors. (a) $SnO_2$ nanoparticle/$Nb_2CT_x$ nanosheet composite electrode. Reprinted with permission from Ankitha et al. (2024). Copyright 2024, Elsevier. (b) $Nb_2C$ MXene/$MnFe_2O_4$. Reprinted with permission from Wu et al. (2023). Copyright 2023, Elsevier.

**TABLE 8.1**
**MXene–metal oxide electrochemical sensors for the detection of various molecules**

| Sensor probe | Analyte | Technique | Linear range | LOD | Ref. |
|---|---|---|---|---|---|
| $SnO_2/Nb_2CT_x/CC$ | Propofol | CA | 1–300 μM | 0.24 μM | Ranjith et al. (2021) |
| $Sm_2O_3/TiC/GCE$ | Nimodipine | DPV | 0.001–30 μM and 57–950 μM | 0.0042 μM | Anupriya, Senthilkumar, and Chen (2022) |
| $MnFe_2O_4/Nb_2C/GCE$ | Acetaminophen | DPV | 0.1–1000 μM | 0.079 μM | Anupriya, Senthilkumar, and Chen (2022) |
| | Dopamine | | 0.1–60 μM | 0.070 μM | |
| $MXene/Y_2O_3/CS\text{-}PPO/Pt$ | Catechol | Amperometry | 0.04–3.2 μM | 0.007 μM | Chen et al. (2024) |
| $Ti_3C_2T_x/CuO/GCE$ | Diclofenac sodium | SWV | 0.34–31.67 | 0.0083 μM | Mini et al. (2024) |
| | | DPV | | 0.0085 μM | |
| $MCH/cDNA/PDA/Ti_3C_2T_x/MWCNTs/NiCo_2O_4/SGPGE$ | Aflatoxin B1 | DPV | 2.5–200 ng/mL | 1.890 ng/mL | Shen et al. (2023) |
| $Cu_xO\text{-}MWCNTs\text{-}COOH/Ti_3C_2T_x/GCE$ | Benomyl | DPV | 10.0 nM to 10.0 μM | 3 nM | Zhong et al. (2022) |
| $AChE/SnO_2/Nb_2CT_x/GCE$ | Chlorpyrifos | DPV | $5.1 \times 10^{-14}$ to $5.1 \times 10^{-7}$ M | $5.1 \times 10^{-14}$ M | Guo et al. (2023) |
| $AChE\text{-}Chit/Ti_3C_2/Au\ NPs/MnO_2/Mn_3O_4/GCE$ | Methamidophos | DPV | $10^{-12}$ to $10^{-6}$ M | $1.34 \times 10^{-13}$ M | Song et al. (2019) |
| $V_2O_5/Ti_3C_2T_x/GCE$ | Bisphenol A | DPV | 411 nm to 31.2 μM | 87 nM | Thukkaram et al. (2023) |
| MXene aerogel-CuO/CC | $Cd^{2+}$ | DPV | 4–800 μg/L | 0.3 μg/L | Wen et al. (2022) |
| | $Pb^{2+}$ | | 4–1200 μg/L | 0.2 μg/L | |
| $TiO_2@Ti_3C_2T_x/GCE$ | Imidacloprid | DPV | 0.07–10.0 μM and 10.0–70.0 μM | 23.3 nM | Yang et al. (2023) |

*(Continued)*

**TABLE 8.1 (Continued)**
**MXene–metal oxide electrochemical sensors for the detection of various molecules**

| Sensor probe | Analyte | Technique | Linear range | LOD | Ref. |
|---|---|---|---|---|---|
| Ag–$TiO_2$/$Ti_3C_2T_x$/GCE | Paraquat | DPV | 0.05–1 μM and 1–15 μM | 0.01 μM | Hou et al. (2023) |
| $SnO_2$ QDs-$Ti_3C_2$/GCE | Luteolin | DPV | 0.1–1200 nM | 0.14 nM | Gao et al. (2024) |
| $Ti_3C_2T_x$@$RuO_2$/CC | Kaempferol | DPV | 0.06–1 and 1–15 μM | 0.039 μM | Wang et al. (2023) |
| $TiO_2$-$Ti_3C_2T_x$/PEDOT:PSS/GCE | Baicalein | DPV | 0.007–10.0 μM | 2.33 nM | Xue et al. (2023) |
| $CuMoO_4$/$Ti_3C_2T_x$-GCE | Rutin | DPV | 1–15 μM | 42.9 nM | Ramadoss et al. (2024) |
| MXene-$FeWO_4$/GCE | Rutin | DPV | 4–147 nM | 0.42 nM | Ranjith et al. (2021) |
| $Ti_2C$–$TiO_2$/GCE | Glucose | CA | 0.1–200 μM | 0.12 μM | Kumar et al. (2022) |
| $Cu_2O$–rGO–$Ti_3C_2T_x$ | Glucose | CA | 0.01–30 mM | – | Gopal et al. (2022) |
| ZnO tetrapods/$Ti_3C_2T_x$/GCE | Glucose | CA | 0.05–0.7 mM | 17 μM | Myndrul et al. (2022) |
| $Cu_2O$/$Ti_3C_2T_x$/rGO/GCE | Glucose | CA | 0.1–14 mM and 15–40 mM | 1.1 μM | Alanazi et al. (2023) |
| $Co_3O_4$/MXene@CFE | Glucose | Amperometry | 0.05 μM to 7.44 mM | 10 nM | Manoj et al. (2022) |
| Au@CuO/$V_2CT_x$ on LIG | Glucose | CA | 0.005–5.0 mM | 1.8 μM | Cui et al. (2023) |
| $NiCo_2O4$/$Ti_2NbC_2$/GCE | Glucose | CA | 0.025–1 mM | 298 μM | Subramania et al. (2023) |
| MXene@$CeO_2$/chitosan | Glucose | DPV | 0.01–2.5 μM | 0.004 μM | Jiang et al. (2023) |
| $Ti_3C_2T_x$-$Fe_2O_3$/GCE | $H_2O_2$ | CV | 10–1000 nM | 7.46 nM | Nagarajan, Sundaramurthy, and Sundramoorthy (2022) |
| CuO–$CeO_2$/$Ti_3C_2T_x$/GCE | $H_2O_2$ | CA | 5.0–100 μM | 1.67 μM | Zhou et al. (2022) |
| MXene–$Co_3O_4$/SPE | $H_2O_2$ | LSV | 0–75 μm | 0.5 μm | Singh et al. (2023) |
| NiO/$Ti_3C_2T_x$/GCE | $H_2O_2$ | DPV | 10 μM to 4.5 mM | 0.34 μM | Ramachandran et al. (2019) |

propofol sensing by using chronoamperometry. Wu et al. fabricated an MXene($Nb_2C$)/ bimetallic oxide ($MnFe_2O_4$) composite with Schottky junctions by electrostatic interaction for the sensing of acetaminophen and dopamine using the DPV technique (Wu et al. 2023). The $MnFe_2O_4$ nanoparticles chemically self-assemble onto $Nb_2C$ MXene, facilitated by the cross-linking action of glutaraldehyde, which enhances the binding strength through electrostatic interactions (Figure 8.3(b)). This process also facilitates charge transfer across the interface and influences the electronic properties of the $MnFe_2O_4$/$Nb_2C$ MXene. The $MnFe_2O_4$/$Nb_2C$/GCE sensor exhibited a linear range from 0.1 to 1000 μM and 0.1 to 60 μM for acetaminophen and dopamine, respectively. LOD calculated for acetaminophen is 0.079 μM and for dopamine is 0.070 μM.

Anupriya et al. successfully synthesized a nanocomposite with rare-earth metal oxide, $Sm_2O_3$ nanoparticles, and two-dimensional TiC MXene by a controllable hydrothermal method (Anupriya, Senthilkumar, and Chen 2022). $Sm_2O_3$/2D TiC/GCE modified electrode was further utilized for the electrochemical sensing of nimodipine by using DPV analysis. This electrochemical sensor has demonstrated a notable level of sensitivity of 1.5 μA/μM/$cm^2$ and an LOD of 4.2 nM toward detecting nimodipine in biological samples. Table 8.1 shows the comparison of MXene–metal oxide electrochemical sensors for the detection of various molecules.

## 8.4 CONCLUSION

In summary, MXene–metal oxide composites exhibit significant potential in improving the efficiency of electrochemical biosensors, which are used for the detection of diverse biomolecules, environmental pollutants, and medicinal drugs. The distinct blend of exceptional electrical conductivity, large surface area, and customizable functionalization enhances sensitivity and selectivity across a range of biosensing applications. These composites not only improve the catalytic activity but also provide benefits in terms of durability and reusability, making them appropriate for practical applications in medical diagnostics and environmental monitoring. The continuous research and development in this area illustrate the capacity of MXene–metal oxide composites to have a substantial influence on future electrochemical biosensing technologies. Although MXene–metal oxide composites provide numerous benefits, there are still various obstacles that need to be overcome for their effective use in biosensing applications. Further work is needed to address issues such as the repeatability of sensor production, the long-term stability of the composites in different environmental conditions, and the development of effective ways to prevent potential interference in complex biological samples. Addressing these challenges is vital for the successful implementation of these advanced biosensing technologies in real-world settings.

## REFERENCES

Alanazi, Nadyah, Tamil S. Gopal, Muthumareeswaran Muthuramamoorthy, Amani A. E. Alobaidi, Reem A. Alsaigh, Mohammed H. Aldosary, Saravanan Pandiaraj, Maram Almutairi, Andrews N. Grace, and Abdullah Alodhayb. 2023. "$Cu_2O$/MXene/rGO ternary nanocomposites as sensing electrodes for nonenzymatic glucose sensors." *ACS Applied Nano Materials* 6 (13):12271–12281. doi: 10.1021/acsanm.3c01959.

Ali, Adnan. 2016. "Effect of surface oxidation on performance of $Ti_3C_2T_x$/MO composite as anode materials for lithium ion batteries." Qatar Foundation Annual Research Conference Proceedings, Hamad bin Khalifa University Press (HBKU Press) Qatar.

Anasori, Babak, Maria R. Lukatskaya, and Yury Gogotsi. 2017. "2D metal carbides and nitrides (MXenes) for energy storage." *Nature Reviews Materials* 2 (2):16098. doi: 10.1038/natrevmats.2016.98.

Ankitha, Menon, Ajith M. Arjun, Neermunda Shabana, and Pathath A. Rasheed. 2023. "A mini review on recent advances in MXene based electrochemical wearable sensing devices." *Biomedical Materials & Devices* 1 (1):339–350. doi: 10.1007/s44174-022-00010-7.

Ankitha, Menon, Pushpalatha V. Vaishag, Punnoli Muhsin, and Pathath A. Rasheed. 2024. "Flexible and disposable electrodes based on $SnO_2$ nanoparticle/$Nb_2CT_x$ nanosheet composites for selective and ultrasensitive detection of propofol." *Journal of Industrial and Engineering Chemistry* 131:449–458. doi: 10.1016/j.jiec.2023.10.049.

Anupriya, Jeyaraman, Thirumurugan Senthilkumar, and Shen M. Chen. 2022. "A precise electrochemical sensor based on $Sm_2O_3$/2D TiC hybrid for highly sensitive and selective detection of antihypertensive drug nimodipine." *Colloids and Surfaces A: Physicochemical and Engineering Aspects* 641:128531. doi: 10.1016/j.colsurfa.2022.128531.

Aravind SS Jyothirmayee, and Sundara Ramaprabhu. 2012. "Dopamine biosensor with metal oxide nanoparticles decorated multi-walled carbon nanotubes." *Nanoscience Methods* 1 (1):102–114. doi: 10.1080/17458080.2010.520136.

Arjun, A Mohan, Megha, Shinde, and Gymama, Slaughter. 2023. "Application of MXene in the electrochemical detection of neurotransmitters: A review." *IEEE Sensors Journal* 23 (15):16456–16466. doi: 10.1109/JSEN.2023.3285594.

Ataca, Can, Hasan Şahin, and Salim Ciraci. 2012. "Stable, single-layer $MX_2$ transition-metal oxides and dichalcogenides in a honeycomb-like structure." *The Journal of Physical Chemistry C* 116 (16):8983–8999. doi: 10.1021/jp212558p.

Aydoğdu, Gözde, Derya K. Zeybek, Bülent Zeybek, and Şule Pekyardımcı. 2013. "Electro chemical sensing of NADH on NiO nanoparticles-modified carbon paste electrode and fabrication of ethanol dehydrogenase-based biosensor." *Journal of Applied Electro chemistry* 43 (5):523–531. doi: 10.1007/s10800-013-0536-3.

Cai, Guofa, Sangbaek Park, Xing Cheng, Alice L.-S. Eh, and Pooi S. Lee. 2018. "Inkjet-printed metal oxide nanoparticles on elastomer for strain-adaptive transmissive electrochromic energy storage systems." *Science and Technology of Advanced Materials* 19 (1):759–770. doi: 10.1080/14686996.2018.1526031.

Chauhan, Nidhi, and Chandra S. Pundir. 2011. "An amperometric biosensor based on acetylcholinesterase immobilized onto iron oxide nanoparticles/multi-walled carbon nanotubes modified gold electrode for measurement of organophosphorus insecticides." *Analytica Chimica Acta* 701 (1):66–74. doi: 10.1016/j.aca.2011.06.014.

Chen, Haifan, Gongwei Xia, Jianfang Chen, and Hao Zheng. 2024"Biosensor based on Mxene-$Y_2O_3$ composite for ultrasensitive detection of catechol in water samples." *International Journal of Electrochemical Science* 19(11):100823. doi: 10.1016/j.ijoes.2024.100823.

Cui, Feiyun, Haiming Sun, Xueqiong Yang, Hang Zhou, Yiqi Wu, Jing Li, Hongxia Li, Jianlei Liu, Chijia Zeng, Bin Qu, Jufan Zhang, and Qin Zhou. 2023. "Laser-induced graphene (LIG)-based Au@CuO/$V_2CT_x$ MXene non-enzymatic electrochemical sensors for the urine glucose test." *Chemical Engineering Journal* 457:141303. doi: 10.1016/j.cej.2023.141303.

Dong, Xiao-Chen, Hang Xu, Xue-Wan Wang, Yin-Xi Huang, Mary B. Chan-Park, Hua Zhang, Lian-Hui Wang, Wei Huang, and Peng Chen. 2012. "3D graphene–cobalt oxide electrode for high-performance supercapacitor and enzymeless glucose detection." *ACS Nano* 6 (4):3206–3213. doi: 10.1021/nn300097q.

Vilian, AT Ezhil, Rajesh Madhu, Shen-Ming Chen, Vediyappan Veeramani, Mani Sivakumar, Yun S. Huh, and Young-Kyu Han. 2015. "Facile synthesis of $MnO_2$/carbon nanotubes decorated with a nanocomposite of Pt nanoparticles as a new platform for the electrochemical detection of catechin in red wine and green tea samples." *Journal of Materials Chemistry B* 3 (30):6285–6292. doi: 10.1039/C5TB00508F.

Gao, Feng, Weihua Hong, Tao Yang, Chenhui Qiao, Jingjia Li, Xi Xiao, Ziying Zhao, Chao Zhang, and Junyuan Tang. 2024. "Expanded interlayer spacing of $SnO_2$ QDs-decorated MXene for highly selective luteolin detection with Ultra-Low limit of detection." *Journal of Colloid and Interface Science* 653:561–569. doi: 10.1016/j.jcis.2023.09.091.

Gopal, Tamil S., Khalid E. Alzahrani, Abdulaziz K. Assaifan, Hamad Albrithen, Abdullah Alodhayb, Muthumareeswaran Muthuramamoorthy, Saravanan Pandiaraj, and Andrews N. Grace. 2022. "Reduced graphene oxide supported MXene based metal oxide ternary composite electrodes for non-enzymatic glucose sensor applications." *Scientific Reports* 12 (1):20583. doi: 10.1038/s41598-022-24700-w.

Guo, Wei, Lijun Liang, Yaxu Zhao, Congyi Zhao, Xiong Lu, Yunpeng Cao, and Faming Gao. 2023. "In-situ growth of $SnO_2$ nanoparticles on $Nb_2CT_x$ nanosheets as highly sensitive electrochemical sensing platform for organophosphorus pesticide detection." *Colloids and Surfaces B: Biointerfaces* 224:113238. doi: 10.1016/j.colsurfb.2023.113238.

Hou, Enhui, Zhiguo Kong, Jianfeng Wu, Hairui Wang, Ping Nie, Ming Lu, and Limin Chang. 2023. "Preparation of Ag–$TiO_2$/MXene composite material for electrochemical detection of paraquat." *Journal of Materials Science: Materials in Electronics* 34 (14):1177. doi: 10.1007/s10854-023-10603-7.

Hussain, Sajjad, Hailiang Liu, Dhanasekaran Vikraman, Syed H. A. Jaffery, Ghazanfar Nazir, Faisal Shahzad, Khalid M. Batoo, Jongwan Jung, Jungwon Kang, and Hyun-Seok Kim. 2023. "Tuning of electron transport layers using MXene/metal–oxide nanocomposites for perovskite solar cells and X-ray detectors." *Nanoscale* 15 (16):7329–7343. doi: 10.1039/D3NR01196H.

Jiang Shuyue, Haifeng Zhang, Zhaoxin Li, Liang Chen, Liang Yin, and Xiaowei Liu. 2023. "Sponge-inspired MXene@$CeO_2$ detector for ultra-sensitive detection of glucose." *Materials Today Chemistry* 32:101638. doi: 10.1016/j.mtchem.2023.101638.

Kumar, Rajesh, and Lakhveer Singh. 2022. "$Ti_3C_2T_x$ MXene as electrocatalyst for designing robust glucose biosensors." *Advanced Materials Technologies* 7 (12):2200151. doi: 10.1002/admt.202200151

Kumar, Vinod, Sudheesh K. Shukla, Meenakshi Choudhary, Jalaj Gupta, Priyanka Chaudhary, Saurabh Srivastava, Mukesh Kumar, Manoj Kumar, Devojit K. Sarma, Bal C. Yadav, and Vinod Verma. 2022. "$Ti_2C$-$TiO_2$ MXene nanocomposite-based high-efficiency non-enzymatic glucose sensing platform for diabetes monitoring." *Sensors* 22 (15):5589. doi: 10.3390/s22155589

Li, Tao, Dawei Shang, Shouwu Gao, Bo Wang, Hao Kong, Guozheng Yang, Weidong Shu, Peilong Xu, and Gang Wei. 2022. "Two-dimensional material-based electrochemical sensors/biosensors for food safety and biomolecular detection." *Biosensors* 12 (5):314. doi: 10.3390/bios12050314

Liao, Jianjun, Shiwei Lin, Yue Yang, Kai Liu, and Wencai Du. 2015. "Highly selective and sensitive glucose sensors based on organic electrochemical transistors using $TiO_2$ nanotube arrays-based gate electrodes." *Sensors and Actuators B: Chemical* 208:457–463. doi: 10.1016/j.snb.2014.11.038.

Liu, Hui, Congyue Duan, Chenhui Yang, Wanqiu Shen, Fen Wang, and Zhenfeng Zhu. 2015. "A novel nitrite biosensor based on the direct electrochemistry of hemoglobin immobilized on MXene-$Ti_3C_2$." *Sensors and Actuators B: Chemical* 218:60–66. doi: 10.1016/j.snb.2015.04.090.

Lu, Decheng, Huijuan Zhao, Xinying Zhang, Yingying Chen, and Lingyan Feng. 2022. "New horizons for MXenes in biosensing applications." *Biosensors* 12 (10):820. doi: 10.3390/bios12100820.

Ma, Junjie, Fang Huang, Aihao Xu, Dong Wei, Xiangyu Chen, Wencan Zhao, Zhengjun Chen, Xucai Yin, Jinliang Zhu, Huibing He, and Jing Xu. 2024. "Three-phase-heterojunction $Cu/Cu_2O–Sb_2O_3$ catalyst enables efficient $CO_2$ electroreduction to CO and high-performance aqueous $Zn–CO_2$ battery." *Advanced Science* 11 (16):2306858. doi: 10.1002/advs.202306858.

Manoj, Devarajan, Ayesha Aziz, Nadeem Muhammad, Zhanpeng Wang, Fei Xiao, Muhammad Asif, and Yimin Sun. 2022. "Integrating $Co_3O_4$ nanocubes on MXene anchored CFE for improved electrocatalytic activity: Freestanding flexible electrode for glucose sensing." *Journal of Environmental Chemical Engineering* 10 (5):108433. doi: 10.1016/j.jece.2022.108433.

Mini, Aarcha A., Arghya Chakravorty, Sudip Das, Shikha Awasthi, Andrews N. Grace, Sarvesh K. Pandey, and Vimala Raghavan. 2024. "CuO nanoparticles passivated 2D MXene-based voltammetric sensor for detecting environmental hazardous pollutant." *Microchemical Journal* 201:110648. doi: 10.1016/j.microc.2024.110648.

Munteanu, Raluca-Elena, Paola S. Moreno, Mattia Bramini, and Szilveszter Gáspár. 2021. "2D materials in electrochemical sensors for in vitro or in vivo use." *Analytical and Bioanalytical Chemistry* 413 (3):701–725. doi: 10.1007/s00216-020-02831-1.

Myndrul, Valerii, Emerson Coy, Nataliya Babayevska, Veronika Zahorodna, Vitalii Balitskyi, Ivan Baginskiy, Oleksiy Gogotsi, Mikhael Bechelany, Maria T. Giardi, and Igor Iatsunskyi. 2022. "MXene nanoflakes decorating ZnO tetrapods for enhanced performance of skin-attachable stretchable enzymatic electrochemical glucose sensor." *Biosensors and Bioelectronics* 207:114141. doi: 10.1016/j.bios.2022.114141.

Nagarajan, Ramila D., Anandhakumar Sundaramurthy, and Ashok K. Sundramoorthy. 2022. "Synthesis and characterization of MXene ($Ti_3C_2T_x$)/iron oxide composite for ultrasensitive electrochemical detection of hydrogen peroxide." *Chemosphere* 286:131478. doi: 10.1016/j.chemosphere.2021.131478.

Natu, Varun, James L. Hart, Maxim Sokol, Helen Chiang, Mitra L. Taheri, and Michel W. Barsoum. 2019. "Edge capping of 2D-MXene sheets with polyanionic salts to mitigate oxidation in aqueous colloidal suspensions." *Angewandte Chemie* 131 (36):12785–12790. doi: 10.1002/anie.201906138

Ouikhalfan, Mohammed, Ahmet Sari, Hassan Chehouani, Brahim Benhamou, and Alper Biçer. 2019. "Preparation and characterization of nano-enhanced myristic acid using metal oxide nanoparticles for thermal energy storage." *International Journal of Energy Research* 43 (14):8592–8607. doi: 10.1002/er.4856

Poizot, Philippe. Laruelle **Stéphane**. Grugeon Sylvie, Loi cDupont, and Jean-Marie. Tarascon. 2000. "Nano-sized transition-metal oxides as negative-electrode materials for lithium-ion batteries." *Nature* 407 (6803):496 499. doi: 10.1038/35035045.

Prabhakar, Nikhil, Amir Humayun, Ponpandian Nagamony, Umamaheswari Rajaji, Razan A. Alshgari, Ting-Yu Liu, and Viswanathan Chinnuswamy. 2024. "Effect on delamination of $Nb_4C_3T_x$ MXene supported tungsten oxide for high-performance supercapacitor." *Journal of Energy Storage* 98:112830. doi: 10.1016/j.est.2024.112830.

Rahman, Md. Mahbubur, Ahammad. AJ. Saleh, Joon-Hyung Jin, Sang J. Ahn, and Jae-Joon Lee. 2010. "A comprehensive review of glucose biosensors based on nanostructured metal-oxides." *Sensors* 10 (5):4855–4886. doi: 10.3390/s100504855

Ramachandran, Rajendran, Changhui Zhao, Muniyandi Rajkumar, Krishnamoorthy Rajavel, Pengli Zhu, Wenlu Xuan, Zong-Xiang Xu, and Fei Wang. 2019. "Porous nickel oxide microsphere and $Ti_3C_2T_x$ hybrid derived from metal-organic framework for battery-type supercapacitor electrode and non-enzymatic $H_2O_2$ sensor." *Electrochimica Acta* 322:134771. doi: 10.1016/j.electacta.2019.134771.

Ramadoss, Jagadeesh, Mani Govindasamy, Arumugam Sonachalam, Chi-Hsien Huang, and Asma A. Alothman. 2024. "$CuMoO_{4/Ti3}C_2T_x$ nanocomposite layers perform as an ultrasensitive electrochemical sensor for the detection of antioxidant rutin." *Microchimica Acta* 191 (4):226. doi: 10.1007/s00604-024-06267-7.

Ranjith, Kugalur S., Vilian AT. Ezhil, Seyed M. Ghoreishian, Reddicherla Umapathi, Yun S. Huh, and Young-Kyu Han. 2021. "An ultrasensitive electrochemical sensing platform for rapid detection of rutin with a hybridized 2D-1D MXene-$FeWO_4$ nanocomposite." *Sensors and Actuators B: Chemical* 344:130202. doi: 10.1016/j.snb.2021.130202.

Rhouati, Amina, Mohammed Berkani, Yasser Vasseghian, and Nasrin Golzadeh. 2022. "MXene-based electrochemical sensors for detection of environmental pollutants: A comprehensive review." *Chemosphere* 291:132921. doi: 10.1016/j.chemosphere.2021.132921.

Shen, Jinhui, Juan Liu, Siyi Yang, Xin Yao, Huanbao Fa, Changjun Hou, and Mei Yang. 2023. "Novel electrochemical sensor based on PDA/MXene/MWCNTs/$NiCo_2O_4$ nanocomposites for rapid, sensitive, and selective detection of Aflatoxin B1." *Food Analytical Methods* 16 (6):1055–1068. doi: 10.1007/s12161-023-02464-x.

Shirsat, Mahendra D., and Tibor Hianik. 2023. "Electrochemical detection of heavy metal ions based on nanocomposite materials." *Journal of Composites Science* 7 (11):473. doi: 10.3390/jcs7110473

Singh, Sima, Arshid Numan, Mohammad Khalid, Ivana Bello, Elisabetta Panza, and Stefano Cinti. 2023. "Facile and affordable design of MXene-$Co_3O_4$-based nanocomposites for detection of hydrogen peroxide in cancer cells: Toward portable tool for cancer management." *Small* 19 (51):2208209. doi: 10.1002/smll.202208209.

Song, Dandan, Xinyu Jiang, Yanshan Li, Xiong Lu, Sunrui Luan, Yuanzhe Wang, Yan Li, and Faming Gao. 2019. "Metal–organic frameworks-derived $MnO_2$/$Mn_3O_4$ microcuboids with hierarchically ordered nanosheets and $Ti_3C_2$ MXene/Au NPs composites for electrochemical pesticide detection." *Journal of Hazardous Materials* 373:367–376. doi: 10.1016/j.jhazmat.2019.03.083.

Subramania, Ashok K., Sivanandam Sugumaran, Prakash Sethuramalingam, Rajasekaran Ramesh, Preethi Dhandapani, and Subramania Angaiah. 2023. "$NiCo_2O_4$/$Ti_2NbC_2$ (double MXene) nanohybrid-based non-enzymatic electrochemical biosensor for the detection of glucose in sweat." *Bioprocess and Biosystems Engineering* 46 (12):1755–1763. doi: 10.1007/s00449-023-02930-0.

Thakur, Anupma, Nithin Chandran BS, Karis Davidson, Annabelle Bedford, Hui Fang, Yooran Im, Vaishnavi Kanduri, Brian C. Wyatt, Srinivasa K. Nemani, and Valeriia Poliukhova. 2023. "Step-by-step guide for synthesis and delamination of $Ti_3C_2T_x$ MXene." *Small Methods* 7 (8):2300030.

Thukkaram, Mano P., Arghya Chakravorty, Aarcha A. Mini, Karthikeyan Ramesh, Andrews N. Grace, and Vimala Raghavan. 2023. "Titanium carbide MXene and $V_2O_5$ composite-based electrochemical sensor for detection of bisphenol A." *Microchemical Journal* 193:109004. doi: 10.1016/j.microc.2023.109004.

Ustad, Ruhan E., Somnath S. Kundale, Kasturi A. Rokade, Snehal L. Patil, Vijay D. Chavan, Kalyani D. Kadam, Harshada S. Patil, Sarita P. Patil, Rajanish K. Kamat, Deok-kee Kim, and Tukaram D. Dongale. 2023. "Recent progress in energy, environment, and electronic applications of MXene nanomaterials." *Nanoscale* 15 (23):9891–9926. doi: 10.1039/D2NR06162G.

Wang, Bingrong, Yang Cao, Yong Chen, Xiaoyong Lai, Juan Peng, Jinchun Tu, and Xiaotian Li. 2016. "Rapid synthesis of rGO conjugated hierarchical $NiCo_2O_4$ hollow mesoporous nanospheres with enhanced glucose sensitivity." *Nanotechnology* 28 (2):025501.

Wang, Jing, Qingbin Xu, Jinxin Liu, Weijun Kong, and Linchun Shi. 2023. "Electrostatic self-assembly of MXene on ruthenium dioxide-modified carbon cloth for electrochemical detection of kaempferol." *Small* 19 (34):2301709. doi: 10.1002/smll.202301709.

Wee, Shianlin, Xiliang Lian, Evgeniya Vorobyeva, Akhil Tayal, Vladimir Roddatis, Fabio La Mattina, Dario G. Vazquez, Netanel Shpigel, Mathieu Salanne, and Maria R. Lukatskaya. 2024. "Tuning MXene Properties through Cu Intercalation: Coupled Guest/Host Redox and Pseudocapacitance." *ACS nano 18*(14):10124–10132. doi: 10.1021/acsnano.3c12989

Wen, Li, Jiangbo Dong, Huisi Yang, Jiaying Zhao, Zhikun Hu, Haiyan Han, Changjun Hou, Xiaogang Luo, and Danqun Huo. 2022. "A novel electrochemical sensor for simultaneous detection of $Cd^{2+}$ and $Pb^{2+}$ by MXene aerogel-CuO/carbon cloth flexible electrode based on oxygen vacancy and bismuth film." *Science of The Total Environment* 851:158325. doi: 10.1016/j.scitotenv.2022.158325.

Wu, Peng, Xueling Zhao, Yaxin Ding, Yufu Huang, Donghai Lin, Lili Xie, Zhanhong Li, Zhigang Zhu, Hongli Zhao, and Minbo Lan. 2023. "Self-assembled multilayer $Nb_2C$ MXene/$MnFe_2O_4$ electrochemical sensor with Schottky junctions for the detection of acetaminophen and dopamine." *Colloids and Surfaces A: Physicochemical and Engineering Aspects* 667:131377. doi: 10.1016/j.colsurfa.2023.131377.

Xiao, Fei, Yuanqing Li, Xiaoli Zan, Kin Liao, Rong Xu, and Hongwei Duan. 2012. "Growth of metal–metal oxide nanostructures on freestanding graphene paper for flexible biosensors." *Advanced Functional Materials* 22 (12):2487–2494.

Xue, Shuya, Min Shi, Jinye Wang, Jiapeng Li, Guanwei Peng, Jingkun Xu, Yansha Gao, Xuemin Duan, and Limin Lu. 2023. "$TiO_2$-MXene/PEDOT: PSS composite as a novel electrochemical sensing platform for sensitive detection of baicalein." *Molecules* 28 (7): 3262.

Yang, Jing, Changxi Deng, Wei Zhong, Guanwei Peng, Jin Zou, Yan Lu, Yansha Gao, Mingfang Li, Songbai Zhang, and Limin Lu. 2023. "Electrochemical activation of oxygen vacancy-rich $TiO_2$@MXene as high-performance electrochemical sensing platform for detecting imidacloprid in fruits and vegetables." *Microchimica Acta* 190 (4):146. doi: 10.1007/s00604-023-05734-x.

Yang, Meiqing, Haozi Lu, and Song Liu. 2022a. "Recent advances of MXene-based electrochemical immunosensors." *Applied Sciences* 12 (11):5630.

Yang, Qinqin, Ying Qin, Tiantian Lei, Yaping Zhong, Xipeng Cui, Yu He, and Gongwu Song. 2022b. "$Co(OH)_2$/MXene-$Ti_3C_2$ nanocomposites with triple-enzyme mimic activities as hydrogel sensing platform for on-site detection of hypoxanthine." *Microchimica Acta* 189 (12):481. doi: 10.1007/s00604-022-05584-z.

Zhao, Wei, Pei-Yu Ge, Jing-Juan Xu, and Hong-Yuan Chen. 2009. "Selective detection of hypertoxic organophosphates pesticides via PDMS composite based acetylcholinesterase-inhibition biosensor." *Environmental Science & Technology* 43 (17):6724–6729. doi: 10.1021/es900841n.

Zhong, Wei, Jin Zou, Yanyu Xie, Jing Yang, Mingfang Li, Shuwu Liu, Yansha Gao, Xiaoqiang Wang, and Limin Lu. 2022. "Three-dimensional nano-$Cu_xO$-MWCNTs-COOH/MXene heterostructure: An efficient electrochemical platform for highly sensitive and selective sensing of benomyl in fruit samples." *Journal of Electroanalytical Chemistry* 920:116586. doi: 10.1016/j.jelechem.2022.116586.

Zhou, Kaiwei, Yang Li, Shujuan Zhuang, Jie Ren, Feng Tang, Jinglin Mu, and Ping Wang. 2022. "A novel electrochemical sensor based on CuO-$CeO_2$/MXene nanocomposite for quantitative and continuous detection of $H_2O_2$." *Journal of Electroanalytical Chemistry* 921:116655. doi: 10.1016/j.jelechem.2022.116655.

# 9 MXene–Metal Chalcogenide Composite–Modified Electrochemical Sensors

*Keerthi Kodakat*

## 9.1 INTRODUCTION

Given the remarkable benefits of electrochemical techniques, including their ease of use, miniaturisation potential, reproducibility, sensitivity and economic nature, it is not surprising that electrochemical sensing technology has made a significant contribution to the development of electroanalytical chemistry to its current state (Hernández-Rodríguez, Rojas, and Escarpa 2021). Moreover, an increasing demand for the fast and selective detection and quantification of molecules in clinical, environmental and agricultural fields has led to a true renaissance in this area. Recent advances in electrochemical sensors are fuelled by the search for highly selective, sensitive and stable materials that can effectively transduce chemical or biological interactions into electrical signals (Umapathi et al. 2022). The electrochemical studies of 2D materials have presented a substantial growth during the last few years with notable implications in the field of sensors (Raza et al. 2024). Among them, MXenes are 2D transition metal carbides or nitrides which are derived from layered compounds called MAX phases with a general formula $M_{n+1}AX_n$ ($n = 1, 2, 3, 4$) where M represents any early transition metal, A can be an element from group 13–16 (mostly Al or Si) and X is C and/or N (Deysher et al. 2020). MXenes are represented by a general formula $M_{n+1}X_nT_x$ where T represent the surface terminations such as F, O, OH or Cl. MXenes are synthesised by etching the A element from the MAX phase. After their discovery in 2011, MXenes garnered significant attention in the electrochemical sensing field as they exhibit characteristic electrochemical properties owing to their ultra-thin structure, high surface area to volume ratio and tuneable electronic structure that could allow control of their electrochemical properties (Raza et al. 2024). MXene gained wide attention in a short time in electrochemical sensing research due to its exceptional electrical conductivity (up to 20–24 kS/cm) and rich and tuneable surface chemistry arising from its terminal functional groups, such as –OH, =O and –F. Their large surface area allows facile electron transfer, and they can also act as an effective immobilisation matrix for application in biosensors (Gayathri et al. 2025; Rhouati et al. 2022). Also, the hydrophilic nature of MXenes plays a vital role in their biocompatibility, which makes them a promising


DOI: 10.1201/9781003736318-9

material in biosensing applications (Dmytriv and Lushchak 2024). Till date, MXenes have been widely explored in the electrochemical sensing of metal ions, environmental pollutants, organic compounds, pharmaceutical drugs, biomarkers, etc. (Rhouati et al. 2022). More than 50 MXenes have been reported to date, and the number goes beyond a hundred when considering the distinct surface terminations and elemental ratios (Khan et al. 2024). Ti-based MXenes (e.g. $Ti_3C_2T_x$), the first MXene to be synthesised, are the most widely studied in electrochemical sensing applications (Pei et al. 2021). Other classes of MXenes, such as $Mo_2C$ and $Nb_2C$, are also being investigated nowadays (Gogotsi 2023). Bare, unmodified MXenes have been used in electrochemical sensors for the determination of hydrogen peroxide (Lorencova et al. 2017), glucose (Chia et al. 2020), dopamine (Shahzad et al. 2019), nitrite (Liu et al. 2015), etc.

Two critical features of MXenes that significantly affect their electrochemical sensing performance are their electronic structure and surface functionalities (Ezzah Ab Latif et al. 2022). The electronic properties, which include band gap and conductivity, play a remarkable role in the interaction of MXene with the target molecule during sensing. These properties become highly influential in predicting the possible sensing mechanism, which in turn has a significant effect on its selectivity and sensitivity (Ezzah Ab Latif et al. 2022). Defect engineering in MXenes involves introduction, alteration or control of the structural or compositional defects to enhance their electronic and catalytic properties (Tang et al. 2022). For example, density functional theory (DFT) and Born–Oppenheimer molecular dynamic (BOMD) simulations on single vacancy defects in MXenes suggest that surface terminations play a significant role in defect formation (Bandyopadhyay, Ghosh, and Pati 2018). Moreover, depending on the type of defect and the accumulation of charges near the defect sites, MXenes may exhibit notable metal-to-semiconductor or semiconductor-to-metal transitions. Studies have also shown that unpaired electrons in spin-split d-orbitals in defective MXenes can also induce magnetic behaviours (Bandyopadhyay, Ghosh, and Pati 2018). Atomistic insights gained from in-depth theoretical calculations on heteroatom-doped or hybrid structures of MXenes have also shown that the strong electronic coupling between the components of the hybrids holds great promise for their electrocatalytic applications (Tang et al. 2022; Peng et al. 2019). For example, the composite of $Ti_3C_2T_x$ with the metal-organic framework (MOF) (cobalt 1,4 benzenedicarboxylate (CoBDC)) synthesised by Zhao et al. outperformed the standard $IrO_2$-based catalyst in oxygen evolution reaction (OER) activity (Zhao et al. 2017). The hybrid catalyst demonstrated better OER reaction kinetics compared to the standard catalyst. These studies point out the critical role of interface control of MXene-based hybrids for robust electrocatalytic applications (Zhao et al. 2017).

The surface functionalities of MXene represent the active sites on the 2D structure that are responsible for the binding interaction with the target molecules. These surface functionalities can be either the inherent functionalities formed during the synthesis of the MXenes or they can be introduced at a later stage, as in a post-synthetic modification procedure (Lee et al. 2021). The inherent active sites of MXene may have a favourable or adverse effect on the sensing performance, depending entirely on the nature of the target analyte, and this is where the post-synthetic modification of the MXene becomes significant. It allows tailoring of the surface properties by carefully controlling the active sites with the help of covalent or non-covalent alternatives to get a structure with the desired properties (Ezzah Ab Latif et al. 2022). This allows selective interaction of the analyte with the MXene modified electrode surface, contributing to better selectivity of the developed sensor (Mozafari and Soroush 2021). The non-covalent modification of the MXene implies van der Waals, H-bonding or electrostatic interactions with the hetero species. For example, the non-covalent functionalisation of $Ti_3C_2T_x$ with a cationic porphyrin is reported to show better stability of the MXene in water against oxidation, which proposes a useful strategy for promising applications in sensors, optoelectronics and biomedicines (Thurakkal and Zhang 2022). In contrast, covalent modification involves binding of MXenes with other small molecules, heteroatoms, other 2D structures or polymers by exploiting covalent interactions. An example is the covalent linkage of polyethyleneglycol carboxylic acid onto $Ti_3C_2T_x$, which results in highly conductive and ordered microstructure films (Lee et al. 2021).

Thus, incorporating suitable material can tune the properties of the pristine MXene to the desired ones. One of the widely explored post-synthetic modification MXenes is the incorporation of other 2D materials such as graphene, MOFs and transition metal dichalcogenides (TMDs). The following section discusses the various MXene-based hybrid architectures used in electrochemical sensing.

## 9.2 MXENE-BASED HYBRID ARCHITECTURES FOR ELECTROCHEMICAL SENSING

One of the major compelling reasons behind the modification of MXenes to hybrid architectures is their unstable nature in ambient conditions and aqueous environments. The agglomeration tendency of nanosheets, low mechanical stability, and their weak stability in oxygen atmosphere are the downsides of MXenes, which subside their performance in sensing applications (Ankitha et al. 2023). Intercalation of small molecules and metal ions such as urea, ethanolamine, $Na^+$ and $K^+$ has been proven to prevent aggregation by increasing the interlayer distance of the MXene nanosheets (Al-Temimy et al. 2020; Kim et al. 2021). In addition, formation of MXene hydrogels or aerogels with 3D porous structure or stable 3D network formed by the integration of MXenes with polymers, including polyvinyl alcohol or chitosan are also found to increase the life of MXene (Zhang et al. 2020; Orangi et al. 2021; Li et al. 2024; Kallingal et al. 2022). Another possible method is the hybridisation of MXene with foreign nanostructures like carbon nanotubes (CNTs) and other 2D materials such as graphene, MOFs, hexagonal boron nitride (hBN) and TMDs (Figure 9.1) (Lim et al. 2020). Recent investigations have demonstrated that integration of the 2D structures with MXenes leads to the high performance of the resulting heterostructures in electrochemical sensing applications (Wang et al. 2024b). The growing prominence of the 2D heterostructure of MXenes in electrochemical sensing applications is due to the opportunity to change the active sites on the surface and electronic states of the material that could render much better electrochemical properties than the pristine MXenes (Ezzah Ab Latif et al. 2022). More explicitly, the multifold application of 2D heterostructure of MXenes arises from their customisable surface chemistry and

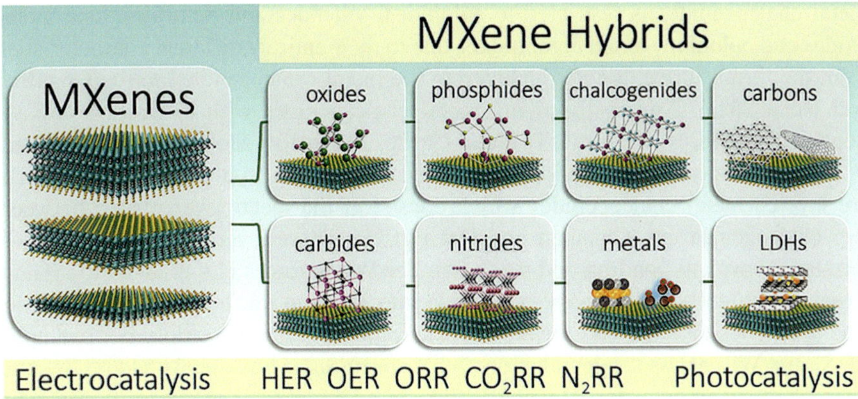


**FIGURE 9.1** Schematic representation of the application of MXene hybrids in various electrocatalytic applications. Reprinted with permission from Lim et al. (2020).

interlayer structure, and the possibility of upgradation of the heterostructure to 3D nanoporous structures (Liu et al. 2021; Gu et al. 2019). Graphene in combination with MXenes has shown considerable potential in enhancing the catalytic property and electroactive surface area, thereby acting as an ideal sensing interface for the sensitive detection of numerous analytes. Wang et al. employed a self-assembled $Ti_3C_2T_x$/graphene composite for the sensitive determination of p-nitrophenol in aqueous media with a detection limit of 0.16 μM (Wang et al. 2022). Chen et al. reported another pioneering research where a unique 3D interconnected micro mesoporous $Ti_3C_2T_x$-graphene aerogel was used for the dynamic quantification of glucose in sweat and for the real-time monitoring and adjustment of the pH of sweat (Chen et al. 2024). Another class of 2D materials used in combination with MXenes are layered double hydroxides (LDHs). They are 2D materials consisting of positive metal hydroxide layers and interlayers of anions. Li et al. synthesised NiCo-LDH on the $Ti_3C_2T_x$ surface, and the resulting 3D porous NiCo-LDH/$Ti_3C_2T_x$ nanocomposite was applied for the non-enzymatic sensing of glucose (Li et al. 2019). High redox activity and surface area of the LDH facilitated the diffusion of glucose onto the electrode surface (Li et al. 2019). Meanwhile, MXene, due to its high conductivity and hydrophilic nature, accelerated the rate of oxidation of glucose at the electrode–electrolyte interface (Li et al. 2019).

In recent years, significant research has focused on the synthesis and application of MXene hybrid structures based on metal chalcogenides ($MoS_2$, $WSe_2$, etc.). This chapter aims to give a comprehensive outline of the reports on MXene–metal chalcogenide heterostructures for electrochemical sensing, highlighting their advancements and potential for various analytical applications.

### 9.2.1 Emergence of MXene–Metal Chalcogenide Heterostructures

Metal chalcogenides constitute an important class of inorganic compounds that consist of one or more chalcogenide anions and electropositive metal ions (Chauhan and Deka 2021). Although all group 16 elements are referred to as chalcogens, the term metal chalcogenides is restricted to sulphides, selenides and tellurides. The metal oxides and polonium-based compounds, due to their entirely different characteristics from the former three, are not included in the metal chalcogenide family (Chauhan and Deka 2021). Among the many metal chalcogenides with different compositions and structures like $Na_2S$, $Ti_2S$ and $Cu_2Se$, TMDs like $MoS_2$, $MoTe_2$, $WSe_2$ and $WS_2$ gained significant attention in the field of research in a short period. TMDs are layered metal chalcogenides which consist of the metal layer placed between two chalcogen layers (Chauhan and Deka 2021). These layers are held by strong intralayer covalent bonding and weak van der Waals forces (Lv et al. 2015). Their structural properties allow easy exfoliation into 2D nano sheets similar to graphene. The properties of TMDs are diverse depending on their composition. For example, $MoS_2$ and $WS_2$ act as semiconductors, $NbS_2$ and $VSe_2$ act as metals and $NbSe_2$ and $TaS_2$ act as superconductors (Lv et al. 2015). In addition, their properties can also be tuned by varying crystal structures and the number of stacking layers in the crystals. Therefore, proper engineering of their structure and composition is essential in exploiting their properties for applications in energy storage, sensing, catalysis and various other applications (Figure 9.2) (Jin et al. 2021).

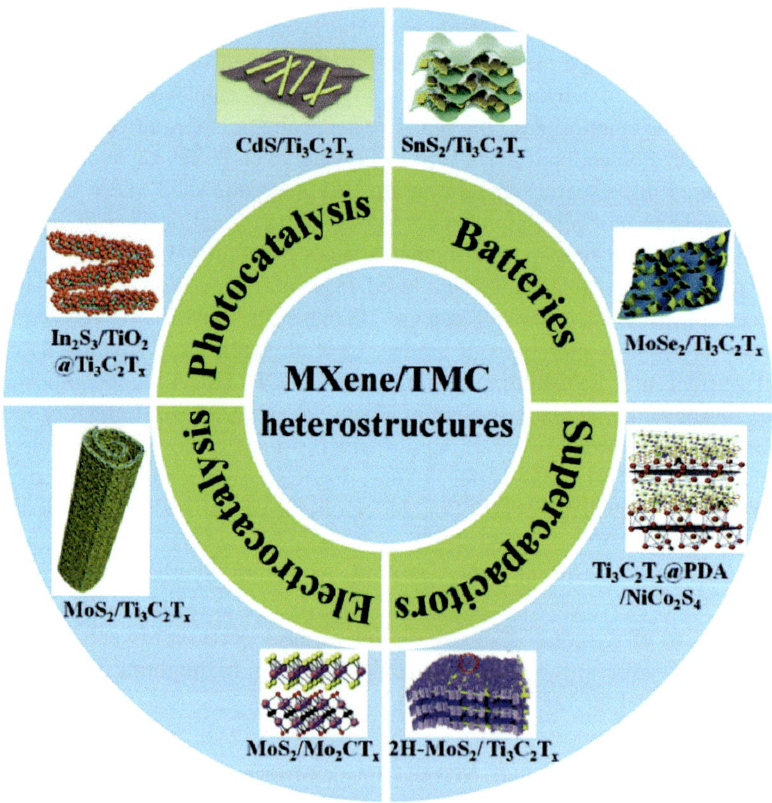


**FIGURE 9.2** Schematic representation of MXene–metal chalcogenide composites for various applications. Reprinted with permission from Jin et al. (2021).

MXene–metal chalcogenide composites have gained attention among researchers due to the unique combination of their properties. MXene with rich surface functionalities and charge transfer capabilities serves as a suitable material for anchoring metal chalcogenides such as $MoS_2$ and ZnS. One of the general synthetic strategies adopted for the synthesis of the composite is the hydrothermal method. Such a hybrid modification can alter morphologies and crystal phases. They can also induce edge-plane defects and influence the surface functional group densities (Ezhil Vilian et al. 2023). The hybrid structures also influence the stability of the MXene by preventing the agglomeration of the nanosheet (Sheikh et al. 2023). In addition, surface functionalisation in MXene can increase the wettability of metal chalcogenides, contributing to better electrolyte accessibility. The following section discusses the recent reports on the application of composites of MXene with TMDs and other metal chalcogenides in electrochemical sensing applications.

### 9.2.2 Application of MXene-TMD Heterostructures

The application of TMDs in electrochemical sensors has gained wide attention due to the higher number of active sites at their edges. Combining 2D TMDs with 2D MXenes is an attractive way of increasing the sensor's performance, as it endows the excellent properties of both materials (Zhao et al. 2022a). The nanosheets of TMDs can be uniformly embedded on the MXene surface, mitigating the restacking issue by increasing the interlayer spacing. Apart from that, the heterostructure can facilitate fast electron transfer, thereby increasing the diffusion of analytes, contributing to the better performance of the sensor (Rajaji et al. 2022). $MoS_2$ is one of the widely used TMDs in electrochemical sensor applications (Tajik et al. 2022).

Zhao et al., in 2022, reported the development of a label-free biosensor for miRNA-21 based on $Ti_3C_2T_x$-$MoS_2$ heterostructure coupled with catalytic hairpin assembly (Zhao et al. 2022a). The MXene-$MoS_2$ composite was prepared by a hydrothermal method followed by calcination at high temperature. Thionine and AuNPs were anchored onto the surface of the $Ti_3C_2T_x$-$MoS_2$ composite. Thionine served as the internal electrochemical signal source for the label-free sensor, whereas the AuNPs acted as the immobilisation matrix for the capture probes. The quantification of miRNA-21 was carried out by noting the decrease in the response of thionine, followed by the hybridisation of target miRNA-21 with the capture probes. The high surface area of MXene-$MoS_2$ composite amplified the hybridisation efficiency of miRNA-21 with the capture probe at the electrode surface. The synergistic effect that arose upon combining MXene-$MoS_2$ composite with the relatively mature technology of catalytic hairpin assembly enabled the determination of miRNA-21 over a wide dynamic range from 100 fM to 100 nM with a limit of detection of 26 fM. Kheyrabadi et al. developed an aptasensor for the hormone, thyroxine, based on a 3D gold nanostructure deposited on $MoS_2$@$Ti_3C_2T_x$ nanohybrid (3DGN/$MoS_2$@$Ti_3C_2T_x$/SPCE) modified screen-printed carbon electrode (Kashefi-Kheyrabadi et al. 2021). The thiol-terminated ferrocene labelled RNA aptamer was immobilised on 3DGN/$MoS_2$@$Ti_3C_2T_x$/SPCE by exploiting gold-thiol chemistry. The quantification of thyroxine was realised by noting the decrease in the electrochemical response of the redox label upon the binding of thyroxine to the aptamer. The sensor enabled a very low detection limit of 0.3 pg/mL over a linear range from 0.78 to 7.8 × $10^6$ pg/mL. An aptasensor for the transmembrane glycoprotein mucin 1 was developed by Xu et al. (2024). The mucin 1 aptamer and the complementary ssDNA were immobilised on MXene/$MoS_2$/PdPtNPs-based sensing platform. A manganese porphyrin complex intercalated into the dsDNA was used for signal amplification. The binding of mucin to its specific aptamer resulted in the enhanced current response from hydroquinone oxidation in the presence of $H_2O_2$ decomposition assisted by PdPtNPs and porphyrin complex. The proposed sensor enabled the detection of mucin 1 over a linear range from 1 pg/mL to 10 ng/mL with a detection limit of 0.36 pg/mL. Zhao et al. fabricated a carbon fibre paper-$Ti_3C_2T_x$ MXene-$MoS_2$ platform (CFP-MXene-$MoS_2$), which served as a flexible biosensing interface for the detection of multi-analytes in biofluids (Zhao et al. 2022b). The proposed sensor was employed for the determination of molecules such as ascorbic acid, uric acid, dopamine and miRNA. It is worth mentioning that the robust and disposable sensing platform allowed the ultrasensitive determination of miRNA with a detection limit of 3.16 aM.

Ankitha et al. used a carbon cloth modified with a hydrothermally synthesised $Nb_2CT_x$–$MoS_2$ composite for the ultrasensitive determination of dopamine (Ankitha et al. 2023). Under optimised experimental conditions, the determination of dopamine was possible in the range from 1.0 fM to 100.0 μM with a detection limit of 0.23 fM. Ankitha et al. synthesised a novel composite, $ReS_2$-$Nb_2CT_x$, using a hydrothermal method (Ankitha, Shabana, and Rasheed 2023). The composite was used to modify the carbon cloth electrode via drop casting, and the modified electrode was employed for the determination of the drug dipyridamole in human serum. The atomic ratio of $ReS_2$ in the composite was optimised, and under the optimised conditions, the sensor presented a linear range from 100 pM to 1 μM for the chronoamperometric determination of dipyridamole. The LOD obtained was 28 pM. Electrochemical determination of the stress hormone, serotonin, has been carried out by Richard et al., by using a single-stranded carbon yarn electrode modified with 1T $SnS_2$-$Nb_2CT_x$ composite (Richard et al. 2024). The enhanced electrochemical behaviour of serotonin at the composite-modified electrode is explained by the heterojunction formation due to the interfacial internal electric field (IIEF) formation. The linear range and LOD obtained for the sensor were 1–100 μM and 81 nM, respectively. Aiming towards point-of-care applications, a prototype for the sensor was developed on a syringe. The successful application of the sensor for the determination of serotonin in artificial cerebrospinal fluid, interstitial fluid, human serum and sweat samples validates the real-time applicability of the sensor.

Ranjith et al. synthesised MWCNT bridged $MoS_2$/$Ti_3C_2T_x$ composite via a multistep process and applied it for the quantification of the antibiotic chloramphenicol (Ranjith et al. 2024). The superior catalytic property exhibited by the composite, due to the synergistic effect arising from the individual properties of the composite, allowed excellent sensing of chloramphenicol by chronoamperometry at −0.6 V. The linear range of concentration obtained was from 8 to 152 nM with a limit of detection as low as 0.32 nM. In another promising work by Rajaji et al., a hydrothermally synthesised $MoS_2$ sphere combined with sulphur-doped $Ti_3C_2T_x$ nano catalysts deposited on a laser-induced graphene electrode was successfully applied for

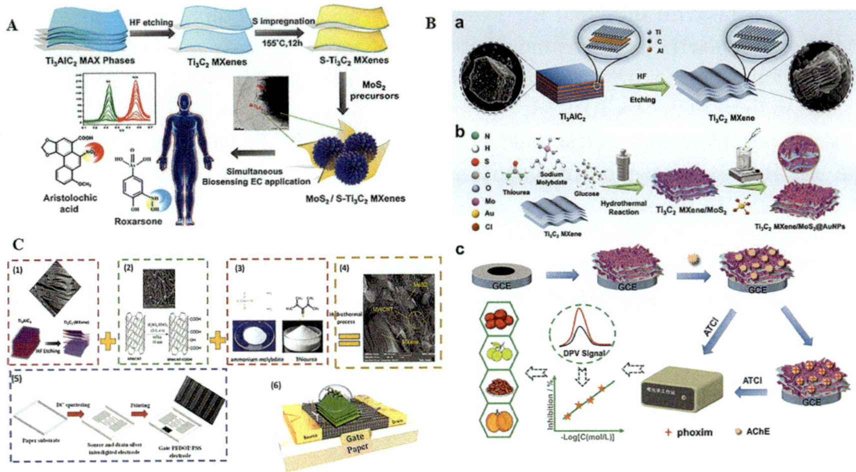

**FIGURE 9.3** Schematic representation of (A) synthesis and application of $Ti_3C_2T_x$/$MoS_2$ nanocatalyst for the sensing of aristolochic acid and roxarsone, (B) application of AChE-$Ti_3C_2T_x$/$MoS_2$@AuNPs for the determination of phoxim and (C) material synthesis process and fabrication of FET sensor for glucose. Reprinted with permission from Rajaji et al. (2022), Zhang et al. (2025), and Farahmandpour et al. (2025).

the electrochemical determination of hazardous molecules such as aristolochic acid and roxarsone (Figure 9.3(A)) (Rajaji et al. 2022). The sensor was able to detect aristolochic acid in the range from 0.01 to 875.1 μM, with a detection limit of 1.65 nM. The linear range obtained for roxarsone was the same as that of aristolochic acid with a detection limit of 2.31 nM. In a similar work reported by Vilian et al., Pd nanoparticles were loaded onto $Ti_3C_2T_x$-$MoS_2$ heterostructure for the quantification of the toxic organic compound bisphenol A in an environmental sample (Ezhil Vilian et al. 2023). The heterostructure-modified screen-printed electrode allowed the determination of bisphenol A over a wide linear concentration range from 5.0 to 175.0 nM with a detection limit of 0.12 nM. The practical applicability of the sensor was also verified in milk samples, river and lake water samples with satisfactory recovery values. In a report by Wang et al., Pt-doped $Ti_3C_2T_x$–$MoS_2$ composite was used to immobilise acetylcholine esterase (AChE) chitosan mixture on the glassy carbon electrode surface (Wang et al. 2024a). The modified electrode was used for the determination of the organic pesticide, chlorpyrifos. Acetylcholine is hydrolysed by the action of AChE to electroactive thiocholine, which can be detected by DPV measurement. In the presence of chlorpyrifos, the formation of thiocholine is prevented by the inhibition of the action of AChE, due to the formation of phosphorylated AChE upon binding with chlorpyrifos. The sensor exhibited analytical performance over the range of concentration from 1.0 pM to 1.0 μM with a detection limit of 0.47 pM. In a similar work done by Zhang et al., the $Ti_3C_2T_x$/$MoS_2$@AuNPs heterostructure was used to immobilise AChE for the sensitive determination of the organothiophosphate insecticide, phoxim (Figure 9.3(B)) (Zhang et al. 2025). The presence of phoxim inhibited the hydrolysis of acetylthiocholine by AChE, which led to a decreased electrochemical response of the hydrolysis product thiocholine. The response of phoxim was linear in the range from $1 \times 10^{-13}$ to $1 \times 10^{-7}$ M with a detection limit of $5.29 \times 10^{-15}$ M.

A disposable and wireless field effect transistor (FET) based paper sensor for the non-invasive detection of glucose in saliva was developed by Farahmandpour et al. (2025). In the sensor set-up, silver served as the source and drain electrodes, PEDOT-PSS as the gain electrode and the 2D material composite of MWCNT/$Ti_3C_2T_x$/$MoS_2$ as the FET channel (Figure 9.3(C)). The sensor was further developed into a fully integrated wireless device, using which glucose monitoring can be done

**TABLE 9.1**
**Comparison of recent reports of MXene-TMD heterostructure-based electrochemical sensors**

| Material | Target analyte | Range of determination | LOD | Ref. |
|---|---|---|---|---|
| $Ti_3C_2T_x$-$MoS_2$-Thi-AuNPs | miRNA-21 | 100 fM to 100 nM | 26 fM | Zhao et al. (2022a) |
| CFP-$Ti_3C_2T_x$-$MoS_2$ | Ascorbic acid | 10–100 μM | 0.89 μM | Zhao et al. (2022b) |
| | Uric acid | 0.5–200 μM | 0.35 μM | |
| | Dopamine | 0.5–150 μM | 0.23 μM | |
| | miRNA | 100 aM to 10 nm | 3.16 aM | |
| $Nb_2CT_x$–$MoS_2$ | Dopamine | 1.0 fM to 100 μM | 0.23 fM | Ankitha et al. (2023) |
| MWCNT/$Ti_3C_2T_x$/$MoS_2$ | Glucose | 0.01 μM to 1.3 mM | 500 pM | Farahmandpour et al. (2025) |
| Aptamer/3DGN/$MoS_2$@$Ti_3C_2T_x$ | Thyroxine | $0.78–7.8 \times 10^6$ pg/mL | 0.30 pg/mL | Kashefi-Kheyrabadi et al. (2021) |
| MXene/$MoS_2$/PdPtNPs | Mucin 1 | 1 pg/mL to 10 ng/mL | 0.36 pg/mL | Xu et al. (2024) |
| $ReS_2$-$Nb_2CT_x$ | Dipyridamole | 100 pM to 1 μM | 28 pM | Ankitha, Shabana, and Rasheed (2023) |
| 1T $SnS_2$-Nb2$CT_x$ | Serotonin | 1–100 μM | 81 nM | Richard et al. (2024) |
| $MoS_2$/S-$Ti_3C_2T_x$ | Aristolochic acid | –875.10 μM | 1.65 nM | Rajaji et al. (2022) |
| | Roxarsone | 0.01–875.10 μM | 2.31 nM | |
| Pd-$MoS_2$-S-$Ti_3C_2T_x$ | Bisphenol A | 5.0–175.0 nM | 0.12 nM | Ezhil Vilian et al. (2023) |
| AChE-chit/Pt-$MoS_2$-$Ti_3C_2T_x$ | Chlorpyrifos | 1.0 pM to 1.0 μM | 0.47 pM | Wang et al. (2024a) |
| AchE-$Ti_3C_2T_x$/$MoS_2$@AuNPs | Phoxim | $1 \times 10^{-13}$ to $1 \times 10^{-7}$ M | $5.29 \times 10^{-15}$ M | Zhang et al. (2025) |
| MWCNT/$Ti_3C_2T_x$/$MoS_2$ | Chloramphenicol | 8.0–152.0 nM | 0.32 nM | Ranjith et al. (2024) |

via a smartphone. The linear range obtained for glucose using the FET device was from 0.01 μM to 1.3 mM, and the detection limit was 500 pM. A comparison of the recent reports on MXene-TMD heterostructure-based electrochemical sensors is depicted in Table 9.1.

### 9.2.3 Application of MXene in Combination with Other Metal Chalcogenides

While TMDs have been most extensively employed in combination with MXenes for electrochemical sensing applications, other metal chalcogenides have also been reported to make significant contributions in this domain (Kotteeswaran et al. 2025). Metal sulphides, selenides and tellurides with different morphologies such as nanoparticles, nanorods, quantum dots and hierarchical structures exhibit unique electronic and catalytic properties, which can complement the higher conductivity, surface area and versatile functionalisation in MXenes (Fard et al. 2024). Although significant research on these metal chalcogenide–MXene combinations has been explored in electrochemical energy storage and conversion applications, recent studies on their sensing application have proven that by leveraging the interfacial interaction between MXene sheets and metal chalcogenide heterostructures, facile charge transfer processes can be achieved, which are critical in the sensing performance (Table 9.2) (Fard et al. 2024).

Arif et al. synthesised $Nb_2CT_x$/ZnS nanocomposite for their application in the selective determination of dopamine. Under optimised analytical conditions, the composite-modified glassy carbon electrode enabled the determination of dopamine in the range from 0.09 to 0.82 mM with a limit of detection of 1.39 μM (Arif et al. 2021). In 2022, Mi et al. synthesised a nanorod/nanoplate/nanosheet hybrid structure consisting of $CuCo_2S_4$/CuS/$Ti_3C_2T_x$ ternary structure via a one-step, low-temperature solvothermal method (Mi et al. 2022). The well-constructed 1D/2D/2D

**TABLE 9.2**
**Comparison of reports of MXene–metal chalcogenide composite-based electrochemical sensors**

| Material | Target analyte | Range of determination | LOD | Ref. |
|---|---|---|---|---|
| $Nb_2CT_x$/ZnS | Dopamine | 0.09–0.82 mM | 1.39 μM | Arif et al. (2021) |
| $Ag_2Se@V_2CT_x$ | L-Fucose | 2.5–25 μM, 25–500 μM | 1.85 μM | Panneer Selvam and Cho (2023) |
| $CuCo_2S_4$/CuS/$Ti_3C_2T_x$ | Chlorpyrifos | $2.85 \times 10^{-12}$ to $2.85 \times 10^{-6}$ M | 0.10 pM | Mi et al. (2022) |
| NMPG/CuS/$Ti_3C_2T_x$ | Nalbuphine | 5–150 μM | 1.60 μM | Hossein Mashhadizadeh, Ghalkhani, and Sohouli (2024) |
| CuZnS/$Ti_3C_2T_x$ | Chloramphenicol | 0.002–0.02 μM | 0.18 nM | Kotteeswaran et al. (2025) |
| $AgBiS_2$/$Ti_3C_2T_x$ | 4-Nitrophenol | 0.02–1869.0 μM | 0.0025 μM | Gopi et al. (2024) |

heterostructure was used as the AChE biosensor for the determination of chlorpyrifos. A wide detection range from 2.85 pM to 2.85 μM was obtained with a detection limit as low as 0.1 pM.

Masshadizadeh et al. synthesised a ternary composite of flower-like CuS structures, N-doped porous graphene and $Ti_3C_2T_x$ MXene by following a multistep process (Hossein Mashhadizadeh, Ghalkhani, and Sohouli 2024). The composite incorporated screen-printed carbon electrode was used for the determination of a semi-synthetic opioid, nalbuphine (Figure 9.4(A)). The synergistic effect arising from the surface area and high conductivity of the flower-like CuS structure and the surface functional groups of MXene enabled the electrooxidation of nalbuphine at a lower peak potential and with higher peak current. The sensor reported a wide linear range from 5 to 150 μM with a limit of detection of 1.60 μM. Hydrothermally synthesised CuZnS nanospheres and $Ti_3C_2T_x$ MXene composite were used by Kotteeswaran et al. for the picomolar detection of the antibiotic, chloramphenicol (Kotteeswaran et al. 2025). The composite was characterised using field emission scanning electron

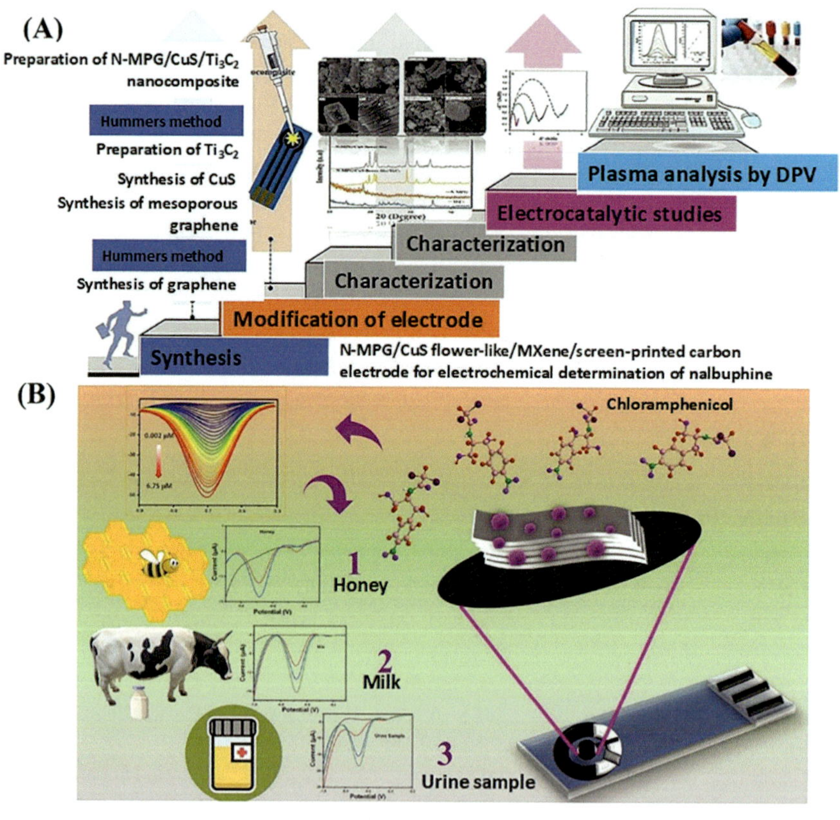


FIGURE 9.4 Schematic representation of (A) application of NMPG/CuS/$Ti_3C_2T_x$ composite for the sensing of nalbuphine and (B) application of CuZnS/$Ti_3C_2T_x$-modified screen-printed electrode for the determination of chloramphenicol in food and biological samples. Reprinted with permission from Mashhadizadeh et al. (2024) and Kotteeswaran et al. (2025).

microscopy (FE-SEM), X-ray diffraction (XRD), X-ray photoelectron spectroscopy (XPS) and electrochemical techniques. The excellent electrocatalytic efficiency of the composite for the electro-reduction of chloramphenicol allowed its determination using DPV in the concentration range from 0.002 to 0.02 μM with a detection limit of 0.18 nM (Figure 9.4(B)). Accuracy in the performance of the sensor was ascertained by carrying out the determination using the amperometric technique, and the detection limit obtained (0.15 nM) was comparable to that obtained from DPV. The real-time applicability of the developed sensor was also studied by spike recovery analysis in samples of urine, milk, honey and eye drops. Relatively good recovery values and relative standard deviation (RSD) values obtained suggested the practical utility of the sensor.

Gopi et al. synthesised an amalgam of silver bismuth sulphide ($AgBiS_2$) and $Ti_3C_2T_x$ MXene through a simple stirring process (Gopi et al. 2024). The composite-modified GCE was used for the selective determination of the environmental pollutant, 4-nitrophenol. Under optimised experimental conditions, the sensor exhibited a wide linear range from 0.02 to 1869.0 μM with a detection limit of 0.0025 μM. The practical applicability of the sensor was also successfully verified in tap water samples.

Selvam et al. reported a molecularly imprinted polymer-based electrochemical sensor to determine the early cancer biomarker L-fucose (Panneer Selvam and Cho 2023). p-Aminobenzoic acid was selected as the monomer via theoretical optimisation. The composite of $Ag_2Se$ and $V_2CT_x$ MXene ($Ag_2Se@V_2CT_x$) deposited on the screen-printed electrode surface served to amplify the signal response. A ratiometric sensing strategy was adopted based on the decrease in oxidation current of the $Ag_2Se@V_2CT_x$ composite with an increase in the current response of L-fucose upon its binding to the cavities. The ratiometric strategy, together with the inherent selectivity of the imprinted polymer, provided extraordinary selectivity with a low detection limit of 1.85 μM. In addition, the practical applicability of the sensor was also analysed in urine and serum with good recovery values.

## 9.3 CONCLUSION AND FUTURE PERSPECTIVES

MXene heterostructures have emerged as bright materials in electrochemical sensing applications. Among them, MXene–metal chalcogenide hybrids leverage their high conductivity, tuneable surface chemistry and hydrophilicity along with their unique catalytic activity and semiconducting properties, resulting in enhanced electrochemical performance. In recent years, a significant amount of research has been carried out in engineering MXene–metal chalcogenide hybrids for their application in diverse areas of electrochemical sensing, including the detection and quantification of environmental pollutants, biomolecules and pharmaceuticals. Tuning the properties of these heterostructures by adopting various synthesis and modification procedures has been highly influential in optimising their electrochemical performance. Doping with metal atoms, intercalation and hybridisation with other nanomaterials were some of the widely used strategies to advance the selectivity and sensitivity of the sensors. Furthermore, their high surface area and versatile surface functionalisation allowed facile immobilisation of biorecognition elements such as aptamers and DNAs onto these heterostructures, which boosted their application in electrochemical biosensors. The successful application of these sensors in real samples and advancements in flexible and disposable electrochemical sensors have given the way for their application in point-of-care monitoring. Among the numerous possible combinations of MXene and metal chalcogenides, $Ti_3C_2T_x$-$MoS_2$ heterostructure-based electrodes are the most widely explored in the field of electrochemical sensing owing to the unparalleled properties arising from the 2D/2D heterostructures.

Despite the numerous achievements, major hurdles that need to be addressed before the application of MXene heterostructures in commercial utilisation of the sensors include their reproducibility and long-term stability, as degradation in ambient and aqueous conditions may adversely affect their sensing performance. Exploration of MXenes beyond Ti MXenes and the development of heterostructures with other emerging metal chalcogenides may uncover novel sensing mechanisms with improved stability, selectivity and practical applicability. In summary, MXene–metal chalcogenide hybrids hold immense potential for the advancement of electrochemical sensing technologies. Although challenges remain in terms of practical applicability, ongoing research in related interdisciplinary areas is expected to give breakthroughs leading to their application in clinical, environmental and industrial fields.

## REFERENCES

Al-Temimy, Ameer, Babak Anasori, Katherine A. Mazzio, Florian Kronast, Mykola Seredych, Narendra Kurra, Mohamad-Assaad Mawass, Simone Raoux, Yury Gogotsi, and Tristan Petit. 2020. "Enhancement of $Ti_3C_2$ MXene pseudocapacitance after urea intercalation studied by soft X-ray absorption spectroscopy." *The Journal of Physical Chemistry C* 124 (9):5079–5086. doi: 10.1021/acs.jpcc.9b11766.

Ankitha, Menon, Neermunda Shabana, Ajith M. Arjun, Punnoli Muhsin, and Pathath A. Rasheed. 2023. "Ultrasensitive electrochemical detection of dopamine from human serum samples by $Nb_2CT_x$-$MoS_2$ hetero structures." *Microchemical Journal* 187:108424. doi: 10.1016/j.microc.2023.108424.

Ankitha, Menon, Neermunda Shabana, and Pathath. A. Rasheed. 2023. "A novel $ReS_2$–$Nb_2CT_x$ composite as a sensing platform for ultrasensitive and selective electrochemical detection of dipyramidole from human serum." *Graphene and 2D Materials* 8 (1):27–37. doi: 10.1007/s41127-022-00055-x.

Arif, Nimrah, Sundus Gul, Manzar Sohail, Syed Rizwan, and Mudassir Iqbal. 2021. "Synthesis and characterization of layered $Nb_2C$ MXene/ZnS nanocomposites for highly selective electrochemical sensing of dopamine." *Ceramics International* 47 (2):2388–2396. doi: 10.1016/j.ceramint.2020.09.081.

Bandyopadhyay, Arkamita, Dibyajyoti Ghosh, and Swapan K. Pati. 2018. "Effects of point defects on the magnetoelectronic structures of MXenes from first principles." *Physical Chemistry Chemical Physics* 20 (6):4012–4019. doi: 10.1039/C7CP07165E.

Chauhan, Himani, and Sasanka Deka. 2021. "Chapter 6 - Supercapacitors based on two-dimensional transition metal dichalcogenides and their hybrids." In *Fundamentals and Supercapacitor Applications of 2D Materials*, edited by Chandra S. Rout and Dattatray J. Late, 159–191. Elsevier.

Chen, Yuxian, Haoyu Xiao, Qiaolin Fan, Weilong Tu, Shiqi Zhang, Xiao Li, and Tao Hu. 2024. "Fully integrated biosensing system for dynamic monitoring of sweat glucose and real-time pH adjustment based on 3D graphene MXene aerogel." *ACS Applied Materials & Interfaces* 16 (41):55155–55165. doi: 10.1021/acsami.4c13013.

Chia, Hui L., Carmen C. Mayorga-Martinez, Nikolas Antonatos, Zdeněk Sofer, Jesus J. Gonzalez-Julian, Richard D. Webster, and Martin Pumera. 2020. "MXene titanium carbide-based biosensor: Strong dependence of exfoliation method on performance." *Analytical Chemistry* 92 (3):2452–2459. doi: 10.1021/acs.analchem.9b03634.

Deysher, Grayson, Christopher E. Shuck, Kanit Hantanasirisakul, Nathan C. Frey, Alexandre C. Foucher, Kathleen Maleski, Asia Sarycheva, Vivek B. Shenoy, Eric A. Stach, Babak Anasori, and Yury Gogotsi. 2020. "Synthesis of $Mo_4VAlC_4$ MAX phase and two-dimensional $Mo_{4VC4}$ MXene with five atomic layers of transition metals." *ACS Nano* 14 (1):204–217. doi: 10.1021/acsnano.9b07708.

Dmytriv, Tetiana R., and Volodymyr I. Lushchak. 2024. "Potential biosafety of Mxenes: Stability, biodegradability, toxicity and biocompatibility." *The Chemical Record* 24 (4):e202300338. doi: 10.1002/tcr.202300338.

Ezhil Vilian, A. T., Jitendra N. Tiwari, Munirah Alhammadi, Gokul Bhaskaran, Seung-Kyu Hwang, Suheon Kim, Krishan Kumar, Annamalai S. Kumar, Yun S. Huh, and Young-Kyu Han. 2023. "Palladium nanoparticles anchored $MoS_2$-MXene composite modified electrode for rapid sensing of toxic bisphenol A in aqueous media." *Chemical Engineering Journal* 469:144017. doi: 10.1016/j.cej.2023.144017.

Ezzah Ab Latif, Farah, Arshid Numan, Nabisab M. Mubarak, Mohammad Khalid, Ezzat C. Abdullah, Norhuda A. Manaf, and Rashmi Walvekar. 2022. "Evolution of MXene and its 2D heterostructure in electrochemical sensor applications." *Coordination Chemistry Reviews* 471:214755. doi: 10.1016/j.ccr.2022.214755.

Farahmandpour, Milad, Daryoosh Dideban, Masoomeh M. Dehbali, and Zoheir Kordrostami. 2025. "Disposable wireless FET paper sensor based on MWCNT/MXene/$MoS_2$ composite for non-invasive glucose detection in saliva." *Journal of Alloys and Compounds* 1014:178715. doi: 10.1016/j.jallcom.2025.178715.

Fard, Narges E., Ilnaz Raeisi, Zivar Yousefipour, and Rashid Mosavin. 2024. "Review of metal sulfide-based MXene nanocomposites for environmental applications, gas sensing, energy storage, and photothermal therapy." *ACS Applied Nano Materials* 7 (6):5698–5728. doi: 10.1021/acsanm.3c05455.

Gayathri, Vijayakumar G., Bartholomew Richard, Jithin T. Chacko, Jagadeesh Bayry, and Pathath A. Rasheed. 2025. "Non-Ti MXenes: New biocompatible and biodegradable candidates for biomedical applications." *Journal of Materials Chemistry B* 13 (4):1212–1228. doi: 10.1039/D4TB01904K.

Gogotsi, Yury. 2023. "The future of MXenes." *Chemistry of Materials* 35 (21):8767–8770. doi: 10.1021/acs.chemmater.3c02491.

Gopi, Praveen K., C. G. Sanjayan, S. Akhil, Chandan H. Ravikumar, Siripong Thitamadee, Supornchai Kongpatanakul, R. G. Balakrishna, and Werasak Surareungchai. 2024. "Silver bismuth sulphide ($AgBiS_2$)-MXene composite as high-performance electrochemical sensing platform for sensitive detection of pollutant 4-nitrophenol." *Electrochimica Acta* 498:144616. doi: 10.1016/j.electacta.2024.144616.

Gu, Hui, Yidan Xing, Ping Xiong, Huiling Tang, Chenchen Li, Shu Chen, Rongjin Zeng, Kai Han, and Guoyue Shi. 2019. "Three-dimensional porous $Ti_3C_2T_x$ MXene–graphene hybrid films for glucose biosensing." *ACS Applied Nano Materials* 2 (10):6537–6545. doi: 10.1021/acsanm.9b01465.

Hernández-Rodríguez, Juan F., Daniel Rojas, and Alberto Escarpa. 2021. "Electrochemical Sensing directions for next-generation healthcare: Trends, challenges, and frontiers." *Analytical Chemistry* 93 (1):167–183. doi: 10.1021/acs.analchem.0c04378.

Hossein Mashhadizadeh, Mohammad, Masoumeh Ghalkhani, and Esmail Sohouli. 2024. "Synthesis and characterization of N-MPG/CuS flower-like/MXene to modify a screen-printed carbon electrode for electrochemical determination of nalbuphine." *Journal of Electroanalytical Chemistry* 957:118130. doi: 10.1016/j.jelechem.2024.118130.

Jin, Jun, Tuo Xiao, You-fang Zhang, Han Zheng, Huanwen Wang, Rui Wang, Yansheng Gong, Beibei He, Xianhu Liu, and Kun Zhou. 2021. "Hierarchical MXene/transition metal chalcogenide heterostructures for electrochemical energy storage and conversion." *Nanoscale* 13 (47):19740–19770. doi: 10.1039/D1NR05799E.

Kallingal, Nithusha, Muni R. Maurya, M. S. Sajna, Huseyin C. Yalcin, Hassen M. Ouakad, Issam Bahadur, Somaya Al-Maadeed, and Kishor K. Sadasivuni. 2022. "A highly sensitive wearable pressure sensor capsule based on PVA/Mxene composite gel." *3 Biotech* 12 (8):171. doi: 10.1007/s13205-022-03221-3.

Kashefi-Kheyrabadi, Leila, Aneesh Koyappayil, Taeeun Kim, Yong-Pil Cheon, and Min-Ho Lee. 2021. "A $MoS_2$@$Ti_3C_2T_x$ MXene hybrid-based electrochemical aptasensor (MEA) for sensitive and rapid detection of thyroxine." *Bioelectrochemistry* 137:107674. doi: 10.1016/j.bioelechem.2020.107674.

Khan, Karim, Ayesha K. Tareen, Waqas Ahmad, Iftikhar Hussain, Mujeeb U. Chaudhry, Asif Mahmood, Muhammad F. Khan, Han Zhang, and Zhongjian Xie. 2024. "Recent advances in non-Ti MXenes: Synthesis, properties, and novel applications." *Advanced Science* 11 (36):2303998. doi: 10.1002/advs.202303998.

Kim, Soobin, Juyun Lee, Sehyun Doo, Yun C. Kang, Chong M. Koo, and Seon J. Kim. 2021. "Metal-ion-intercalated MXene nanosheet films for $NH_3$ gas detection." *ACS Applied Nano Materials* 4 (12):14249–14257. doi: 10.1021/acsanm.1c03814.

Kotteeswaran, Sindhuja, Sarawut Kondee, Weeraphat Pon-On, Wahidah H. Al-Qahtani, Chatchawal Wongchoosuk, and Mani Govindasamy. 2025. "CuZnS microflowers anchored on MXene nanosheets based electrochemical sensor for pico-molar determination of chloramphenicol in biological samples and dairy products." *Microchemical Journal* 212:113291. doi: 10.1016/j.microc.2025.113291.

Lee, Jacob T., Brian C. Wyatt, Gregory A. Davis, Jr., Adrianna N. Masterson, Amber L. Pagan, Archit Shah, Babak Anasori, and Rajesh Sardar. 2021. "Covalent surface modification of $Ti_3C_2T_x$ MXene with chemically active polymeric ligands producing highly conductive and ordered microstructure films." *ACS Nano* 15 (12):19600–19612. doi: 10.1021/acsnano.1c06670.

Li, Menghui, Liang Fang, Hua Zhou, Fang Wu, Yi Lu, Haijun Luo, Yuxin Zhang, and Baoshan Hu. 2019. "Three-dimensional porous MXene/NiCo-LDH composite for high performance non-enzymatic glucose sensor." *Applied Surface Science* 495:143554. doi: 10.1016/j.apsusc.2019.143554.

Li, Yan, Han Zhang, Guo Qu, Lingying Xie, Sisi Tang, Hang Lei, Yuxia Zhong, and Yue-Fei Zhang. 2024. "Efficient removal of antibiotics from wastewater by chitosan/polyethyleneimine/$Ti_3C_2$ MXene composite hydrogels: Synthesis, adsorption, kinetics and mechanisms." *Colloids and Surfaces A: Physicochemical and Engineering Aspects* 702:135111. doi: 10.1016/j.colsurfa.2024.135111.

Lim, Kang R. G., Albertus D. Handoko, Srinivasa K. Nemani, Brian Wyatt, Hai-Ying Jiang, Junwang Tang, Babak Anasori, and Zhi W. Seh. 2020. "Rational design of two-dimensional transition metal carbide/nitride (MXene) hybrids and nanocomposites for catalytic energy storage and conversion." *ACS Nano* 14 (9):10834–10864. doi: 10.1021/acsnano.0c05482.

Liu, Hui, Congyue Duan, Chenhui Yang, Wanqiu Shen, Fen Wang, and Zhenfeng Zhu. 2015. "A novel nitrite biosensor based on the direct electrochemistry of hemoglobin immobilized on MXene-$Ti_3C_2$." *Sensors and Actuators B: Chemical* 218:60–66. doi: 10.1016/j.snb.2015.04.090.

Liu, Miao, Zeyu Wang, Peng Song, Zhongxi Yang, and Qi Wang. 2021. "Flexible MXene/rGO/CuO hybrid aerogels for high performance acetone sensing at room temperature." *Sensors and Actuators B: Chemical* 340:129946. doi: 10.1016/j.snb.2021.129946.

Lorencova, Lenka, Tomas Bertok, Erika Dosekova, Alena Holazova, Darina Paprckova, Alica Vikartovska, Vlasta Sasinkova, Jaroslav Filip, Peter Kasak, Monika Jerigova, Dusan Velic, Khaled A. Mahmoud, and Jan Tkac. 2017. "Electrochemical performance of $Ti_3C_2T_x$ MXene in aqueous media: Towards ultrasensitive $H_2O_2$ sensing." *Electrochimica Acta* 235:471–479. doi: 10.1016/j.electacta.2017.03.073.

Lv, Ruitao, Joshua A. Robinson, Raymond E. Schaak, Du Sun, Yifan Sun, Thomas E. Mallouk, and Mauricio Terrones. 2015. "Transition metal dichalcogenides and beyond: Synthesis, properties, and applications of single- and few-layer nanosheets." *Accounts of Chemical Research* 48 (1):56–64. doi: 10.1021/ar5002846.

Mi, Yuping, Yisong Zhao, Jianmin Chen, Xiaolu Li, Yunxia Yang, and Faming Gao. 2022. "Ternary heterostructures of 1D/2D/2D $CuCo_2S_4$/CuS/$Ti_3C_2$ MXene: Boosted amperometric sensing for chlorpyrifos." *Journal of Hazardous Materials* 438:129419. doi: 10.1016/j.jhazmat.2022.129419.

Mozafari, Mohammad, and Masoud Soroush. 2021. "Surface functionalization of MXenes." *Materials Advances* 2 (22):7277–7307. doi: 10.1039/D1MA00625H.

Orangi, J., H. Tetik, P. Parandoush, E. Kayali, D. Lin, and M. Beidaghi. 2021. "Conductive and highly compressible MXene aerogels with ordered microstructures as high-capacity electrodes for Li-ion capacitors." *Materials Today Advances* 9:100135. doi: 10.1016/j.mtadv.2021.100135.

Panneer Selvam, Sathish, and Sungbo Cho. 2023. "Silver chalcogenide loaded $V_2CT_x$ MXene-molecularly imprinted polymer-based novel ratiometric sensor for the early predictive cancer marker: L-Fucose." *Chemical Engineering Journal* 469:144016. doi: 10.1016/j.cej.2023.144016.

Pei, Yangyang, Xiaoli Zhang, Zengyu Hui, Jinyuan Zhou, Xiao Huang, Gengzhi Sun, and Wei Huang. 2021. "$Ti_3C_2T_X$ MXene for sensing applications: Recent progress, design principles, and future perspectives." *ACS Nano* 15 (3):3996–4017. doi: 10.1021/acsnano.1c00248.

Peng, Jiahe, Xingzhu Chen, Wee-Jun Ong, Xiujian Zhao, and Neng Li. 2019. "Surface and heterointerface engineering of 2D MXenes and their nanocomposites: Insights into electro- and photocatalysis." *Chem* 5 (1):18–50. doi: 10.1016/j.chempr.2018.08.037.

Rajaji, Umamaheswari, Pattan-Siddappa Ganesh, Sang-Youn Kim, Mani Govindasamy, Razan A. Alshgari, and Ting-Yu Liu. 2022. "$MoS_2$ sphere/2D S-$Ti_3C_2$ MXene nanocatalysts on laser-induced graphene electrodes for hazardous aristolochic acid and roxarsone electrochemical detection." *ACS Applied Nano Materials* 5 (3):3252–3264. doi: 10.1021/acsanm.1c03680.

Ranjith, Kugalur S., Ali Mohammadi, A. T. Ezhil Vilian, Soobin Han, Yun S. Huh, and Young-Kyu Han. 2024. "Synergistic effects of CNT-bridged dual-phase $MoS_2$ on MXene as a ternary hybrid electrode for rapid sensing of chloramphenicol in aqueous media." *Chemical Engineering Journal* 500:157487. doi: 10.1016/j.cej.2024.157487.

Raza, Saleem, Asif Hayat, Tariq Bashir, Cheng Chen, Liguo Shen, Yasin Orooji, and Hongjun Lin. 2024. "Electrochemistry of 2D-materials for the remediation of environmental pollutants and alternative energy storage/conversion materials and devices, a comprehensive review." *Sustainable Materials and Technologies* 40:e00963. doi: 10.1016/j.susmat.2024.e00963.

Rhouati, Amina, Mohammed Berkani, Yasser Vasseghian, and Nasrin Golzadeh. 2022. "MXene-based electrochemical sensors for detection of environmental pollutants: A comprehensive review." *Chemosphere* 291:132921. doi: 10.1016/j.chemosphere.2021.132921.

Richard, Bartholomew, Kannankuzhiyan Niyas, Manjusha Mathew, and Pathath A. Rasheed. 2024. "Lab on a syringe: Strategic design of 1T $SnS_2$/MXene embedded single-stranded carbon yarn for real-time stress biomarker detection." *ACS Applied Electronic Materials* 6 (9):6905–6915. doi: 10.1021/acsaelm.4c01265.

Shahzad, Faisal, Aamir Iqbal, Shabi A. Zaidi, Suk-Won Hwang, and Chong M. Koo. 2019. "Nafion-stabilized two-dimensional transition metal carbide ($Ti_3C_2T_x$ MXene) as a high-performance electrochemical sensor for neurotransmitter." *Journal of Industrial and Engineering Chemistry* 79:338–344. doi: 10.1016/j.jiec.2019.03.061.

Sheikh, Zulfqar A., Pranav K. Katkar, Honggyun Kim, Shania Rehman, Karim Khan, Vijay D. Chavan, Rajan Jose, Muhammad F. Khan, and Deok-kee Kim. 2023. "Transition metal chalcogenides, MXene, and their hybrids: An emerging electrochemical capacitor electrodes." *Journal of Energy Storage* 71:107997. doi: 10.1016/j.est.2023.107997.

Tajik, Somayeh, Zahra Dourandish, Fariba G. Nejad, Hadi Beitollahi, Peyman M. Jahani, and Antonio Di Bartolomeo. 2022. "Transition metal dichalcogenides: Synthesis and use in the development of electrochemical sensors and biosensors." *Biosensors and Bioelectronics* 216:114674. doi: 10.1016/j.bios.2022.114674.

Tang, Yi, Chenhui Yang, Xingtao Xu, Yunqing Kang, Joel Henzie, Wenxiu Que, and Yusuke Yamauchi. 2022. "MXene nanoarchitectonics: Defect-engineered 2D MXenes towards enhanced electrochemical water splitting." *Advanced Energy Materials* 12 (12):2103867. doi: 10.1002/aenm.202103867.

Thurakkal, Shameel, and Xiaoyan Zhang. 2022. "Noncovalent functionalization of $Ti_3C_2T_X$ using cationic porphyrins with enhanced stability against oxidation." *Materials Chemistry Frontiers* 6 (5):561–569. doi: 10.1039/D1QM01326B.

Umapathi, Reddicherla, Seyed M. Ghoreishian, Sonam Sonwal, Gokana M. Rani, and Yun S. Huh. 2022. "Portable electrochemical sensing methodologies for on-site detection of pesticide residues in fruits and vegetables." *Coordination Chemistry Reviews* 453: 214305. doi: 10.1016/j.ccr.2021.214305.

Wang, Fei, Yao Zhu, Long Qian, Yuhao Yin, Ziyu Yuan, Yuting Dai, Tao Zhang, Dongya Yang, and Fengxian Qiu. 2024a. "Lamellar $Ti_3C_2$ MXene composite decorated with platinum-doped $MoS_2$ nanosheets as electrochemical sensing functional platform for highly sensitive analysis of organophosphorus pesticides." *Food Chemistry* 459:140379. doi: 10.1016/j.foodchem.2024.140379.

Wang, Jing, Qingbin Xu, Yang Yang, Jinxin Liu, Weijun Kong, and Linchun Shi. 2024b. "An electrochemical sensor founded on heterogeneous MXene & MOF composite for tanshinol sensing." *Talanta* 268:125344. doi: 10.1016/j.talanta.2023.125344.

Wang, Xue, Mengjia Li, Shuang Yang, Xuting Bai, and Jiajia Shan. 2022. "Self-assembled $Ti_3C_2T_X$ MXene/graphene composite for the electrochemical reduction and detection of p-nitrophenol." *Microchemical Journal* 179:107473. doi: 10.1016/j.microc.2022.107473.

Xu, Xiaohansi, Na Li, Jiangjian Huang, Tong Zhu, Xin Su, Yi Ma, Renxiang Yang, Jia Ruan, and Huilan Su. 2024. "The novel few layered MXene/$MoS_2$/PdPtNPs nanomaterials combined with hybrid chain reaction for electrochemical detection of mucin 1." *Microchemical Journal* 197:109827. doi: 10.1016/j.microc.2023.109827.

Zhang, Jiani, Xiaohui Guo, Jian Zhang, Xiaobing Guo, Yuying Xu, and Lijuan Chen. 2025. "$Ti_3C_2$ MXene/$MoS_2$@AuNPs ternary nanocomposite for highly sensitive electrochemical detection of phoxim residues in fruits." *Food Chemistry* 462:140939. doi: 10.1016/j.foodchem.2024.140939.

Zhang, Yi-Zhou, Jehad K. El-Demellawi, Qiu Jiang, Gang Ge, Hanfeng Liang, Kanghyuck Lee, Xiaochen Dong, and Husam N. Alshareef. 2020. "MXene hydrogels: Fundamentals and applications." *Chemical Society Reviews* 49 (20):7229–7251. doi: 10.1039/D0CS00022A.

Zhao, Jiaying, Congjuan He, Weixuan Wu, Huisi Yang, Jiangbo Dong, Li Wen, Zhikun Hu, Mei Yang, Changjun Hou, and Danqun Huo. 2022a. "MXene-$MoS_2$ heterostructure collaborated with catalyzed hairpin assembly for label-free electrochemical detection of microRNA-21." *Talanta* 237:122927. doi: 10.1016/j.talanta.2021.122927.

Zhao, Jiaying, Congjuan He, Weixuan Wu, Huisi Yang, Lan Peng, Li Wen, Zhikun Hu, Changjun Hou, and Danqun Huo. 2022b. "MXene-$MoS_2$ carbon-fiber-based flexible electrochemical interface for multiple bioanalysis in biofluids." *Chemical Engineering Journal* 446:136841. doi: 10.1016/j.cej.2022.136841.

Zhao, Li, Biliang Dong, Shaozhou Li, Lijun Zhou, Linfei Lai, Zhiwei Wang, Shulin Zhao, Min Han, Kai Gao, Min Lu, Xiaoji Xie, Bo Chen, Zhengdong Liu, Xiangjing Wang, Hao Zhang, Hai Li, Juqing Liu, Hua Zhang, Xiao Huang, and Wei Huang. 2017. "Interdiffusion reaction-assisted hybridization of two-dimensional metal–organic frameworks and $Ti_3C_2T_x$ nanosheets for electrocatalytic oxygen evolution." *ACS Nano* 11 (6): 5800–5807. doi: 10.1021/acsnano.7b01409.

# 10 MXene–Carbon-Modified Electrochemical Sensors

*Jerome Rajendran, Shingirirai Chakoma, and Rahim Esfandyarpour*

## 10.1 INTRODUCTION

Since the isolation of graphene in 2004 (Novoselov et al. 2004), the exploration of two-dimensional (2D) nanomaterials has surged across multiple disciplines due to their exceptional properties. These materials are pivotal in sectors ranging from photovoltaics (Rogdakis, Karakostas, and Kymakis 2021) to electrochemical energy production (Yi et al. 2022), sensors (Hu et al. 2022), catalysts (Nagalingam and Grace 2022), and beyond. In 2011, Barsoum et al. introduced MXenes, a new class of 2D materials composed of transition metal carbides, nitrides, and carbonitrides, sharing a structural resemblance with graphene (Naguib et al. 2011). These materials are typically represented by the formula $M_{n+1}X_nT_x$, where n ranges from 1 to 4, M denotes an early transition metal (e.g., Ti, V, Nb, Mo), X is either carbon or nitrogen, and $T_x$ symbolizes surface-terminating groups such as =O, –OH, and –F. The most common method for producing MXenes involves selectively etching the A layer from MAX phases ($M_{n+1}AX_n$, with A being a group 13 or 14 element like Al). The MAX phases are characterized by their layered hexagonal crystal structures, where A layers and $M_{n+1}X_n$ units are stacked alternately (Jolly, Paranthaman, and Naguib 2021). Due to weaker M–A bonds compared to M–X bonds, the A layers can be selectively removed, yielding $M_{n+1}X_n$ layers with specific surface functionalization after mild exfoliation (Nagalingam and Grace 2022). Furthermore, MXenes with two different transition metals in the M layers are referred to as double-transition-metal MXenes. Depending on whether these metals are arranged in an out-of-plane or in-plane order, these materials are known as ordered double-transition-metal MXenes (Khaledialidusti, Mishra, and Barnoush 2020). To date, around 30 varieties of MXenes have been synthesized experimentally, with expectations for further additions as novel synthesis methods evolve, utilizing polar organic solvents and Lewis acidic molten salts, among others (Dong et al. 2021).

MXenes exhibit numerous advantageous properties attributable to their unique 2D structures and adaptable surface chemistry (Ghandehari et al. 2024). These include metallic conductivity, exemplified by $Ti_3C_2T_x$ thin films, which exhibit conductivities up to ~10,000 S/cm (Luo et al. 2020), and a significant specific surface area, with a porous $Ti_3C_2T_x$ aerogel displaying more than 170 $m^2/g$ (Li et al. 2017). Their mechanical properties are also notable, with a high strain energy at failure,


DOI: 10.1201/9781003736318-10

approximately 1500 kJ/m$^3$ for $Ti_3C_2T_x$ films post-intercalant removal by proton acid colloidal treatment, and an excellent Young's modulus in the range of 80–100 GPa in directions perpendicular to the basal plane (Firestein et al. 2020). Furthermore, their magnetic properties can be modified through strain application, as seen with a 2% strain inducing a 203 μeV shift per Zr atom in $Zr_2N$ (Yue et al. 2021), and through temperature variations, like in $Cr_2TiC_2T_x$, which undergoes a magnetic transition around 30 K (Hantanasirisakul et al. 2020). However, there are several challenges to be addressed before MXenes can be fully utilized in technological applications. These include their tendency to restack or aggregate due to van der Waals forces and hydrogen bonding, which limits surface accessibility and active site availability when used in electrochemical application (Wu et al. 2020); swelling in aqueous solutions caused by their hydrophilic nature, which affects their performance in filtration membranes by allowing monovalent cations to pass through (Ding et al. 2020); high reflectivity of densely packed ultrathin films, which can lead to significant secondary electromagnetic pollution (Zhang et al. 2021c); and gradual oxidation by atmospheric moisture and oxygen, reducing their overall conductivity (Iqbal et al. 2021).

Despite these challenges, the customizable composition, versatile structures, and surface chemistries of MXenes offer substantial opportunities for integration with other low-dimensional materials. In general, the MXenes' rich variety of reactive surface groups and large specific area facilitate numerous anchoring sites for active catalysts (Rajendran et al. 2022b). However, the inherent interlayer attractions and intermolecular interactions among MXene nanosheets frequently lead to their clustering and organization during electrode setup, which adversely affects their electrochemical capabilities (Cao et al. 2022). In response, ongoing research efforts focus on integrating MXenes with structurally stable materials to mitigate aggregation within the MXene matrix (Chen et al. 2017). Notable among these materials are molybdenum carbide (Wei et al. 2021), various carbon structures (Liao et al. 2021), polymers (Nagalingam and Grace 2022), and metal oxides (Liu et al. 2018), among others (Zhang et al. 2023).

Carbon materials, in particular, stand out due to their ubiquity, diverse nanostructures, and robust chemical properties. These include three-dimensional graphene (like foam, fiber, and paper) and two-dimensional graphene oxide; one-dimensional entities like carbon nanotubes (CNTs) and carbon nanofibers (CNFs); zero-dimensional carbon quantum dots. Carbon structures not only boast a vast active surface area and numerous pores (Gao et al. 2020; Feng et al. 2022; Rajendran et al. 2022a; Rajendran et al. 2021a) but also offer essential structural stability, making them superior catalysts for electrochemical sensing. When compared to various MXene-based materials, hybrid MXene/carbon systems uniquely address weaknesses in MXene structures by enhancing support and creating spacing between MXene nanosheets. Moreover, collaborative effects between carbon-based substances and MXenes impart superior electron transport properties and enhance overall electrochemical performance.

Recent advancements in MXene/carbon composites have been remarkable. The ensuing sections of this review will delve into these advances, particularly focusing on their controllable synthesis, which encompasses the methods of production and adjustments in morphology. Additionally, the importance of structural integration in overcoming the stability challenges of MXene will be highlighted. This section will also investigate the range of structural forms found in MXene/carbon hybrids and how they relate to specific functions in electrochemical systems. This chapter wraps up by identifying current barriers in MXene/carbon composite use and suggesting potential future directions to support continued innovation in this rapidly developing area of materials science.

## 10.2 SYNTHESIS OF MXENE–CARBON COMPOSITES

Recent studies have introduced a variety of creative MXene/carbon systems produced by merging MXenes with special carbon-based substances (Zhao et al. 2014). The synthesis of these composites typically involves a two-stage process. First, MXenes are synthesized by etching atomic layers selectively from MAX phase compounds (Fu et al. 2021). Afterward, the MXenes are combined with carbon-based materials using a broad range of fabrication methods, including but not limited to spray drying, ball milling, and electrospinning (Chen et al. 2020; Lee et al. 2023). Figure 10.1 illustrates that carbon materials, which are structurally stable, such as 1D structures like CNFs and CNTs, and 2D graphene, are particularly effective. The integration of

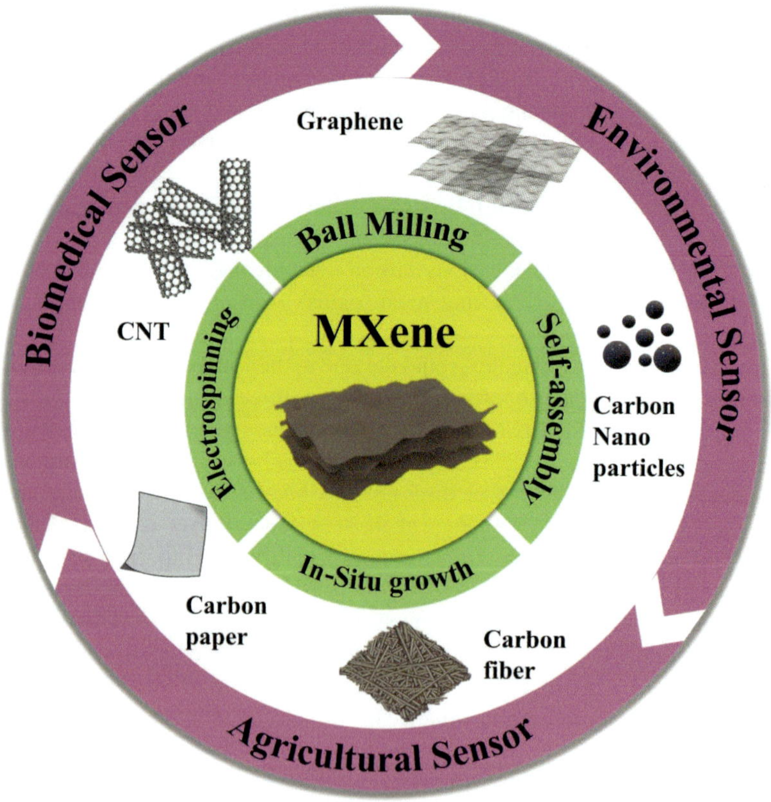


**FIGURE 10.1** Overview of MXene/carbon composite uses in electrochemical sensor development.

MXene with carbon materials has predominantly been achieved through mechanical methods, like self-assembly (Quan et al. 2024), spray drying (Chen et al. 2020), ball milling (Kalambate et al. 2023), in situ growth (Gao et al. 2020), electrospinning (Cheng et al. 2023), and vacuum-assisted filtration (Zhang et al. 2021a). Mechanical strategies like self-assembly, in situ growth, and ball milling primarily leverage intermolecular interactions to form novel three-dimensional structures. The spray drying technique is notable for its ability to produce composite microspheres with a controllable size distribution. Moreover, vacuum-assisted filtration and electrospinning are particularly utilized for creating flexible MXene/carbon electrodes, catering to their specific applications in electrochemical sensing.

### 10.2.1 Self-Assembly Techniques

Self-assembly is a process where nanomaterials autonomously organize into well-ordered structures. This occurs when foundational units coalesce into a stable and regularly shaped configuration, driven primarily by electrostatic forces. MXene surfaces typically carry a negative charge because of the adsorption of chloride and hydroxide ions. This phenomenon is especially significant in such contexts (Zhou et al. 2022b). Illustrating this, Gogotsi and colleagues (Xie et al. 2016) demonstrated the production of porous $Ti_3C_2$/CNT hybrid paper electrodes using self-assembly (Figure 10.2(A)).

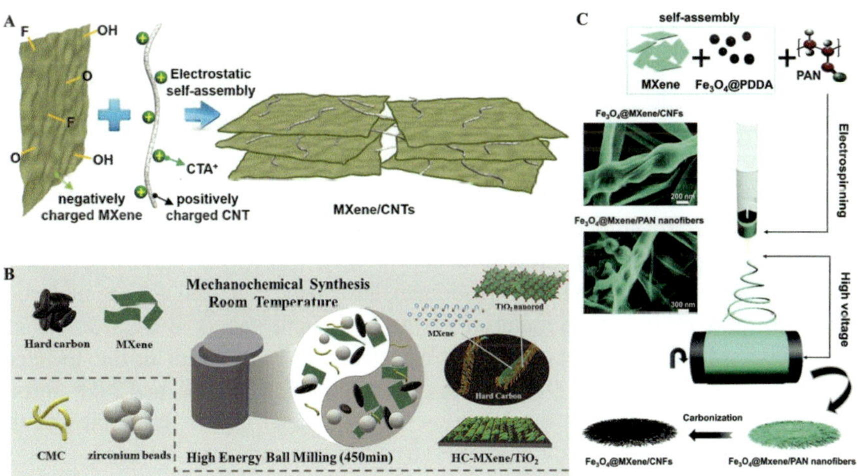


**FIGURE 10.2** (A) Illustration of the assembly process for creating a porous MXene/carbon nanotubes (CNT) electrode through self-assembly techniques. This diagram highlights the interaction between negatively charged MXene nanosheets, attributed to their surface functional groups, and positively charged CNTs that are modified with cetyltrimethylammonium bromide. (B) Diagram depicting the mechanochemical production of hybrid carbon HC-MXene/$TiO_2$ electrodes, utilizing $Ti_3C_2$ as a multifunctional conductive component. (C) Detailed schematic of the step-by-step fabrication of $Fe_3O_4$@MXene/carbon nanofibers (CNFs) nanocomposites, showcasing each stage of the complex synthesis process.

### 10.2.2 Ball Milling Methodology

The ball milling technique leverages the kinetic energy of milling balls to effectively grind and homogenize various components, noted for its simplicity and consistent performance (Wei et al. 2022). Specifically, for MXene/carbon composites, this method facilitates a chemical interaction between the oxygen-containing reactive groups on MXene surfaces and the carbon-based sites in carbon structures. This interaction serves as a crucial binding mechanism within the composites.

Illustratively, Gao et al. (2021) successfully fabricated HC-MXene composites via ball milling, a process that enables the formation of Ti–O–C covalent linkages between MXene ($Ti_3C_2T_x$) and HC (Figure 10.2(B)). In their procedure, HC, MXene, and carboxymethyl cellulose were mixed at a mass ratio of 7:2:1 and dissolved in purified water. This blend underwent ball milling using 5 mm zirconia balls with a 1:20 ratio of material to ball mass. The treatment caused MXene edge oxidation, which led to $TiO_2$ nanorod development and the formation of a distinct MXene/$TiO_2$ junction interface. The resulting Ti–O–C bonds help stabilize the junction by protecting it from structural disruptions during ion intercalation and removal. Despite these benefits, ball milling is sometimes limited by low operational efficiency and significant energy demands (Zhou et al. 2022a). Additionally, abrasion-induced contamination during milling may limit its viability for large-scale production.

### 10.2.3 Electrospinning Techniques

Electrospinning is a method characterized by the electrostatic atomization of polymer solutions. Under an electric field, these solutions are transformed into fine jets, which solidify into fibers over a distance. This approach is appreciated for being straightforward, versatile, and economically efficient (Wang et al. 2019). Polyacrylonitrile (PAN) is frequently used as a carbon precursor in electrospinning because of its extreme carbon yield and excellent spinnability (Yan et al. 2021). Gogotsi et al. employed this technique to create independent $Ti_3C_2T_x$/CNF electrode structures by first applying electrospinning a mixture containing PAN and $Ti_3C_2T_x$ MXene flakes, followed by a carbonization process (Levitt et al. 2019). Other polymers, such as poly(ether amide) and poly(vinyl-pyrrolidone), have also been used as base materials for electrospinning in carbon material synthesis. Enhancements in electrochemical properties have been achieved by integrating active metal nanoparticles into the electrospinning solution, such as the process depicted for $Fe_3O_4$@MXene/CNFs fibrous film (Figure 10.2(C)). In this setup, hollow $Fe_3O_4$ nanoparticles are mixed into the PAN/$Ti_3C_2$ solution, resulting in a fibrous structure that combines 2D $Ti_3C_2$-encased $Fe_3O_4$ spheres within N-doped carbon nanofibers after pre-oxidation and carbonization (Guo et al. 2021). This configuration not only facilitates electron transport but also provides a buffer against the volumetric expansion of $Fe_3O_4$, thereby enhancing the electrical conductivity and structural stability of the film.

### 10.2.4 In Situ Growth Techniques

The in situ growth method is a cutting-edge technique for fabricating MXene/carbon composites, often involving chemical vapor deposition. This process can occur under or close to atmospheric conditions, making the conditions relatively gentle. By altering the coating's composition, it is possible to create composites with gradient densities and controlled purity. A notable application of this technique was presented by Ko (Li et al. 2020), who managed to uniformly cultivate multi-walled carbon nanotubes (MWCNTs) on a carbon cloth-supported MXene sheet (MWCNTs-MXene@CC). The initial step involved anchoring a binary metal catalyst to the MXene surface to facilitate the growth of MWCNTs at 700°C under low pressure. Further advancements were seen when metal-organic framework (MOFs) like zinc imidazolate framework-67 (ZIF-67) and ZIF-8 were incorporated into the MXene matrix,

**TABLE 10.1**
**Comparison of the merits and demerits of different synthesis methods for MXene–carbon composites**

| Method | Merits | Demerits |
|---|---|---|
| Self-assembly techniques | • **High structural control**: enables precise control over the arrangement of MXene and carbon.<br>• **Versatility**: applicable to a broad range of materials and structures.<br>• **Mild conditions**: typically operates under mild conditions, preserving material integrity. | • **Scalability issues**: difficult to scale up for industrial applications.<br>• **Time-consuming**: assembly processes can be slow.<br>• **Environmental sensitivity**: performance can be influenced via environmental conditions like temperature and pH. |
| Ball milling methodology | • **Scalability**: suitable for large-scale production.<br>• **Cost-effective**: uses relatively inexpensive equipment.<br>• **Effective mixing**: can achieve homogeneous dispersion of MXene and carbon. | • **Impurity introduction**: mechanical milling can introduce impurities from the milling media.<br>• **Limited control**: limited control over particle size distribution and morphology.<br>• **High energy consumption**: process is energy-intensive. |
| Electrospinning techniques | • **High surface area**: produces nanofibers with a high surface area, beneficial for electrochemical applications.<br>• **Uniformity**: can produce uniform and continuous fibers.<br>• **Flexibility**: versatile for creating different composite structures. | • **Material limitations**: limited to polymers that can be electrospun.<br>• **High voltage requirement**: requires high voltage, posing safety concerns.<br>• **Scalability challenges**: difficult to scale up for mass production. |
| In situ growth techniques | • **High crystallinity**: produces composites with high crystallinity and strong interfacial bonding.<br>• **Integration**: combines synthesis and deposition in a single step, improving efficiency.<br>• **Complex structures**: suitable for fabricating complex hierarchical structures. | • **Complexity and cost**: requires precise control of synthesis conditions, making it complex and expensive.<br>• **Material specificity**: limited to specific precursor materials and conditions.<br>• **Reproducibility issues**: can have reproducibility challenges due to the sensitivity of the process parameters. |

leading to the formation of a MOF/MXene composite upon which CNTs were grown in situ, culminating in a ternary Co/MXene/CNT composite after carbonization (Wang et al. 2021).

The various synthesis methods for MXene–carbon composites each have distinct advantages and challenges. Table 10.1 provides a systematic comparison of these methods, including self-assembly techniques, ball milling methodology, electrospinning techniques, and in situ growth techniques. Each method is evaluated based on its scientific merits, such as structural control, scalability, and cost-effectiveness, as well as its demerits, including potential impurities, process complexity, and scalability challenges. This detailed analysis serves as a valuable resource for researchers and engineers focused on optimizing MXene–carbon composite synthesis methodologies. Table 10.1 shows the comparison of the merits and demerits of different synthesis methods for MXene–carbon composites.

## 10.3 PROPERTIES OF MXENE–CARBON COMPOSITES AND THEIR DISTINCTION FROM PARENT MATERIALS

The integration of MXenes with carbon-based materials such as graphene and carbon nanotubes (CNTs) has garnered significant attention in the field of electrochemical sensors, as these composites combine the exceptional properties of both materials, leading to enhanced sensor performance. MXenes, particularly $Ti_3C_2T_x$, are known for their high electrical conductivity, with values ranging from 2000 to 10,000 S/cm depending on synthesis conditions and surface terminations (Khazaei et al. 2013). Graphene, with a conductivity of 200,000 S/cm, and CNTs, which can achieve up to 10,000 S/cm in single-walled configurations, also contribute significantly to the conductivity of these composites (Novoselov et al. 2004; Tans, Verschueren, and Dekker 1998). The combination of MXenes with graphene or CNTs in composites results in a synergistic improvement in conductivity. For instance, MXene-graphene composites have shown conductivity as high as 12,000 S/cm (Pandey et al. 2024), attributed to the prevention of MXene sheet restacking by carbon materials, which preserves conductive pathways and enhances electron transport. This enhanced conductivity directly translates to improved sensitivity and response times in electrochemical sensors, as the efficient electron transfer facilitates rapid and accurate detection of analytes. Furthermore, the mechanical properties of MXenes, characterized by a Young's modulus of up to 330 GPa, are enhanced when combined with the superior mechanical strength of graphene and CNTs, which exhibit Young's moduli around 1 TPa and tensile strengths up to 63 GPa (Lee et al. 2008; Yu et al. 2000). The resulting MXene–carbon composites offer improved mechanical stability and flexibility, making them ideal for use in flexible and wearable sensors where mechanical resilience is critical. Importantly, these composites also exhibit superior electrochemical performance in sensing applications. MXenes alone are known for their excellent electrochemical properties, such as rapid electron transfer kinetics, which are essential for high-performance sensors. When combined with carbon materials, the composite structure promotes better surface area utilization and enhances the active sites available for analyte interaction. This leads to significant improvements in sensor selectivity and sensitivity. For example, MXene-graphene composites have demonstrated enhanced detection capabilities for various biomarkers and environmental pollutants, with lower detection limits and higher response rates compared to sensors made from individual components. Additionally, the stability and durability of MXenes, which are often challenged by susceptibility to oxidation, are substantially improved in MXene–carbon composites. The carbon components in these composites act as protective barriers, reducing the rate of oxidation and extending the material's operational lifespan, thus enhancing the long-term reliability of the sensors (Zhan et al. 2020). This enhanced stability is particularly beneficial in sensors that require long-term deployment in harsh environments. Overall, MXene–carbon composites represent a significant advancement over their parent materials, offering a range of enhanced properties including improved electrical conductivity, mechanical strength, sensing performance, and stability. These improvements enable the advance of next-generation electrochemical sensors with faster response times, higher sensitivity, and greater durability, making them highly effective for applications in environmental monitoring, biomedical diagnostics, and industrial process control.

## 10.4 STATE-OF-THE-ART PROGRESS OF MXENE/CARBON COMPOSITES IN ELECTROCHEMICAL SENSORS

Electrochemical sensors have become pivotal in monitoring environmental changes, detecting biomolecules, and analyzing pharmaceutical and food quality. In this domain, carbon materials, owing to their excellent conductivity, large specific surface area, and stable nanotube structures, serve as a robust foundation for sensor technology. These materials are highly sought after not only for their biocompatibility and modifiable surfaces, which allow for various functional groups through pretreatments, but also for their dispersibility in solutions, making them ideal for constructing diverse sensor types.

While carbon-based materials share some electrochemical properties with edge plane pyrolytic graphite electrodes, MXene composites bring unique advantages to electrochemical sensor applications. MXenes themselves are distinguished by their high conductivity, remarkable mechanical strength, and favorable hydrophilic properties, setting the stage for their integration into electrochemical sensor development (Lee et al. 2023; Rajendran 2023; Yi et al. 2022). However, the lamellar structure of MXenes is susceptible to collapse, presenting a significant barrier to their long-term functional stability. To counteract this limitation and enhance electron transport efficiency, MXene/carbon composites have been developed, leveraging the inherent structural stability of carbon materials (Wang et al. 2020).

Recent advancements in MXene/carbon composites have been profound, with extensive research focusing on their application in sensor fabrication. These composites have been successfully implemented across a broad spectrum of sensor technologies, including environmental sensors and those designed for biomolecular, pharmaceutical, and agricultural/food analysis. The combination of MXene and carbon materials not only mitigates the structural challenges posed by MXenes but also augments the electrochemical capabilities of the sensors, thereby broadening their application potential and improving their performance and reliability in various analytical scenarios.

### 10.4.1 Development of MXene/Carbon Composites for Biomedical and Pharmaceutical Sensor Applications

The burgeoning field of biomedical and pharmaceutical sensing stands as a pivotal component of modern healthcare, enhancing diagnostic accuracy and enabling effective treatment strategies. Sensors in these fields detect and quantify biological or chemical events, offering crucial insights into patient health and the efficacy of therapeutic interventions. MXene/carbon composites, with their unique electrochemical properties, have emerged as ideal candidates for advancing sensor technology in these critical areas. In biomedical applications, the early detection and monitoring of biomarkers can significantly influence outcomes in disease management and treatment. Sensors that can provide real-time, accurate measurements of biological

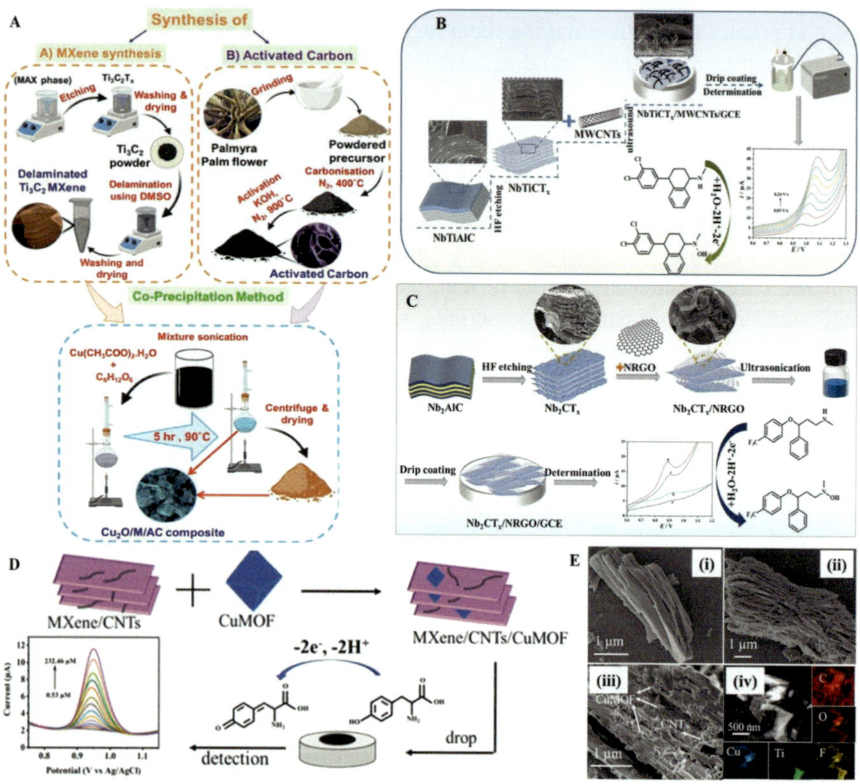

**FIGURE 10.3** (A) Illustration depicting the synthesis process of the MXene-activated carbon-based $Cu_2O$ ($Cu_2O$/M/AC) composite. (B) Overview of the formation of $NbTiCT_x$/multi-walled carbon nanotubes (MWCNTs)/glassy carbon electrode (GCE) and its application in detecting Sertraline (SER). (C) Diagram showing the assembly of $Nb_2CT_x$/nitrogen-doped reduced graphene oxide (NRGO)/GCE and its utilization in fluoxetine detection. (D) Schematic representation of the MXene/CNTs/copper metal-organic framework (CuMOF) synthesis procedure and the mechanism involved in tyrosine detection. (E) SEM images of MXene (i), MXene integrated with CNTs (ii), and MXene/CNTs/CuMOF (iii), accompanied by an elemental mapping of MXene/CNTs/CuMOF (iv).

markers are indispensable for conditions such as diabetes, cancer, and cardiovascular diseases. Pharmaceutical sensing, however, plays a critical role in drug development and management, ensuring the safety and effectiveness of medications through precise dosage and monitoring of drug metabolism.

In recent advancements within the realm of biomedical sensors, Tamil Selvi Gopal's research (Gopal et al. 2024) introduced a MXene-activated carbon-based $Cu_2O$ ($Cu_2O$/M/AC) nanocomposite, created through the coprecipitation method, aimed at glucose detection (Figure 10.3(A)). This sensor displays an extensive linear range of 0.004–13.3 mM and a further range of 15.3–28.4 mM, with a low limit of detection (LOD) of 1.96 μM. The sensor's capabilities are further emphasized by its sensitivity, with values recorded at 430.3 and 240.5 $\mu A/mM/cm^2$, respectively. These characteristics underscore the sensor's potential in non-enzymatic glucose monitoring, which is crucial for diabetes management. In the study conducted by Nana Tang et al. (2024), they developed a novel electrochemical sensor using a bimetallic $NbTiCT_x$/MWCNTs nanocomposite for the detection of Sertraline (SER) (Figure 10.3(B)). The nanocomposite, produced through ultrasonic treatment of $NbTiCT_x$ combined with MWCNTs, was applied to modify glassy carbon electrodes ($NbTiCT_x$/MWCNTs/GCE).

The sensor demonstrated excellent electrochemical properties. This setup enabled a sensitive detection of Sertraline across two linear ranges: 10.0–100.0 μM and 5.0–10.0 μM, with a detection limit of 0.16 μM. The sensor enables the sensitive determination of Sertraline in both urine and serum samples, highlighting its potential for broader biomedical applications. Aiting Chen et al. (2024) reported on the $Nb_2CT_x$/nitrogen-doped reduced graphene oxide (NRGO) composite, prepared via ultrasonic treatment, for the electrochemical determination of fluoxetine at micromolar concentrations (Figure 10.3(C)). The sensor operates with a dual linear range of 1.0–100.0 μM and an LOD of 0.34 μM, which offers a new avenue for monitoring therapeutic levels of fluoxetine in human serum and urine. Further, Jing Chen et al. (2022) developed an enzyme-free electrochemical sensor using a carbon–MXene based composite, specifically MXene/CNTs/CuMOF, synthesized through ultrasonic treatment (Figure 10.3(D)). In Figure 10.3(E)(i), the MXene structure is depicted as having an accordion-like layered configuration. The incorporation of CNTs into MXene (Figure 10.3(E)(ii)) results in a more expanded multilayer structure due to the CNTs serving as spacers that increase the separation between MXene layers. Furthermore, SEM imagery in Figure 10.3(E)(iii) illustrates the successful attachment of Cu MOF octahedral particles onto the MXene/CNTs framework. The elemental mapping in Figure 10.3(E)(iv) confirms the presence of elements verifying the successful synthesis of the composite. This novel sensor efficiently detects tyrosine, a neurotransmitter precursor, showcasing a broad linear detection range of 0.53–232.46 μM and an impressive LOD at 0.19 μM. Menon Ankitha et al. (2024) developed a $Ti_3C_2T_x$/conductive carbon yarn (CCY) composite through ultrasonic treatment, utilized in a wearable sensor patch for on-site monitoring of dopamine levels. This cutting-edge sensor offers a linear range from 1 nM to 1 μM and a LOD of 316 pM.

Shipeng Zhang and his team (Zhang et al. 2021b) have pioneered the development of a MXene/MWCNTs composite, expertly crafted through ultrasonic treatment. This advanced composite is at the core of a highly sensitive, flexible, wireless, and battery-free electrochemical sensing patch designed for the real-time monitoring of potassium ions ([$K^+$]) in human sweat. The integration of a microfluidic unit for sweat collection enhances the sensor's efficiency and functionality, facilitating a seamless and continuous monitoring experience (Figure 10.4(A) and (B)). The hybrid, multi-dimensional network structure of the sensor delivers a great surface activation area and accelerates charge transfer, essential for the effective capture and analysis of [$K^+$] across a linear response range of 1–32 mM. Utilizing near field communication (NFC), the device ensures uninterrupted data transmission to smartphones.

Figure 10.4(C) and (D) displays a photograph of the sensor's sophisticated flexible antennas and circuits, which are meticulously fabricated using photolithography and wet etching techniques on a copper foil-coated polyimide film atop a silicon wafer. The ultrathin flexible antenna and circuit components, which are assembled using low-temperature solder paste for the surface-mounted components. To safeguard the NFC electronics from sweat and external moisture, the circuits are encased in Polydimethylsiloxane (PDMS), as depicted in Figure 10.4(E). The microfluidic system, constructed from PDMS, is adeptly attached to the sensor to optimize the collection and handling of sweat, showcased in Figure 10.4(F). Figure 10.4(G) presents the sensor's remarkably compact and lightweight design, with a diameter of just 3.3 cm, making it exceptionally portable and user-friendly. Finally, Figure 10.4(H) captures the real-time interaction between the wireless NFC sensor

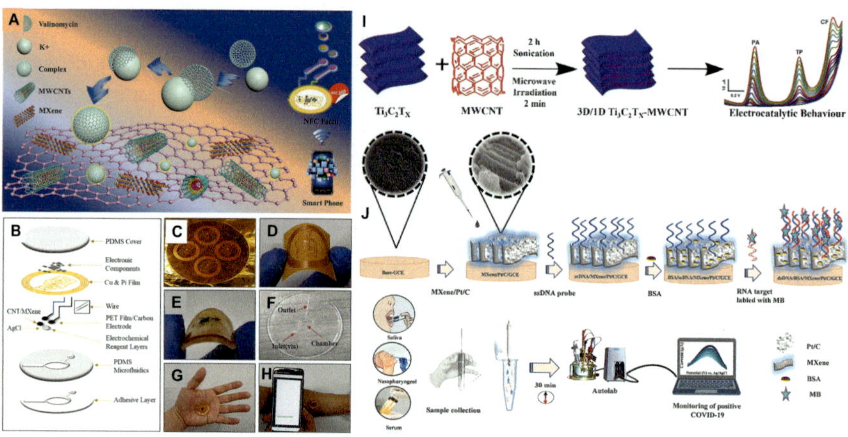


**FIGURE 10.4** (A) Illustrative depiction of the operational mechanism of a wireless, battery-free electrochemical sensing patch incorporating MWCNTs-MXene hybrid multi-dimensional networks. (B) Detailed exploded view schematic of the flexible, wireless, battery-free microfluidic and electronic system. (C) and (D) Fabrication steps involving lithography and etching to construct the flexible antenna and circuits. (E) Sealing and protection of flexible circuits using PDMS. (F) Overview of the microfluidic system design. (G) Compact and lightweight circular patch sensor system with a diameter of 3.3 cm. (H) Diagram showing real-time communication between the wireless NFC sensor and a smartphone. (I) Detailed schematic of the 3D/1D $Ti_3C_2T_x$-MWCNT/screen-printed carbon electrode (SPE) nanohybrid-modified electrode demonstrating electrocatalytic activity. (J) Process diagram for constructing a DNA/nano/biosensor and the electrochemical detection method using an Autolab device.

and a smartphone, demonstrating the system's capability for immediate communication and data relay. Next, Elancheziyan Mari et al. (2024) have also contributed to the field with their development of a MXene/MWCNT nanocomposite, synthesized via microwave reaction. This material forms the basis of a screen-printed carbon electrode (SPE) that simultaneously detects paracetamol, theophylline, and caffeine in human blood samples (Figure 10.4(I)). The sensor's linear range is remarkably broad, detecting paracetamol from 1.0 to 90.1 μM, theophylline from 2.0 to 62.0 μM, and caffeine from 2.0 to 90.9 μM, with respective LODs of 0.23, 0.57, and 0.43 μM. Monireh Bolourinezhad et al.'s (2023) research introduces a groundbreaking MXene/Pt/C nanocomposite, utilizing ultrasonic synthesis, tailored for the detection of COVID-19 (Figure 10.4(J)). This sensor uses MXene nanosheets and carbon platinum (Pt/C) to amplify the detection signal in response to hybridization with the virus's specific oligonucleotide target. The electrochemical sensor displays a broad dynamic range from 1 aM to 100 nM and an extremely low LOD of 0.4 aM, making it highly effective for early COVID-19 diagnostics. The sensor's accuracy and specificity were validated through tests on 192 clinical samples, showing 100% accuracy and sensitivity, and 97.87% specificity. This highlights its suitability for rapid, reliable COVID-19 testing across various biological matrices, including saliva and nasopharyngeal swabs.

In biomedical applications, particularly for wearable sensors, the biocompatibility and stability of materials are paramount. MXenes, despite their advantageous properties, face challenges in these areas. To address these issues, researchers have commonly modified MXenes with polymers, a strategy that enhances both their stability and biocompatibility (Chakoma et al. 2024a; Chakoma et al. 2025). For instance, Geng et al. (Xuan et al. 2016) successfully enhanced the stability of MXenes by modifying them with polyethylene glycol (PEG), while Chen et al. (Ren et al. 2019) employed polyvinyl pyrrolidone (PVP) to improve both stability and biocompatibility. This focus on biocompatibility is especially crucial in wearable sensing technologies, where materials must remain safe and effective during prolonged contact with biological systems (NajafiKhoshnoo et al. 2023; Chakoma et al. 2024c; Rajendran and Esfandyarpour 2024; Chakoma et al. 2024b). Despite the progress made with polymer modifications, ongoing research is essential to develop new strategies that further optimize MXene performance in biomedical fields.

### 10.4.2 Development of MXene–Carbon Composites for Environmental Sensing Applications

Environmental sensing is critical for monitoring the health of ecosystems, ensuring public safety, and enforcing environmental regulations. It encompasses the detection and quantification of pollutants, monitoring of water quality, and assessment of soil contamination. Such sensors not only provide essential data for environmental management and policy-making but also help in the early detection of potential environmental hazards, which can prevent or mitigate adverse impacts on human health and biodiversity. In this context, Xue Wang et al. (2022) developed a sensitive electrochemical sensor for p-nitrophenol detection using a $Ti_3C_2T_x$ MXene/graphene composite (D-$Ti_3C_2T_x$/GR), prepared via an optimized minimally intensive layer delamination method (Figure 10.5(A)). The composite was applied to a glassy carbon electrode, providing a linear response to p-nitrophenol concentrations ranging from 1 to 175 μmol/L and achieving a LOD of 0.16 μmol/L. This sensor underscores the potential for rapid and effective environmental pollutant monitoring in water samples. Expanding the scope to antibiotic detection, Kunling Wei et al. (2024) synthesized a poly(diallyldimethyl-ammonium chloride) assembled graphene oxide and MXene composite. This composite was utilized to construct a capable electrochemical sensor with an extensive detection scale of 0.04–200 μM and a LOD of 1.22 nM for furazolidone, demonstrating successful detection in both tap and river water (Figure 10.5(B)). Building on the theme of advanced composites, Guangfei Qu et al. (2023) introduced a novel Pt@SWCNTs-MXene-reduced graphene oxide (rGO) composite, synthesized via ultrasonic treatment. This sensor system, designed for the

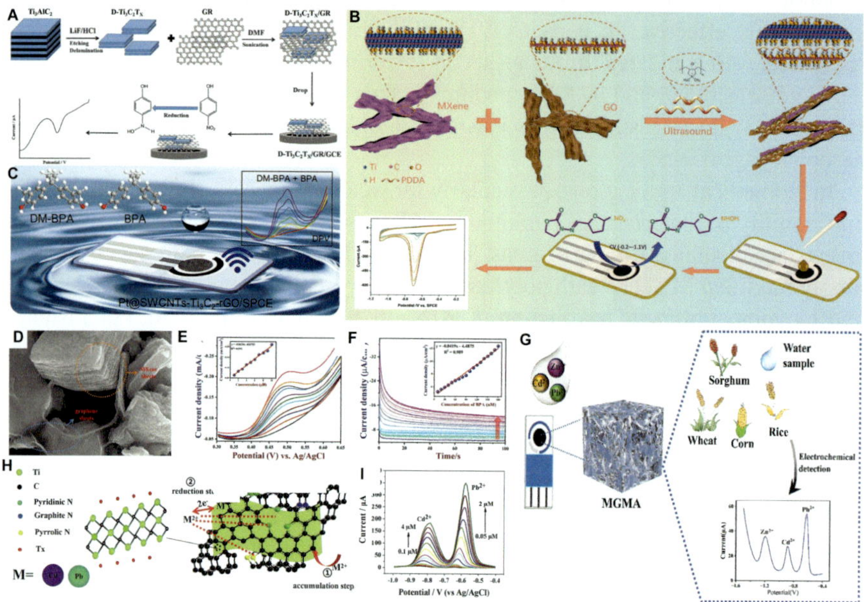


**FIGURE 10.5** (A) Diagram illustrating the creation of D-$Ti_3C_2T_x$/GR/GCE and its use in the electrochemical detection of p-NP. (B) Depiction of the method for synthesizing the CNF electrocatalyst aimed at detecting sulfadiazine. (C) Visualization of the simultaneous and direct detection process for dimethyl bisphenol A (DM-BPA) and bisphenol A (BPA) utilizing Pt@ single-walled carbon nanotubes (SWCNTs)-MXene-rGO hybrid materials. (D) SEM images showing the detailed structure of the graphene/MXene nanocomposite. (E) Differential pulse voltammetry (DPV) profiles and (F) amperometric detection responses at varying concentrations of BPA in a 0.1 M phosphate buffer solution (PBS) using a graphene/MXene nanocomposite-modified electrode. (G) Conceptual diagram of 3D melamine-doped graphene oxide/MXene composite aerogel (MGMA)/screen-printed carbon electrode (SPCE) for detecting $Zn^{2+}$, $Cd^{2+}$, and $Pb^{2+}$ in various environmental contexts. (H) Illustration of the detection mechanism for $Cd^{2+}$ and $Pb^{2+}$ and the interaction between nitrogen-doped sites and heavy metal ions (HMIs). (I) Square wave anodic stripping voltammetry (SWASV) curves recorded from $Ti_3C_2$@N-C/GCE for the detection range of $Pb^{2+}$ (0.05–2 μM) and $Cd^{2+}$ (0.1–4 μM).

simultaneous detection of bisphenol A (BPA) and dimethyl bisphenol A (DM-BPA), featured rGO produced by the electrochemical reduction of graphene oxide under an electric field of −1.2 V (Figure 10.5(C)). It demonstrated exceptional electrochemical properties, with a detection range of 0.006–7.4 μmol/L and remarkably low detection limits of 2.8 and 3 nmol/L for BPA and DM-BPA, respectively, showcasing its robust reproducibility, stability, and interference resistance, particularly in thermal paper samples. However, Jerome Rajendran et al. (2022b) also developed a highly sensitive and selective electrochemical sensor for BPA detection, crucial for ensuring environmental and food safety. Utilizing a top-down approach, a 2D hybrid graphene/$Ti_3C_2T_x$ nanocomposite (Gr/MXene) was synthesized.

The arrangement of the MXene and Gr/MXene nanohybrid was confirmed through SEM as depicted in Figure 10.5(D). Further innovations include the preparation of a modified GCE with the 2D hybrid Gr/MXene nanocomposite for BPA oxidation in 100 mM PBS. Under optimized conditions, the Gr/MXene/GCE displayed a wide linear detection range spanning 10–180 nM and 1–10 μM BPA, with detection limits of 4.08 nM and 0.35 μM achieved by amperometry and differential pulse voltammetry (DPV) methods respectively, as illustrated in Figure 10.5(E) and (F). Hao Cheng et al. (2023) synthesized a $Ti_3C_2$ MXene/$Co_3O_4$/carbon nanofibers ($Ti_3C_2$ MXene/$Co_3O_4$/CNFs) composite using a HF exfoliating $Ti_3AlC_2$ strategy, combined with electrospinning and thermal annealing to incorporate $Co_3O_4$ and $Ti_3C_2$ MXene into the CNFs. This innovative composite was used to develop a hybrid modified electrode that facilitated ultra-sensitive and rapid analysis of 4-aminophenol (4-AP), achieving a broad detection range from 0.5 to 150 μM and a low detection limit of 0.018 μM. The electrode's effectiveness was validated through successful applications in analyzing 4-AP spiked in tap water and blood serum samples. Next, Yuanyuan Chen et al. (2023a) developed a novel sensor using a three-dimensional melamine-doped graphene oxide/MXene composite aerogel for the simultaneous and sensitive detection of $Zn^{2+}$, $Cd^{2+}$, and $Pb^{2+}$ (Figure 10.5(G)). Fabricated using a self-assembly method, this sensor enhanced electrical conductivity and facilitated metal ion uptake, achieving detection limits of 0.48 μg/L, 0.45 μg/L, and 0.29 μg/L for the respective metals, demonstrating its capability in monitoring heavy metal contamination effectively.

Transitioning to pollution monitoring, Xiao Zhang et al. (2022) developed an innovative electrochemical sensor for the simultaneous detection of $Cd^{2+}$ and $Pb^{2+}$ ions, crucial pollutants in environmental monitoring. Utilizing a nitrogen-doped carbon-coated $Ti_3C_2$-MXene heterostructure ($Ti_3C_2$@N-C), prepared via a hydrothermal reaction, the sensor showcased exceptional sensitivity and selectivity (Figure 10.5(H)). The sensor displayed linear detection ranges of 0.1–4 μM for $Cd^{2+}$ and 0.05–2 μM for $Pb^{2+}$, with impressively low detection limits of 2.55 nM for $Cd^{2+}$ and 1.10 nM for $Pb^{2+}$ (Figure 10.5(I)). The enhanced electrical conductivity of the heterostructure and the nitrogen doping provided abundant adsorbent sites for metal ions, facilitating rapid and efficient electron transfer. This setup not only improved the detection capabilities but also exhibited excellent performance in the presence of interfering ions. Xiao Zhang's work presents a significant advancement in using MXene-based composites for effective and reliable heavy metal monitoring in both seawater and tap water, contributing to the broader application of these materials in environmental safety assessments.

### 10.4.3 MXene–Carbon Hybrid Materials in Agricultural Sensor Development

Agricultural sensing technology is pivotal in modern agriculture, significantly enhancing efficiency, productivity, and environmental sustainability. Sensors used in agriculture can provide critical information on soil conditions, plant health, and environmental factors, enabling farmers to make informed decisions about irrigation, fertilization, and pest management. This targeted approach helps in optimizing resource use, reducing waste, and increasing crop yields. Furthermore, agricultural sensors play a crucial role in precision farming techniques, adapting agricultural practices to the specific conditions of each plot of land, thereby fostering more sustainable and environmentally friendly farming methods. Yanmei Shi et al. (2024) employed a graphene/MXene composite, also synthesized through ultrasonic treatment, to detect rutin in buckwheat tea (Figure 10.6(A)). The composite, rGO-polystyrene sulfonate

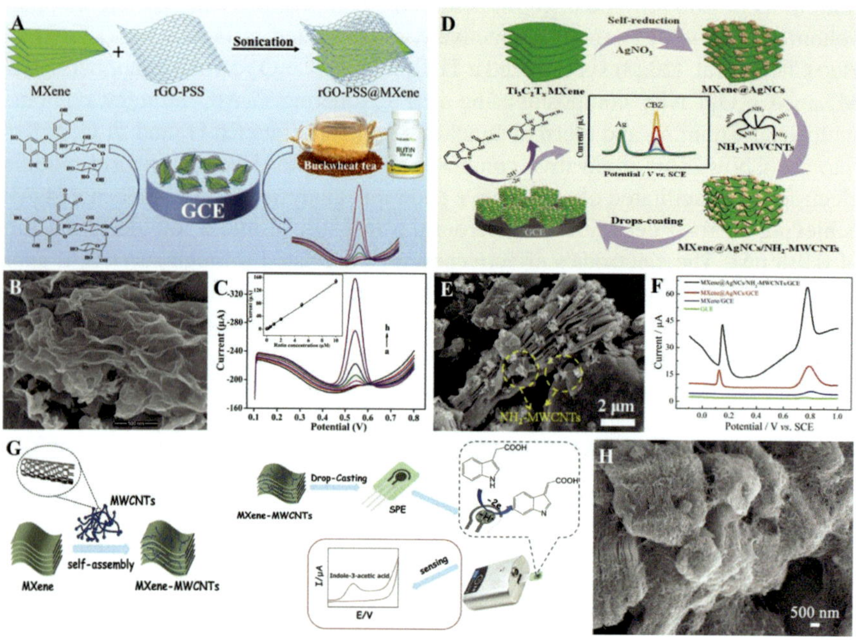


**FIGURE 10.6** (A) Illustration of the synthesis process for rGO-PSS@MXene/GCE and its use in electrochemical rutin detection. (B) SEM imagery depicting the rGO-PSS@MXene heterostructure. (C) DPV traces at varying rutin concentrations on rGO-polystyrene sulfonate (PSS)@MXene/GCE, with an inset showing the linear relationship between response current and rutin concentration. (D) Diagram showing the synthesis of MXene@AgNCs/$NH_2$-MWCNTs/GCE and its application in detecting carbendazim (CBZ). (E) SEM visuals of the MXene@AgNCs/$NH_2$-MWCNTs composite. (F) DPV responses of bare GCE, MXene/GCE, MXene@AgNCs/GCE, and MXene@AgNCs/$NH_2$-MWCNTs/GCE in PBS (0.1 M, pH 6.0) containing 5.0 μM CBZ. (G) Detailed synthesis route for MXene-MWCNTs composite and preparation process for MXene-MWCNTs/SPE used in the electrochemical detection of indole-3-acetic acid (IAA). (H) SEM images of the MXene/MWCNTs composite.

(PSS)@MXene, addresses the self-stacking tendencies of MXene and reduced graphene oxide, enhancing electron transfer and electrocatalytic activity. The SEM image (Figure 10.6(B)) reveals that the rGO-PSS@MXene heterostructure has successfully formed, exhibiting a distinct interconnected honeycomb multilayer network. This structure resembles crumpled silk shapes, with rGO enveloping the surface of MXene, indicating the effective assembly of the rGO-PSS with MXene. This sensor not only demonstrated a sensitive detection range from 0.005 to 10.0 μM and a detection limit of 1.8 nM, as shown in Figure 10.6(C), but also validated its accuracy against HPLC results, confirming its reliability and precision for quality control in herbal products. Building on the theme of heterostructure utility, Wei Zhong et al. (2021) introduced a novel ratiometric electrochemical sensor for detecting carbendazim, utilizing a composite of MXene@Ag nanoclusters and amino-functionalized multi-walled carbon nanotubes (MXene@AgNCs/$NH_2$-MWCNTs) (Figure 10.6(D)). This sensor, synthesized via ultrasonic treatment, effectively inhibits the aggregation of MXene flakes and enhances electrocatalytic activity. In the structure of MXene@AgNCs/$NH_2$-MWCNTs (Figure 10.6(E)), the $NH_2$-MWCNTs are observed surrounding the MXene@AgNCs, serving as conductive 'bridges' that facilitate improved electron transfer (Figure 10.6(F)). The sensor's superior design achieves a remarkably low detection limit of 0.1 nM and has shown excellent performance in testing vegetable samples, highlighting its potential for real-world agricultural applications.

Yuxue Chen et al.'s (2023b) creation of a wireless portable electrochemical sensor for indole-3-acetic acid (IAA) detection using an MXene/MWCNTs composite exemplifies the application of these materials in plant physiology Figure 10.6(G). The sensor's ability to analyze IAA content across various parts of pea seedlings offers valuable insights into plant growth and development, supporting precise agricultural practices. Further expanding the scope of MXene/carbon composites, Meng Gao et al. (2020) developed an MXene/MWCNTs composite sensor for the detection of synephrine in fruit samples. This composite leverages the large electrochemically active surface area and the combined physical and chemical properties of MXene and MWCNTs, as shown in SEM Figure 10.6(H), to offer high sensitivity and excellent electrocatalytic activity. With a linear response to synephrine concentrations ranging from 0.5 to 70 μM and a low detection limit of 0.167 μM, the sensor exhibits outstanding repeatability and anti-interference capabilities, making it ideal for monitoring compounds in complex food matrices.

## 10.5 CONCLUSION AND FUTURE OUTLOOK

In the emerging field of MXene–carbon-modified electrochemical sensors, the integration of MXenes with carbon materials has unleashed remarkable electrochemical and electrocatalytic capabilities. These hybrid materials, distinguished by their high specific surface area and abundant functional groups, have demonstrated significant potential in detecting a broad spectrum of analytes. Recently reported representative MXene–carbon-based hybrid electrochemical sensors are summarized in Table 10.2. The strategic incorporation of secondary carbon materials such as graphene, CNT, carbon dots with MXenes enhances the selectivity, sensitivity, and overall stability of the sensors. Carbon materials improve sensor

**TABLE 10.2**
**Overview of hybrid electrochemical sensors based on MXene–carbon composites for applications in biomedical, environmental, and agricultural detection**

| Application | Carbon–MXene-based composite | Synthesis method | Sensing analyte | Linear range | LOD | Real sample | Ref. |
|---|---|---|---|---|---|---|---|
| Biomedical and pharmaceutical | MXene/CNTs/CuMOF | Ultrasonic treatment | Tyrosine | 0.53–232.46 µm | 0.19 µm | Serum | Chen et al. (2022) |
| | Graphene/MXene nanocomposite | Self-assembled under ultrasound | Tlanine aminotransferase | 0.5–400 U/L | 0.16 U/L | Serum | Quan et al. (2024) |
| | N, S-CDs/$Ti_3C_2T_x$ composites | Condensation reaction | Dopamine | 1–1000 µM | 0.91 µM; 3.71 µM | Urine | Hou et al. (2024) |
| | $TiO_2$-MXene/MWCNTs/PDA/GCE | Ultrasonic treatment | Xanthine/guanine | 0.1–100 µmol/L | 0.018 µmol/L | HFF-1 cells | Liu et al. (2024) |
| | $Nb_2CT_x$/NRGO | Ultrasonic treatment | Fluoxetine | 1.0–100.0 µM | 0.34 µM | Serum | Chen et al. (2024) |
| | $Cu_2O$/M/AC | Coprecipitation method | Glucose | 0.004–13.3 mM | 1.96 µM | Serum | Gopal et al. (2024) |
| | $Ti_3C_2T_x$/CCY | Ultrasonic treatment | Glucose | 1 nM/µM | 316 pM | Serum | Ankitha, Shamsheera, and Rasheed (2024) |
| | Au/rGO-$Ti_3C_2$ | Hydrothermal reaction | Glucose | 0.01–21 mM | 3.1 µM | Saliva | Shang et al. (2024) |
| | $NbTiCT_x$/MWCNTs nanocomposite | Ultrasonic treatment | Sertraline | 5.0–100.0 µM | 0.16 µM | Serum | Tang et al. (2024) |
| | MXene and graphene composite | Ultrasonic treatment | Cortisol | 1 fM to 10 µM | 0.4 fM | Saliva | Liu et al. (2023) |
| | MXene/graphene/AuNPs 3D porous composites | Ultrasonic treatment | Glucose | 2 µM to 6 mM | 2 µM | Sweat | Feng et al. (2023) |
| | MXene/Pt/C nanocomposite | Ultrasonic treatment | COVID-19 | 1 aM to 100 nM | 0.4 aM | COVID-19 | Bolourinezhad et al. (2023) |
| | $Cu_2O$/MXene/rGO ternary nanocomposites | Hydrothermal reaction | Glucose | 0.1–14 and 15–40 mM | 1.1 µM | Biological sample | Alanazi et al. (2023) |
| | MXene/graphene nanocomposite | Ultrasonic treatment | Nicotine | 1–55 µM | 290 nM | Saliva | Rajendran et al. (2022a) |

*(Continued)*

**TABLE 10.2 (Continued)**
**Overview of hybrid electrochemical sensors based on MXene–carbon composites for applications in biomedical, environmental, and agricultural detection**

| Application | Carbon–MXene-based composite | Synthesis method | Sensing analyte | Linear range | LOD | Real sample | Ref. |
|---|---|---|---|---|---|---|---|
| | MXene/graphene nanoplatelets | Ball milling | Imiquimod (IMQ) | 0.071–92.0 µM | 0.487 nM | Imiquimod (IMQ) cream | Kalambate et al. (2023) |
| | Au–Pd/MXene/LSG | Self-assembly | Ascorbic acid (AA), uric acid (UA), and dopamine (DA) | AA, 10–1600 µM, DA, 12–240 µM, and UA, 8–100, 200–800 µM | 3 µM AA, 0.13 µM DA, and 1.47 µM UA | Urine | Wang et al. (2023) |
| | MXene/MWCNT | Ultrasonic treatment | Amyloid-β protein | 1.0–100.0 fg/mL | 0.3 fg/mL | Plasma | Özcan et al. (2020) |
| | MXene/CNH | Ultrasonic treatment | Adrenaline | 1.0 nM to 60.0 µM | 0.3 nM | Urine | Chen et al. (2021a) |
| | RGO/MXene composite | Ultrasonic treatment | $H_2O_2$ | 0.1–300 µM | 0.028 µM | Cancer cell | Zhao et al. (2021) |
| | MXene/nitrogen-doped rGO | Hydrothermal reaction | Adrenaline | 10.0 nM to 90.0 µM | 3.0 nM | Urine | Chen et al. (2021b) |
| | MXene/MWCNTs | Ultrasonic treatment | $K^+$ | 1–32 mM | – | Sweat | Zhang et al. (2021b) |
| | MXene/g-$C_3N_4$ | Electrostatic self-assembly ultrasonic treatment | Ciprofloxacin | 0.4–1000 nM | 0.13 nM | Tablet | Yuan et al. (2021) |
| | $Ti_3C_2$/G-MWCNTs/ZnO | Ultrasonic treatment | Dopamine | 0.01–30 µM | 3.3 nM | Serum | Ni et al. (2022) |
| | MXene/MWCNT nanocomposite | Microwave reaction | Paracetamol, theophylline, and caffeine | 1.0–90.1, 2.0–62.0, and 2.0–90.9 µM | 0.23, 0.57, and 0.43 µM | Blood | Mari et al. (2024) |
| | MXene/$Co_3O_4$/carbon nanofibers composite | Electrospinning | 4-Aminophenol | 0.5–150 µM | 0.018 µM | Serum, water | Cheng et al. (2023) |
| Environmental | Pt@SWCNTs-MXene-rGO | Ultrasonic treatment | Bisphenol A | 0.006–7.4 µmol/L | 2.8 and 3 nmol/L | Thermal paper samples | Qu et al. (2023) |

*(Continued)*

**TABLE 10.2 (Continued)**
**Overview of hybrid electrochemical sensors based on MXene–carbon composites for applications in biomedical, environmental, and agricultural detection**

| Application | Carbon–MXene-based composite | Synthesis method | Sensing analyte | Linear range | LOD | Real sample | Ref. |
|---|---|---|---|---|---|---|---|
| | D-$Ti_3C_2T_x$/GR/GCE | Ultrasonic treatment | p-Nitrophenol | 1–175 μmol/L | 0.16 μmol/L | Water | Wang et al. (2022) |
| | Poly(diallyldimethyl-ammonium chloride) (PDDA) assembled graphene oxide and MXene composites | Ultrasonic treatment | Furazolidone | 0.04–200 μM | 1.22 nM | Water | Wei et al. (2024) |
| | UiO-66-$NH_2$-MXene@rGO | Ultrasonic treatment | Cadmium and lead | 2–800 ppb | 0.3 ppb | Water | Dong et al. (2024) |
| | Melamine-doped rGO/MXene | Ultrasonic treatment | $Zn^{2+}$, $Cd^{2+}$, and $Pb^{2+}$ | 1.0 and 1000 μg/L | 0.48 μg/L, 0.45 μg/L and 0.29 μg/L | Tap water | Chen et al. (2023a) |
| | AuNP/carbon quantum dots-MXene | Microwave reaction | Nitrite | 1–500 μM | 0.078 μM | Tap water | Feng et al. (2022) |
| | $Ti_3C_2$@N-C | Hydrothermal reaction | $Cd^{2+}$ and $Pb^{2+}$ | 0.1–4 μM; 0.05–2 μM | $Cd^{2+}$ 2.55 nM, $Pb^{2+}$ 1.10 nM | Water | Zhang et al. (2022) |
| | Graphene/MXene nanocomposite | Ultrasonic treatment | Bisphenol A | 1–10 μM | 0.35 μM | Water | Rajendran et al. (2022b) |
| Agricultural | MXene@AgNCs/ $NH_2$-MWCNTs/GCE | Ultrasonic treatment | Carbendazim | 0.3 nM to 10 μM | 0.1 nM | Food | Zhong et al. (2021) |
| | Graphene/MXene | Ultrasonic treatment | Rutin | 0.005–10.0 μM | 1.8 nM | Buckwheat tea | Shi et al. (2024) |
| | MXene/MWCNTs | Ultrasonic treatment | Synephrine | 0.5–70 μM | 0.167 μM | Fruit | Gao et al. (2020) |
| | MXene-MWCNTs composite | Ultrasonic treatment | Indole-3-acetic acid | 0.05–125.0 μmol/L | 16.7 nmol/L | Plant leaves | Chen et al. (2023b) |
| | MXene-modified carbon cloth | Electrodeposition | $H_2O_2$ | 1–100 μM | 200 nM | Milk | Arjun et al. (2023) |
| | PDA/MXene/MWCNTs/ $NiCo_2O_4$ nanocomposites | Ultrasonic treatment | Aflatoxin B1 | 2.5–200 ng/mL | 1.890 ng/mL | Food samples | Shen et al. (2023) |

performance through several key mechanisms. Graphene, with its high surface area and excellent electrical conductivity, facilitates efficient electron transport and provides numerous active sites for analyte interaction, enhancing sensitivity (Geim and Novoselov 2007; Iro et al. 2021; Rajendran, Reshetilov, and Sundramoorthy 2021a, b; Jerome and Sundramoorthy 2020; Govindaraj et al. 2023). CNTs enhance performance due to their high aspect ratio and unique electronic properties, which promote rapid electron transfer and increase the density of reactive sites available for analyte binding, thereby improving both sensitivity and selectivity (Kumar et al. 2021; Baughman, Zakhidov, and De Heer 2002; Vignesh Kumar and Rajendran 2024). Additionally, carbon dots offer tunable photoluminescence and superior biocompatibility, enabling highly specific and sensitive detection of various analytes through fluorescence resonance energy transfer (FRET) mechanisms and other photophysical interactions (Jerome and Sundramoorthy 2019; Li et al. 2012). These properties, combined with the intrinsic advantages of MXenes, result in sensors with superior performance characteristics. This multicomponent synergy is not only improving the sensory performance but also introducing capabilities for multifunctional and multimodal detection platforms.

CNTs, carbon nanotubes; CuMOF, copper metal-organic framework; N, S-CDs, nitrogen–sulfur-doped carbon dots; PDA/GCE, polydopamine/glassy carbon electrode; $Nb_2CT_x$/NRGO, niobium carbide/nitrogen-doped reduced graphene oxide; $Cu_2O$/M/AC, cuprous oxide/MXene/activated carbon; $Ti_3C_2T_x$/CCY, MXene/conductive carbon yarn; Au/rGO-$Ti_3C_2$, gold/reduced graphene oxide-MXene; $NbTiCT_x$/MWCNTs, MXene/multi-walled carbon nanotubes; AuNPs, gold nanoparticles; MXene/Pt/C, MXene/carbon platinum; Au–Pd/MXene/LSG, bimetallic palladium–gold/MXene/laser scribed porous graphene; CNH, carbon nanohorn; g-$C_3N_4$, graphitic carbon nitride; $Ti_3C_2$/G-MWCNTs/ZnO, MXene/graphene modified multi-walled carbon nanotubes/zinc oxide; SWCNTs, single-walled carbon nanotubes; D-$Ti_3C_2T_x$/GR/GCE, delaminated titanium carbide/graphene/glassy carbon electrode; $Ti_3C_2$@N-C, nitrogen doped carbon-coated MXene heterostructure; AgNCs, silver nanoclusters; $NH_2$-MWCNTs, amino-functionalized multi-walled carbon nanotubes; GCE, glass carbon electrode; PDA, poly dopamine.

However, future development of these sensors faces critical challenges, particularly concerning the chemical stability and the propensity for MXene nanosheets to restack in ambient environments. The principal concern lies in irreversible oxidation, which can degrade MXenes, thereby curtailing the practical deployment of these materials in electronic sensing devices. Recent explorations into functionalizing MXene with other nanomaterials and merging MXene electrochemical sensors with photoelectrochemical technologies signify a promising direction. These novel sensor designs offer exciting prospects for enhancing sensitivity and selectivity. The escalating urgency for environmental monitoring and public health safeguards highlights the need for sensors capable of rapid, on-site detection of trace pollutants like perfluorinated compounds, antibiotics, and disinfection by-products in diverse environmental matrices. While MXene–carbon hybrid sensors have proven effective in laboratory settings, their performance in real environmental conditions, especially in terms of resisting interference, requires further investigation and validation. Additionally, the development of high-throughput sensor technology that can adapt to complex, multicomponent dynamics of real-world environments is an essential research trajectory.

In conclusion, MXene–carbon-modified electrochemical sensors represent a cutting-edge advancement in sensor technology. Due to their distinct architecture and multifunctional behavior, these composites are poised to revolutionize the field of electrochemical sensing, promoting enhanced capabilities for environmental and biomedical applications.

## ACKNOWLEDGMENTS

This research was funded through start-up support awarded to R.E. by the Henry Samueli School of Engineering and the Department of Electrical Engineering and Computer Science at the University of California, Irvine.

## REFERENCES

Alanazi, Nadyah, Tamil S. Gopal, Muthumareeswaran Muthuramamoorthy, Amani A. E Alobaidi, Reem A. Alsaigh, Mohammed H. Aldosary, Saravanan Pandiaraj, Maram Almutairi, Andrews N. Grace, and Abdullah Alodhayb. 2023. "$Cu_2O$/MXene/rGO ternary nanocomposites as sensing electrodes for nonenzymatic glucose sensors." *ACS Applied Nano Materials* 6 (13):12271–12281.

Ankitha, Menon, Fathima Shamsheera, and P. Abdul Rasheed. 2024. "MXene-integrated single-stranded carbon yarn-based wearable sensor patch for on-site monitoring of dopamine." *ACS Applied Electronic Materials* 6 (1):599–610.

Arjun, Ajith M., Neermunda Shabana, Menon Ankitha, and P. Abdul Rasheed. 2023. "Electrochemical deposition of Prussian blue on $Nb_2CT_x$ MXene modified carbon cloth for the non-enzymatic electrochemical detection of hydrogen peroxide." *Microchemical Journal* 185:108301.

Baughman, Ray H., Anvar A. Zakhidov, and Walt A. De Heer. 2002. "Carbon nanotubes–The route toward applications." *Science* 297 (5582):787–792.

Bolourinezhad, Monireh, Majid Rezayi, Zahra Meshkat, Saman Soleimanpour, Majid Mojarrad, Seyed Hamid Aghaee-Bakhtiari, and Seyed M. Taghdisi. 2023. "Design of a rapid electrochemical biosensor based on MXene/Pt/C nanocomposite and DNA/RNA hybridization for the detection of COVID-19." *Talanta* 265:124804.

Cao, Fangcheng, Ye Zhang, Hongqing Wang, Karim Khan, Ayesha K. Tareen, Wenjing Qian, Han Zhang, and Hans Ågren. 2022. "Recent advances in oxidation stable chemistry of 2D MXenes." *Advanced Materials* 34 (13):2107554.

Chakoma, Shingirirai, Jerome Rajendran, Xiaochang Pei, Anita Ghandehari, Jorge A. Tavares Negrete, and Rahim Esfandyarpour. 2024a. "Pioneering 3D-printed devices with triboelectric nanogenerators for advanced self-powered force sensing." *IEEE Sensors Letters,* 9 (1):1–4.

Chakoma, Shingirirai, Jerome Rajendran, Xiaochang Pei, Anita Ghandehari, Jorge A. Tavares Negrete, and Rahim Esfandyarpour. 2024b. "Reusable and wireless sensor for on-demand molecular stress biomarker assessment." *2024 IEEE 20th International Conference on Body Sensor Networks (BSN),* Chicago, IL, USA.

Chakoma, Shingirirai, Xiaochang Pei, Huiting Qin, Anita Ghandehari, Sahar NajafiKhoshnoo, Jerome Rajendran, and Rahim Esfandyarpour. 2024c. "A passive, reusable, and resonating wearable sensing system for on-demand, non-invasive, and wireless molecular stress biomarker detection." *Nano Research,* 17 (8):7542–7556.

Chakoma, Shingirirai, Jerome Rajendran, Xiaochang Pei, Anita Ghandehari, Jorge A. T. Negrete, and Rahim Esfandyarpour. 2025. "A flexible self-powered multi-nanomaterials 3D-printed triboelectric nanogenerator-based force sensor." In *Oxide-based Materials and Devices XVI*, San Francisco, California, United States.

Chen, Aiting, Yanping Wei, Du Tuo, Chuanqin Zhou, Shuting Shi, Nana Tang, Quanguo He, and Jun Liu. 2024. "Submicromolar electrochemical sensing platform for fast fluoxetine quantification based on $Nb_2CT_x$ MXene and nitrogen-doped graphene oxide nanocomposites." *Journal of Alloys and Compounds* 970:172557.

Chen, Chi, Muhammad Boota, Xiuqiang Xie, Mengqiang Zhao, Babak Anasori, Chang E. Ren, Ling Miao, Jianjun Jiang, and Yury Gogotsi. 2017. "Charge transfer induced polymerization of EDOT confined between 2D titanium carbide layers." *Journal of Materials Chemistry A* 5 (11):5260–5265.

Chen, Jing, Ying Chen, Shuying Li, Jiao Yang, Jianbin Dong, and Xiaoquan Lu. 2022. "MXene/CNTs/Cu-MOF electrochemical probe for detecting tyrosine." *Carbon* 199:110–118. doi: 10.1016/j.carbon.2022.07.021.

Chen, Mengmeng, Xiaoyu Hu, Kun Li, Jinkun Sun, Zhuangjian Liu, Baigang An, Xiang Zhou, and Zunfeng Liu. 2020. "Self-assembly of dendritic-lamellar MXene/carbon nanotube conductive films for wearable tactile sensors and artificial skin." *Carbon* 164:111–120.

Chen, Shuxian, Min Shi, Jing Yang, Yongfang Yu, Quan Xu, Jingkun Xu, Xuemin Duan, Yansha Gao, and Limin Lu. 2021a. "MXene/carbon nanohorns decorated with conductive molecularly imprinted poly (hydroxymethyl-3,4-ethylenedioxythiophene) for voltammetric detection of adrenaline." *Microchimica Acta* 188:1–11.

Chen, Shuxian, Min Shi, Quan Xu, Jingkun Xu, Xuemin Duan, Yansha Gao, Limin Lu, Feng Gao, Xiaoqiang Wang, and Yongfang Yu. 2021b. "$Ti_3C_2T_x$ MXene/nitrogen-doped reduced graphene oxide composite: A high-performance electrochemical sensing platform for adrenaline detection." *Nanotechnology* 32 (26):265501.

Chen, Yuanyuan, Peng Zhao, Yi Liang, Yi Ma, Yiyi Liu, Jinsong Zhao, Jingzhou Hou, Changjun Hou, and Danqun Huo. 2023a. "A sensitive electrochemical sensor based on 3D porous melamine-doped rGO/MXene composite aerogel for the detection of heavy metal ions in the environment." *Talanta* 256:124294.

Chen, Yuxue, Yunxiu Sun, Yanyan Niu, Baoli Wang, Zejun Zhang, Lina Zeng, Lin Li, and Wei Sun. 2023b. "Portable electrochemical sensing of indole-3-acetic acid based on self-assembled MXene and multi-walled carbon nanotubes composite modified screen-printed electrode." *Electroanalysis* 35 (3):e202200279.

Cheng, Hao, Liwen Zhang, Jun Feng, Tingfan Tang, and Danfeng Qin. 2023. "A novel sensor based on $Ti_3C_2$ MXene/$Co_3O_4$/carbon nanofibers composite for the sensitive detection of 4-aminophenol." *Chemosphere* 341:139981.

Ding, Li, Libo Li, Yanchang Liu, Yi Wu, Zong Lu, Junjie Deng, Yanying Wei, Jürgen Caro, and Haihui Wang. 2020. "Effective ion sieving with $Ti_3C_2T_x$ MXene membranes for production of drinking water from seawater." *Nature Sustainability* 3 (4):296–302.

Dong, Hanyu, Peng Xiao, Na Jin, Binbin Wang, Ying Liu, and Zifeng Lin. 2021. "Molten salt derived $Nb_2CT_x$ MXene anode for Li-ion batteries." *ChemElectroChem* 8 (5):957–962.

Dong, Jiangbo, Xinyao Li, Li Wen, Yi Ma, Jingbing Xu, Huibo Luo, Jingzhou Hou, Changjun Hou, and Danqun Huo. 2024. "A novel electrochemical strategy based on MXene@rGO composite aerogel-doped UiO-66-$NH_2$ for simultaneous detection of cadmium and lead in grain and water samples." *Food Chemistry* 437:137835.

Feng, Lei, Wenjing Qin, Yanli Wang, Changshun Gu, Xinxin Li, Junji Chen, Jianxing Chen, Hanying Qiao, Min Yang, and Zhenhao Tian. 2023. "$Ti_3C_2T_x$ MXene/graphene/AuNPs 3D porous composites for high sensitivity and fast response glucose biosensing." *Microchemical Journal* 184:108142.

Feng, Xiwen, Guangda Han, Jihai Cai, and Xiaoying Wang. 2022. "Au@carbon quantum dots-MXene nanocomposite as an electrochemical sensor for sensitive detection of nitrite." *Journal of Colloid and Interface Science* 607:1313–1322.

Firestein, Konstantin L., Joel E. von Treifeldt, Dmitry G. Kvashnin, Joseph F. S. Fernando, Chao Zhang, Alexander G. Kvashnin, Evgeny V. Podryabinkin, Alexander V. Shapeev, Dumindu P. Siriwardena, and Pavel B. Sorokin. 2020. "Young's modulus and tensile strength of $Ti_3C_2$ MXene nanosheets as revealed by in situ TEM probing, AFM nanomechanical mapping, and theoretical calculations." *Nano Letters* 20 (8):5900–5908.

Fu, Bo, Jingxuan Sun, Cong Wang, Ce Shang, Lijun Xu, Jiebo Li, and Han Zhang. 2021. "MXenes: Synthesis, optical properties, and applications in ultrafast photonics." *Small* 17 (11):2006054.

Gao, Meng, Yu Xie, Wuliang Yang, and Limin Lu. 2020. "Fabrication of novel electrochemical sensor based on MXene/MWCNTs composite for sensitive detection of synephrine." *International Journal of Electrochemical Science* 15 (5):4619–4630.

Gao, Pan, Haiting Shi, Tianshuai Ma, Shuaitong Liang, Yuanhua Xia, Zhiwei Xu, Shuo Wang, Chunying Min, and Liyan Liu. 2021. "MXene/$TiO_2$ heterostructure-decorated hard carbon with stable Ti–O–C bonding for enhanced sodium-ion storage." *ACS Applied Materials & Interfaces* 13 (43):51028–51038.

Geim, Andre K., and Konstantin S. Novoselov. 2007. "The rise of graphene." *Nature Materials* 6 (3):183–191.

Ghandehari, Anita, Jorge A. Tavares-Negrete, Jerome Rajendran, Qian Yi, and Rahim Esfandyarpour. 2024. "Optimization of process parameters in 3D-nanomaterials printing for enhanced uniformity, quality, and dimensional precision using physics-guided artificial neural network." *Discover Nano* 19 (1):1–17.

Gopal, Tami S., Jaimson T. James, Bharath Gunaseelan, Karthikeyan Ramesh, Vimala Raghavan, K. Amarnath, V. Ganesh Kumar, Sofia Jennifer Rajasekaran, Saravanan Pandiaraj, and M. R. Muthumareeswaran. 2024. "MXene-embedded porous carbon-based $Cu_2O$ nanocomposites for non-enzymatic glucose sensors." *ACS Omega* 9 (7):8448.

Govindaraj, Muthukumar, Jerome Rajendran, Udhaya Ganesh P. K., Magesh Kumar Muthukumaran, and Balamurugan Jayaraman. 2023. "Graphitic carbon nitride nanosheets decorated with strontium tungstate nanospheres as an electrochemical transducer for sulfamethazine sensing." *ACS Applied Nano Materials* 6 (2):930–945.

Guo, Ying, Deyang Zhang, Ya Yang, Yangbo Wang, Zuxue Bai, Paul K. Chu, and Yongsong Luo. 2021. "MXene-encapsulated hollow $Fe_3O_4$ nanochains embedded in N-doped carbon nanofibers with dual electronic pathways as flexible anodes for high-performance Li-ion batteries." *Nanoscale* 13 (8):4624–4633.

Hantanasirisakul, Kanit, Babak Anasori, Slavomir Nemsak, James L. Hart, Jiabin Wu, Yizhou Yang, Rajesh V. Chopdekar, Padraic Shafer, Andrew F. May, and Eun J. Moon. 2020. "Evidence of a magnetic transition in atomically thin $Cr_2TiC_2T_x$ MXene." *Nanoscale Horizons* 5 (12):1557–1565.

Hou, Yaru, Fanyong Yan, Xinyi Bai, Xin Wei, Liang Chen, and Yang Fu. 2024. "Designing advanced 0D/2D heterojunctions of N, S-co-doped carbon dots/$Ti_3C_2T_x$ MXene nanosheets in electrochemical sensor applications." *Ceramics International* 50 (7): 10735–10745.

Hu, Luhing, Beom J. Kim, Seunghyeon Ji, Juyeong Hong, Ajit K. Katiyar, and Jong-Hyun Ahn. 2022. "Smart electronics based on 2D materials for wireless healthcare monitoring." *Applied Physics Reviews* 9 (4):041308.

Iqbal, Aamir, Junpyo Hong, Tae Y. Ko, and Chong M. Koo. 2021. "Improving oxidation stability of 2D MXenes: Synthesis, storage media, and conditions." *Nano Convergence* 8 (1):9.

Iro, Zaharaddeen S., C. Subramani, Jerome Rajendran, and Ashok Kumar Sundramoorthy. 2021. "Promising nature-based activated carbon derived from flowers of Borassus flabellifer for supercapacitor applications." *Carbon Letters* 31(6):1145–1153.

Jerome Rajendran, and Ashok Kumar Sundramoorthy. 2019. "Hydrothermal synthesis of boron nitride quantum dots/poly(luminol) nanocomposite for selective detection of ascorbic acid." *Journal of the Electrochemical Society* 166 (9):B3017.

Jerome, Rajendran, and Ashok K. Sundramoorthy. 2020. "Preparation of hexagonal boron nitride doped graphene film modified sensor for selective electrochemical detection of nicotine in tobacco sample." *Analytica Chimica Acta* 1132:110–120.

Jolly, Shae, M. Parans Paranthaman, and Michael Naguib. 2021. "Synthesis of $Ti_3C_2T_z$ MXene from low-cost and environmentally friendly precursors." *Materials Today Advances* 10:100139.

Kalambate, Pramod K., Vitsarut Primpray, Wichayaporn Kamsong, Chanpen Karuwan, and Wanida Laiwattanapaisal. 2023. "Disposable electrochemical sensor based on MXene/graphene nanoplatelets/ionic liquid for the determination of anticancer drug imiquimod." *Journal of the Electrochemical Society* 170 (11):117508.

Khaledialidusti, Rasoul, Abhishek K. Mishra, and Afrooz Barnoush. 2020. "Atomic defects in monolayer ordered double transition metal carbide ($Mo_2TiC_2T_x$) MXene and $CO_2$ adsorption." *Journal of Materials Chemistry C* 8 (14):4771–4779.

Khazaei, Mohammad, Masao Arai, Taizo Sasaki, Chan-Yeup Chung, Natarajan S. Venkataramanan, Mehdi Estili, Yoshio Sakka, and Yoshiyuki Kawazoe. 2013. "Novel electronic and magnetic properties of two-dimensional transition metal carbides and nitrides." *Advanced Functional Materials* 23 (17):2185–2192.

Kumar, Thomaati H. V., Jerome Rajendran, Ramila D. Nagarajan, Gayathri Jeevanandam, Anatoly N. Reshetilov, and Ashok K. Sundramoorthy. 2021. "Selective chemistry-based separation of semiconducting single-walled carbon nanotubes and alignment of the nanotube array network under electric field for field-effect transistor applications." *ACS Omega* 6 (8):5146–5157.

Lee, Changgu, Xiaoding Wei, Jeffrey W. Kysar, and James Hone. 2008. "Measurement of the elastic properties and intrinsic strength of monolayer graphene." *Science* 321 (5887): 385–388.

Lee, Sang W., Xiaochang Pei, Jerome Rajendran, and Rahim Esfandyarpour. 2023. "A wireless and battery-free wearable pressure sensing system for human-machine interaction and health monitoring." *IEEE Journal on Flexible Electronics* 2(6):439–447.

Levitt, Ariana S., Mohamed Alhabeb, Christine B. Hatter, Asia Sarycheva, Genevieve Dion, and Yury Gogotsi. 2019. "Electrospun MXene/carbon nanofibers as supercapacitor electrodes." *Journal of Materials Chemistry A* 7 (1):269–277.

Li, Haitao, Zhenhui Kang, Yang Liu, and Shuit-Tong Lee. 2012. "Carbon nanodots: synthesis, properties and applications." *Journal of Materials Chemistry* 22 (46):24230–24253.

Li, Hui, Rui Chen, Mumtaz Ali, Haiwon Lee, and Min J. Ko. 2020. "In situ grown MWCNTs/MXenes nanocomposites on carbon cloth for high-performance flexible supercapacitors." *Advanced Functional Materials* 30 (47):2002739.

Li, Lu, Mingyi Zhang, Xitian Zhang, and Zhiguo Zhang. 2017. "New $Ti_3C_2$ aerogel as promising negative electrode materials for asymmetric supercapacitors." *Journal of Power Sources* 364:234–241.

Liao, Leiping, Aitang Zhang, Kun Zheng, Rui Liu, Yujun Cheng, Lihua Wang, Aihua Li, and Jingquan Liu. 2021. "Fabrication of cobaltous sulfide nanoparticle-modified 3D MXene/carbon foam hybrid aerogels for all-solid-state supercapacitors." *ACS Applied Materials & Interfaces* 13 (24):28222–28230.

Liu, Hengchao, Wenjing Qin, XinXin Li, Lei Feng, Changshun Gu, Junji Chen, Zhenhao Tian, Jianxing Chen, Min Yang, and Hanying Qiao. 2023. "Molecularly imprinted electrochemical sensors based on $Ti_3C_2T_x$-MXene and graphene composite modifications for ultrasensitive cortisol detection." *Analytical Chemistry* 95 (44):16079–16088.

Liu, Junqi, Meng Wang, Saijin Wang, Yanli Zhao, Shi Zhou, Jinlian Li, and Dongmei Wu. 2024. "Assessment of oxygenated polycyclic aromatic hydrocarbon cytotoxicity using $TiO_2$-MXene/MWCNTs/PDA/GCE electrochemical sensor." *Electrochimica Acta* 474: 143515.

Liu, Yi-Tao, Peng Zhang, Ning Sun, Babak Anasori, Qi-Zhen Zhu, Huan Liu, Yury Gogotsi, and Bin Xu. 2018. "Self-assembly of transition metal oxide nanostructures on MXene nanosheets for fast and stable lithium storage." *Advanced Materials* 30 (23):1707334.

Luo, Jianmin, Edward Matios, Huan Wang, Xinyong Tao, and Weiyang Li. 2020. "Interfacial structure design of MXene-based nanomaterials for electrochemical energy storage and conversion." *InfoMat* 2 (6):1057–1076.

Mari, Elancheziyan, Murugesan Duraisamy, Muthusankar Eswaran, Senthilkumar Sellappan, Keehoon Won, Pranjal Chandra, Pei-Chien Tsai, Po-Chin Huang, Yi-Hsun Chen, and Yuan-Chung Lin. 2024. "Highly electrochemically active $Ti_3C_2T_x$ MXene/MWCNT nanocomposite for the simultaneous sensing of paracetamol, theophylline, and caffeine in human blood samples." *Microchimica Acta* 191 (4):1–14.

Nagalingam, Suruthi P., and Andrews N. Grace. 2022. "Poly(3,4-ethylenedioxythiophene) decorated MXene as an alternative counter electrode for dye-sensitized solar cells." *Materials Today Chemistry* 26:101113.

Naguib, Michael, Murat Kurtoglu, Volker Presser, Jun Lu, Junjie Niu, Min Heon, Lars Hultman, Yury Gogotsi, and Michel W. Barsoum. 2011. "Two-dimensional nanocrystals produced by exfoliation of $Ti_3AlC_2$." In *MXenes,* edited by Yury Gogotsi, 15–29. Jenny Stanford Publishing.

NajafiKhoshnoo, Sahar, Taeil Kim, Jorge A. Tavares-Negrete, Xiaochang Pei, Prativa Das, Sang W. Lee, Jerome Rajendran, and Rahim Esfandyarpour. 2023. "A 3D nanomaterials-printed wearable, battery-free, biocompatible, flexible, and wireless pH sensor system for real-time health monitoring." *Advanced Materials Technologies* 8 (8):2201655.

Ni, Meijun, Jia Chen, Chenxi Wang, Yilin Wang, Linzi Huang, Weicheng Xiong, Pengcheng Zhao, Yixi Xie, and Junjie Fei. 2022. "A high-sensitive dopamine electrochemical sensor based on multilayer $Ti_3C_2$ MXene, graphitized multi-walled carbon nanotubes and ZnO nanospheres." *Microchemical Journal* 178:107410.

Novoselov, Kostya S., Andre K. Geim, Sergei V. Morozov, De-eng Jiang, Yanshui Zhang, Sergey V. Dubonos, Irina V. Grigorieva, and Alexandr A. Firsov. 2004. "Electric field effect in atomically thin carbon films." *Science* 306 (5696):666–669.

Özcan, Nermin, Hilal Medetalibeyoglu, Onur Akyıldırım, Necip Atar, and Mehmet L. Yola. 2020. "Electrochemical detection of amyloid-β protein by delaminated titanium carbide MXene/multi-walled carbon nanotubes composite with molecularly imprinted polymer." *Materials Today Communications* 23:101097.

Pandey, Mayank, Kalim Deshmukh, Akhila Raman, Aparna Asok, Saritha Appukuttan, and G. Ramanath Suman. 2024. "Prospects of MXene and graphene for energy storage and conversion." *Renewable and Sustainable Energy Reviews* 189:114030.

Qu, Guangfei, Yun Zhang, Junhong Zhou, Huimin Tang, Wei Ji, Zhoupeng Yan, Keheng Pan, and Ping Ning. 2023. "Simultaneous electrochemical detection of dimethyl bisphenol A and bisphenol A using a novel Pt@SWCNTs-MXene-rGO modified screen-printed sensor." *Chemosphere* 337:139315.

Quan, Changyun, Lingling Quan, Qinying Wen, Minghui Yang, and Ting Li. 2024. "Alanine aminotransferase electrochemical sensor based on graphene@MXene composite nanomaterials." *Microchimica Acta* 191 (1):45.

Rajendran, Jerome. 2023. "Amperometric determination of salivary thiocyanate using electrochemically fabricated poly(3,4-ethylenedioxythiophene)/MXene hybrid film." *Journal of Hazardous Materials* 449:130979.

Rajendran, Jerome, Anatoly N. Reshetilov, and Ashok K. Sundramoorthy. 2021a. "An electrochemically exfoliated graphene/poly(3,4-ethylenedioxythiophene) nanocomposite-based electrochemical sensor for the detection of nicotine." *Materials Advances* 2 (10): 3336–3345.

Rajendran, Jerome, Anatoly N. Reshetilov, and Ashok K. Sundramoorthy. 2021b. "Preparation of hybrid paper electrode based on hexagonal boron nitride integrated graphene nanocomposite for free-standing flexible supercapacitors." *RSC Advances* 11 (6):3445–3451.

Rajendran, Jerome, Ashok K. Sundramoorthy, Dhanraj Ganapathy, Raji Atchudan, Mohamed A. Habila, and Deepak Nallaswamy. 2022a. "2D MXene/graphene nanocomposite preparation and its electrochemical performance towards the identification of nicotine level in human saliva." *Journal of Hazardous Materials* 440:129705.

Rajendran, Jerome, and Rahim Esfandyarpour. 2024. "Revolutionizing personalized health: The frontier of wearable biomolecule sensors through 3D printing innovation." *Biomedical Materials & Devices* 3(2):818–834.

Rajendran, Jerome, Tamil S. Kannan, Lokhendra S. Dhanasekaran, Preethika Murugan, Raji Atchudan, Zeid A. ALOthman, Mohamed Ouladsmane, and Ashok K. Sundramoorthy. 2022b. "Preparation of 2D graphene/MXene nanocomposite for the electrochemical determination of hazardous bisphenol A in plastic products." *Chemosphere* 287:132106.

Ren, Xiangyi, Minfeng Huo, Mengmeng Wang, Han Lin, Xuxia Zhang, Jun Yin, Yu Chen, and Honghong Chen. 2019. "Highly catalytic niobium carbide (MXene) promotes hematopoietic recovery after radiation by free radical scavenging." *ACS Nano* 13 (6):6438–6454.

Rogdakis, Konstantinos, Nikolaos Karakostas, and Emmanuel Kymakis. 2021. "Up-scalable emerging energy conversion technologies enabled by 2D materials: From miniature power harvesters towards grid-connected energy systems." *Energy & Environmental Science* 14 (6):3352–3392.

Shang, Ling-Jun, Shuai-Qun Yu, Xin-Wei Shang, Xiang-Yu Wei, Hui-Ying Wang, Wen-Shuai Jiang, and Qiong-Qiong Ren. 2024. "A non-invasive glucose sensor based on 3D reduced graphene oxide-MXene and AuNPs composite electrode for the detection of saliva glucose." *Journal of Applied Electrochemistry* 54 (8):1–11.

Shen, Jinhui, Juan Liu, Siyi Yang, Xin Yao, Huanbao Fa, Changjun Hou, and Mei Yang. 2023. "Novel electrochemical sensor based on PDA/MXene/MWCNTs/$NiCo_2O_4$ nanocomposites for rapid, sensitive, and selective detection of aflatoxin B1." *Food Analytical Methods* 16 (6):1055 1068.

Shi, Yanmei, Kai Hu, Lin Mei, Liqin Chao, Mingxia Wu, Zhihong Chen, Xiangxiang Wu, Jingyi Qiao, Pingsheng Zhu, and Mingsan Miao. 2024. "Platforms of graphene/MXene heterostructure for electrochemical monitoring of rutin in drug and Tartary buckwheat tea." *Talanta* 270:125548.

Tang, Nana, Yanping Wei, Quanguo He, Shuting Shi, Chuanqin Zhou, Aiting Chen, and Jun Liu. 2024. "Facile electrochemical determination of sertraline at nanomolar concentration using bimetallic MXene $NbTiCT_x$ and MWCNTs hybridized electrodes as sensing platform." *Materials Today Chemistry* 36:101901.

Tans, Sander J., Alwin R. M. Verschueren, and Cees Dekker. 1998. "Room-temperature transistor based on a single carbon nanotube." *Nature* 393 (6680):49–52.

Vignesh Kumar, Thoomatti Haridass, and Jerome Rajendran. 2024. "Recent progress in electrochemical methods for microplastics detection." In *Microplastics and Pollutants*, edited by V. Sivasankar and T. G. Sunitha:249–263. Springer, Cham.

Wang, Chunmei, Renfei Cheng, Peng-Xiang Hou, Yonghui Ma, Abdul Majeed, Xiaohui Wang, and Chang Liu. 2020. "MXene-carbon nanotube hybrid membrane for robust recovery of Au from trace-level solution." *ACS Applied Materials & Interfaces* 12 (38): 43032–43041.

Wang, Jingtao, Tongkun Zhao, Zhihao Yang, Yang Chen, Yong Liu, Junxiao Wang, Pengfei Zhai, and Wenjia Wu. 2019. "MXene-based Co, N-codoped porous carbon nanosheets regulating polysulfides for high-performance lithium–sulfur batteries." *ACS Applied Materials & Interfaces* 11 (42):38654–38662.

Wang, Qiufan, Jiaheng Liu, Guofu Tian, and Daohong Zhang. 2021. "Co@N-CNT/MXenes in situ grown on carbon nanotube film for multifunctional sensors and flexible supercapacitors." *Nanoscale* 13 (34):14460–14468.

Wang, Xue, Mengjia Li, Shuang Yang, Xuting Bai, and Jiajia Shan. 2022. "Self-assembled $Ti_3C_2T_X$ MXene/graphene composite for the electrochemical reduction and detection of p-nitrophenol." *Microchemical Journal* 179:107473.

Wang, Ying, Peng Zhao, Bo Gao, Min Yuan, Jiabing Yu, Zeping Wang, and Xianping Chen. 2023. "Self-reduction of bimetallic nanoparticles on flexible MXene-graphene electrodes for simultaneous detection of ascorbic acid, dopamine, and uric acid." *Microchemical Journal* 185:108177.

Wei, Kunling, Keheng Pan, Guangfei Qu, Jin Qin, Jiaxin Lv, and Yuqi Liang. 2024. "Self-assembly of GO-MXene hybrid composites by polyelectrolyte: As an efficient electrocatalyst for selective electrochemical detection of the antibiotic furazolidone." *Journal of Environmental Chemical Engineering* 12(2):112319.

Wei, Ming, Bing Wang, Miao Chen, Honghong Lyu, Xinqing Lee, Shengsen Wang, Zebin Yu, and Xueyang Zhang. 2022. "Recent advances in the treatment of contaminated soils by ball milling technology: Classification, mechanisms, and applications." *Journal of Cleaner Production* 340:130821.

Wei, Yi, Peng Zhang, Razium A. Soomro, Qizhen Zhu, and Bin Xu. 2021. "Advances in the synthesis of 2D MXenes." *Advanced Materials* 33 (39):2103148.

Wu, Zhitan, Tongxin Shang, Yaqian Deng, Ying Tao, and Quan-Hong Yang. 2020. "The assembly of MXenes from 2D to 3D." *Advanced Science* 7 (7):1903077.

Xie, Xiuqiang, Meng-Qiang Zhao, Babak Anasori, Kathleen Maleski, Chang E. Ren, Jingwen Li, Bryan W. Byles, Ekaterina Pomerantseva, Guoxiu Wang, and Yury Gogotsi. 2016. "Porous heterostructured MXene/carbon nanotube composite paper with high volumetric capacity for sodium-based energy storage devices." *Nano Energy* 26:513–523.

Xuan, Jinnan, Zhiqiang Wang, Yuyan Chen, Dujuan Liang, Liang Cheng, Xiaojing Yang, Zhuang Liu, Renzhi Ma, Takayoshi Sasaki, and Fengxia Geng. 2016. "Organic-base-driven intercalation and delamination for the production of functionalized titanium carbide nanosheets with superior photothermal therapeutic performance." *Angewandte Chemie* 128 (47):14789–14794.

Yan, Sheng-xue, Shao-hua Luo, Qing Wang, Ya-hui Zhang, and Xin Liu. 2021. "Rational design of hierarchically sulfide and MXene-reinforced porous carbon nanofibers as advanced electrode for high energy density flexible supercapacitors." *Composites Part B: Engineering* 224:109246.

Yi, Qian, Xiaochang Pei, Prativa Das, Huiting Qin, Sang W. Lee, and Rahim Esfandyarpour. 2022. "A self-powered triboelectric MXene-based 3D-printed wearable physiological biosignal sensing system for on-demand, wireless, and real-time health monitoring." *Nano Energy* 101:107511.

Yu, Min-Feng, Oleg Lourie, Mark J. Dyer, Katerina Moloni, Thomas F. Kelly, and Rodney S. Ruoff. 2000. "Strength and breaking mechanism of multiwalled carbon nanotubes under tensile load." *Science* 287 (5453):637–640.

Yuan, Chen, Zhilong He, Qiaowei Chen, Xuandong Wang, Chunyang Zhai, and Mingshan Zhu. 2021. "Selective and efficacious photoelectrochemical detection of ciprofloxacin based on the self-assembly of 2D/2D g-$C_3N_4$/$Ti_3C_2$ composites." *Applied Surface Science* 539:148241.

Yue, Yunliang, Buwei Wang, Nanxi Miao, Chao Jiang, Hongwei Lu, Bowen Zhang, Yankai Wu, Jie Ren, and Min Wang. 2021. "Tuning the magnetic properties of $Zr_2N$ MXene by biaxial strain." *Ceramics International* 47 (2):2367–2373.

Zhan, Xiaoxue, Chen Si, Jian Zhou, and Zhimei Sun. 2020. "MXene and MXene-based composites: Synthesis, properties and environment-related applications." *Nanoscale Horizons* 5 (2):235–258.

Zhang, Lu, Yao Lu, Shaowei Lu, Hao Zhang, Ziping Zhao, Chengkun Ma, Keming Ma, and Xiaoqiang Wang. 2021a. "Lifetime health monitoring of fiber reinforced composites using highly flexible and sensitive MXene/CNT film sensor." *Sensors and Actuators A: Physical* 332:113148.

Zhang, Shipeng, Md Abu Zahed, Md Sharifuzzaman, Sanghyuk Yoon, Xue Hui, Sharat C. Barman, Sudeep Sharma, Hyo S. Yoon, Chani Park, and Jae Y. Park. 2021b. "A wearable battery-free wireless and skin-interfaced microfluidics integrated electrochemical sensing patch for on-site biomarkers monitoring in human perspiration." *Biosensors and Bioelectronics* 175:112844.

Zhang, Xiao, Dong An, Zhaoshun Bi, Wei Shan, Beibei Zhu, Long Zhou, Liangmin Yu, Han Zhang, Shuwei Xia, and Meng Qiu. 2022. "$Ti_3C_2$-MXene@N-doped carbon heterostructure-based electrochemical sensor for simultaneous detection of heavy metals." *Journal of Electroanalytical Chemistry* 911:116239.

Zhang, Yunlong, Zhaoxuan Feng, Xiaoshan Wang, Han Hu, and Mingbo Wu. 2023. "MXene/carbon composites for electrochemical energy storage and conversion." *Materials Today Sustainability* 22:100350.

Zhang, Zhiwei, Zhihao Cai, Yi Zhang, Yaling Peng, Ziyuan Wang, Lun Xia, Suping Ma, Zhanzhao Yin, Ruofeng Wang, and Yishu Cao. 2021c. "The recent progress of MXene-based microwave absorption materials." *Carbon* 174:484–499.

Zhao, Meng-Qiang, Chang E. Ren, Zheng Ling, Maria R. Lukatskaya, Chuanfang Zhang, Katherine L. Van Aken, Michel W. Barsoum, and Yury Gogotsi. 2014. "Flexible MXene/carbon nanotube composite paper with high volumetric capacitance." *Advanced Materials* 27 (2):339–345.

Zhao, Peng, Yuanyuan Jia, Yi Liang, Jilin Zheng, Mei Yang, Danqun Huo, Yongzhong Wang, and Changjun Hou. 2021. "A Prussian blue-doped RGO/MXene composite aerogel with peroxidase-like activity for real-time monitoring of $H_2O_2$ secretion from living cells." *Chemical Communications* 57 (77):9870–9873.

Zhong, Wei, Feng Gao, Jin Zou, Shuwu Liu, Mingfang Li, Yansha Gao, Yongfang Yu, Xiaoqiang Wang, and Limin Lu. 2021. "MXene@Ag-based ratiometric electrochemical sensing strategy for effective detection of carbendazim in vegetable samples." *Food Chemistry* 360:130006.

Zhou, Fengchen, Shaobo Li, Liuzhang Ouyang, Jiangwen Liu, Jun Liu, Zhenguo Huang, and Min Zhu. 2022a. "Facile synthesis of black phosphorene via a low melting media assisted ball milling." *Chemical Engineering Journal* 444:136593.

Zhou, Rui, Bo Tu, Dacheng Xia, Hanping He, Zhiwei Cai, Nan Gao, Gang Chang, and Yunbin He. 2022b. "High-performance Pt/$Ti_3C_2T_x$ MXene based graphene electrochemical transistor for selective detection of dopamine." *Analytica Chimica Acta* 1201:339653.

# 11 MXene–Metal-Organic Framework-Modified Electrochemical Sensors

*Misheal Merlin V. S., Rijo Rajeev, and Anitha Varghese*

## 11.1 INTRODUCTION

MXenes are two-dimensional transition metal carbides, nitrides, or carbonitrides demonstrating properties including dispersibility in water and high conductivity (Rajeev and Varghese 2024). They exhibit properties such as hydrophilicity, excellent electrochemical activity, nanoscale thickness, ion intercalation performance, mechanical strength, environmental benignity, tunable band gap, and large surface-to-volume ratio with numerous active sites. The structural and electronic properties of MXenes render them suitable for applications such as electrochemical sensing (Rajeev et al. 2024; Rajeev, Thadathil, and Varghese 2023). MXene-based materials have diverse applications in electrochemical sensors (Sariga et al. 2023). MXenes such as titanium carbide are used as electrochemical biosensors to detect glucose with a low LOD of 2.6 μM (Wu et al. 2021a). Lei and co-workers developed non-invasive MXene-based electrochemical biosensors. These sensors monitor biomarkers such as glucose and lactate with a sensitivity of 35.3 $\mu A/mm/cm^2$ and 11.4 $\mu A/mm/cm^2$ for glucose and lactate, respectively (Lei et al. 2019). Rakhi et al. developed MXene-based sensors for glucose detection. These sensors exhibited selectivity, stability, and high sensitivity of 4.2 $\mu A/mM/cm^2$ and LOD of 5.9 μM (Rakhi et al. 2016). MXenes can be directly used as signal transducers or as electrocatalysts to enhance the detection of small molecules, drugs, pollutants, and biomarkers.

MOFs are porous crystalline polymers of inorganic metal ions and organic ligands (Jose, Sariga, and Varghese 2023). They exhibit high catalytic activity, large specific surface area, tuneability, and well-defined pore size. These properties are used for the modification of suitable applications, redox active sites, and antimicrobial properties (Mathew et al. 2024). MOFs promote rapid electron transfer through redox-hopping, π–π interactions, π–d interactions, electroactive guest-promoted transfer, and defect engineering (Xie, Skorupskii, and Dincă 2020) in the structure. The chemical functionalities that can interact with specific analytes are tunable. This makes them suitable for electrochemical sensing (Zhou et al. 2021; Ma et al. 2020). MXenes and MOFs have been investigated for various electrocatalytic-based applications. However, MXenes and MOFs utilized individually pose limitations that hurdle their potential. MXene sheets tend to agglomerate due to weak van der Waals


DOI: 10.1201/9781003736318-11

interactions. This agglomeration can hinder their ability to perform effectively in various applications due to inefficient electron transfer (Safarkhani et al. 2023; Liang et al. 2024). Pristine MOFs are susceptible to moisture, undergoing degradation and poor conduction of heat and electricity, which limits their applications. However, by forming MXene–MOF composites, each component's limitations can be overcome. The resulting composite possesses tailored properties such as high sensitivity, selectivity, low LOD, tunable electronic structure, high surface area, high dielectric losses, and facile synthetic methods (Saini et al. 2021).

Detection of various external stimuli for monitoring biological, physical, and chemical changes is achieved using sensors. Among different techniques used for sensing, electrochemical sensing has been reported to be superior and efficient. The principle of these sensors is based on the interaction between the analyte of interest and the sensing layer, which generates an electrical signal that is measured to obtain valuable information. These sensors monitor their environment and provide real-time data on the surrounding chemical composition (Baranwal et al. 2022). Electrochemical sensors detect various environmental pollutants, biomarkers, and food adulterants (Benny et al. 2021). These electrochemical sensors are classified into five categories: voltammetric, amperometric, impedimetric, potentiometric, and conductive. Voltammetric and amperometric methods are found to be superior for electrochemical sensing (Kajal et al. 2022). Electrochemical sensing of environmental pollutants, pesticides, food adulterants, and biomarkers is efficiently detected using composites such as MXene–MOF (Rodrigues et al. 2024). This chapter comprehensively outlines the properties, synthesis methods, and applicability of MXene–MOF composites for electrochemical sensing.

## 11.2 FUNDAMENTALS OF MXENE–MOF COMPOSITES

MXenes are 2D transition metal carbides, nitrides or carbonitrides with the general formula $M_{n+1}X_nT$ where M = early transition metal, $n + 1$ layers interleaved by $n$ layers, $n = 1, 2, 3, 4$ (Nidhi et al. 2024); T = surface termination such as –OH, –O, –F, or –I on the outermost M layers; X = carbon, nitrogen, or carbonitride; A = 13 or 14 group elements. In MXenes, the X element atoms are located at the octahedral interstitial sites within the hexagonal crystal lattice of the M elements. This configuration creates subunits composed of edge-sharing $M_6X$ octahedra. Meanwhile, the surface groups are positioned at various sites on the M elements, with their specific locations influenced by the chemistry of both the surface groups ($T_x$) and the M elements. The diversity, consistency, and precise control over the $T_x$ surface groups significantly affect the behavior of MXenes, which determines their performance (Lim et al. 2022; Riazi, Taghizadeh, and Soroush 2021). For MXenes containing two types of transition metals (double transition metal MXenes), the metal atoms (M) can be arranged in two ways: either randomly distributed (forming a solid solution) or in an ordered pattern. The ordered arrangement is energetically more favorable, meaning it has a lower energy state and is thus more stable (Pei et al. 2021). MXenes were discovered by Gogotsi and his team in 2011 using MAX phase (Naguib et al. 2011) (a layered material containing early transition metals, carbon, and nitrogen). The MAX phase was made to intercalate, where guest atoms or molecules were inserted between its layers. This weakens the bonding between the layers, followed by strong acid etching, which removes specific elements from the MAX phase and leaves behind the desired two-dimensional MXene structure. The properties of MXenes are governed by their composition, lateral size, and stacking order, which leads to chemical stability and mechanical and optical properties (Pei et al. 2021). MXenes demonstrate exceptional electrical conductivity, significantly accelerating heterogeneous electron transfer rates in electrochemical sensors. They are also easily synthesized with good solution dispersibility and stability, which is crucial for electrode modification via drop-casting. MXenes exhibit high stretchability and biocompatibility, making them ideal for developing flexible, wearable health monitoring and clinical analysis sensors (Wu et al. 2021b).

MOFs demonstrate large surface area, tunable porosity, and high porosity, which improve the sensitivity of the sensors through rapid mass transfer and the concentration of analytes. The unsaturated metal coordination sites and uniform network structure in MOFs enhance the catalytic activity. Small molecules and heavy metals are selectively detected due to the presence of specific channel sizes in MOFs (Kajal et al. 2022; Daniel et al. 2022). The formation of composites with MXenes and MOFs enhances the properties of various analytes for efficient sensing. MXene nanosheets are stabilized by the intercalation of MOFs between the individual nanosheets. These MOFs act as spacers between the MXene layers, which keep them separated (Saini et al. 2021). Therefore, the formation of composites with MXenes and MOF has been reported to demonstrate enhanced stability and performance for electrochemical sensing.

## 11.3 SYNTHESIS OF MXENE–MOF-BASED COMPOSITES

MXene–MOF hybrids are reported to be synthesized using various methods, including in situ growth, ex situ, solvothermal, hydrothermal, self-assembly, and electrophoretic deposition. These techniques influence the morphology, composition, and targeted applications of the hybrid material. The precursors, solvents, and reaction conditions are utilized to tailor the properties of these hybrids for specific applications.

### 11.3.1 In situ Methods of Synthesis for MXene–MOF Composites

In situ growth strategies have been utilized to prepare MXene–MOF hybrids with a uniform structure by directly growing MOFs on MXene surfaces or within their layers. This approach integrates both the materials, which maximizes their synergistic effects. The desired properties in the final MXene–MOF hybrid are achieved by precisely controlling the reaction parameters. Wen et al. synthesized CoNi-ZIF-67 (MOF), grown directly on $Ti_3C_2T_x$ MXene. This composite material was prepared through a relatively simple in situ coprecipitation method. The pre-synthesized $Ti_3C_2T_x$ MXene flakes acted as a scaffold for the growth of CoNi-ZIF-67. Due to etching-induced functional groups, the negatively charged surface of $Ti_3C_2T_x$ attracted positively charged $Co^{2+}$ and $Ni^{2+}$ ions. These ions interacted with the organic linker (2-methylimidazole) in solution, forming CoNi-ZIF-67 crystals directly on the MXene surface and within its interlayers. This in situ approach exhibited several advantages, such as eliminating separate MOF preparation and simplifying the overall process. A strong interaction between the MOF and MXene components was also developed, which improved the material's properties. This method was facile and had the potential for good interfacial contact with desirable characteristics in the final MXene–MOF composite (Wen et al. 2019). Xiao et al. fabricated a 2D/2D heterostructure electrode material, $Ti_3C_2T_x$ MXene nanosheets coupled with an ultrathin NiCo-MOF. Lignin was then introduced to modify the MXene surface. The NiCo-MOF nanosheets were grown directly onto the MXene sheets via the in situ process. The stability and uniform structure were obtained due to the bonding between the MOF's carboxylate and hydroxyl groups on the MXene and lignin. The NiCo-MOF offered abundant active sites, and excellent electron transport was facilitated by the conductive network of the MXene (Xiao et al. 2022b).

### 11.3.2 Solvothermal and Hydrothermal

Solvothermal and hydrothermal methods are employed for the synthesis of MXene–MOF-based composites. These methods offer excellent control over the structural properties of the product. Composites synthesized by these methods demonstrate high stability and orderly arrangement of the material (Mathew et al. 2024). Wu et al. developed a facile method to synthesize MOFs derived from MXenes. A delamination process using TMAOH intercalation was employed for the preparation of the MOF, and $H_2$TCPP (meso-tetra(4-carboxyl-phenyl) porphyrin) was chosen as the ligand due to its extensive planar aromatic ring and protonated carboxylic groups, which can effectively bind to MXene surface atoms. Later, the solvothermal technique was used to transform the $V_2CT_x$ MXene and $H_2$TCPP ligand into a V-PMOF ($V_2CT_x$-MOF) composite. The resultant MOF displayed a 2D nanosheet morphology. In this design, the guest molecules were incorporated (like non-volatile $H_3PO_4$) in such a way that they can interact with the porphyrin core's nitrogen atoms, aiming to endow the MOF with proton-conducting properties. The prepared $V_2CT_x$-PMOF nanosheets were stacked to form oriented thin films through spin coating. This approach created MOF thin films with proton conductivity suitable for high-performance electrochemical sensor devices (Wu et al. 2019). Ravipatti et al. synthesized Zr-MOF/$V_2C$ MXene composite on nickel foam for the detection of epinephrine from blood serum. These composites were synthesized via a solvothermal method. The synthesized compound exhibited excellent selectivity and a low detection limit (Ravipati, Sai Deepak, and Badhulika 2024). Wang et al. synthesized a composite material, which demonstrated excellent electrochemical properties. Nickel–cobalt-based MOF (NCM) was grown on nickel foam using a hydrothermal method, which was then used as a template for electrodeposition. A solution containing MXene flakes, nickel ($Ni^{2+}$), and cobalt ($Co^{2+}$) ions was used as the electrolyte. The metal cations were redistributed within the MXene layers during electrodeposition, forming unique hexagonal MXene–Ni–Co nanosheets directly on the NCM template (M-NC@NCM/NF). This particular nanosheet morphology enhanced the electrochemical performance of the electrode. This structure of the composite promotes efficient electron and ion transport. The flattened, two-dimensional nature of the sheets minimizes the distance charged ions and electrons need to travel within the material. The numerous nanosheets created abundant active sites for electrochemical reactions, providing a high surface area. The unique morphology of the composite helps prevent the material from agglomerating, which can hinder its effectiveness (Wang et al. 2021). Xiao and co-workers developed a novel electrochemical sensor detecting As(III) in water. The sensor was based on a Fe-MOF/MXene composite synthesized using the hydrothermal method. The presence of MXene nanosheets provided a conductive scaffold and abundant active sites for As(III) adsorption on which the Fe-MOF was grown directly. This Fe-MOF/MXene composite exhibited strong As-hydroxyl bonding, substantiated using X-ray photoelectron spectroscopy, and was highly selective toward As(III) detection. Compared to pristine electrodes or electrodes modified with just Fe-MOF or MXene, the Fe-MOF/MXene sensor demonstrated a superior sensitivity of 8.94 $\mu A/(ng/L)/cm^2$ and a low LOD of 0.58 ng/L. This sensor successfully detected As(III) in samples taken from lake and river water samples, highlighting its potential applicability in environmental monitoring (Xiao et al. 2022a). Cancer maker exosomes were directly sensed with MXene–MOF-based composites prepared by Wang and co-workers. This composite was prepared via hydrothermal synthesis. The synthesized composite demonstrated anti-fouling properties and high selectivity (Wang et al. 2024b).

### 11.3.3 Ex situ Methods of Synthesis of MXene–MOF Composites

A facile and convenient synthesis of MXene–MOF-based composites can be obtained through ex situ synthesis. Self-assembly, electrochemical synthesis, and direct mixing are ex situ methods used to prepare MXene–MOF composites. Employing these methods, highly efficient MXene–MOF composites with enhanced active sites have been prepared.

### 11.3.4 Self-assembly

Self-assembly synthesis for MXene–MOF composites is recommended due to its facile operation and scale-up ability. Optimizing several parameters, such as precursor, pH, temperature, and solvent selection, MXene–MOF composites with excellent morphological regulation and uniform structural arrangement can be achieved through self-assembly synthesis (Aldhaher, Rabiee, and Iravani 2024).

Liu et al. developed a flexible gas sensor to detect harmful $NO_2$ gas. The sensor design addressed the limitations of conventional sensors by combining an MOF with a conductive MXene aerogel. A thermal and solvent-based process created selective pores within a rigid Cu-based MOF. Self-assembly technique was utilized for the synthesis, using hydrogen bonding between a polymer grafted onto the Cu-MOF and the surface hydroxyl groups of MXene to create a conductive Cu-MOF/MXene aerogel. The aerogel was embedded within a flexible sodium alginate hydrogel for high-performance sensors for $NO_2$ gas detection (Liu et al. 2024b). Tu et al. developed MXene/CNHs/β-CD-MOFs for sensitive carbendazim (CBZ) pesticide detection. This composite combined MXenes with good electrical conductivity and ample surface-active sites, positively charged carbon nanohorns (CNHs), and β-cyclodextrin-MOF (β-CD-MOFs), demonstrating high CBZ adsorption. The MXene/CNHs structure was prepared to employ self-assembly via electrostatic interaction between positively charged CNHs and negatively charged MXene, which provided a scaffold for β-CD-MOFs. This composite provided a conductive scaffold with abundant active sites, high porosity, and strong target molecule recognition. This synergy led to the designed sensor possessing high sensitivity, wide detection range, and stability and was successfully utilized for environmental and food safety monitoring. This composite electrode achieved a low LOD of 1.0 nM for CBZ, demonstrating the potential for real-world applications (Tu et al. 2020). Cheng et al. synthesized Ni-MOF/$Ti_3C_2$ nanosheets via facile electrostatic self-assembly. This composite demonstrated high surface area, numerous active sites, and excellent electron transfer efficiency. The layered structure of the composite exhibited high conductivity and stability (Cheng et al. 2021b). MXene–MOF-based composites were developed to detect tanshinol. The self-assembly method was utilized to prepare these compounds. Enhanced electron transfer and selectivity were exhibited by this composite (Wang et al. 2024a).

### 11.3.5 Direct Mixing

Direct mixing of MXene and MOF solutions is a simple method of preparation of the combined composites. Chen et al. developed a Ce-MOF/MXene composite with heightened electrochemical performance toward L-Tryptophan (L-Trp) detection. The Ce-MOF was then synthesized via a facile direct mixing method. The composite was prepared by combining Ce-MOF and $Ti_3C_2T_x$ MXene in a 1:1 ratio. The Ce-MOF/$Ti_3C_2T_x$ MXene exhibited several desirable qualities, such as a wide linear range and a low LOD for L-Trp concentrations, signifying its ability to detect a broad range of concentrations. The fabricated sensor also showcased high selectivity and sensitivity for L-Trp, effectively distinguishing it from other substances. The composite also demonstrated long-term stability and good reproducibility (Chen et al. 2022c).

### 11.3.6 Electrochemical Synthesis

Electrochemical synthesis of MXene–MOF composites is an advantageous synthetic method. The milder reaction conditions and straightforward operation make this method suitable for the preparation of these composites. Hong et al. developed an MXene-ZIF-8 composite via electrophoretic deposition, which was used to create a conductive MXene layer. This MXene layer was used as the base for the ZIF-8 layer growth. Utilizing a fast current-driven synthesis method resulted in the rapid growth of the ZIF-8 layer on the MXene layer. This composite exhibited impressive electrochemical activity (Hong et al. 2022).

## 11.4 ELECTROCHEMICAL SENSING APPLICATIONS OF MXENE–MOF COMPOSITES

High sensitivity, extremely low LOD, enhanced stability, large surface area, surface functionalization, and high performance make MXene–MOF composites suitable for electrochemical sensing. Environmental pollutants in water, soil, and air, as well as food adulterants and pesticides, must be detected to prevent toxicity due to such contaminants. Safe levels of various biomarkers must be maintained; concentration abnormalities can cause diseases, as shown in Figure 11.1. Therefore, detecting these components is essential and can be performed efficiently with MXene–MOF electrochemical sensors, as shown in Table 11.1.

### 11.4.1 MXene–MOF-based Electrochemical Sensors for the Detection of Environmental Pollutants

Electrochemical sensing of environmental pollutants, including heavy metals, phenolic compounds, and pesticides, is necessary due to their adverse effects. When food and water contaminated by these pollutants are consumed by humans and other life forms, they are susceptible to disrupting the homeostasis of the body. Pollutants in air, water, and soil must be detected efficiently, which can be performed via MXene–MOF composites. Trivalent As is considered a threat to the human body, which affects the reproductive, respiratory, and renal systems. Therefore, these pollutants must be effectively detected, which can be used for monitoring their levels in soil, water, and food, thereby

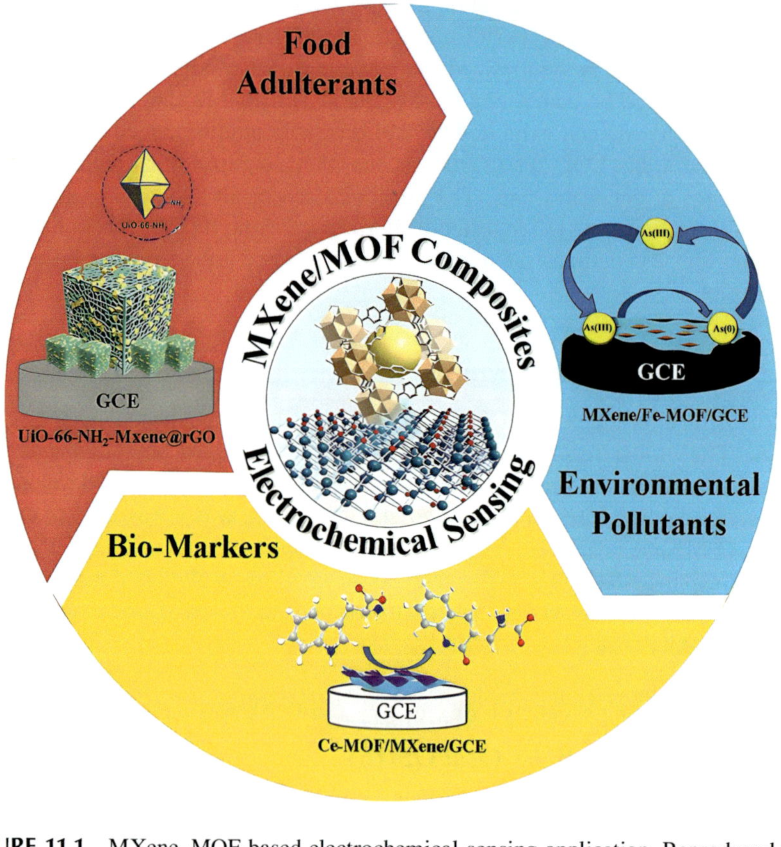


**FIGURE 11.1** MXene–MOF-based electrochemical sensing application. Reproduced with permission from Dong et al. (2024) Copyright 2024, Elsevier, Y. Chen et al. (2022) Copyright 2022, Elsevier, and P. Xiao et al. (2022) Copyright 2022, Elsevier.

**TABLE 11.1**
**Analytical parameters of MXene–MOF-based electrochemical sensors for detecting various analytes**

| Electrode modification and electrolyte | Synthesis | Analyte | Analytical parameters | Ref. |
|---|---|---|---|---|
| Fe-MOF/MXene/GCE 0.1 M PBS, pH = 5 | Hydrothermal method | As(III) | LOD – 0.58 ng/L Linear range – 10–100 ng/L | Xiao et al. (2022a) |
| MOF-derived-Alk-$Ti_3C_2$/N-PC/GCE 0.1 M PBS | Self-assembling method | Hydroquinone Catechol | LOD – 4.8 nM for hydroquinone and 3.1 nM for catechol Linear range – 0.5–150 μM | Huang et al. (2020) |
| MOF@MXene $V_2CT_x$-PMOF/SPE PBS, pH = 7.4 | Solvothermal method | NO | Linear range – 1–13 μm | Tan et al. (2022) |
| Cu-MOF/MXene/GCE 0.1 M PBS, pH = 7.0 | One-step solvothermal method | $H_2O_2$ | LOD – 0.35 μM Linear range – 1 μM to 6.12 mM | Cheng et al. (2021a) |
| AuNPs@ZIF-L-$Ti_3C_2T_x$/GCE 0.1 M PBS, pH = 7.0 | | Carbendazim | LOD – 1.2 nM Linear range – 0.01–0.20 μM | Liu et al. (2024a) |
| MXene@rGO composite aerogel-doped UiO-66-$NH_2$/GCE 0.1 M KCl | Hydrothermal | $Cd^{2+}$ | LOD – 0.46 ppb Linear range – 2–800 ppb | Dong et al. (2024) |
| | | $Pb^{2+}$ | LOD – 0.40 ppb Linear range – 1–600 ppb | |
| Ce-MOF/$Ti_3C_2T_x$/GCE 0.1 M PBS, pH = 7.0 | | L-Tryptophan | LOD – 0.19 μM Linear range – 0.2–139 μM | Chen et al. (2022c) |
| MXene/CNTs/CuMOF/GCE 0.1 M KCl | Hydrothermal | Tyrosine | LOD – 0.19 μM Linear range – 0.53–232.46 μM | Chen et al. (2022a) |
| PCN-222@MXene/GCE | Solvothermal | Oxytetracycline | LOD – 0.44 fg/mL Linear range – 0.01–100 pg/mL | Wei et al. (2024) |
| Eu-MOF/PDA@MXene/MGCE PBS, pH = 7.4 | Hydrothermal method | Trimethylamine oxide | LOD – $1.25 \times 10^{-18}$ mol/L Linear range – $9 \times 10^{-18}$ 9 $\times 10^{-15}$ mol/L | Liang et al. (2024) |
| $NH_2$-MIL-53(Fe)/CS/MXene/GCE 0.1 M PBS, pH = 5 | Solvothermal synthesis | Adenine | LOD – 0.57 μM Linear range – 3–118 μM | Chen et al. (2022b) |
| | | Guanine | LOD – 0.17 μM Linear range – 2–120 μM | |
| NiCo-MOF/$Ti_3C_2$ | | Acetaminophen | LOD – 0.01 μM Linear range – 0.01–400 μM | Lu et al. (2021) |
| | | Dopamine | Linear range – 0.01–300 μM | |
| | | Uric acid | Linear range – 0.01–350 μM | |

reducing their impact (Kumunda et al. 2021). Highly sensitive sensors for detecting As (III) in water samples have been developed with low LOD and high sensitivity. Xiao et al. developed Fe-MOF/MXene sensors that effectively determined the As ions due to the bonding between As and hydroxyl. The –OH groups on MXene and Fe-MOF acted like anchors that attract As(III) ions through coordination and facilitate their adsorption onto the electrode surface. The adsorption process enhanced the transformation of As(III) to As(0). This composite was excellent for detecting As(III) due to a synergistic effect between the MXene and MOF. The Fe-MOF/MXene composite sensor demonstrated impressive electrochemical performance, several surface-active sites, and good electrical conductivity, which improved the overall electroanalytical activity. Superior sensitivity of 1.367 μA/(ng/L)/cm$^2$ at a linear range of 10–100 ng/L was demonstrated by the composite for the precise detection of even trace amounts of As(III) with a remarkably low LOD of 0.58 ng/L. The synergistic effect between Fe-MOF and MXene further enhanced the capturing and detection of As (Xiao et al. 2022a). Huang et al. synthesized alkalization-intercalated $Ti_3C_2$ MXene and MOF-derived N-doped porous carbon (alk-$Ti_3C_2$/N-PC) electrochemical sensors for efficient detection of hydroquinone and catechol, as shown in Figure 11.2. These phenolic compounds are harmful environmental pollutants that cause health hazards to humans and other life forms. Regular $Ti_3C_2$ material undergoes restacking, a limitation in applications such

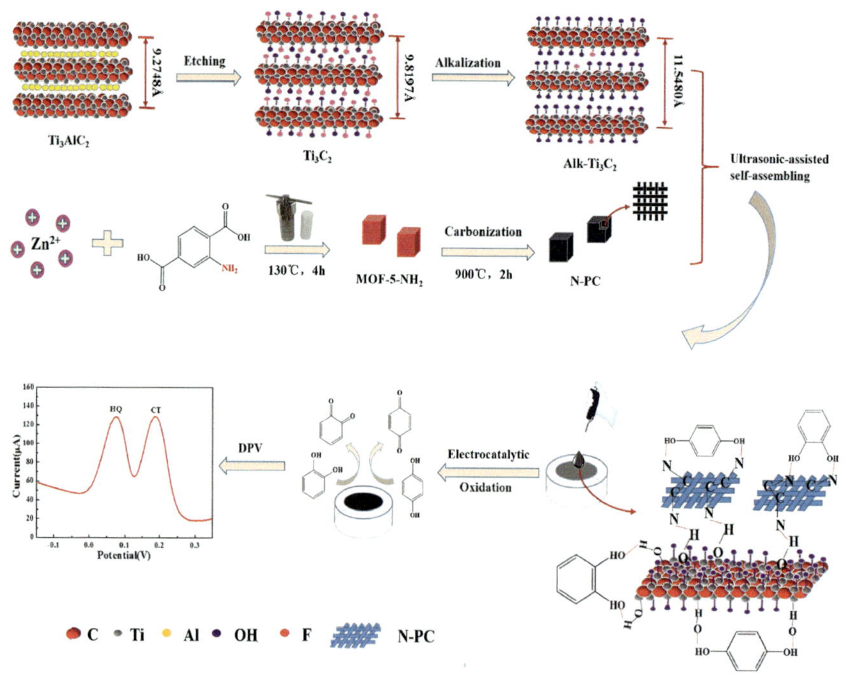


**FIGURE 11.2** Schematic of synthesis of modified electrodes and detection of analytes for environmental pollutants. (a) Schematic illustration of the preparation and application of alk-$Ti_3C_2$/N-PC/GCE. Adapted with permission from Huang et al. (2020). Copyright (2020) Elsevier.

as electrochemical sensing. This composite was self-assembled by combining N-doped porous carbon (N-PC) with specially treated $Ti_3C_2$ (alk-$Ti_3C_2$). This design prevented restacking and allowed hydroquinone and catechol detection via hydrogen bonding. The excellent conductivity associated with the composite is due to the alk-$Ti_3C_2$ and high surface area from N-PC, leading to ultrasensitive detection with a low LOD for hydroquinone (4.8 nM) and catechol (3.1 nM). In addition, a linear detection range of 0.5–150 μM was observed for hydroquinone and catechol, thereby making alk-$Ti_3C_2$/N-PC a promising tool for environmental monitoring of phenolic compounds (Huang et al. 2020). Nitric oxide (NO) is an environmental pollutant hazardous to human health and ecosystems. High NO levels in the air can cause respiratory problems and ozone depletion. The nitrogen cycles are affected by the abnormal concentrations of NO in water. Therefore, high-performance sensors for NO are essential. MXene/MOF composites are formed by partially converting MXene into MOF. This incomplete conversion creates an interface between the two materials that enhances the poor conductivity of MOFs and the weak NO adsorption of MXenes. This composite demonstrated improved conductivity from MXene with a built-in electric field at the interface for better charge transfer and preferential NO adsorption compared to pristine MOF or MXene. This design resulted in a sensor with high NO sensitivity with a linear detection window of 1–13 μM. The sensitivity was reported as 9.4 $\mu A/\mu m/cm^2$, which makes it suitable for NO detection (Tan et al. 2022). Liao et al. developed an N-$Ti_3C_2$/PC composite sensitive to detecting pollutants like 4-aminophenol and acetaminophen. These pollutants were reported as a threat to the aquatic environment. This composite combined MXene (N-$Ti_3C_2$) with a porous carbon derived from ZIF-8. The aggregation of MXene sheets was counteracted by incorporating ZIF-8, which was reported to improve conductivity and performance. This N-$Ti_3C_2$/PC composite exhibited a detection range of 1–150 μM, excellent stability, a low LOD of 0.059 μM for 4-aminophenol and 0.050 μM for acetaminophen, and the ability to distinguish between the two pollutants (Liao et al. 2022).

### 11.4.2 MXene–MOF-based Electrochemical Sensors for Food Safety Monitoring

Adulterants, pesticides, and harmful chemicals in food are a growing concern that demands the detection of adulterants and pesticides, which is helpful for the mitigation of resulting health issues. MXene–MOF electrochemical sensors are highly efficient for detecting such food adulterants. $H_2O_2$ in exceeding concentrations in food threatens human health, necessitating the detection of its safe limit concentrations. Novel electrochemical sensors for detecting hydrogen peroxide have been developed with high sensitivity and low LOD. Cheng et al. developed an electrochemical sensor that combines a three-dimensional flower-like copper-based MOF (Cu-MOF) with MXene nanosheets. The Cu-MOF assisted by offering several metal active sites and a large surface area, while the MXene contributes high electrical conductivity. The catalytic activity of the sensor toward $H_2O_2$ reduction was enhanced due to the synergistic effect. Cu-MOF/MXene/glassy carbon electrode (GCE) sensor showcased high sensitivity, with a broad detection range (1 μmol/L to 6.12 mmol/L) and a low LOD of 0.35 μmol/L. The designed sensor demonstrated good selectivity due to

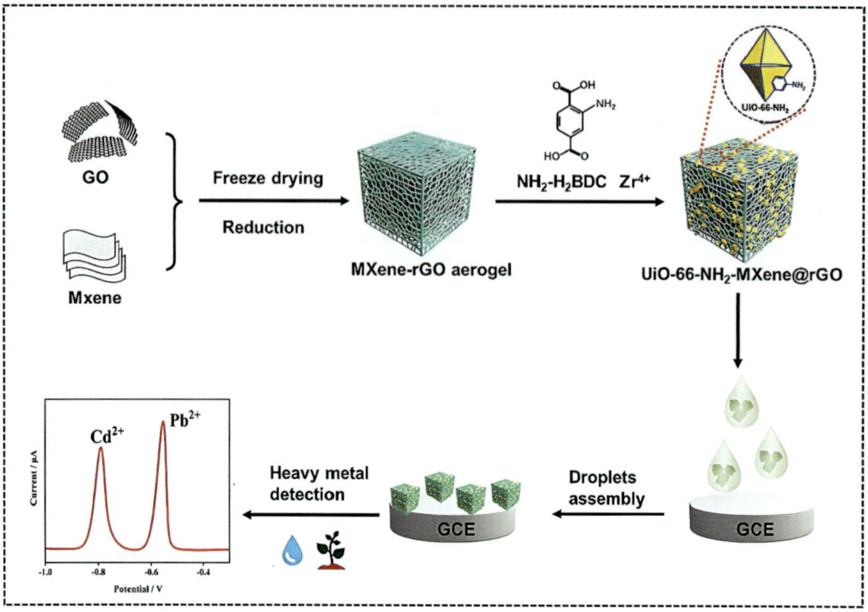


**FIGURE 11.3** Schematic of synthesis and detection of analytes for food safety. (a) Construction process of UiO-66-$NH_2$-MXene@rGO/GCE sensor electrode and schematic diagram of simultaneous detection of $Cd^{+2}$ and $Pb^{2+}$. Adapted with permission from Dong et al. (2024). Copyright (2024) Elsevier.

the high catalytic activity of Cu-MOF toward $H_2O_2$ and the low detection potential. This was suitable for selective $H_2O_2$ detection in complex samples like milk and serum. Therefore, various properties of this sensor, including ultra-sensitivity, rapid detection, and simple operation, make it suitable for real-time analysis of $H_2O_2$ in food and biological samples (Cheng et al. 2021a). Residues of carbendazim (CBZ) in agricultural products consumed have increased the demand to develop ultrasensitive sensors. Liu et al. developed a new composite material, AuNPs@ZIF-L-$Ti_3C_2T_x$, for use in the electrochemical detection of CBZ. This composite was synthesized using a few-layer $Ti_3C_2T_x$ and AuNPs@ZnO as precursors. A key feature of this method was using L-histidine, which aided the synthesis of a ZnO layer on the AuNPs before ZIF-L was grown, leading to a unique salix leaf-like structure with dart-like and carambola-like features. This structure was reported to be critical for the sensor performance. When CBZ was introduced, it was oxidized at the electrode surface through a two-electron transfer process. The sensor exhibited a low LOD of 1.2 nM and a high sensitivity of 66,684 μA/mM/$cm^2$. It also exhibited good reproducibility, repeatability, and storage stability (Liu et al. 2024a). Heavy metals such as $Cd^{2+}$ and $Pb^{2+}$ can accumulate in crops and water bodies due to contamination in soil, which can cause serious health issues for humans and other life forms that consume them. These are extremely toxic and hazardous, which can cause kidney disease, osteoporosis, and cancer. Therefore, detecting $Cd^{2+}$ and $Pb^{2+}$ levels in water and food is vital for safeguarding public health and ensuring food safety. Dong et al. fabricated a novel sensor for detecting $Cd^{2+}$ and $Pb^{2+}$ in water and food (Figure 11.3). A combination of

MXene@rGO aerogel and UiO-66-$NH_2$ was used to develop this sensor. MXene@rGO acted as a scaffold, which provided a conductive network for fast electron transfer and a porous structure for the formation of UiO-66-$NH_2$. Several amino groups in the MOF were utilized to attract and capture $Cd^{2+}$ and $Pb^{2+}$ ions. $PhNH_2$/$PhNH^{3+}$ promoted the detection process by reducing and oxidizing these metal ions. This unique combination developed sensors with very low LODs for $Cd^{2+}$ (0.46 ppb) and $Pb^{2+}$ (0.40 ppb). These sensors could accurately detect $Cd^{2+}$ and $Pb^{2+}$ simultaneously in samples like grain and water suitable for environmental and food safety monitoring (Dong et al. 2024). Aflatoxins produced by fungus-like *Aspergillus flavus* are highly toxic and carcinogenic. They can damage the DNA, causing mutations that lead to cancer. Therefore, aflatoxins must be effectively detected and eliminated from food to prevent harmful consumer effects. Kaur et al. developed a nanocomposite from $NH_2$-Mo-MOF and $Ti_3C_2$, a screen-printed electrode (SPE) for detecting Aflatoxin B1 (AFB1). The $NH_2$-Mo-MOF provided abundant surface area and amine groups for antibody immobilization, preventing the $Ti_3C_2$ sheets from aggregating. $Ti_3C_2$ provided conductivity to the composite and improved its structural stability. The positive charge on the composite developed due to the amine functionality was suitable for binding the antibody. The fabricated sensor could detect AFB1 with a wide concentration range of 0.06–50 ng/mL and a low LOD of 8 pg/mL (Kaur et al. 2024).

### 11.4.3 MXene–MOF-based Electrochemical Sensors for Detection of Various Biomarkers

Electrochemical sensing with MXene–MOF composites is utilized to detect biomarkers effectively. This is essential for the early detection and prevention of diseases caused by them. MXene–MOF-based electrochemical sensors are susceptible to such biomarkers with very low LOD and the ability to simultaneously detect different types of analytes. Therefore, these composites are considered superior for this application (Timilsina et al. 2021). Important metabolisms in the body are determined by an essential amino acid called L-Tryptophan. Imbalances in this amino acid must be monitored to prevent various disorders. Chen and co-workers developed a novel sensor for detecting L-Tryptophan, an essential amino acid in the human body. $Ti_3C_2T_x$ MXene nanosheets were combined with MOF to form these sensors that demonstrated high conductivity and efficiency. The aggregation of the nanosheets was prevented by the distinct framework of the composite, thereby improving the electrochemical performance. This modification also exhibited exceptional detection capabilities, with a wide range of sensitivity of 0.2 μM to 139 μM and a remarkably low LOD of 0.19 μM. This composite demonstrated excellent sensitivity, selectivity, and stability in samples such as blood serum. This Ce-MOF/$Ti_3C_2T_x$ MXene sensor could be utilized for sensitive and reliable L-Tryptophan detection in biomedical and clinical settings (Chen et al. 2022c).

Tyrosine is an amino acid that regulates emotions, memory, and thyroid function by producing neurotransmitters and hormones. Imbalances in tyrosine levels cause various diseases like depression, dementia, and Parkinson's. Therefore, a fast, accurate, cost-effective method to detect tyrosine levels in human serum is essential. Chen et al. reported a novel sensor for the detection of tyrosine sensor employing

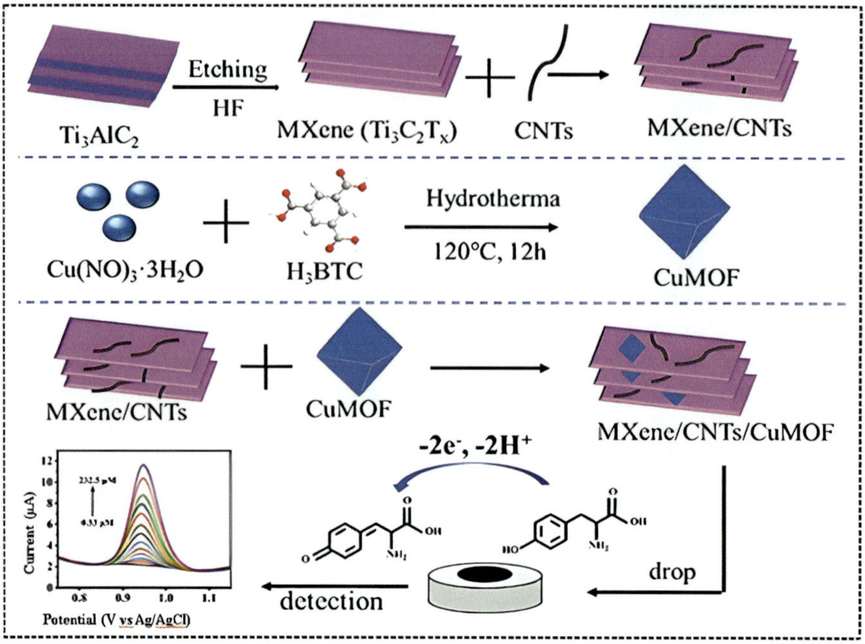


**FIGURE 11.4** Schematic illustration of synthesis and detection of biomarkers. (a) Schematic diagram of the MXene/CNTs/Cu-MOF synthesizing procedure and the sensing mechanism for detecting tyrosine. Adapted with permission from J. Chen et al. (2022a). Copyright (2022) Elsevier.

a composite material called MXene/CNTs/Cu-MOF, as shown in Figure 11.4. This composite combined MXene, carbon nanotubes (CNTs), and copper-based MOF (Cu-MOF). CNTs were embedded within MXene, which prevented MXene sheets from clumping together. This composite was enhanced by adding octahedral and porous Cu-MOF particles, which increased its overall porosity and catalytic ability. This sensor demonstrated a linear range of 0.53 μM to 232.46 μM and a low LOD of 0.19 μM (Chen et al. 2022a). Wei and co-workers developed a PCN-222@MXene composite material suitable for electrochemical aptasensors. The composite was synthesized by combining a zirconium-based MOF (PCN-222) with MXene nanosheets. PCN-222 was first constructed using a Zr cluster and meso-tetra(4-carboxyphenyl) porphine, which was then linked with the MXene nanosheets via an in situ assembly process. The synergy between the two materials was utilized, where the MOF contributed abundant Zr(IV) sites, high surface area, and good stability, while MXene nanosheets provided excellent electrical conductivity. These combined properties made PCN-222@MXene suitable for electrochemical aptasensors due to its potential for high sensitivity, selectivity, and stability. This aptasensor was utilized in detecting oxytetracycline (OTC), which causes damage to the liver and is also known to cause allergies. OTC binds to the aptamers, which hinders the electron transfer and reduces the current. The LOD for OTC was found to be 0.44 fg/mL, which is more sensitive than the literature. The aptasensor also showed good selectivity for OTC over other substances and stability over time and could detect OTC in diluted milk samples (Wei et al. 2024). Trimethylamine oxide (TMAO) is reported to induce cardiovascular and other chronic diseases. The detection of TMAO is essential to prevent these diseases. For this, Liang et al. developed a novel composite material, Eu-MOF/PDA@MXene, applicable in electrochemical sensors to detect ultra-trace trimethylamine oxide (TMAO). This composite was synthesized by etching $Ti_3AlC_2$ with LiF and HCl to prepare $Ti_3AlC_2$-MXene. PDA was modified on its surface, which was introduced to prevent the aggregation of MXenes and increase the spacing between MXene layers. Eu-MOF/PDA@MXene composites were in situ synthesized using a hydrothermal technique. $Fe_3O_4$ nanoparticles were introduced to Eu-MOF/PDA@MXene composite using self-assembly to enhance the conductivity. The sensor was modified using a magnetic glassy carbon electrode (GCE) with Eu-MOF/PDA@MXene/$Fe_3O_4$. A molecularly imprinted polymer (MIP) selective to TMAO was also employed, and the fabricated sensor exhibited high sensitivity and selectivity for TMAO detection. The composite exhibited a low LOD of $1.25 \times 10^{-18}$ mol/L (Liang et al. 2024). Chen et al. developed $NH_2$-MIL-53(Fe)/CS/MXene nanocomposite electrochemical sensors for sensing adenine and guanine found in DNA. The composite was synthesized by combining MXene, $NH_2$-MIL-53(Fe) (iron MOF), and chitosan. The $Ti_3AlC_2$ powder was etched with hydrofluoric acid to obtain MXene nanosheets. A solvothermal reaction with ferric chloride and a linker molecule prepared $NH_2$-MIL-53(Fe). These were combined in N,N-Dimethylformamide (DMF) solution under ultrasound and stirring, which allowed them to interact and form the $NH_2$-MIL-53(Fe)/CS/MXene composite after drying. The sensor displayed low LOD toward detecting guanine (0.17 μM) and adenine (0.57 μM). $NH_2$-MIL-53(Fe)/CS/MXene exhibited a better electrochemical response to adenine and guanine than Fe-MOF and MXene. The sensor displayed good anti-interference ability and maintained a stable detection performance over 50 minutes of continuous testing. Therefore, the developed sensor demonstrated promising potential for detecting adenine and guanine with good sensitivity, anti-interference ability, and stability (Chen et al. 2022b). Lu and co-workers developed a novel biosensor made using a composite based on nickel–cobalt-based MOF (NiCo-MOF) and MXene ($Ti_3C_2$) for simultaneous quantification of acetaminophen (APAP), dopamine (DA), and uric acid (UA) in biological fluids. This biosensor was superior due to the combined properties of the composite, such as heightened electrocatalytic activity offered by NiCo-MOF and excellent conductivity of $Ti_3C_2$. This material also demonstrated a large surface area and improved detection performance. The NiCo-MOF/$Ti_3C_2$ composite effectively detected a wide range of concentrations for each target molecule, such as APAP (0.01–400 μM), DA (0.01–300 μM), and UA (0.01–350 μM) with good sensitivity. The composite exhibited a low LOD of 0.01 μM and was utilized for simultaneous detection. This biosensor demonstrated excellent stability, reproducibility, and resistance to interference from other molecules. Thus, making it suitable for practical applications in real biological samples like serum and urine with 100% accuracy (Lu et al. 2021).

## 11.5 CONCLUSION AND FUTURE PERSPECTIVES

MXene–MOF composites are promising materials for electrochemical sensors. MXenes exhibit high conductivity, high optoelectronic properties, antibacterial properties, and large surfaces, whereas MOFs demonstrate tunability, porosity, crystallinity, high density, and high selectivity, which can be combined to prepare MXene–MOF sensors. The synergistic effects between these composites improve the surface area, conductivity, electrocatalytic properties, and surface functionality. These composites also exhibit anti-fouling and high electron transfer properties. Enhanced selectivity of analytes, repeatability, strong structural and mechanical stability, a wide range of detection, low LOD, and stability are achieved by utilizing these composites for electrochemical sensing. However, these composites must overcome certain limitations for various applications. The existing synthetic methods result in defective MXene sheets that are prone to depleting the conductivity of MXenes. Milder synthetic methods are required to develop MXene sheets with higher stability and enhanced properties. The pore size of most MOFs is a limitation for detecting larger biomolecules. Methods to increase the pore size to the mesoporous range must be developed, which is crucial for broader applications. Most studies are focused on a limited range of MOFs. A wider variety of MOFs must be explored that can be compatible with MXenes. This can be beneficial for the detection of a wide variety of analytes. New types of MXene–MOF composites can be developed, opening doors to novel research. Examination of MXene derivatives, like hollow spheres or aerogels, can be utilized to improve sensor performance. Improved theoretical simulations and characterization techniques are needed to understand these materials better. Also, studies based on MXenes-MOFs interactions are crucial for rational design, and advanced characterization techniques are required. Thereby, designing electrochemical sensors with precisely controlled properties. Further development of MXene–MOF composites represents a promising avenue for developing future NEMS and MEMS devices, offering a platform for emerging advancements in sensing.

## REFERENCES

Aldhaher, Abdullah, Navid Rabiee, and Siavash Iravani. 2024. "Exploring the Synergistic Potential of MXene-MOF Hybrid Composites: A Perspective on Synthesis, Properties, and Applications." *Hybrid Advances* 5 (April): 100131. doi:10.1016/j.hybadv.2023.100131.

Baranwal, Jaya, Brajesh Barse, Gianluca Gatto, Gabriela Broncova, and Amit Kumar. 2022. "Electrochemical Sensors and Their Applications: A Review." *Chemosensors* 10 (9): 363. doi:10.3390/chemosensors10090363.

Benny, Libina, Anjali John, Anitha Varghese, Gurumurthy Hegde, and Louis George. 2021. "Waste Elimination to Porous Carbonaceous Materials for the Application of Electrochemical Sensors: Recent Developments." *Journal of Cleaner Production* 290 (March): 125759. doi:10.1016/j.jclepro.2020.125759.

Chen, Jing, Ying Chen, Shuying Li, Jiao Yang, Jianbin Dong, and Xiaoquan Lu. 2022a. "MXene/CNTs/Cu-MOF Electrochemical Probe for Detecting Tyrosine." *Carbon* 199 (October): 110–18. doi:10.1016/j.carbon.2022.07.021.

Chen, Jing, Shuying Li, Ying Chen, Jiao Yang, and Jianbin Dong. 2022b. "Highly Selective Detection of Adenine and Guanine by $NH_2$-MIL-53(Fe)/CS/MXene Nanocomposites with Excellent Electrochemical Performance." *Microchimica Acta* 189 (9): 328. doi: 10.1007/s00604-022-05376-5.

Chen, Yue, Shaobin Li, Li Zhang, Tao Jing, Jianxin Wang, Lijie Zhao, Fengbo Li, Chao Li, and Jingyu Sun. 2022c. "Facile and Fast Synthesis of Three-Dimensional Ce-MOF/$Ti_3C_2T_X$ MXene Composite for High Performance Electrochemical Sensing of L-Tryptophan." *Journal of Solid State Chemistry* 308 (April): 122919. doi:10.1016/j.jssc.2022.122919.

Cheng, Dan, Peipei Li, Xiaohua Zhu, Meiling Liu, Youyu Zhang, and Yang Liu. 2021a. "Enzyme-free Electrochemical Detection of Hydrogen Peroxide Based on the Three-Dimensional Flower-like Cu-based Metal Organic Frameworks and MXene Nanosheets." *Chinese Journal of Chemistry* 39 (8): 2181–87. doi:10.1002/cjoc.202100158.

Cheng, Lin, Yuncheng Tang, Min Xie, Yitong Sun, and Hong Liu. 2021b. "2D Ultrathin NiMOF Decorated by $Ti_3C_2$ MXene for Highly Improved Photocatalytic Performance." *Journal of Alloys and Compounds* 864 (May): 158913. doi:10.1016/j.jallcom.2021.158913.

Daniel, Miriam, Georgeena Mathew, Masakazu Anpo, and Bernaurdshaw Neppolian. 2022. "MOF Based Electrochemical Sensors for the Detection of Physiologically Relevant Biomolecules: An Overview." *Coordination Chemistry Reviews* 468 (October): 214627. doi:10.1016/j.ccr.2022.214627.

Dong, Jiangbo, Xinyao Li, Li Wen, Yi Ma, Jingbing Xu, Huibo Luo, Jingzhou Hou, Changjun Hou, and Danqun Huo. 2024. "A Novel Electrochemical Strategy Based on MXene@ rGO Composite Aerogel-Doped UiO-66-$NH_2$ for Simultaneous Detection of Cadmium and Lead in Grain and Water Samples." *Food Chemistry* 437 (March): 137835. doi: 10.1016/j.foodchem.2023.137835.

Hong, Xilu, Zong Lu, Yali Zhao, Luxi Lyu, Li Ding, Yanying Wei, and Haihui Wang. 2022. "Fast Fabrication of Freestanding MXene-ZIF-8 Dual-Layered Membranes for $H_2$/$CO_2$ Separation." *Journal of Membrane Science* 642 (February): 119982. doi:10.1016/j.memsci.2021.119982.

Huang, Runmin, Dan Liao, Sisi Chen, Jingang Yu, and Xinyu Jiang. 2020. "A Strategy for Effective Electrochemical Detection of Hydroquinone and Catechol: Decoration of Alkalization-Intercalated $Ti_3C_2$ with MOF-Derived N-Doped Porous Carbon." *Sensors and Actuators B: Chemical* 320 (October): 128386. doi:10.1016/j.snb.2020.128386.

Jose, Sandra, Sariga, and Anitha Varghese. 2023. "Design and Structural Characteristics of Conducting Polymer-Metal Organic Framework Composites for Energy Storage Devices." *Synthetic Metals* 297 (August): 117421. doi:10.1016/j.synthmet.2023.117421.

Kajal, Navdeep, Vishavjeet Singh, Ritu Gupta, and Sanjeev Gautam. 2022. "Metal Organic Frameworks for Electrochemical Sensor Applications: A Review." *Environmental Research* 204 (March): 112320. doi:10.1016/j.envres.2021.112320.

Kaur, Gurjeet, Saloni Sharma, Neha Bhardwaj, Manoj K. Nayak, and Akash Deep. 2024. "Highly Sensitive and Selective Electrochemical Detection of Aflatoxin B1 in Water and Pistachio Samples with MOF/MXene Composite Based Sensor." *Food Control* 165 (November): 110694. doi:10.1016/j.foodcont.2024.110694.

Kumunda, Coster, Abolanle S. Adekunle, Bhekie B. Mamba, Ntuthuko W. Hlongwa, and Thabo T. I. Nkambule. 2021. "Electrochemical Detection of Environmental Pollutants Based on Graphene Derivatives: A Review." *Frontiers in Materials* 7 (February): 616787. doi:10.3389/fmats.2020.616787.

Lei, Yongjiu, Wenli Zhao, Yizhou Zhang, Qiu Jiang, Jr-Hau He, Antje J. Baeumner, Otto S. Wolfbeis, Zhong Lin Wang, Khaled N. Salama, and Husam N. Alshareef. 2019. "A MXene-Based Wearable Biosensor System for High-Performance In Vitro Perspiration Analysis." *Small* 15 (19): 1901190. doi:10.1002/smll.201901190.

Liang, Xilin, Yanmei Yan, Guangyan Liu, Shifeng Hou, and Shili Hou. 2024. "Constructing Magnetically Functionalized Sandwich-like Eu-MOF/PDA@MXene Hybrids for Extremely Sensitive Molecular Imprinted Sensor toward Trace Detecting Trimethylamine Oxide." *Chemical Engineering Journal* 483 (March): 149301. doi:10.1016/j.cej.2024.149301.

Liao, Dan, Zhenhua Liu, Runmin Huang, Jingang Yu, and Xinyu Jiang. 2022. "In-Situ Construction of Porous Carbon on Embedded N-Doped MXene Nanosheets Composite for Simultaneous Determination of 4-Aminophenol and Acetaminophen." *Microchemical Journal* 175 (April): 107067. doi:10.1016/j.microc.2021.107067.

Lim, Kang R. G., Mikhail Shekhirev, Brian C. Wyatt, Babak Anasori, Yury Gogotsi, and Zhi W. Seh. 2022. "Fundamentals of MXene Synthesis." *Nature Synthesis* 1 (8): 601–14. doi:10.1038/s44160-022-00104-6.

Liu, Xuan, Yong Zhang, Yun-Qing Liu, Kanru Fu, Yuanqi Wang, and Na Wei. 2024a. "Topological Structure of Au Nanoparticles@salix Leaf-like Metal–Organic Framework-MXene for Highly Sensitive Determination of Carbendazim." *Sensors and Actuators B: Chemical* 404 (April): 135251. doi:10.1016/j.snb.2023.135251.

Liu, Zhuo, He Lv, Yuchi Zhang, Jian Wang He, Le Han, Shuang Li, Lin Yang, and Yan Xu. 2024b. "Synergistic Etching and Hydrogen Bonding-Induced Self-Assembly of MXene/MOF Hybrid Aerogel for Flexible Room-Temperature Gas Sensing." *ACS Sensors* 9: 3641–3651. doi:10.1021/acssensors.4c00745.

Lu, Xing, Feng Zhang, Yimeng Sun, Kai Yu, Wei Guo, and Fengyu Qu. 2021. "A 2D/2D NiCo-MOF/$Ti_3$ $C_2$ Heterostructure for the Simultaneous Detection of Acetaminophen, Dopamine and Uric Acid by Differential Pulse Voltammetry." *Dalton Transactions* 50 (45): 16593–600. doi:10.1039/D1DT03208A.

Ma, Teng, Huibo Li, Jian-Gong Ma, and Peng Cheng. 2020. "Application of MOF-Based Materials in Electrochemical Sensing." *Dalton Transactions* 49 (47): 17121–29. doi:10.1039/D0DT03388J.

Mathew, Aleena E., Sandra Jose, Ann M. Babu, and Anitha Varghese. 2024. "State of the Art MOF-Composites and MXene-Composites: Synthesis, Fabrication and Diverse Applications." *Materials Today Chemistry* 36 (March): 101927. doi:10.1016/j.mtchem.2024.101927.

Naguib, Michael, Murat Kurtoglu, Volker Presser, Jun Lu, Junjie Niu, Min Heon, Lars Hultman, Yury Gogotsi, and Michel W. Barsoum. 2011. "Two-Dimensional Nanocrystals: Two-Dimensional Nanocrystals Produced by Exfoliation of $Ti_3$ $AlC_2$." *Advanced Materials* 23 (37): 4207–4207. doi:10.1002/adma.201190147.

Nidhi, Nahid T., Vipul Bhardwaj, Sudheshna Moka, Manoj K. Singh, Manika Khanuja, and Gaurav Sharma. 2024. "An Overview on Synthesis of MXene and MXene Based Nanocomposites for Supercapacitors." *Materials Today Communications* 41 (December): 110223. doi:10.1016/j.mtcomm.2024.110223.

Pei, Yangyang, Xiaoli Zhang, Zengyu Hui, Jinyuan Zhou, Xiao Huang, Gengzhi Sun, and Wei Huang. 2021. "$Ti_3C_2T_X$ MXene for Sensing Applications: Recent Progress, Design Principles, and Future Perspectives." *ACS Nano* 15 (3): 3996–4017. doi:10.1021/acsnano.1c00248.

Rajeev, Rijo, Anila R. Cherian, Ditto A. Thadathil, and Anitha Varghese. 2024. "Electrochemical Determination of Vitamin B6 Using Coral-like $MnO_2$-Pi on $Ti_3C_2T_x$ MXene." *Materials Research Bulletin* 169 (January): 112523. doi:10.1016/j.materresbull.2023.112523.

Rajeev, Rijo, and Anitha Varghese. 2024. "High-Performance β-Cyclodextrin-$Ti_3C_2T_x$ MXene-Based Electrochemical Sensor for the Detection of Neurological Disorder Biomarker." *Journal of the Electrochemical Society* 171 (2): 027504. doi:10.1149/1945–7111/ad2316.

Rajeev, Rijo, Ditto A. Thadathil, and Anitha Varghese. 2023. "New Horizons in Surface Topography Modulation of MXenes for Electrochemical Sensing toward Potential Biomarkers of Chronic Disorders." *Critical Reviews in Solid State and Materials Sciences* 48 (5): 580–622. doi:10.1080/10408436.2022.2078789.

Rakhi Baby, Pranati Nayak, Chuan Xia, and Husam N. Alshareef. 2016. "Novel Amperometric Glucose Biosensor Based on MXene Nanocomposite." *Scientific Reports* 6 (1): 36422. doi:10.1038/srep36422.

Ravipati, Manaswini, Mavuri S. Deepak, and Sushmee Badhulika. 2024. "Solvothermal Synthesis of Zr-Metal-Organic Framework/$V_2C$ MXene Based Composite on Nickel Foam for Highly Selective Detection of Epinephrine in Simulated Blood Serum." *Microchemical Journal* 207 (December): 112120. doi:10.1016/j.microc.2024.112120.

Riazi, Hossein, Golnoush Taghizadeh, and Masoud Soroush. 2021. "MXene-Based Nano composite Sensors." *ACS Omega* 6 (17): 11103–12. doi:10.1021/acsomega.0c05828.

Rodrigues, Paul, Harun Bangali, Ebraheem A. M. Saleh, Srwa R. Hamza, Bakhadir Suyunovich Mirzaev, Fatima Ghali, Beneen M. Hussien, Sahar B. Hussein, Ruqayah T. Habash, and Yasser F. Mustafa. 2024. "Metal-Organic Framework/MXene Nanohybrid Composites as an Emerging Electrochemical Sensing Platform for Food Safety and Biomedical Monitoring: From Synthesis to Application." *Electrochimica Acta* 494 (August): 144424. doi:10.1016/j.electacta.2024.144424.

Safarkhani, Moein, Abdullah Aldhaher, Eder C. Lima, Masoumeh Zargar, Erica E. Jung, YukSuk Huh, and Navid Rabiee. 2023. "Engineering MXene@MOF Composites for a Wide Range of Applications: A Perspective." *ACS Applied Engineering Materials* 1 (11): 3080–98. doi:10.1021/acsaenm.3c00529.

Saini, Haneesh, Nikitha Srinivasan, Veronika Šedajová, Mandira Majumder, Deepak P. Dubal, Michal Otyepka, Radek Zbořil, Narendra Kurra, Roland A. Fischer, and Kolleboyina Jayaramulu. 2021. "Emerging MXene@Metal–Organic Framework Hybrids: Design Strategies toward Versatile Applications." *ACS Nano* 15 (12): 18742–76. doi:10.1021/acsnano.1c06402.

Sariga, Ann M. B., Shashi Kumar, Rijo Rajeev, Ditto A. Thadathil, and Anitha Varghese. 2023. "New Horizons in the Synthesis, Properties, and Applications of MXene Quantum Dots." *Advanced Materials Interfaces* 10 (5): 2202139. doi:10.1002/admi.202202139.

Tan, Yi, Li Yang, Dong Zhai, Lanju Sun, Shengliang Zhai, Wei Zhou, Xiao Wang, Wei-Qiao Deng, and Hao Wu. 2022. "MXene-Derived Metal-Organic Framework@MXene Heterostructures toward Electrochemical NO Sensing." *Small* 18 (50): 2204942. doi:10.1002/smll.202204942.

Timilsina, Sanjay S., Pawan Jolly, Nolan Durr, Mohamed Yafia, and Donald E. Ingber. 2021. "Enabling Multiplexed Electrochemical Detection of Biomarkers with High Sensitivity in Complex Biological Samples." *Accounts of Chemical Research* 54 (18): 3529–39. doi:10.1021/acs.accounts.1c00382.

Tu, Xiaolong, Feng Gao, Xue Ma, Jin Zou, Yongfang Yu, Minfang Li, Fengli Qu, Xigen Huang, and Limin Lu. 2020. "Mxene/Carbon Nanohorn/β-Cyclodextrin-Metal-Organic Frameworks as High-Performance Electrochemical Sensing Platform for Sensitive Detection of Carbendazim Pesticide." *Journal of Hazardous Materials* 396 (September): 122776. doi:10.1016/j.jhazmat.2020.122776.

Wang, Jiaheng, Jiaxu Gong, Huan Zhang, Linlin Lv, Yuxing Liu, and Yatang Dai. 2021. "Construction of Hexagonal Nickel-Cobalt Oxide Nanosheets on Metal-Organic Frameworks Based on MXene Interlayer Ion Effect for Hybrid Supercapacitors." *Journal of Alloys and Compounds* 870 (July): 159466. doi:10.1016/j.jallcom.2021.159466.

Wang, Jing, Qingbin Xu, Yang Yang, Jinxin Liu, Weijun Kong, and Linchun Shi. 2024a. "An Electrochemical Sensor Founded on Heterogeneous MXene & MOF Composite for Tanshinol Sensing." *Talanta* 268 (February): 125344. doi:10.1016/j.talanta.2023.125344.

Wang, Yindian, Chengcheng Xu, Yangyang Zhou, Jaebeom Lee, Qiang Chen, and Hongxia Chen. 2024b. "Interface-Engineered 2D Heterojunction with Photoelectric Dual Gain: Mxene@MOF-Enhanced SPR Spectroscopy for Direct Sensing of Exosomes." *Small* 20 (23): 2308897. doi:10.1002/smll.202308897.

Wei, Cheng, Chuanbin Fan, Guanghui Tian, Ziao Zong, Junli Wang, Xinying Kong, Hongming He, and Feng Guo. 2024. "Rational Construction of a Zirconium–Organic framework@MXene Nanocomposite for Constructing a Sensitive Electrochemical Aptasensor." *Microchemical Journal* 202 (July): 110778. doi:10.1016/j.microc.2024.110778.

Wen, Yangyang, Zhiting Wei, Chang Ma, Xiaofei Xing, Zhenxing Li, and Dan Luo. 2019. "MXene Boosted CoNi-ZIF-67 as Highly Efficient Electrocatalysts for Oxygen Evolution." *Nanomaterials* 9 (5): 775. doi:10.3390/nano9050775.

Wu, Hao, Maram Almalki, Xiangming Xu, Yongjiu Lei, Fangwang Ming, Arijit Mallick, Vladimir Roddatis, Sergei Lopatin, Osama Shekhah, Mohamed Eddaoudi, and Husam N. Alshareef. 2019. "MXene Derived Metal–Organic Frameworks." *Journal of the American Chemical Society* 141 (51): 20037–42. doi:10.1021/jacs.9b11446.

Wu, Minying, Qian Zhang, Yuanyuan Fang, Chao Deng, Fangzhou Zhou, Yi Zhang, Xingdong Wang, Yi Tang, and Yajun Wang. 2021a. "Polylysine-Modified MXene Nanosheets with Highly Loaded Glucose Oxidase as Cascade Nanoreactor for Glucose Decomposition and Electrochemical Sensing." *Journal of Colloid and Interface Science* 586 (March): 20–29. doi:10.1016/j.jcis.2020.10.065.

Wu, Xinzhao, Pinyi Ma, Ying Sun, Fangxin Du, Daqian Song, and Guobao Xu. 2021b. "Application of MXene in Electrochemical Sensors: A Review." *Electroanalysis* 33 (8): 1827–51. doi:10.1002/elan.202100192.

Xiao, Ping, Guodong Zhu, Xiaohong Shang, Bin Hu, Boshuang Zhang, Zhaoyu Tang, Jianmao Yang, and Jianyun Liu. 2022a. "An Fe-MOF/MXene-Based Ultra-Sensitive Electrochemical Sensor for Arsenic(III) Measurement." *Journal of Electroanalytical Chemistry* 916 (July): 116382. doi:10.1016/j.jelechem.2022.116382.

Xiao, Tuo, Jun Jin, Youfang Zhang, Wen Xi, Rui Wang, Yanshen Gong, Beibei He, and Huanwen Wang. 2022b. "Rational Construction of 2D/2D $Ti_3C_2T_x$/NiCo MOF Heterostructure for Highly Efficient $Li^+$ Storage." *Electrochimica Acta* 427 (September): 140851. doi:10.1016/j.electacta.2022.140851.

Xie, Lilia S., Grigorii Skorupskii, and Mircea Dincă. 2020. "Electrically Conductive Metal–Organic Frameworks." *Chemical Reviews* 120 (16): 8536–80. doi:10.1021/acs.chemrev.9b00766.

Zhou, Zhenyu, Soumya Mukherjee, Shujin Hou, Weijin Li, Martin Elsner, and Roland A. Fischer. 2021. "Porphyrinic MOF Film for Multifaceted Electrochemical Sensing." *Angewandte Chemie International Edition* 60 (37): 20551–57. doi:10.1002/anie.202107860.

# 12 MXene-based Devices for Wearable Electrochemical Sensors

*Ajith Mohan Arjun and Sanjiv Sharma*

## 12.1 INTRODUCTION

Wearable sensors have been used for managing metabolic diseases (Vulpe et al. 2024b), diagnosing life-threatening diseases (Dincer et al. 2019), drug monitoring (Arjun et al. 2022), and prognosis of disease states (Chen, Kim, and Lu 2024). The market for wearable sensors has shown great potential with a projected market size of $10.19 billion in 2032. This huge market size is an indicator of the utility that this technology holds in the healthcare sector. Additionally, the COVID pandemic has cemented the need for a robust remote monitoring and diagnostic system, which could be made possible only with the use of highly sensitive sensors to gather data on various physiological states. Surface modifications are key to realizing a robust system of highly sensitive sensors (Arjun et al. 2023). Additionally, materials also form an important part of realizing flexible wearable sensors that can be used for real-time monitoring, diagnosis, and prognosis applications (Khan et al. 2023).

The realization of these surface modifications is possible using a variety of materials including carbon-based materials (Chipangura et al. 2024; Kirchner and Hirsch 2020), metals (Islam et al. 2020), transition metal dichalcogenides (Tajik et al. 2022; Wang, Huang, and Wu 2017), and transition metal carbides and nitrides (Wang, Huang, and Wu 2017; Wu et al. 2021; Kalambate et al. 2019). Among these materials, transition metal carbides and nitrides, also known as MXenes, have been gaining popularity. MXenes distinguish themselves from other materials with a unique combination of properties. MXenes are materials that are derived from MAX phase precursors that follow the general formula $M_{n+1}AX_n$ where M is a transition metal (Ti, V, Nb, Ta, Cr, and Mo), A is an element of group 13 or 14 (Al, Ge, As, Sn Ga, In, Si, and Pb), and X is carbon or nitrogen and $n$ can be any integer between 1 and 4. The synthesis of MXenes is carried out in two steps, viz., etching and delamination. The etching step involves the removal of the A from MAX resulting in the formation of MXenes. The etching step has been carried out using acids like HF or by introducing HF in the reaction mixture using compounds like LiF (Downes et al. 2024). Due to the harmful nature of HF, other methods like molten salt etching (Zhang et al. 2024b), ball milling method (Xue et al. 2020), and use of hydrogen halides barring HF have also been reported (Kumar 2024). Post etching, the delaminated MXenes are dispersed in organic solvents due to the presence of terminal groups including F and O, which are formed as a result of the etching process. This dispersion

DOI: 10.1201/9781003736318-12

ability of MXenes has made it an attractive material for solution-based processing using high-throughput fabrication methods like blade (Zhang et al. 2020; Noriega et al. 2024), spin (Liu et al. 2024; Kim et al. 2024), spray coating (Wang et al. 2024), and filtration (Khan et al. 2024), which have great potential for the fabrication of electrode materials for sensing applications (Figure 12.1(A) and (B)).

The family of MXenes can be broadly classified into single transition metal and double transition metal MXenes (Hussain et al. 2024). Single transition metal MXenes have been extensively studied for multiple applications including electrochemical sensing (Kalambate et al. 2019), catalysis (Li and Wu 2019), energy storage (Anasori, Lukatskaya, and Gogotsi 2017), environmental remediation (Rasool et al. 2019), and biomedical applications (Huang et al. 2018). However, the use of double transition metal MXenes is still being explored actively in the energy storage arena

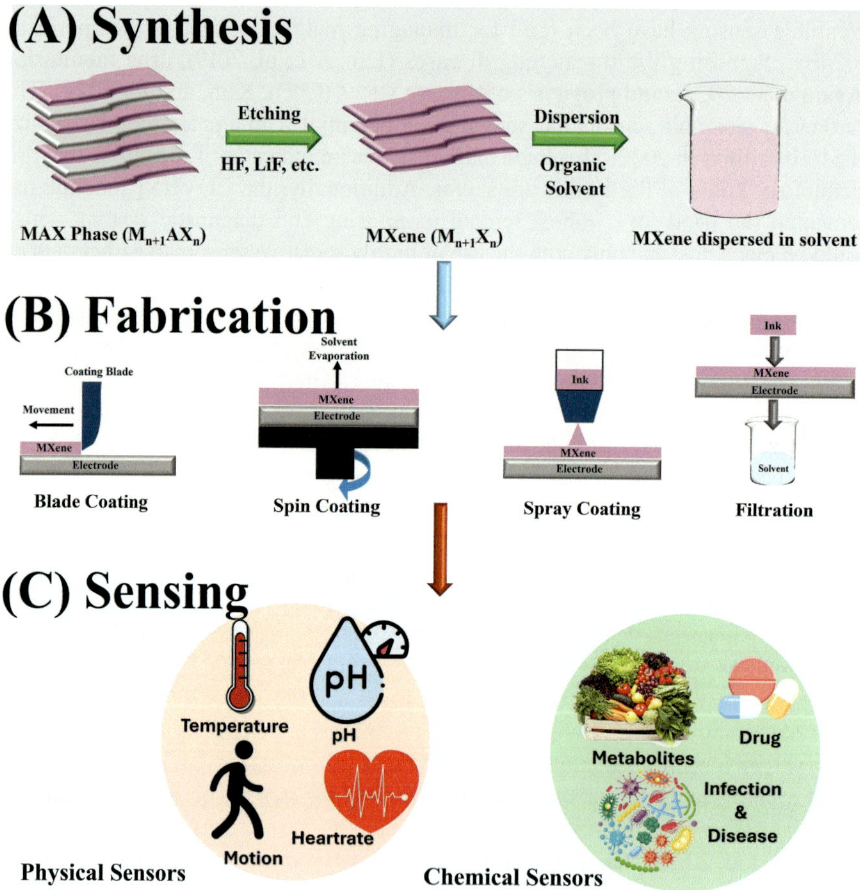


**FIGURE 12.1** Schematic showing: (A) synthesis of MXenes, (B) fabrication of MXene-based electrodes for electrochemical sensing, and (C) applications of these electrodes in physical and chemical sensing.

(Hong et al. 2020). Unlike most carbon-based materials, MXenes exhibit metallic conductivity and hydrophilicity, making them suitable for electronics and biomedical applications (Rasheed et al. 2022; Ankitha et al. 2023). Their two-dimensional structure and lower density compared to metals offer advantages in flexible and lightweight designs. While some transition metal dichalcogenides are semiconductors, MXenes are inherently metallic, with tunable surface chemistry for diverse applications (Gogotsi and Anasori 2019). Compared to bulk carbides and nitrides, the two-dimensional morphology and surface terminations of MXenes enable higher surface area and property fine-tuning (Gogotsi and Anasori 2019). Overall, MXenes merge the benefits of multiple material classes, offering a versatile platform for advancing various technologies (Zhou et al. 2023). Additionally, MXenes also possess an array of exceptional properties that make them prime candidates for the next generation of wearable sensor technologies (Ma et al. 2021). Their high electrical conductivity ensures efficient and rapid signal transduction, essential for real-time monitoring of physiological parameters. The expansive surface area of MXenes provides abundant active sites for the adsorption and interaction with target analytes, leading to enhanced sensitivity and lower detection limits, crucial for detecting subtle changes in biomarker concentrations (Mathew and Rout 2021). Moreover, the remarkable tunability of their surface chemistry allows for precise tailoring of MXene properties and selective targeting of specific biomarkers relevant to health and wellness, such as glucose, lactate, or cortisol (Gürbüz and Ciftci 2024). Their hydrophilic nature facilitates seamless integration with biological fluids like sweat or tears, enabling non-invasive and continuous monitoring. Additionally, the electrochemical stability of MXenes ensures reliable and long-term performance, a critical requirement for wearable devices. The potential biocompatibility of certain MXenes further supports their safe and sustainable integration into wearable sensors for continuous health monitoring (Driscoll et al. 2018). Finally, the inherent mechanical flexibility of MXenes allows them to conform to the contours of the body, ensuring comfort and uninterrupted data collection, vital for accurate and personalized health insights (Driscoll et al. 2018). The sensing applications involving MXenes can be broadly classified into physical and chemical sensors (Liu et al. 2023; Ganesan et al. 2023). While physical sensors measure physical parameters like temperature (Cao et al. 2019), humidity (Waheed et al. 2024), heart rate (Xie et al. 2023), and motion signal (Fan et al. 2022), chemical sensors sense parameters like changes in the levels of metabolites (Chen et al. 2024), biomarkers of disease (Hermawan et al. 2021), levels of infection (Hu et al. 2024), and drug molecules in the physiology of the user (Kalambate et al. 2019) (Figure 12.1(C)).

This chapter objectivizes the presentation of benchmark literature reports involving the application of MXenes in electrochemical sensing, specifically in the context of wearable sensing applications. It will present a future direction in the applications of this interesting material for wearable biosensing applications.

## 12.2 BIOCOMPATIBILITY OF MXENES

The use of MXenes for wearable sensing applications is an attractive avenue, which is being explored continuously. MXenes exhibit promising biocompatibility, making them suitable for wearable sensing applications (Li et al. 2023a). Their two-dimensional structure, hydrophilic surfaces, and tunable surface chemistry allow for good interaction with biological systems, minimizing adverse reactions and enabling long-term wearability (Chen et al. 2022). Studies have shown that MXenes are generally non-toxic to cells and tissues, with minimal inflammatory responses and no significant adverse effects on cell viability or function. A recent study by Unal et al. concluded that MXenes did not induce any significant loss of T lymphocytes or cell viability. Additionally, this study also showed that the levels of some activation markers significantly reduced for cell samples treated with MXenes, which indicated a potential anti-inflammatory effect (Unal et al. 2021). In addition to this, Huang et al. also reported on the use of electrospun MXene incorporated nanofibrous membrane for cutaneous regeneration which provided a surface for cutaneous regeneration and promotion of antibacterial activity. This membrane in addition to providing a surface for cutaneous regeneration and promoting antibacterial activity (Huang et al. 2024b). Another study by Liao et al. showed that MXene sheets coated with epigallocatechin gallate and iron metal phenol nanodots inhibited the expression of heat shock proteins. This study helped in proving the biocompatibility of this material (Liao et al. 2023). Additionally, Zhao et al. used MXene polyvinyl pyrrolidone sheets for the treatment of acute kidney injury (Zhao et al. 2021). There have also been reports on the use of MXenes for high-resolution neural interfaces. Vitale and coworkers used MXenes for this application because they offer high volumetric capacitance, are electrically conductive, possess surfaces that can be functionalized, and can be readily processed via aqueous dispersions (Driscoll et al. 2018). Although most of these reports dealt with the use of Ti-based MXenes, Delogu and coworkers studied the inhibition of T cell–dendritic cell interactions, which are pivotal for immune activation. This study confirmed that vanadium-based MXenes similar to Ti MXenes possessed good biocompatibility (Fusco et al. 2023). The use of wearable devices mainly revolves around the skin as the skin is the matrix for accessing matrices like sweat, interstitial fluid, and blood (Vulpe et al. 2024a). Zheng et al. fabricated an MXene-based scaffold, which showed good anti-inflammatory, antibacterial, and conductive properties.

This MXene-based layer also promoted cell proliferation, collagen deposition, and fibroblast migration (Zheng et al. 2021). Due to a multitude of benign properties related to therapeutics (Soleymaniha et al. 2019), wound healing (Zarepour et al. 2024), antibacterial properties (Seidi et al. 2023), and the ability to promote cell proliferation (Ma, Zhang, and Lei 2023), MXenes are ideal candidates to be used as electrode materials for wearable electrochemical sensing applications. While the biocompatibility of MXene is a crucial factor for its integration into wearable devices, a detailed discussion on this topic is beyond the scope of this chapter, which focuses on electrochemical wearable MXene-based sensors. Nevertheless, it is important to acknowledge that the promising biocompatibility of MXenes makes them attractive candidates for such applications. Further exploration of their biocompatibility in the context of long-term wearability and interaction with biological systems will be essential for the continued development and advancement of MXene-based wearable electrochemical sensors.

## 12.3 MXENE-BASED WEARABLE ELECTROCHEMICAL SENSORS

The biocompatibility of MXenes is an important aspect, which supports its application in wearable electrochemical sensors. Depending on the bioreceptor used for sensing, MXene-based electrochemical sensors can be divided into catalytic, enzymatic, antibody-based, and aptamer-based sensing applications. As the body's most extensive organ, the skin provides a valuable source for sampling a wide array of biomarkers that are indicative of both general health and clinical conditions. It comprises three distinct layers: epidermis, dermis, and subcutaneous tissue. The epidermis, the outermost layer, acts as a barrier and interacts with the external environment. The dermis, beneath the epidermis, provides structural support and contains blood vessels, nerves, and sweat glands. As the deepest layer, the subcutaneous tissue is predominantly made up of fat and connective tissue, which serves to insulate and protect underlying structures. Understanding the microanatomy of the skin is crucial for the development and integration of wearable sensors, as it dictates the accessibility of biomarkers and the optimal placement of sensing devices (Friedel et al. 2023). Epidermal wearable sensors, primarily focused on sweat analysis, offer a non-invasive window into various health parameters by detecting electrolytes, metabolites, and small biomolecules present in sweat. This approach is advantageous due to its ease of access and continuous monitoring capabilities. However, it grapples with challenges such as fluctuating sweat rates influenced by factors like exercise and ambient temperature, a limited range of detectable biomarkers compared to blood or interstitial fluid (ISF), and the risk of contamination from external substances on the skin's surface. In contrast, subdermal wearable sensors delve deeper, targeting the skin's ISF, a fluid that closely mirrors blood composition. This allows for a broader spectrum of biomarker detection, including glucose, hormones, and larger proteins, providing a more comprehensive insight into an individual's physiological state. ISF-based sensing boasts stable collection conditions, as ISF levels and composition are less susceptible to external fluctuations. Additionally, the minimally invasive extraction methods employed for ISF collection reduce the risk of contamination (Vulpe et al. 2024a). However, challenges remain in achieving continuous and user-friendly monitoring due to the technical complexities and costs associated with ISF sensing devices. Despite these challenges, the future of wearable sensing is increasingly leaning toward ISF-based systems due to their potential for more accurate and comprehensive health monitoring. Advancements in microneedle technology and skin-interfacing sensors are paving the way for less invasive and more user-friendly ISF extraction methods. The engineering of trustworthy, long-term wearable biosensors benefits significantly from using ISF as a matrix, owing to its robustness and reduced contamination risk. Although obstacles need addressing, the potential of ISF-based sensing to reveal more detailed information about individual health marks an important step forward for personalized medicine and continuous health tracking, potentially revolutionizing future healthcare paradigms (Teymourian et al. 2020). An analysis of the literature related to MXene-based wearable sensors reveals that work on epidermal wearable sensors has been an active area. However, further research is required to understand the nuances of related to the use of MXenes for subdermal wearable sensing. On carrying out an analysis of MXene-based electrochemical sensors in the literature, it is evident that reports exist on the use of single and multiplexed detection of biomarkers and physiologically relevant parameters. On the one hand, single sensors are designed to detect a specific analyte and offer high sensitivity and specificity for targeted monitoring (Macchia et al. 2024). On the other hand, multiplexed sensors simultaneously detect multiple analytes, providing a comprehensive overview of physiological states and disease progression (Irimes et al. 2024). This multi-analyte capability enables early disease detection, personalized treatment plans, and accurate prognoses. As technology advances, both single and multiplexed sensors are becoming increasingly miniaturized, wearable, and non-invasive, paving the way for continuous, real-time health monitoring in various settings, from hospitals to personal homes. The succeeding sections will present a literature overview of the use of MXenes for wearable electrochemical sensing.

### 12.3.1 Catalytic Sensing

Catalytic sensing with MXenes or their composites leverages their inherent electrocatalytic properties to facilitate the direct oxidation or reduction of target analytes. This interaction induces a measurable electrochemical response, typically a change in current or potential, proportional to the analyte concentration. By establishing a calibration curve relating the electrochemical signal to known analyte concentrations, this response can be accurately translated into a quantitative measure of the analyte concentration within the sample. Glucose sensing is a common sensor application, which has been reported using the catalytic sensing mechanism. Li et al. fabricated a microfluidic wearable sensor using a Pt, MXene composite for glucose sensing (Li et al. 2023b). The validation of the sensor was carried out using monitoring of glucose generated in sweat during different times, and it was found that the sensor was able to measure glucose in the range of 0–8 mM, which was carried out by applying a potential of −0.1 V (Figure 12.2(A)). Creatinine and sodium ions are important indicators for kidney health monitoring. A MXene-based sensor was realized for this application by Kalasin et al. This sensor was fabricated by using poly(3,4-ethylenedioxythiophene) polystyrenesulfonate (PEDOT:PSS) and polypyrrole (PPy) for potentiometric sensing of sodium ions and voltametric sensing of creatinine (Kalasin and Sangnuang 2023). These active layers were sandwiched between sodiated nanoporous-layer carbon–polyethylene glycol and sodiated $Ti_3C_2T_x$ MXene nanosheet. These sensors were able to detect sodium ions and creatinine in the range of 0.001–50 mM and 0.6–2800 μM, respectively. The validation of this sensor was carried out by a volunteer wearing these sensors (Figure 12.2(B)). Ankitha et al. reported on the development of a wearable yarn-based sensor for the detection of dopamine from sweat (Ankitha, Shamsheera, and Rasheed 2024). The sensor was composed of MXene-coated yarns integrated into the absorbent pad of an antiseptic layer that was used for the collection of sweat (Figure 12.2(C)). The sensor was able to detect dopamine in the range from 1 nM to 1 μM with a limit of detection of 316 pM.

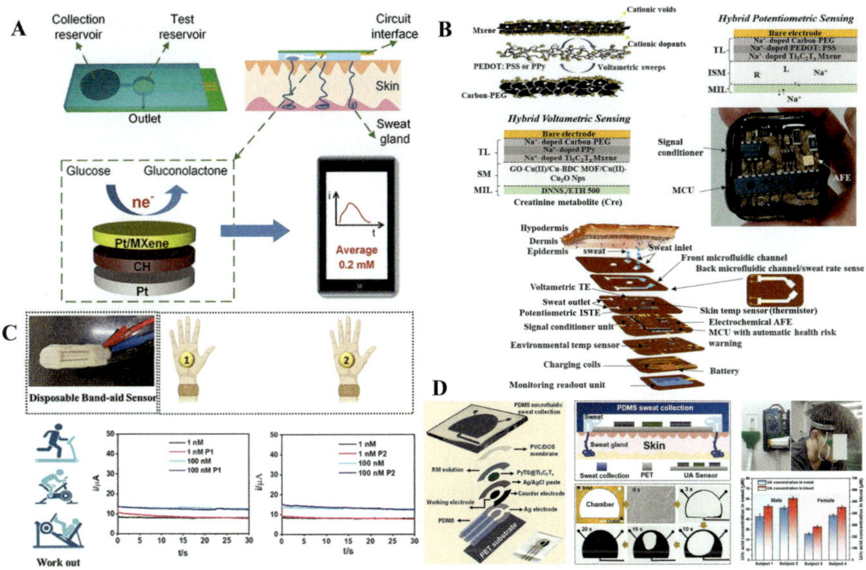


**FIGURE 12.2** (A) Microfluidics integrated non-enzymatic wearable electrochemical sensors reported by Li, Chen, et al. (2023). Reprinted with permission from Li, Chen, et al. (2023). (B) Multiplexed wearable sensor for the detection of sodium ions and creatinine developed by Kalasin and Sangnuang (2023). Reprinted with permission from Kalasin and Sangnuang (2023). (C) MXene-based disposable band aid sensor developed by Ankitha, Shamsheera, and Rasheed (2024). Reprinted with permission from Ankitha, Shamsheera, and Rasheed (2024). (D) MXene-PyTS-based uric acid sensor developed by Chen et al. (2024) for the wearable detection of sweat-based uric acid. Reprinted with permission from Chen et al. (2024).

Another important metabolite is uric acid, which is important from the point of view of many physiological conditions like kidney diseases and gout. Chen et al. fabricated a MXene 1,3,6,8-pyrene tetrasulfonic acid sodium salt for the detection of uric acid (Chen et al. 2024). This composite provides many sites for the redox of uric acid. The sensor was able to detect uric acid in the range of 5–100 μM (Figure 12.2(D)). The sensor was validated on volunteers after the intake of purine-rich food and was benchmarked by using a commercial blood-based uric acid analyzer. Melatonin is a hormone that is associated with sleep. Variations in the levels of this hormone can lead to disorders like Alzheimer's disease. Hence, a monitoring system for the detection of the levels of this molecule is highly crucial. Richard et al. developed a wearable system employing an $Nb_2CT_x$ MXene Zn MOF composite for the detection of the levels of this molecule in cerebrospinal fluid, human serum, and human sweat (Richard et al. 2024). This composite was coated on a carbon yarn. The sensor was able to detect this hormone in the range 1–100 μM with a limit of detection of 215 nM. Boobphahom utilized a composite of MXene, polyvinyl alcohol, $TiO_2$, and graphene oxide coated on a screen-printed electrode (Boobphahom et al. 2021a). This material was coated on an absorbent material and used for the detection of urinary norepinephrine, which is an important biomarker for screening brain tumors. This sensor was able to detect norepinephrine in the range of 1–60 μM with a limit of detection of 6 nM. Zhang et al. reported on the use of a MXene reduced graphene oxide composite for the deposition of $Zn^{2+}$ ions (Zhang et al. 2024a). This sensor was used for the detection of ascorbic acid from biofluids such as sweat. The detection range of this sensor was determined to be 0–5 mM in real matrices.

### 12.3.2 Enzymatic Sensing

Enzymatic sensing involves the immobilization of biomolecule-specific enzymes on MXene-modified surfaces for wearable sensing of important physiological metabolites. Myndrul et al. fabricated a skin-attachable sensor for the monitoring of glucose from sweat. The active component of this sensor was a composite of MXene/ZnO tetrapods incorporated with GOx. The sensor was able to detect glucose achieving a limit of detection of 17 μM in sweat. The detection was carried out at a potential of −0.24 V (Myndrul et al. 2022). A prime example of the use of MXene was reported by Lei et al., who fabricated a multiplexed enzymatic sensor for the detection of glucose and lactate from sweat (Lei et al. 2019). In this report, the MXene was modified using Prussian blue and the oxidation of glucose and lactate was carried out using glucose and lactate oxidase, respectively. The sensor was able to detect glucose and lactate in the range of 10 μM to 1.5 mM and 0–22 mM, respectively. Wearable electrochemical sensors are important for the multiplexed monitoring of multiple parameters like glucose, pH, and $Ca^{2+}$ ions (Figure 12.3(A)). A sensor for the measurement of these parameters was reported by Zahed et al. who used a nanoporous carbon MXene composited with platinum nanoparticles for the detection of glucose from sweat. The enzymatic glucose biosensor exhibited a good detection range of 3 μM to 1.5 mM (Zahed et al. 2022). Potassium ions play a crucial role in maintaining a balance of physiological fluids and regulating body functions. The detection of potassium ions was carried out by using a MXene–carbon nanotube composite coated with valinomycin to protect the underlying electrode from effects caused by the immobilization of the potassium ions (Zhang et al. 2021). The sensor was integrated into an NFC unit to transmit data to a smartphone. A microfluidic channel was also integrated into the sensor surface for effective collection of sweat from the surface of the skin. The sensor was able to detect potassium ions in the range from 1 to 32 mM from human sweat and was demonstrated to work well when worn by a human subject (Figure 12.3(B)). Real-time nutrition management is highly crucial for monitoring abnormalities in the health of the wearer. Skin-adhering fibrous electrodes are attractive avenues for wearable sensors. Sharifuzzaman et al. reported on the development of a soft wearable patch incorporating MXene/fluoropolymer nanofiber-derived hierarchical porous $TiO_2$ nanocatalyst interconnected 3D fibrous carbon nanohybrid electrodes (Sharifuzzaman et al. 2023). This electrode coated with platinum nanoparticles was used for the loading of GOx enzyme (Figure 12.3(C)). The sensor was able to detect glucose in the range from 10 μM to 2 mM. The sensor was used for on-body analysis of sweat-based glucose. Additionally, the bare electrode without the enzyme was also used for the detection of sweat pH.

Li et al. employed a highly integrated sensing paper modified with an MXene methylene blue composite (Li et al. 2021a). This electrode was used for the enzymatic detection of glucose and lactate using glucose oxidase and lactate oxidase.

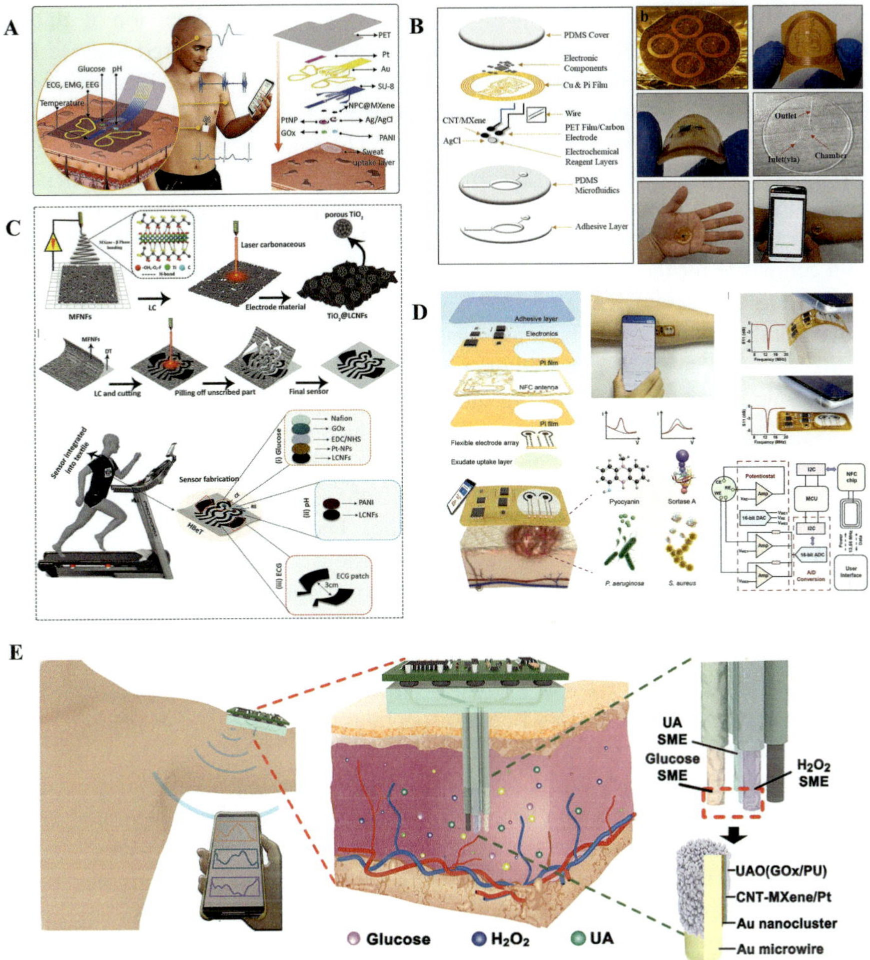


**FIGURE 12.3** (A) Multiplexed electrochemical wearable sensor reported by Zahed et al. Reprinted with permission from Zahed et al. (2022). (B) Wearable electrochemical potassium sensor developed by Zhang et al. Reprinted with permission from Zhang et al. (2022). (C) MXene/Fluoropolymer-derived carbonaceous soft wearable sensor for sweat-based metabolite detection developed by Sharifuzzaman et al. Reprinted with permission from Sharifuzzaman et al. (2023). (D) Wearable sensor for wound healing monitoring developed by Shi et al. Reprinted with permission from Shi et al. (2023). (E) Wearable sensor for the detection of metabolites from the interstitial space developed by Huang et al. Reprinted with permission from Huang, Yao, et al. (2024).

The range of detection for glucose and lactate exhibited by this sensor was found to be 0.08–1.25 mM and 0.3–20.3 mM, respectively. This sensor was integrated with a wearable electronics unit and used for the detection of sweat-based glucose and lactate of a volunteer. Reem Khan et al. reported on the use of an MXene ceria composite as a catalytic amplifier of reduction of peroxide that is produced by the immobilized glucose oxidase (Khan and Andreescu 2024). This sensing surface was transferred onto a wearable platform and was used for the detection of glucose from sweat. Pan et al. reported on the use of a MXene functionalized PEDOT:PSS composite for the enzymatic detection of glucose (Pan et al. 2024). The sensor was incorporated onto a screen-printed carbon electrode and employed for the on-body non-invasive detection of sweat-based glucose. The range of detection of this sensor was reported to be 0.003–1.29 mM from sweat. Wounds that do not heal on time cause an immense burden to the healthcare system. Uric acid is an important biomarker for wound status. Sharifuzzaman et al. reported on the use of an MXene gold composite for the immobilization of uricase, which is used for the detection of uric acid (Sharifuzzaman et al. 2020). The sensor was able to detect uric acid in the range from 50 to 1200 μM. The sensor was also able to detect the pH and temperature of the wound. Song et al. reported on a 3D printed epifluidic electronic skin, which incorporated a multiplexed sensing unit for the detection of glucose, alcohol, pH, temperature, and pressure (Song et al. 2023). The active element of these sensors was 3D printed using layers comprising porous CNT–styrene–butadiene–styrene (CNT–SBS) as the working electrode and MXene–Prussian blue as the mediator for the enzymatic sensing of glucose and alcohol. The data from the sensor was fed into a machine learning algorithm to predict various aspects of the physiology of the wearer. The detection of infections from wounds is an important step toward wound healing and its associated treatment protocol. To diagnose bacterial infections in wounds, Shi et al. reported on the detection of gram-positive *Staphylococcus aureus* and gram-negative *Pseudomonas aeruginosa*. The detection was carried out by detection of virulence factors like transpeptidase sortase A and pyocyanin using a composite of MXenes and gold nanoparticles (Shi et al. 2023). The sensor was able to detect these biomarkers in the range from 1 pg/mL to 100 ng/mL for sortase A, and 1 to 100 μM for pyocyanin (Figure 12.3(D)). This sensor was integrated into a wearable unit and into a mouse model for the detection of these biomarkers.

The interstitial space, the fluid-filled area surrounding cells and tissues, holds a wealth of biomarkers reflecting physiological and pathological states. This space acts as a dynamic reservoir for molecules secreted or leaked from cells, making it an ideal target for biosensing technologies. Analyzing ISF biomarkers offers a non-invasive window into real-time health monitoring and disease diagnosis. Huang et al. reported on the use of functionalized gold microwire-based bundle electrodes modified with a composite of Au nanocluster, carbon nanotubes, MXene, and Pt nanoparticles (Figure 12.3(E)). This surface was used for the immobilization of glucose oxidase and uricase for the detection of glucose and uric acid, respectively. The sensor was able to detect glucose in the range of 0–22 mM and uric acid in the range 0–900 μM. The pristine composite electrode was also used for the detection of peroxide from 0 to 10 mM in the interstitial space. This sensor performed exceptionally well when incorporated into the body of a mouse model (Huang et al. 2024a).

### 12.3.3 Antibody, Aptamer, and miRNA-based Sensing

Antibody-, aptamer-, and miRNA-based electrochemical sensing constitutes an affinity-based sensing mechanism that either utilizes an impedimetric sensing method or a voltametric sensing method for the detection of physiologically relevant molecules like hormones or disease biomarkers. Cortisol is a biomarker, which can be used to monitor psychological and emotional stress. San Nah et al. developed an MXene-modified laser-burned graphene electrode for the immobilization of cortisol antibody (Nah et al. 2021). This was then used for the realization of an impedimetric cortisol sensor, which was integrated onto a PDMS-based microfluidic sensor for the detection of cortisol from sweat. This sensor was able to detect cortisol in the range from 0.01 mM to 100 nM. Laochai et al. designed a flexible thread-based immunosensor for the detection of cortisol from sweat (Laochai et al. 2022). The immobilization

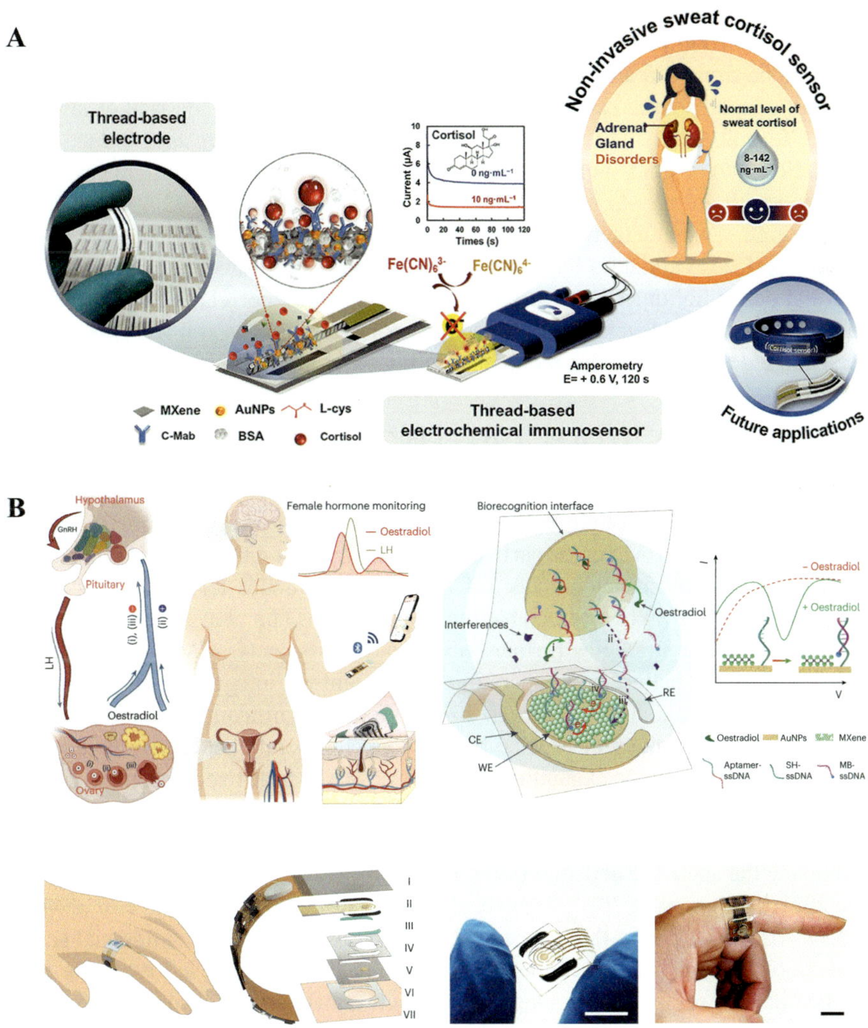


**FIGURE 12.4** (A) Wearable immunosensor for the detection of cortisol developed by Laochai et al. Reprinted with permission from Laochai et al. (2022). (B) Wearable sweat-based aptasensor developed by Ye et al. for real-time monitoring of oestradiol in sweat. Reprinted with permission from Ye et al. (2024).

surface of the cortisol antibody in this work was a composite of MXene, gold nanoparticles, and L-cysteine. This sensor was able to detect cortisol in the range of 5–180 ng/mL achieving a limit of detection of 0.54 ng/mL (Figure 12.4(A)). Zhao et al. reported on the use of a carbon fiber modified with MXene $MoS_2$ for the simultaneous detection of ascorbic acid, uric acid, and dopamine (Zhao et al. 2022). The sensor was also used for the detection of miRNA, which are important biomarkers for a variety of diseases. This sensor employed a hybridization chain reaction (HCR) cascade mechanism, enabling ultrasensitive microRNA detection through signal amplification catalyzed by horseradish peroxidase (HRP). The sensor was able to detect ascorbic acid, uric acid, and dopamine in the range of 10–3000 μM, 0.5–1000 μM, and 0.5–1000 μM, respectively. The range of detection of miRNA was determined to be from 100 am to 10 nM. Yi et al. developed a wearable, skin-interfaced aptamer nanobiosensor capable of non-invasive monitoring of oestradiol levels in sweat (Ye et al. 2024). This innovative device utilizes iontophoresis to induce sweat production and a gold nanoparticle–MXene-based electrode for highly sensitive detection with an ultra-low limit of detection of 0.14 pM (Figure 12.4(B)). The system integrates sweat induction, sampling, analysis, and wireless communication with a user interface. Validation in human participants demonstrated cyclical fluctuation of sweat oestradiol during menstrual cycles and a strong correlation between sweat and blood oestradiol levels, highlighting the potential of this wearable technology for personalized reproductive hormone monitoring.

## 12.4 FUTURE OUTLOOK AND CONCLUSIONS

MXene-based wearable electrochemical sensors exhibit immense potential to revolutionize healthcare through real-time, continuous monitoring of biomarkers. The tunable properties of MXenes, particularly $Ti_3C_2T_x$, in conjunction with their compatibility with printing techniques, offer promising avenues for developing highly sensitive, selective, and customizable sensing platforms (Jia et al. 2023). The advent of passive sensing strategies further enhances the prospects of MXene-based sensors, facilitating battery-free operation and seamless integration into wearable devices (Aziz et al. 2022; Li et al. 2021b). While significant progress has been made in the development of MXene-based sensors, several challenges remain to be addressed. These include expanding the range of MXenes utilized beyond $Ti_3C_2T_x$ to Nb- and V-based MXenes optimizing synthesis methods for improved yield and uniform functionalization, exploring the potential of non-functionalized MXenes, and enhancing the shelf life and stability of MXene-based electrodes (Wang, Wang, and Qiu 2021; Downes et al. 2024; Arjun et al. 2023). Furthermore, the high cost of MAX phases poses a barrier to the widespread adoption of MXenes in commercial sensor applications. Despite these challenges, the future of MXene-based wearable electrochemical sensors is bright. Ongoing research efforts are focused on overcoming these limitations and developing novel MXene composites with enhanced properties. The integration of microfluidics (Safarkhani et al. 2024), multimodal sensing, and artificial intelligence algorithms promises to further enhance the capabilities of these sensors, enabling comprehensive health monitoring and personalized healthcare delivery (Safarkhani et al. 2024; Ballard et al. 2021).

**TABLE 12.1**
**Table of comparison showing MXene-based wearable devices**

| Material modification | Analyte | Type (C/E/A) | Mode (ED/SD) | Detection range | LOD | Ref. |
|---|---|---|---|---|---|---|
| 1. $Ti_3C_2T_x$/PyTS | Uric acid | C | ED | 5–100 µM | 0.48 µM | Chen et al. (2024) |
| Pt/$Ti_3C_2T_x$ | Glucose | C | ED | 0–8 mM | 29.15 µM | Li et al. (2023b) |
| a. $Ti_3C_2T_x$/PEDOT:PSS<br>b. $Ti_3C_2T_x$/PPy | a. Sodium ion<br>b. Creatinine | C | ED | a. 0.12–120 mM<br>b. 0.6–2800 µM | a. 1 µM<br>b. 0.12 µM | Kalasin and Sangnuang (2023) |
| Zinc MOF/$Nb_2CT_x$ | Melatonin | C | ED | 1–100 µM | 215 nM | Richard et al. (2024) |
| $TiO_2$/$Ti_3C_2T_x$-PVA/GO | Norepinephrine | C | ED | 0.01–60 µM | 6 nM | Boobphahom et al. (2021b) |
| $Ti_3C_2T_x$-rGO/$Zn^{2+}$ | Ascorbic acid | C | ED | 0–5 mM | 1.98 µM | Zhang et al. 2024a) |
| ZnO TP/$Ti_3C_2T_x$/GOx | Glucose | E | ED | 0.05–0.7 mM | 17 µM | Myndrul et al. (2022) |
| a. $Ti_3C_2T_x$/PB/GOx<br>b. $Ti_3C_2T_x$/PB/LOx | a. Glucose<br>b. Lactate | E | ED | a. 10 µM to 1.5 mM<br>b. 0–22 mM | a. 0.33 µM<br>b. 0.67 µM | Lei et al. (2019) |
| NPC@$Ti_3C_2T_x$/GOx | Glucose | E | ED | 0.003–1.5 mM | 7 µM | Zahed et al. (2022) |
| $Ti_3C_2T_x$-MWCNT/valinomycin | Potassium ion | E (enzyme-like) | ED | 1–32 mM | Not mentioned | Zhang et al. (2021) |
| $Ti_3C_2T_x$-derived $TiO_2$/ laser-carbonaceous nanofibrous/ Pt NPs/GOx | Glucose | E | ED | 10 µM to 2 mM | 10 µM | Sharifuzzaman et al. (2023) |
| a. $Ti_3C_2T_x$/MB/GOx<br>b. $Ti_3C_2T_x$/MB/LOx | a. Glucose<br>b. Lactate | E | ED | a. 0.08–1.25 mM<br>b. 0.3–20.3 mM | a. 17.05 µM<br>b. 3.73 µM | Li et al. (2021a) |
| a. $Ti_3C_2T_x$/MB/GOx<br>b. $Ti_3C_2T_x$/MB/Lox<br>c. $Ti_3C_2T_x$/MB/HRP | a. Glucose<br>b. Lactate<br>c. Peroxide | E | ED | a. 100 µM to 10 mM<br>b. 1–60 µM<br>c. 100 µM to 10 mM | a. 0.49 µM<br>b. 3.6 nM<br>c. 0.8 µM | Khan and Andreescu (2024) |

*(Continued)*

**TABLE 12.1 (Continued)**
**Table of comparison showing MXene-based wearable devices**

| Material modification | Analyte | Type (C/E/A) | Mode (ED/SD) | Detection range | LOD | Ref. |
|---|---|---|---|---|---|---|
| PEDOT:PSS/$Ti_3C_2T_x$/GOx | Glucose | E | ED | 94–1294 µM | 1.9 µM | Pan et al. (2024) |
| $Ti_3C_2T_x$/laser-guided graphene/uricase | Uric acid | E | ED | 50–1200 µM | 50 µM | Sharifuzzaman et al. (2020) |
| a. $Ti_3C_2T_x$–PB/GOx<br>b. $Ti_3C_2T_x$–PB/AOX | a. Glucose<br>b. Alcohol | E | SD | a. 0–200 µM<br>b. 0–20 mM | Not mentioned | Song et al. (2023) |
| a. $Ti_3C_2T_x$–Au<br>b. $Ti_3C_2T_x$–Au/ferrocene | a. Pyocyanin<br>b. Sortase A | a. C<br>b. E | ED | a. 1–100 µM<br>b. 1–100 ng/mL | Not mentioned | Shi et al. (2023) |
| a. Au NW/Au NC/$Ti_3C_2T_x$-CNT/GOx<br>b. Au NW/Au NC/$Ti_3C_2T_x$-CNT/uricase<br>c. Au NW/Au NC/$Ti_3C_2T_x$-CNT/Pt | a. Glucose<br>b. Uric acid<br>c. Peroxide | a. E<br>b. E<br>c. C | ED | a. 5–15 mM<br>b. 0–900 µM<br>c. 0–60 µM | Not mentioned | Huang et al. (2024a) |
| $Ti_3C_2T_x$/laser-burned graphene | Cortisol | A | ED | 0.01–100 nM | 88 pM | San Nah et al. (2021) |
| L-cys/AuNPs/$Ti_3C_2T_x$ | Cortisol | A | ED | 5–180 ng/mL | 0.54 ng/mL | Laochai et al. (2022) |
| a. CFP/$Ti_3C_2T_x$-$MoS_2$<br>b. CFP/$Ti_3C_2T_x$-$MoS_2$<br>c. CFP/$Ti_3C_2T_x$-$MoS_2$<br>d. CFP/$Ti_3C_2T_x$-$MoS_2$ | a. Ascorbic acid<br>b. Uric acid<br>c. Dopamine<br>d. miRNA | a. C<br>b. C<br>c. C<br>d. A | ED | a. 10–3000 µM<br>b. 0.5–1000 µM<br>c. 0.5–1000 µM<br>d. 100 aM to 10 nM | a. 1.47 µM<br>b. 0.27 µM<br>c. 0.38 µM<br>d. 3.16 aM | Zhao et al. (2022) |
| $Ti_3C_2T_x$-Au NPs/ssDNA | Oestradiol | A | ED | 0–100 pM | 0.14 pM | Ye et al. (2023) |

*Legend used:* catalytic (C), enzymatic (E), affinity-based such as antibody, aptamer, miRNA (A), epidermal (ED), and subdermal (SD).

In conclusion, MXenes have emerged as a promising material for the development of next-generation wearable electrochemical sensors. Their unique properties, coupled with ongoing research and technological advancements, hold the potential to transform the field of wearable sensing and revolutionize healthcare monitoring and diagnostics. MXenes have been extensively explored for non-invasive monitoring of biomarkers in sweat, including glucose, lactate, uric acid, and cortisol. Wearable patches and devices incorporating MXene-based sensors have been developed for real-time, continuous sweat analysis. However, the exploration of applications for this material in minimally invasive sensing such as detecting biomarkers in ISF using microneedle arrays needs to be explored further (Yang et al. 2021). The biocompatibility of MXenes allows for their safe integration into these minimally invasive approaches, enabling continuous health monitoring with minimal discomfort or risk to the wearer. By addressing these challenges and pursuing these future directions, MXene-based wearable electrochemical sensors can unlock their full potential, paving the way for a new era of personalized healthcare and disease prevention. Table 12.1 shows a detailed comparison of MXene-based wearable devices.

## REFERENCES

Anasori, Babak, Maria R. Lukatskaya, and Yury Gogotsi. 2017. "2D metal carbides and nitrides (MXenes) for energy storage." *Nature Reviews Materials* 2 (2):16098. doi: 10.1038/natrevmats.2016.98.

Ankitha, Menon, Ajith Mohan. Arjun, Neermunda Shabana, and Pathath Abdul Rasheed. 2023. "A mini review on recent advances in MXene based electrochemical wearable sensing devices." *Biomedical Materials & Devices* 1 (1):339–350. doi: 10.1007/s44174-022-00010-7.

Ankitha, Menon, Fathima Shamsheera, and Pathath Abdul Rasheed. 2024. "MXene-integrated single-stranded carbon yarn-based wearable sensor patch for on-site monitoring of dopamine." *ACS Applied Electronic Materials* 6 (1):599–610. doi: 10.1021/acsaelm.3c01668.

Arjun, Ajith. Mohan, Menon Ankitha, Neermunda Shabana, Pushpalatha Vijayakumar Vaishag, Fathima Shamsheera, Mundummal Mufeeda, and Pathath Abdul Rasheed. 2023. "An overview on surface modification of niobium MXenes for diagnostic and prognostic applications." *FlatChem* 41:100538. doi: 10.1016/j.flatc.2023.100538.

Arjun, Ajith M., Prasannakumari H. Krishna, Anish R. Nath, and Pathath Abdul Rasheed. 2022. "A review on advances in the development of electrochemical sensors for the detection of anesthetic drugs." *Analytical Methods* 14 (41):4040–4052. doi: 10.1039/D2AY01290A.

Aziz, Ayesha, Muhammad Asif, Ghazala Ashraf, Tayyaba Iftikhar, Wajid Hussain, and Shenqi Wang. 2022. "Environmental significance of wearable sensors based on MXene and graphene." *Trends in Environmental Analytical Chemistry* 36:e00180. doi: 10.1016/j.teac.2022.e00180.

Ballard, Zachary, Calvin Brown, Asad M. Madni, and Aydogan Ozcan. 2021. "Machine learning and computation-enabled intelligent sensor design." *Nature Machine Intelligence* 3 (7):556–565. doi: 10.1038/s42256-021-00360-9.

Boobphahom, Siraprapa, Tatiya Siripongpreda, DongDong Zhang, Jiaqian Qin, Pranee Rattanawaleedirojn, and Nadnudda Rodthongkum. 2021a. "$TiO_2$/MXene-PVA/GO hydrogel-based electrochemical sensor for neurological disorder screening via urinary norepinephrine detection." *Microchimica Acta* 188 (11):387. doi: 10.1007/s00604-021-04945-4.

Boobphahom, Siraprapa, Tatiya Siripongpreda, DongDong Zhang, Jiaqian Qin, Pranee Rattanawaleedirojn, and Nadnudda Rodthongkum. 2021b. "$TiO_2$/MXene-PVA/GO hydrogel-based electrochemical sensor for neurological disorder screening via urinary norepinephrine detection." *Microchimica Acta* 188 (11):1–12.

Cao, Zherui, Yina Yang, Yinghui Zheng, Wei Wu, Fangfang Xu, Ranran Wang, and Jing Sun. 2019. "Highly flexible and sensitive temperature sensors based on $Ti_3C_2T_x$ (MXene) for electronic skin." *Journal of Materials Chemistry A* 7 (44):25314–25323. doi: 10.1039/C9TA09225K.

Chen, Fan, Jinhao Wang, Lijuan Chen, Haoliang Lin, Dongxue Han, Yu Bao, Wei Wang, and Li Niu. 2024. "A wearable electrochemical biosensor utilizing functionalized $Ti_3C_2T_x$ MXene for the real-time monitoring of uric acid metabolite." *Analytical Chemistry* 96 (9):3914–3924. doi: 10.1021/acs.analchem.3c05672.

Chen, Liang, Xinyue Dai, Wei Feng, and Yu Chen. 2022. "Biomedical applications of MXenes: From nanomedicine to biomaterials." *Accounts of Materials Research* 3 (8):785–798. doi: 10.1021/accountsmr.2c00025.

Chen, Xiaodong, Dae-Hyeong Kim, and Nanshu Lu. 2024. "Introduction: Wearable devices." *Chemical Reviews* 124 (10):6145–6147. doi: 10.1021/acs.chemrev.4c00271.

Chipangura, Yevedzo E., Brian D. Spindler, Philippe Bühlmann, and Andreas Stein. 2024. "Design criteria for nanostructured carbon materials as solid contacts for ion-selective sensors." *Advanced Materials* 36 (8):2309778. doi: 10.1002/adma.202309778.

Dincer, Can, Richard Bruch, Estefanía Costa-Rama, Maria T. Fernández-Abedul, Arben Merkoçi, Andreas Manz, Gerald A. Urban, and Firat Güder. 2019. "Disposable sensors in diagnostics, food, and environmental monitoring." *Advanced Materials* 31 (30):1806739. doi: 10.1002/adma.201806739.

Downes, Marley, Christopher E. Shuck, Bernard McBride, Jeffrey Busa, and Yury Gogotsi. 2024. "Comprehensive synthesis of $Ti_3C_2T_x$ from MAX phase to MXene." *Nature Protocols* 19 (6):1807–1834. doi: 10.1038/s41596-024-00969-1.

Driscoll, Nicolette, Andrew G. Richardson, Kathleen Maleski, Babak Anasori, Oladayo Adewole, Pavel Lelyukh, Lilia Escobedo, D. K. Cullen, Timothy H. Lucas, Yury Gogotsi, and Flavia Vitale. 2018. "Two-dimensional $Ti_3C_2$ MXene for high-resolution neural interfaces." *ACS Nano* 12 (10):10419–10429. doi: 10.1021/acsnano.8b06014.

Fan, Wenjing, Cheng Li, Wei Li, and Jin Yang. 2022. "3D waterproof MXene-based textile electronics for physiology and motion signals monitoring." *Advanced Materials Interfaces* 9 (14):2102553. doi: 10.1002/admi.202102553.

Friedel, Mark, Ian A. P. Thompson, Gerald Kasting, Ronen Polsky, David Cunningham, Hyongsok T. Soh, and Jason Heikenfeld. 2023. "Opportunities and challenges in the diagnostic utility of dermal interstitial fluid." *Nature Biomedical Engineering* 7 (12):1541–1555. doi: 10.1038/s41551-022-00998-9.

Fusco, Laura, Arianna Gazzi, Christopher E. Shuck, Marco Orecchioni, Eiman I. Ahmed, Linda Giro, Barbara Zavan, Açelya Yilmazer, Klaus Ley, Davide Bedognetti, Yury Gogotsi, and Lucia G. Delogu. 2023. "$V_4C_3$ MXene immune profiling and modulation of T cell-dendritic cell function and interaction." *Small Methods* 7 (8):2300197. doi: 10.1002/smtd.202300197.

Ganesan, Selvaganapathy, Kalaipriya Ramajayam, Thangavelu Kokulnathan, and Arunkumar Palaniappan. 2023. Recent advances in two-dimensional MXene-based electrochemical biosensors for sweat analysis. *Molecules* 28 (1-24). doi: 10.3390/molecules 28124617.

Gogotsi, Yury, and Babak Anasori. 2019. "The rise of MXenes." *ACS Nano* 13 (8):8491–8494. doi: 10.1021/acsnano.9b06394.

Gürbüz, Berfin, and Fatih Ciftci. 2024. "Bio-electric-electronics and tissue engineering applications of MXenes wearable materials: A review." *Chemical Engineering Journal* 489:151230. doi: 10.1016/j.cej.2024.151230.

Hermawan, Angga, Tahta Amrillah, Anung Riapanitra, Wee-Jun Ong, and Shu Yin. 2021. "Prospects and challenges of MXenes as emerging sensing materials for flexible and wearable breath-based biomarker diagnosis." *Advanced Healthcare Materials* 10 (20):2100970. doi: 10.1002/adhm.202100970.

Hong, Weichen, Brian C. Wyatt, Srinivasa K. Nemani, and Babak Anasori. 2020. "Double transition-metal MXenes: Atomistic design of two-dimensional carbides and nitrides." *MRS Bulletin* 45 (10):850–861. doi: 10.1557/mrs.2020.251.

Hu, Yanling, Fangfang Wang, Hui Ye, Jingai Jiang, Shengke Li, Baoying Dai, Jiahui Li, Jun Yang, Xuejiao Song, Junjie Zhang, Yannan Xie, Li Gao, and Dongliang Yang. 2024. "MXene-based flexible electronic materials for wound infection detection and treatment." *npj Flexible Electronics* 8 (1):30. doi: 10.1038/s41528-024-00312-4.

Huang, Kai, Zhongjun Li, Jing Lin, Gang Han, and Peng Huang. 2018. "Two-dimensional transition metal carbides and nitrides (MXenes) for biomedical applications." *Chemical Society Reviews* 47 (14):5109–5124. doi: 10.1039/C7CS00838D.

Huang, Shuang, Chuanjie Yao, Mengyi He, Xinshuo Huang, Zhengjie Liu, Jiayi Chen, Lelun Jiang, Hui-jiuan Chen, and Xi Xie. 2024a. "Functionalized microwires-based bundle electrodes for detection of multiplexed metabolites in interstitial space." *Advanced Engineering Materials* 26 (6):2301662. doi: 10.1002/adem.202301662.

Huang, Yixuan, Shuai He, Sheng Yu, Hannah M. Johnson, Yau K. Chan, Zheng Jiao, Shouteng Wang, Zixiang Wu, and Yi Deng. 2024b. "MXene-decorated nanofibrous membrane with programmed antibacterial and anti-inflammatory effects via steering NF-κB pathway for infectious cutaneous regeneration." *Small* 20 (4):2304119. doi: 10.1002/smll.202304119.

Hussain, Iftikhar, Uzair Sajjad, Onkar J. Kewate, Umay Amara, Faiza Bibi, Abdul Hanan, Darshna Potphode, Muhammad Ahmad, Muhammad S. Javed, Pitcheri Rosaiah, Sajjad Hussain, Karim Khan, Zeeshan Ajmal, Saranya Punniyakoti, Saleh S. Alarfaji, Jee-Hyun Kang, Wail Al Zoubi, Sumanta Sahoo, and Kaili Zhang. 2024. "Double transition-metal MXenes: Classification, properties, machine learning, artificial intelligence, and energy storage applications." *Materials Today Physics* 42:101382. doi: 10.1016/j.mtphys.2024.101382.

Irimes, Maria-Bianca, Alexandra Pusta, Andreea Cernat, Bogdan Feier, Mihaela Tertis, Cecilia Cristea, Anca D. Buzoianu, and Radu Oprean. 2024. "Multiplexed electrochemical sensing devices for chronic diseases diagnosis and monitoring." *TrAC Trends in Analytical Chemistry* 172:117560. doi: 10.1016/j.trac.2024.117560.

Islam, Tamanna, Md M. Hasan, Abdul Awal, Md Nurunnabi, and A. J. S. Ahammad. 2020. Metal nanoparticles for electrochemical sensing: Progress and challenges in the clinical transition of point-of-care testing. *Molecules* 25 (24). doi: 10.3390/molecules25245787.

Jia, Lu, Suqin Zhou, Abbas Ahmed, Zhichao Yang, Shuqiang Liu, Hua Wang, Fu Li, Man Zhang, Yujing Zhang, and Luyi Sun. 2023. "Tuning MXene electrical conductivity towards multifunctionality." *Chemical Engineering Journal* 475:146361. doi: 10.1016/j.cej.2023.146361.

Kalambate, Pramod K., Nayan S. Gadhari, Xiang Li, Zhixiang Rao, Sachin T. Navale, Yue Shen, Vishwanath R. Patil, and Yunhui Huang. 2019. "Recent advances in MXene–based electrochemical sensors and biosensors." *TrAC Trends in Analytical Chemistry* 120:115643. doi: 10.1016/j.trac.2019.115643.

Kalasin, Surachate, and Pantawan Sangnuang. 2023. "Multiplex wearable electrochemical sensors fabricated from sodiated polymers and Mxene nanosheet to measure sodium and creatinine levels in sweat." *ACS Applied Nano Materials* 6 (19):18209–18221. doi: 10.1021/acsanm.3c03469.

Khan, Karim, Ayesha K. Tareen, Muhammad Iqbal, Zhang Ye, Zhongjian Xie, Asif Mahmood, Nasir Mahmood, and Han Zhang. 2023. "Recent progress in emerging novel MXenes based materials and their fascinating sensing applications." *Small* 19 (19):2206147. doi: 10.1002/smll.202206147.

Khan, Reem, and Silvana Andreescu. 2024. "Catalytic MXCe$O_2$ for enzyme based electrochemical biosensors: Fabrication, characterization and application towards a wearable sweat biosensor." *Biosensors and Bioelectronics* 248:115975. doi: 10.1016/j.bios.2023.115975.

Khan, Reem, Zihni O. Uygun, Daniel Andreescu, and Silvana Andreescu. 2024. "Sensitive detection of perfluoroalkyl substances using MXene–AgNP-based electrochemical sensors." *ACS Sensors*. 9(6):3403–3412. doi: 10.1021/acssensors.4c00776.

Kim, Hyuk J., Chung W. Lee, Sohyeon Park, Sungkyun Choi, Sung H. Park, Gi B. Nam, Jung-El Ryu, Tae H. Eom, Byungsoo Kim, Cheol-Joo Kim, Soo Y. Kim, and Ho W. Jang. 2024. "MXene-based high performance microfluidic pH sensors for electronic tongue." *Sensors and Actuators B: Chemical* 409:135636. doi: 10.1016/j.snb.2024.135636.

Kirchner, Eva-Maria, and Thomas Hirsch. 2020. "Recent developments in carbon-based two-dimensional materials: Synthesis and modification aspects for electrochemical sensors." *Microchimica Acta* 187 (8):441. doi: 10.1007/s00604-020-04415-3.

Kumar, Sunil. 2024. "Fluorine-free MXenes: Recent advances, synthesis strategies, and mechanisms." *Small* 20 (16):2308225. doi: 10.1002/smll.202308225.

Laochai, Thidarut, Jutiporn Yukird, Nadtinan Promphet, Jiaqian Qin, Orawon Chailapakul, and Nadnudda Rodthongkum. 2022. "Non-invasive electrochemical immunosensor for sweat cortisol based on L-cys/AuNPs/MXene modified thread electrode." *Biosensors and Bioelectronics* 203:114039. doi: 10.1016/j.bios.2022.114039.

Lei, Yongjiu, Wenli Zhao, Yizhou Zhang, Qiu Jiang, Jr-Hau He, Antje J. Baeumner, Otto S. Wolfbeis, Zhong L. Wang, Khaled N. Salama, and Husam N. Alshareef. 2019. "A MXene-based wearable biosensor system for high-performance in vitro perspiration analysis." *Small* 15 (19):1901190. doi: 10.1002/smll.201901190.

Li, Hui, Rangrang Fan, Bingwen Zou, Jiazhen Yan, Qiwu Shi, and Gang Guo. 2023a. "Roles of MXenes in biomedical applications: Recent developments and prospects." *Journal of Nanobiotechnology* 21 (1):73. doi: 10.1186/s12951-023-01809-2.

Li, Meng, Lichao Wang, Rui Liu, Jianmin Li, Qinghong Zhang, Guoying Shi, Yaogang Li, Chengyi Hou, and Hongzhi Wang. 2021a. "A highly integrated sensing paper for wearable electrochemical sweat analysis." *Biosensors and Bioelectronics* 174:112828. doi: 10.1016/j.bios.2020.112828.

Li, Quan-Fu, Xu Chen, Hai Wang, Ming Liu, and Hui-Ling Peng. 2023b. "Pt/MXene-based flexible wearable non-enzymatic electrochemical sensor for continuous glucose detection in sweat." *ACS Applied Materials & Interfaces* 15 (10):13290–13298. doi: 10.1021/acsami.2c20543.

Li, Zehua, Zepeng Kang, Bo Wu, and Zhiguang Zhu. 2021b. "A MXene-based slurry bioanode with potential application in implantable enzymatic biofuel cells." *Journal of Power Sources* 506:230206. doi: 10.1016/j.jpowsour.2021.230206.

Li, Zhe, and Yue Wu. 2019. "2D early transition metal carbides (MXenes) for catalysis." *Small* 15 (29):1804736. doi: 10.1002/smll.201804736.

Liao, Tao, Zhongyin Chen, Ying Kuang, Zhe Ren, Wenqian Yu, Wen Rao, Linwei Li, Yun Liu, Ziqiang Xu, Bingbing Jiang, and Cao Li. 2023. "Small-size $Ti_3C_2T_x$ MXene nanosheets coated with metal-polyphenol nanodots for enhanced cancer photothermal therapy and anti-inflammation." *Acta Biomaterialia* 159:312–323. doi: 10.1016/j.actbio.2023.01.049.

Liu, Liyuan, Jie Yang, Hongli Zhang, Jianhua Ma, Jiaojiao Zheng, and Chen Wang. 2023. "Recent advances of flexible MXene physical sensor to wearable electronics." *Materials Today Communications* 35:106014. doi: 10.1016/j.mtcomm.2023.106014.

Liu, Yu, Xu Zeng, Yulong Wu, Habib U. Khan, Fei Xiao, Bao Y. Xia, and Yimin Sun. 2024. "Functionalized MXene fiber electrode for the electrochemical sensing of urinary ammonia." *Advanced Sensor and Energy Materials* 3(4):100091. doi: 10.1016/j.asems.2024.100091.

Ma, Chang, Ming-Guo Ma, Chuanling Si, Xing-Xiang Ji, and Pengbo Wan. 2021. "Flexible MXene-based composites for wearable devices." *Advanced Functional Materials* 31 (22):2009524. doi: 10.1002/adfm.202009524.

Ma, Junping, Long Zhang, and Bo Lei. 2023. "Multifunctional MXene-based bioactive materials for integrated regeneration therapy." *ACS Nano* 17 (20):19526–19549. doi: 10.1021/acsnano.3c01913.

Macchia, Eleonora, Fabrizio Torricelli, Mariapia Caputo, Lucia Sarcina, Cecilia Scandurra, Paolo Bollella, Michele Catacchio, Matteo Piscitelli, Cinzia Di Franco, Gaetano Scamarcio, and Luisa Torsi. 2024. "Point-of-care ultra-portable single-molecule bioassays for one-health." *Advanced Materials* 36 (13):2309705. doi: 10.1002/adma.202309705.

Mathew, Minu, and Chandra S. Rout. 2021. "Electrochemical biosensors based on $Ti_3C_2T_x$ MXene: Future perspectives for on-site analysis." *Current Opinion in Electrochemistry* 30:100782. doi: 10.1016/j.coelec.2021.100782.

Myndrul, Valerii, Emerson Coy, Nataliya Babayevska, Veronika Zahorodna, Vitalii Balitskyi, Ivan Baginskiy, Oleksiy Gogotsi, Mikhael Bechelany, Maria T. Giardi, and Igor Iatsunskyi. 2022. "MXene nanoflakes decorating ZnO tetrapods for enhanced performance of skin-attachable stretchable enzymatic electrochemical glucose sensor." *Biosensors and Bioelectronics* 207:114141. doi: 10.1016/j.bios.2022.114141.

Nah, Joong S., Sharat C. Barman, Md Abu Zahed, Md Sharifuzzaman, Hyosang Yoon, Chani Park, Sanghyuk Yoon, Shipeng Zhang, and Jae Y. Park. 2021. "A wearable microfluidics-integrated impedimetric immunosensor based on $Ti_3C_2T_x$ MXene incorporated laser-burned graphene for noninvasive sweat cortisol detection." *Sensors and Actuators B: Chemical* 329:129206. doi: 10.1016/j.snb.2020.129206.

Noriega, Natalia, Mikhail Shekhirev, Christopher E. Shuck, Jonathan Salvage, Armin VahidMohammadi, Marcus K. Dymond, Joseph Lacey, Susan Sandeman, Yury Gogotsi, and Bhavik A. Patel. 2024. "Pristine $Ti_3C_2T_x$ MXene enables flexible and transparent electrochemical sensors." *ACS Applied Materials & Interfaces* 16 (5):6569–6578. doi: 10.1021/acsami.3c14842.

Pan, Yawen, Meiru He, Jing Wu, Hongbin Qi, and Yuanyuan Cheng. 2024. "One-step synthesis of MXene-functionalized PEDOT:PSS conductive polymer hydrogels for wearable and noninvasive monitoring of sweat glucose." *Sensors and Actuators B: Chemical* 401:135055. doi: 10.1016/j.snb.2023.135055.

Rasheed, Pathath Abdul, Ravi P. Pandey, Fawzi Banat, and Shadi W. Hasan. 2022. "Recent advances in niobium MXenes: Synthesis, properties, and emerging applications." *Matter* 5 (2):546–572. doi: 10.1016/j.matt.2021.12.021.

Rasool, Kashif, Ravi P. Pandey, Rasheed, Pathath Abdul, Samantha Buczek, Yury Gogotsi, and Khaled A. Mahmoud. 2019. "Water treatment and environmental remediation applications of two-dimensional metal carbides (MXenes)." *Materials Today* 30: 80–102. doi: 10.1016/j.mattod.2019.05.017.

Richard, Bartholomew, Kannankuzhiyan Niyas, Menon Ankitha, and Rasheed, Pathath Abdul. 2024. "Biological metal–organic framework-embedded MXene nanocomposite as a wearable transducer patch for real-time monitoring of the sleep hormone." *ACS Applied Nano Materials* 7 (8):9585–9597. doi: 10.1021/acsanm.4c01062.

Safarkhani, Moein, Bahareh F. Far, Eder C. Lima, Shima Jafarzadeh, Pooyan Makvandi, Rajender S. Varma, YunSuk Huh, Majid E. Warkiani, and Navid Rabiee. 2024. "Integration of MXene and microfluidics: A perspective." *ACS Biomaterials Science & Engineering* 10 (2):657–676. doi: 10.1021/acsbiomaterials.3c01361.

San Nah, Joong, Sharat C. Barman, Md Abu Zahed, Md Sharifuzzaman, Hyosang Yoon, Chani Park, Sanghyuk Yoon, Shipeng Zhang, and Jae Y. Park. 2021. "A wearable microfluidics-integrated impedimetric immunosensor based on $Ti_3C_2T_x$ MXene incorporated laser-burned graphene for noninvasive sweat cortisol detection." *Sensors Actuators B: Chemical* 329:129206.

Seidi, Farzad, Ahmad A. Shamsabadi, Mostafa D. Firouzjaei, Mark Elliott, Mohammad R. Saeb, Yang Huang, Chengcheng Li, Huining Xiao, and Babak Anasori. 2023. "MXenes antibacterial properties and applications: A review and perspective." *Small* 19 (14):2206716. doi: 10.1002/smll.202206716.

Sharifuzzaman, Md, Ashok Chhetry, Md Abu Zahed, Sang H. Yoon, Chan I. Park, Shipeng Zhang, Sharat C. Barman, Sudeep Sharma, Hyosang Yoon, and Jae Y. Park. 2020. "Smart bandage with integrated multifunctional sensors based on MXene-functionalized porous graphene scaffold for chronic wound care management." *Biosensors and Bioelectronics* 169:112637. doi: 10.1016/j.bios.2020.112637.

Sharifuzzaman, Md, Md Abu Zahed, Md Selim Reza, Md Asaduzzaman, SeongHoon Jeong, Hyesu Song, Dong K. Kim, Shipeng Zhang, and Jae Y. Park. 2023. "MXene/fluoropolymer-derived laser-carbonaceous all-fibrous nanohybrid patch for soft wearable bioelectronics." *Advanced Functional Materials* 33 (21):2208894. doi: 10.1002/adfm.202208894.

Shi, Zhenghan, Chaobo Dai, Peixue Deng, Xin Li, Yue Wu, Jingjiang Lv, Chuhan Xiong, Yifan Shuai, Fenni Zhang, Di Wang, Hao Liang, Yan He, Qingmei Chen, Yanli Lu, and Qingjun Liu. 2023. "Wearable battery-free smart bandage with peptide functionalized biosensors based on MXene for bacterial wound infection detection." *Sensors and Actuators B: Chemical* 383:133598. doi: 10.1016/j.snb.2023.133598.

Soleymaniha, Mohammadreza, Mohammad-Ali Shahbazi, Ali R. Rafieerad, Aziz Maleki, and Ahmad Amiri. 2019. "Promoting role of MXene nanosheets in biomedical sciences: Therapeutic and biosensing innovations." *Advanced Healthcare Materials* 8 (1):1801137. doi: 10.1002/adhm.201801137.

Song, Yu, Roland Y. Tay, Jiahong Li, Changhao Xu, Jihong Min, Ehsan S. Sani, Gwangmook Kim, Wenzheng Heng, Inho Kim, and Wei Gao. 2023. "3D-printed epifluidic electronic skin for machine learning–powered multimodal health surveillance." *Science Advances* 9 (37):eadi6492. doi: 10.1126/sciadv.adi6492.

Tajik, Somayeh, Zahra Dourandish, Fariba G. Nejad, Hadi Beitollahi, Peyman M. Jahani, and Antonio Di Bartolomeo. 2022. "Transition metal dichalcogenides: Synthesis and use in the development of electrochemical sensors and biosensors." *Biosensors and Bioelectronics* 216:114674. doi: 10.1016/j.bios.2022.114674.

Teymourian, Hazhir, Marc Parrilla, Juliane R. Sempionatto, Noelia F. Montiel, Abbas Barfidokht, Robin Van Echelpoel, Karolien De Wael, and Joseph Wang. 2020. "Wearable electrochemical sensors for the monitoring and screening of drugs." *ACS Sensors* 5 (9):2679–2700. doi: 10.1021/acssensors.0c01318.

Unal, Mehmet A., Fatma Bayrakdar, Laura Fusco, Omur Besbinar, Christopher E. Shuck, Süleyman Yalcin, Mine T. Erken, Aykut Ozkul, Cansu Gurcan, Oguzhan Panatli, Gokce Y. Summak, Cemile Gokce, Marco Orecchioni, Arianna Gazzi, Flavia Vitale, Julia Somers, Emek Demir, Serap S. Yildiz, Hasan Nazir, Jean-Charles Grivel, Davide Bedognetti, Andrea Crisanti, Kamil C. Akcali, Yury Gogotsi, Lucia G. Delogu, and Açelya Yilmazer. 2021. "2D MXenes with antiviral and immunomodulatory properties: A pilot study against SARS-CoV-2." *Nano Today* 38:101136. doi: 10.1016/j.nantod.2021.101136.

Vulpe, Georgeta, Guoyi Liu, Sam Oakley, Dimitrios Pletsas, Guanghao Yang, Rosa Dutra, Owen Guy, Yufei Liu, Mark Waldron, Joe Neary, Arjun A. Mohan, and Sanjiv Sharma. 2024a. "Wearable technology for one health: Charting the course of dermal biosensing." *Biosensors and Bioelectronics:* X 19:100500. doi: 10.1016/j.biosx.2024.100500.

Vulpe, Georgeta, Guoyi Liu, Sam Oakley, Guanghao Yang, Arjun A. Mohan, Mark Waldron, and Sanjiv Sharma. 2024b. "Lab on skin: Real-time metabolite monitoring with polyphenol film based subdermal wearable patches." *Lab on a Chip* 24 (7):2039–2048. doi: 10.1039/D4LC00073K.

Waheed, Waqas, Shoaib Anwer, Muhammad U. Khan, Muhammad Sajjad, and Anas Alazzam. 2024. "2D $Ti_3C_2T_x$-MXene nanosheets and graphene oxide based highly sensitive humidity sensor for wearable and flexible electronics." *Chemical Engineering Journal* 480:147981. doi: 10.1016/j.cej.2023.147981.

Wang, Xingyu, Zhiyu Wang, and Jieshan Qiu. 2021. "Stabilizing MXene by hydration chemistry in aqueous solution." *Angewandte Chemie International Edition* 60 (51):26587–26591. doi: 10.1002/anie.202113981.

Wang, Yi-Han, Ke-Jing Huang, and Xu Wu. 2017. "Recent advances in transition-metal dichalcogenides based electrochemical biosensors: A review." *Biosensors and Bioelectronics* 97:305–316. doi: 10.1016/j.bios.2017.06.011.

Wang, Zifeng, Feng Yan, Zhichao Yu, Huina Cao, Zhanying Ma, ZuNa YeErKenTai, Zhanhong Li, Yutong Han, and Zhigang Zhu. 2024. "Fully transient 3D origami paper-based ammonia gas sensor obtained by facile MXene spray coating." *ACS Sensors* 9 (3):1447–1457. doi: 10.1021/acssensors.3c02558.

Wu, Xinzhao, Pinyi Ma, Ying Sun, Fangxin Du, Daqian Song, and Guobao Xu. 2021. "Application of MXene in electrochemical sensors: A review." *Electroanalysis* 33 (8): 1827–1851. doi: 10.1002/elan.202100192.

Xie, Chao, Yi Wang, Zhaosheng Xia, Yu Cheng, Xisheng Cui, Shijie Xu, Xingang Ren, Wenhua Yang, and Zhixiang Huang. 2023. "High-performance broadband flexible photodetectors based on $Ti_3C_2T_x$ MXene/pyramidal thin Si heterostructures with light trapping effect for heart rate detection." *Advanced Materials Technologies* 8 (24):2301232. doi: 10.1002/admt.202301232.

Xue, Ni, Xuesong Li, Mengqi Zhang, Liuyuan Han, Yangyang Liu, and Xutang Tao. 2020. "Chemical-combined ball-milling synthesis of fluorine-free porous MXene for high-performance lithium ion batteries." *ACS Applied Energy Materials* 3 (10):10234–10241. doi: 10.1021/acsaem.0c02081.

Yang, Yen-Chang, Yen-Tzu Lin, Jiashing Yu, Huan-Tsung Chang, Ting-Yu Lu, Tzu-Yen Huang, Anant Preet, Ya-Ju Hsu, Ligang Wang, and Tzu-En Lin. 2021. "MXene nanosheet-based microneedles for monitoring muscle contraction and electrostimulation treatment." *ACS Applied Nano Materials* 4 (8):7917–7924. doi: 10.1021/acsanm.1c01237.

Ye, Cui, Minqiang Wang, Jihong Min, Roland Y. Tay, Heather Lukas, Juliane R. Sempionatto, Jiahong Li, Changhao Xu, and Wei Gao. 2023. "A wearable aptamer nanobiosensor for non-invasive female hormone monitoring." *Nature Nanotechnology* 19:330–337. doi: 10.1038/s41565-023-01513-0.

Ye, Cui, Minqiang Wang, Jihong Min, Roland Y. Tay, Heather Lukas, Juliane R. Sempionatto, Jiahong Li, Changhao Xu, and Wei Gao. 2024. "A wearable aptamer nanobiosensor for non-invasive female hormone monitoring." *Nature Nanotechnology* 19 (3):330–337. doi: 10.1038/s41565-023-01513-0.

Zahed, Md Abu, Md Sharifuzzaman, Hyosang Yoon, Md Asaduzzaman, Dong K. Kim, Seonghoon Jeong, Gagan B. Pradhan, Young D. Shin, Sang H. Yoon, Sudeep Sharma, Shipeng Zhang, and Jae Y. Park. 2022. "A nanoporous carbon-MXene heterostructured nanocomposite-based epidermal patch for real-time biopotentials and sweat glucose monitoring." *Advanced Functional Materials* 32 (49):2208344. doi: 10.1002/adfm.202208344.

Zarepour, Atefeh, Nesa Rafati, Arezoo Khosravi, Navid Rabiee, Siavash Iravani, and Ali Zarrabi. 2024. "MXene-based composites in smart wound healing and dressings." *Nanoscale Advances* 6:3513–3532. doi: 10.1039/D4NA00239C.

Zhang, Jianxin, Junlin Ma, Wenrui Zhang, Hongting Ma, Xiaodong Geng, Hanwen Zhang, Quanli Liu, Jing Zhu, Chunbao Li, Yan Su, and Nan Zhu. 2024a. "Wearable sensors deriving from cationic-induced 2D-2D co-assembled films for nutrient monitoring." *Electrochimica Acta* 477:143787. doi: 10.1016/j.electacta.2024.143787.

Zhang, Jizhen, Na Kong, Simge Uzun, Ariana Levitt, Shayan Seyedin, Peter A. Lynch, Si Qin, Meikang Han, Wenrong Yang, Jingquan Liu, Xungai Wang, Yury Gogotsi, and Joselito M. Razal. 2020. "Scalable manufacturing of free-standing, strong $Ti_3C_2T_x$ MXene films with outstanding conductivity." *Advanced Materials* 32 (23):2001093. doi: 10.1002/adma.202001093.

Zhang, Shipeng, Md Abu Zahed, Md Sharifuzzaman, Sanghyuk Yoon, Xue Hui, Sharat C. Barman, Sudeep Sharma, Hyo S. Yoon, Chani Park, and Jae Y. Park. 2021. "A wearable battery-free wireless and skin-interfaced microfluidics integrated electrochemical sensing patch for on-site biomarkers monitoring in human perspiration." *Biosensors and Bioelectronics* 175:112844. doi: 10.1016/j.bios.2020.112844.

Zhang, Teng, Kateryna Shevchuk, Ruocun J. Wang, Hyunho Kim, Jamal Hourani, and Yury Gogotsi. 2024b. "Delamination of chlorine-terminated MXene produced using molten salt etching." *Chemistry of Materials* 36 (4):1998–2006. doi: 10.1021/acs.chemmater.3c02872.

Zhao, Jiaying, Congjuan He, Weixuan Wu, Huisi Yang, Lan Peng, Li Wen, Zhikun Hu, Changjun Hou, and Danqun Huo. 2022. "MXene-$MoS_2$ carbon-fiber-based flexible electrochemical interface for multiple bioanalysis in biofluids." *Chemical Engineering Journal* 446:136841. doi: 10.1016/j.cej.2022.136841.

Zhao, Xing, Li-Ya Wang, Jia-Meng Li, Li-Mei Peng, Chun-Yan Tang, Xiang-Jun Zha, Kai Ke, Ming-Bo Yang, Bai-Hai Su, and Wei Yang. 2021. "Redox-mediated artificial non-enzymatic antioxidant MXene nanoplatforms for acute kidney injury alleviation." *Advanced Science* 8 (18):2101498. doi: 10.1002/advs.202101498.

Zheng, Hua, Shenqiang Wang, Fang Cheng, Xiaowei He, Zongxu Liu, Wenyan Wang, Li Zhou, and Qiuyu Zhang. 2021. "Bioactive anti-inflammatory, antibacterial, conductive multifunctional scaffold based on MXene@$CeO_2$ nanocomposites for infection-impaired skin multimodal therapy." *Chemical Engineering Journal* 424:130148. doi: 10.1016/j.cej.2021.130148.

Zhou, Jie, Martin Dahlqvist, Jonas Björk, and Johanna Rosen. 2023. "Atomic scale design of MXenes and their parent materials—From theoretical and experimental perspectives." *Chemical Reviews* 123 (23):13291–13322. doi: 10.1021/acs.chemrev.3c00241.

# 13 MXene-based Self-Powered Electrochemical Sensors

*Ajith Mohan Arjun*

## 13.1 INTRODUCTION

Electrochemical sensors are indispensable tools, playing a pivotal role across diverse fields including environmental monitoring, healthcare, industrial process control, the food industry, agriculture, safety and security, automotive systems, and biotechnology research. Despite their widespread utility, conventional electrochemical sensors necessitate an external power source. Stationary systems often draw mains power, frequently reliant on supply grids deriving energy from fossil fuels with associated carbon footprints, while portable units typically depend on batteries. The production of these batteries involves raw materials sourced via mining operations – which can be sources of pollution and environmental disruption – and their availability is often contingent upon complex global supply chains susceptible to disruption. This fundamental reliance on external power introduces significant limitations concerning device autonomy, overall sustainability, environmental impact, maintenance logistics, and system resilience. Addressing these challenges, self-powered electrochemical sensors (SPESs) offer a transformative approach and represent a significant evolution in sensor technology (Zeng et al. 2025). These innovative devices operate by generating their own electrical energy, often through processes where the power output itself is directly modulated by the presence and concentration of the target analyte. This integration of sensing and in situ power generation represents a paradigm shift toward more sustainable, autonomous, and potentially simplified sensing platforms, paving the way for next-generation devices. This chapter will delve into the exciting advancements in this area, focusing specifically on the role of novel MXene materials in developing high-performance SPESs.

This paradigm shift toward self-powered systems necessitates innovative device designs capable of harvesting energy directly from their immediate surroundings or from the electrochemical sensing reaction itself. Realizing such autonomous sensors involves exploring diverse energy conversion strategies integrated directly within the device architecture. The key approaches include harnessing biomechanical energy (e.g., using piezoelectric or triboelectric effects to convert vibrations or pressure changes), exploiting biochemical reactions (as seen in enzymatic biofuel cells where analyte consumption generates current), utilizing thermal gradients (via thermoelectric generators), capturing ambient light (through photovoltaic components), or even

DOI: 10.1201/9781003736318-13

converting chemical potential directly into electricity. The specific choice of energy harvesting mechanism dictates crucial aspects of the sensor's design, operation, and potential applications. For instance, a wearable sensor might leverage body movement via triboelectric nanogenerators, while an environmental sensor might utilize a biofuel cell powered by pollutants (Ou-Yang et al. 2024). Consequently, the development focus extends beyond mere analyte detection to the co-design and integration of sensing elements with miniaturized, efficient power generation units, ensuring the generated power signal reliably reflects the analyte concentration while meeting the operational demands of the device. Understanding these varied operational principles and the associated device engineering challenges is central to advancing the field of self-powered electrochemical sensing.

Among the diverse strategies previously outlined, approaches leveraging biofuel cells (BFCs) and photoelectrochemical (PEC) reactions offer particularly compelling routes toward self-powered sensing, aligning well with biological and certain environmental monitoring needs. These systems are distinct in that they generate power directly from chemical energy sources, such as metabolites readily available in physiological fluids (like glucose or lactate), or by converting captured light energy in PEC systems. A refined BFC-based sensing concept involves designing the fuel cell electrode with a primary biocatalyst for power generation, but coupling its activity to a separate bioreceptor element. This bioreceptor selectively captures or metabolizes the specific analyte of interest. The biochemical consequence of this bioreceptor-analyte interaction (e.g., production or consumption of a substance that acts as a fuel, mediator, or inhibitor for the main BFC reaction) directly influences the rate of the power-generating reaction at the biocatalytic electrode. Consequently, the presence and concentration of the analyte are transduced into a measurable change in the fuel cell's power output. This operational principle contrasts sharply with sensors relying on harvesting ambient mechanical (piezo/triboelectric) or thermal energy, where the analyte typically modulates a physical energy conversion process indirectly. In the BFC/PEC scenarios discussed here, the analyte interaction is more intimately linked to the core energy generation mechanism itself. This direct modulation potentially simplifies signal analysis significantly, as the sensing information can be extracted by monitoring the device's electrical power with relatively straightforward instrumentation. Furthermore, this inherent coupling can also streamline device fabrication; the processes at the transducer stage can be designed to integrate both the power production and sensing elements closely together. This potentially simplifies manufacturing compared to other self-powering technologies where distinct power harvesting and sensing units might need separate fabrication processes followed by complex integration steps to link them functionally. This overall approach thus offers an elegant integration of sensing, power generation, and potentially fabrication.

Given the central role of the active component (transducer) in such devices, selecting the appropriate materials for constructing or surface-modifying these transducers is of utmost importance. The successful realization of these elegantly integrated BFC and PEC-based self-powered sensors, where analyte interaction directly modulates power generation, hinges critically upon the selection and engineering of advanced materials. These materials form the core transducer element, needing to efficiently support both the energy conversion process (biocatalysis or photoelectrocatalysis) and the analyte recognition/interaction mechanism, all while ensuring rapid electron transfer for signal readout. While traditional materials like noble metals, carbon allotropes (nanotubes, graphene), and conducting polymers have been explored, achieving optimal synergy between power generation, sensing performance, stability, and biocompatibility remains a significant challenge.

This is where the unique attributes of novel nanomaterials become paramount. Among the emerging candidates, MXenes, a relatively new family of two-dimensional (2D) transition metal carbides, nitrides, and carbonitrides, have garnered substantial interest for electrochemical applications, including self-powered sensing (Arjun et al. 2023). Typically derived from precursor MAX phases via selective etching (often using hydrofluoric acid or alternative etchants), MXenes possess a distinctive combination of properties that make them exceptionally well suited for the demands of integrated sensing and power generation platforms. Represented by the general formula $M_{n+1}X_nT_x$ (where M is an early transition metal like Ti, V, Mo; X is C or N; $n$ is typically 1–4; and $T_x$ denotes surface functional groups like –O, –OH, –F acquired during etching), MXenes exhibit metallic conductivity, ensuring efficient charge transport comparable to or exceeding many traditional electrode materials. Crucially, unlike pristine graphene, their surfaces are inherently functionalized, rendering them hydrophilic and providing abundant anchor points for immobilizing enzymes, antibodies, aptamers, or other biorecognition elements essential for selective sensing in BFCs. Furthermore, their high specific surface area facilitates high catalyst loading and enhances interaction with the surrounding medium (e.g., electrolyte or biological fluid). The tunable surface chemistry and potential for diverse elemental compositions also open avenues for tailoring their electrocatalytic activity toward specific reactions involved in BFC energy generation (like oxygen reduction or fuel oxidation) or PEC processes. These intrinsic advantages position MXenes as a powerful building block for developing next-generation, high-performance SPESs, addressing limitations associated with conventional materials and paving the way for enhanced sensitivity, stability, and miniaturization. The subsequent sections will explore in detail how these properties are being leveraged in the design and fabrication of MXene-based self-powered sensing devices.

MXenes present numerous advantages for the development of SPESs. Their inherent high metallic conductivity minimizes internal resistance, leading to efficient electron collection, maximized power output, and stronger signals. The two-dimensional structure allows for the creation of thin, flexible, and potentially transparent electrodes suitable for diverse applications, including wearables. Furthermore, MXenes possess hydrophilic surfaces rich in functional groups, making them superior supports for biocatalysts like enzymes; this promotes better loading, enhances enzyme stability and activity, and provides a high density of active sites due to the large surface area, ultimately boosting sensor sensitivity and power generation in biofuel cell configurations. They also show promise in facilitating direct electron transfer, simplifying sensor designs, and improving stability. Additionally, some MXenes exhibit intrinsic electrocatalytic or co-catalytic activity, potentially replacing expensive catalysts or enhancing efficiency in photoelectrochemical systems by improving charge separation and accelerating redox reactions. Finally, their ability to form advanced hybrid nanocomposites with other materials enables fine-tuning of properties, allowing for enhanced catalytic activity, mechanical robustness, or specific functionalities tailored to the target application.

Building upon the advantages outlined previously, this chapter delves into the practical application of MXenes within SPESs. To provide a clear focus, the discussion will concentrate specifically on systems where MXenes serve as active components. Emphasis will be placed on their integration into enzyme, microbial, and photoelectrochemical fuel cells, particularly when these configurations are employed as sensors for detecting physiologically and environmentally significant analytes. This defined scope allows for an in-depth exploration of the functional role of MXenes in these specific sensing platforms.

Before examining the specific contributions of MXenes, it is essential to understand the basic operating principles of the self-powered systems central to this chapter: enzymatic biofuel cells (EBFCs), microbial fuel cells (MFCs), and photoelectrochemical (PEC) cells, particularly concerning their use as sensors (Figure 13.1).

EBFCs generate electrical power through highly specific biochemical reactions catalyzed by enzymes immobilized on electrode surfaces. Typically, an oxidoreductase enzyme at the anode oxidizes a particular fuel (e.g., glucose), while another enzyme or catalyst at the cathode reduces an oxidant (e.g., oxygen). The electrical output (current or voltage) of an EBFC is directly dependent on the concentration of the enzyme's target substrate (the analyte). Their high specificity, derived from the enzymes, makes them attractive for detecting particular biomarkers or metabolites (Xiao et al. 2019).

MFCs harness the metabolic activity of electrogenic microorganisms to convert the chemical energy stored in organic matter directly into electricity. Microbes colonizing the anode oxidize diverse substrates (the fuel) and transfer electrons to the electrode. For sensing applications, changes in the power output of MFC can correlate with fluctuations in the concentration of the microbial substrate (e.g., measuring biochemical oxygen demand (BOD) in wastewater) or the presence of substances that inhibit or stimulate microbial activity (e.g., toxins) (Doglioni, Nardello, and Brunelli 2024).

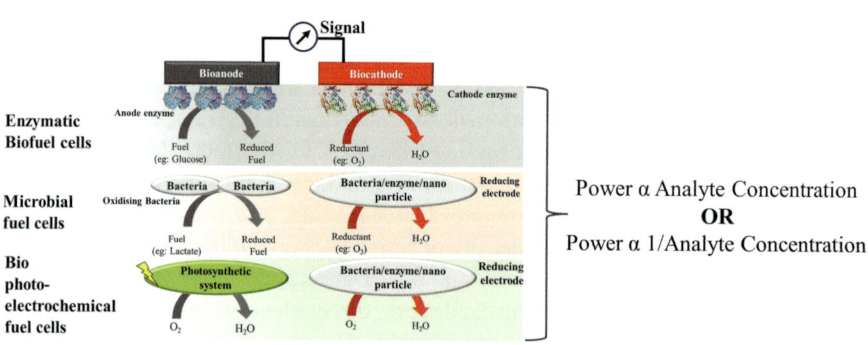


**FIGURE 13.1** Illustration of enzymatic, microbial, and biophotoelectrochemical fuel cells. These systems utilize biological components (enzymes, microbes, photosynthetic systems) to catalyze redox reactions at the anode and cathode, generating an electrical signal (power) that is typically proportional or inversely proportional to the concentration of a target analyte.

Biophotoelectrochemical (PEC) fuel cells/sensor systems utilize semiconductor materials (photoelectrodes) that absorb light to generate electron–hole pairs. These charge carriers drive electrochemical reactions, often oxidizing a fuel at the photoanode or reducing an oxidant at the photocathode, thereby generating electricity assisted by light energy (Zhu et al. 2024a). In a sensing context, the photocurrent or photovoltage can be highly sensitive to the intensity of incident light, or crucially, to the presence of target analytes that interact with the photoelectrode surface, modulate charge transfer kinetics, or participate in the PEC reactions. This allows for light-controlled sensing or detection based on analyte-induced changes in the photo-response.

## 13.2 ENZYME BIOFUEL CELL-BASED SENSORS

The use of enzymes in fuel cells possesses some disadvantages such as the stability of enzymes, lower current production, and issues associated with the immobilization mechanisms used. To mitigate these demerits, Ji et al. implemented an innovative enzyme-free biofuel cell design for their lead ion sensor. They utilized two-dimensional lamellar Au NPs-$Ti_3C_2T_x$ heterostructures for the anode, where ultra-small Au NPs provided glucose oxidase-like catalytic activity, replacing potentially unstable enzymes. Furthermore, the $Ti_3C_2T_x$ MXene component, with its excellent charge storage capacity, was leveraged to provide a large instantaneous current for signal amplification, addressing potential issues of low current production (Ji et al. 2024).

SPESs show great promise but often struggle in practical analytical applications due to low energy conversion efficiency and limited selectivity in complex samples. To overcome these specific limitations. Sensitive and specific detection of biomarkers like soluble CD44 is vital for diagnostic purposes, driving the need for advanced sensor platforms. CD44 is a cell-surface glycoprotein receptor involved in cell adhesion and migration, acting as a key biomarker and therapeutic target, particularly in cancer metastasis and stem cell identification. To meet this demand, Sun et al. reported on the development of a novel self-powered biosensor employing an EBFC specifically for ultrasensitive CD44 detection (Figure 13.2(A)). Its innovative sensing mechanism relies on a bioanode featuring both glucose oxidase and a CD44-specific aptamer. The aptamer initially acts as an insulating layer, suppressing the cell's open-circuit voltage ($E_{OCV}$); however, binding of the target CD44 induces a structural change that facilitates electron transfer, thereby increasing the $E_{OCV}$ in proportion to the analyte concentration. This clever design achieves high sensitivity (LOD of 0.052 ng/mL), excellent selectivity and stability, proving effective in real blood samples without special pretreatment and highlighting its potential for early cancer diagnosis (Sun et al. 2024).

While self-powered EBFCs offer power autonomy, achieving highly precise and sensitive detection of targets like 17β-estradiol (E2) demands sophisticated signal transduction strategies (Figure 13.2(B)). Yi et al. reported the development of such a system by integrating a dual-enzyme approach with an innovative bioconjugate release mechanism within an EBFC. Utilizing advanced carbon/gold nanomaterial composites, the sensor employs a 'bioconjugate' carrying the cathode enzyme (bilirubin oxidase (BOD)) and complementary DNA (csDNA). This bioconjugate is initially anchored near the cathode via hybridization with an E2-specific aptamer; upon E2 binding, this linkage is disrupted, releasing the BOD-bioconjugate. This physical detachment of the essential cathode enzyme hinders electron transfer from the glucose oxidase anode, causing a significant decrease in the open-circuit voltage ($E_{OCV}$) and enabling highly sensitive detection (LOD of 3.3 pg/mL), positioning it as a viable prototype for potential on-site food and environmental safety monitoring (Yi et al. 2024).

Developing ultrasensitive, self-powered sensors suitable for on-site detection of contaminants like 17β-estradiol is crucial for food and environmental safety. Addressing this need, Luo et al. developed a novel biofuel cell (BFC)-based self-powered immunosensing platform (Luo et al. 2024). Their strategy integrates a biogate immunoassay with target-induced biofuel release. Specifically, they used positively charged mesoporous silica nanoparticles (PMSN) loaded with glucose fuel, gated by negatively charged gold nanoparticle-labeled anti-E2 antibodies (AuNPs-Ab) attached via electrostatic interactions. Upon capturing target E2, the antibody binding weakens the gate's adhesion to the PMSN, triggering the release of glucose. This released fuel is then oxidized by glucose oxidase at the bioanode (constructed using carbon nanocage/AuNP composites), leading to a significant *increase* in the BFC's open-circuit voltage ($E_{OCV}$). This innovative BFC-based immunosensor demonstrated exceptional sensitivity for E2 detection (LOD 0.32 pg/mL) and shows significant potential as a viable prototype for mobile bioassay systems.

Sensitive and accurate detection of cancer markers like HER2 is crucial for early diagnosis and treatment monitoring. Addressing this, Hou et al. developed an ultrasensitive self-powered biosensor based on an EBFC. To achieve exceptional sensitivity for HER2 detection, they integrated multiple amplification strategies: advanced electrode materials (Graphdiyne and 1T $MoS_2$), a 3D DNA Walker triggering a hybrid chain reaction (HCR) cascade for significant biochemical signal amplification, and

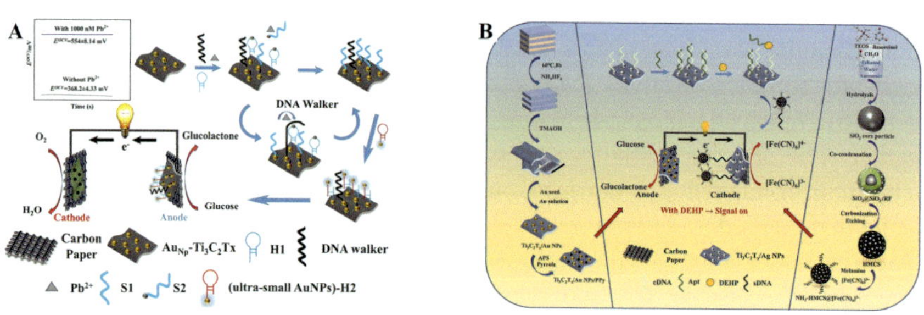


**FIGURE 13.2** (A) A biofuel cell-surface plasmon spectroscopy (BFC-SPS) platform for sensitive lead ($Pb^{2+}$) detection, utilizing an AuNp-$Ti_3C_2T_x$ heterostructure glucose anode, a Co, Mn-NC nanosheet oxygen reduction cathode, and a $Pb^{2+}$ triggered DNA walker for dual electrochemical and optical signal transduction. Adapted with permission from Ji et al. (2024) Copyright (2024). (B) An intra-capacitive self-powered sensor for DEHP detection, featuring a $Ti_3C_2T_x$/Au NPs/PPy ternary heterostructure electrode, aptamer-based recognition of DEHP, and an $NH_2$-HMCS@$[Fe(CN)_6]^{3-}$-based signal tag integrated into a glucose-powered electrochemical cell for signal generation. Adapted with permission from Ji et al. (2025) Copyright (2025) Elsevier.

critically, matched the EBFC with a capacitor to further magnify the output electrical signal (Figure 13.2(C)). This sophisticated design, providing dual-mode electrochemical and colorimetric readouts, resulted in a sensor with an extremely low detection limit (0.03 pg/mL) for HER2 over a very wide dynamic range, offering a powerful new tool for cancer diagnostics (Hou et al. 2023)

Wearable sensors often require external power and reading devices, limiting their utility for on-site detection. Addressing this limitation, Wei et al. developed a reusable, self-powered electrochromic sensor patch based on EBFCs for visualized monitoring directly on the skin (Wei et al. 2025). Their design uses hydrophilic/hydrophobic materials for sweat management and cleverly utilizes lactic acid in sweat as both the target analyte and the fuel for the EBFC. The electrical energy generated by the EBFC reduces Prussian Blue (PB) to Prussian White (PW), providing a visible color change that allows for semiquantitative detection by eye or quantitative analysis via color value or current. Crucially, Wei et al. incorporated reusability by demonstrating that applying an external voltage restores the PB color, enabling multiple uses of the sensor patch and offering a proof-of-concept for practical, visualized wearable monitoring systems.

The widespread presence of endocrine-disrupting phthalates like di(2-ethylhexyl) phthalate (DEHP) necessitates sensitive and portable sensors for environmental and health monitoring, addressing limitations of traditional methods. To address this challenge, Ji et al. developed a novel enzyme-free biofuel cell (BFC)-based self-powered sensor, specifically termed an intra-capacitive BFC self-powered sensor (ICBFC-SPS) (Ji et al. 2025). Their design uniquely integrates a ternary heterostructure capacitive anode and a sensing-interface cathode within a single chamber. The anode, composed of $Ti_3C_2T_x$ MXene, ultra-small Au NPs, and polypyrrole (PPy) NPs, enables efficient enzyme-free glucose oxidation (via Au NPs) while achieving significantly enhanced charge storage through the synergistic combination of MXene's double-layer capacitance and PPy's pseudocapacitance. PPy intercalation also promotes ion diffusion and charge transfer. This sophisticated ICBFC-SPS demonstrated exceptional sensitivity for DEHP detection, achieving an extremely low detection limit (9.51 pg/L) over a vast linear range, highlighting its potential as a transformative tool for rapid, on-site environmental monitoring and public health protection.

Effective immobilization of enzymes like glucose oxidase (GOx) onto conductive supports while maintaining high activity and loading is crucial for developing high-performance electrochemical glucose biosensors. Addressing this, Wu et al. developed 2D nanoreactors by immobilizing GOx onto poly-l-lysine (PLL)-modified $Ti_3C_2$ MXene nanosheets (Wu et al. 2021). Their approach utilized PLL to create a compatible, positively charged surface on the MXene, achieving a high GOx loading capacity (up to 50 wt%). Subsequent cross-linking between GOx and PLL amine groups formed a robust enzyme network. Significantly, Wu et al. demonstrated that the $Ti_3C_2$ MXene itself contributes to superior enzymatic activity by catalytically decomposing the toxic intermediate $H_2O_2$ generated during glucose oxidation, a key advantage over inert supports. When deposited on an electrode, these $Ti_3C_2$-PLL-GOx nanosheets formed an enzymatic electrochemical biosensor exhibiting a low glucose detection limit of 2.6 μM, showcasing the effectiveness of this synergistic material design.

## 13.3 MICROBIAL AND BIOPHOTOELECTROCHEMICAL-BASED SENSORS

Soil microbial fuel cells (SMFCs) offer a route to soil-powered biosensors, but traditional designs often rely on anodic sensing, which Wang et al. (2023) identify as potentially unstable for long-term monitoring (Wang et al. 2023). To address this limitation, their work developed an innovative SMFC biosensor employing cathodic sensing using a carbon-felt cathode, with the anode remaining stable in deep soil. They demonstrated that upon injecting heavy metal ions ($Cd^{2+}$, $Zn^{2+}$, $Pb^{2+}$, or $Hg^{2+}$) near the cathode, the SMFC's output voltage (~400 mV under load) increased sharply and rapidly (2–5 min), allowing quantification in the 0.5–30 mg/L range. Crucially, this cathodic sensing approach enabled efficient monitoring over four months with repeated detections, overcoming the stability concerns associated with anodic sensing and showing clear potential for low-cost, self-powered, long-term environmental monitoring. Existing self-powered photocatalytic fuel cell (PFC) sensors often face limitations due to poor energy conversion efficiency and a lack of features suitable for routine use. Addressing these shortcomings, Qui et al. designed a multifunctional self-powered PFC platform for sensing of Ochratoxin A (OTA), which is a toxic and potentially carcinogenic mycotoxin produced by certain molds (*Aspergillus* and *Penicillium* species) that commonly contaminates food items like cereals, coffee, and dried fruit. In this work, energy conversion performance is targeted using a $Ti_3C_2T_x$ MXene–$TiO_2$ composite photoanode under visible light illumination, yielding a significant reported power output (Pmax of 23.516 μW/cm$^2$) (Qiu et al. 2023) (Figure 13.3(A)). To improve suitability for routine applications, the system cleverly incorporates an electrochromic PB cathode, enabling a visual 'self-checking' capability through color change alongside the primary electrical signal. This platform employs a target-responsive cargo release mechanism, where aptamer binding to OTA triggers glucose fuel release from mesoporous silica nanoparticles, directly

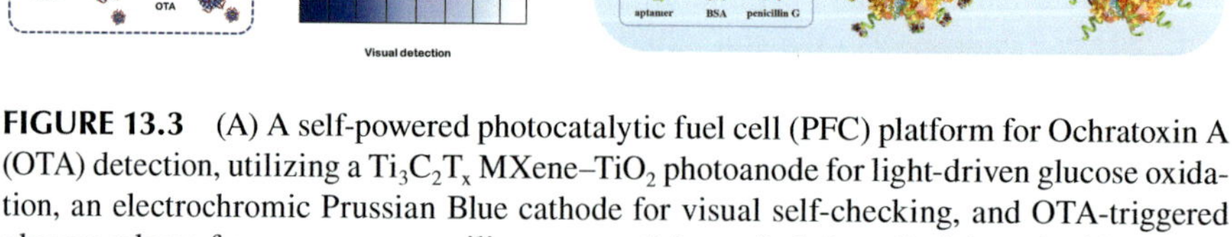


**FIGURE 13.3** (A) A self-powered photocatalytic fuel cell (PFC) platform for Ochratoxin A (OTA) detection, utilizing a $Ti_3C_2T_x$ MXene–$TiO_2$ photoanode for light-driven glucose oxidation, an electrochromic Prussian Blue cathode for visual self-checking, and OTA-triggered glucose release from mesoporous silica nanoparticles to fuel the cell. Adapted with permission from (Qiu et al. 2023) Copyright (2023) Elsevier. (B) A photo-assisted zinc–air battery (ZAB)-driven self-powered aptasensor for penicillin G (PG) detection, employing a Zn-plate photoanode, a Cd-$MoS_2$@$Ti_3C_2T_x$ heterostructure photocathode functionalized with aptamers, and PG-binding-induced modulation of the photoelectrochemical current for sensing. Adapted with permission from Liu et al. (2023) Copyright (2023) Elsevier.

powering the PFC and demonstrating high sensitivity (LOD of 0.0587 ppb) in complex samples like beer.

Jin et al. reported on the construction of a novel self-powered biosensor by uniquely integrating a PFC with a molecular imprinting polymer (MIP) for aflatoxin B1 detection. Aflatoxin B1 (AFB1) is a highly potent carcinogenic mycotoxin produced by Aspergillus molds that contaminates staple crops like corn and peanuts, posing a significant risk for liver cancer. The issue of energy conversion efficiency is tackled by using a $MoS_2$–$Ti_3C_2T_x$ MXene heterojunction as a photoanode within the PFC, which reportedly enhances photocurrent significantly (~3-fold) by harnessing light energy more effectively. Simultaneously, the challenge of poor selectivity is directly addressed by incorporating the MIP, designed to specifically capture the target analyte (AFB1) even amidst interferences, thus enabling sensitive detection (LOD 0.73 pg/mL) without requiring complex sample pretreatment. This dual approach leverages the strengths of both components – MIP for targeted recognition and the MXene-enhanced PFC for efficient, light-driven signal conversion (Jin et al. 2021). Harnessing the photosynthetic power of plant components like thylakoid membranes for bioelectricity generation requires efficient integration with electrode materials, a challenge addressed by Sarode et al., who fabricated a photo-driven bioanode (Sarode et al. 2024). They utilized a facile laser-induced graphene (LIG) substrate, further enhancing it by decorating with $Nb_4C_3T_x$ MXene to provide a high surface area. This modified surface successfully immobilized thylakoid membranes from spinach, leveraging their intrinsic photosynthetic capabilities for light-driven electron generation. Paired with a platinum gas diffusion cathode for efficient oxygen reduction, their resulting photo-biofuel cell demonstrated considerable activity under illumination, such as achieving 0.45 V open-circuit voltage and 7.24 μW/cm$^2$ power density, highlighting their promising strategy that effectively combines the unique features of LIG, MXenes, and thylakoids for plant-based bioelectricity generation.

Ultrasensitive detection of potent toxins like microcystin-RR (MC-RR) is critical for environmental safety, demanding advanced sensors with significant signal amplification, possibly suitable for portable, real-time analysis. Addressing this, Sun et al. constructed a photo-driven self-powered aptasensor (a type of biophotoelectrochemical sensor) incorporating multiple signal amplification strategies (Sun et al. 2023). Their design features a dual-photoelectrode system with a $ZnIn_2S_4$/$Ti_3C_2T_x$ heterojunction photoanode and a $Cu_2O$ photocathode. The $Ti_3C_2T_x$ MXene enhances charge separation and provides a high surface area for immobilizing the MC-RR-specific aptamer. Detection follows a 'signal-off' mechanism where MC-RR binding to the aptamer causes steric hindrance, reducing the photocurrent. Critically, Sun et al. integrated a matching capacitor, boosting sensitivity 22-fold, and proposed a USB-sized micro-workstation for capturing instantaneous signals, enhancing portability. This multi-amplification system achieved ultrasensitive and selective determination of MC-RR with an extremely low detection limit (0.033 pM) and demonstrated applicability in fish samples, paving the way for practical microcystin monitoring. Ultrasensitive and label-free detection of cancer biomarkers like microRNA-155 (miRNA-155) is essential for advancing early diagnostics. Addressing this, Duan et al. constructed a novel impedimetric aptasensing system utilizing a nanohybrid material (Duan et al. 2020). Their approach involved synthesizing $Ti_3C_2T_x$ MXene nanosheets decorated with iron phthalocyanine quantum dots (FePc QDs) via π–π* stacking interaction. This $Ti_3C_2T_x$@FePcQD nanohybrid, featuring mixed-valence states and rich functionality, served as an effective platform for immobilizing the miRNA-155 capture probe. The resulting label-free impedimetric aptasensor demonstrated exceptional sensitivity, achieving an attomolar detection limit (4.3 aM) for miRNA-155 over a broad concentration range (0.01 fM to 10 pM), offering a promising and feasible strategy for diverse biomarker detection.

Improving the detection accuracy of self-powered sensors is crucial for reliable analysis, potentially achievable through multi-signal strategies. Addressing this need, Ji et al. designed a novel PFC-based self-powered aptasensor featuring a unique dual-power-peak output (Ji et al. 2023). Their approach utilized persulfate as an initiator in the PFC system, which not only enhanced charge separation but also uniquely generated a second power peak in the output signal. This dual-peak response enables self-checking within a single readout window, greatly improving accuracy. Furthermore, the incorporation of $V_2C$ MXene accelerated electron transfer, boosting photoelectric conversion efficiency, while an aptamer on the photoanode provided specific detection of chloramphenicol. Quantification based on these dual power peaks allowed for sensitive detection (LOD 0.17 pM) of chloramphenicol over a linear range of 0.5 pM to 3 nM, demonstrating a promising strategy for developing highly accurate and specific multi-signal self-powered sensors (Figure 13.3(B)).

## 13.4 OTHER SELF-POWERED ELECTROCHEMICAL DETECTION TECHNOLOGIES

Developing portable SPESs that integrate power and detection is key for intelligent sensing devices, with applications like monitoring biomarkers such as hydrogen sulfide ($H_2S$). Addressing this, Zhu et al. constructed a novel zinc–air battery (ZAB)-based self-powered sensor for $H_2S$ detection. Their approach utilizes an Fe/$Fe_3C$@graphene foam composite as the oxygen reduction catalyst in the ZAB's air cathode, achieving an open-circuit voltage of 1.30 V (Zhu et al. 2024b). The detection principle relies on the poisoning effect of $H_2S$ on this catalyst; $H_2S$ presence decreases the catalyst's performance, leading to a measurable decrease in the ZAB's discharge voltage. This ZAB-SPES, read using a simple digital multimeter, demonstrated good selectivity and reproducibility for detecting $H_2S$ in colon cancer cell lysate, offering a practical strategy for portable SPES construction (Figure 13.3(C)).

Non-invasive, point-of-care monitoring of stress biomarkers like cortisol in various body fluids requires highly sensitive and selective sensor platforms. Addressing this, Liu et al. developed a novel dual-modal electrochemical aptasensor based on a $V_2CT_x$ MXene and bimetallic CoNi-layered double hydroxide (CoNi-MLDH) heterojunction (Liu et al. 2024). Their work utilized $V_2CT_x$ nanosheets as scaffolds for the in situ growth and subsequent conversion of CoNi-ZIF into CoNi-MLDH nanosheets. This CoNi-MLDH@$V_2CT_x$ composite synergistically combines the excellent electrochemical activity of the MXene with the high surface area, functionality, oxygen reduction activity, and aptamer affinity of the CoNi-MLDH. Leveraging this advanced material, Liu et al. constructed both impedimetric and ZAB-driven self-powered aptasensors, achieving exceptionally low detection limits for cortisol (0.26 and 0.17 fg/mL, respectively) in sweat, urine, and blood, thereby offering a promising approach for sensitive cortisol analysis relevant to mental health monitoring (Figure 13.3(D)).

Sensitive and selective analysis of antibiotics like penicillin G demands advanced analytical methods, particularly portable and self-powered platforms. Addressing this, Liu et al. manufactured a novel photocatalyzed zinc–air battery (ZAB)-driven self-powered aptasensor (Liu et al. 2023). Their approach utilizes a Zn-plate photoanode and a sophisticated 2D/2D Schottky heterojunction photocathode comprising cadmium-doped $MoS_2$ nanosheets grown on $Ti_3C_2T_x$ MXene (Cd–$MoS_2$@$Ti_3C_2T_x$). This heterojunction exhibits enhanced light absorption and charge separation under UV–vis irradiation, boosting the ZAB output voltage (to 1.43 V). An aptamer specific to penicillin G is used for recognition, and binding likely modulates the photocatalytic process at the cathode, altering the ZAB's output power characteristics. The resulting self-powered aptasensor achieved an ultralow detection limit (0.06 fg/mL) for penicillin G, demonstrating high specificity and stability, offering a promising portable analysis method for antibiotics.

Developing fully integrated, flexible, wireless, and battery-free sensing systems is crucial for practical on-site monitoring of analytes like potassium [$K^+$] in sweat, overcoming limitations of traditional wearable sensors. Addressing these requirements, Zhang et al. developed a novel electrochemical sensing patch system (Zhang et al. 2021). Their design incorporates a hybrid multi-dimensional network electrode using multiwalled carbon nanotubes (MWCNT) and MXene-$Ti_3C_2T_x$. This structure provides high surface area and fast charge transfer, effectively immobilizing the $K^+$ selective valinomycin membrane while enhancing electrochemical stability and sensitivity, particularly at low concentrations. The system achieves battery-free operation and wireless data transmission to a smartphone via near field communication (NFC), includes an integrated signal amplification system, and integrates a microfluidic channel for efficient sweat collection and sensor protection. The resulting sensor patch demonstrated high amplified sensitivity (173 mV/dec) and excellent selectivity for [$K^+$], showcasing a complete, practical platform for on-site sweat analysis.

A particularly innovative application leverages the high electrochemical capacitance of certain MXenes, notably $Ti_3C_2T_x$. Ji et al. ingeniously designed enzyme-free ICBFC-SPSs where the $Ti_3C_2T_x$-based anode (often combined with pseudocapacitive materials like PPy) provides significant charge storage capacity (Ji et al. 2025). This stored charge can deliver large instantaneous currents, effectively amplifying the sensing signal upon modulation by the analyte ($Pb^{2+}$ or DEHP), representing a unique approach to boosting SPES sensitivity beyond direct power output modulation. The versatility of MXenes is underscored by the diverse range of SPES platforms they have enabled. Examples span enzyme-free BFCs (Ji et al. 2024), PECs/PFCs (Doglioni, Nardello, and Brunelli 2024), impedimetric aptasensors (San Nah et al. 2021), zinc–air battery (ZAB)-based sensors (Zhu et al. 2024b), and even flexible electrochemical patches. Different

MXene compositions, including $Ti_3C_2T_x$, $V_2CT_x$, and $Nb_4C_3T_x$, have been successfully employed to detect a wide array of analytes relevant to healthcare (cortisol, miRNA-155), environmental monitoring ($Pb^{2+}$, DEHP, 4-NP, MC-RR, PG), and physiological tracking ($K^+$) (San Nah et al. 2021). Furthermore, MXene-based platforms readily integrate with other advanced strategies to push performance boundaries. These include sophisticated molecular recognition using aptamers, signal amplification via DNA nanostructures or matched capacitors (Sun et al. 2023), implementation of self-checking mechanisms using dual signals, and integration into practical device formats incorporating microfluidics and wireless communication (NFC). This demonstrates that MXenes are not just passive components but active enablers of complex, multi-functional sensing systems. Despite these significant advancements, challenges remain in areas such as ensuring the long-term stability of MXene-based SPES in complex real-world samples, developing fully standardized and scalable synthesis methods, and deepening the fundamental understanding of charge transfer and catalytic mechanisms at MXene interfaces within these integrated devices. Continued research focusing on these aspects, exploring new MXene compositions and heterostructures, and optimizing device architecture will be vital.

## 13.4 CONCLUSION

SPESs represent a paradigm shift from conventional power-dependent devices, integrating energy harvesting with analyte detection to enable autonomous, portable, and sustainable sensing solutions. The success of this approach hinges on materials innovation, where 2D MXenes have emerged as exceptionally effective components due to their unique suite of properties: high metallic conductivity enhances charge transfer for efficient energy conversion and signal generation, while their large, inherently functionalized surfaces serve as superior scaffolds for active sensing components, collectively addressing key challenges in both energy conversion and signal transduction. The demonstrated versatility of MXenes across diverse platforms underscores their significant contribution to achieving improved sensitivity and performance. Therefore, MXenes stand out as pivotal materials advancing self-powered sensing technology, paving the way for practical implementation in critical areas like healthcare and environmental protection, with future research promising even greater potential for truly autonomous sensing.

## 13.5 LIMITATIONS AND FUTURE PERSPECTIVES

Conventional electrochemical sensors face significant hurdles regarding autonomy, portability, and sustainability due to their reliance on external power sources, making SPESs a transformative alternative that integrates in situ energy generation with analyte detection, a field where advanced materials are paramount. Among these, 2D MXenes have emerged as exceptionally promising candidates, demonstrating remarkable versatility and performance enhancement across numerous recent studies. Their inherent metallic conductivity, large functionalized surface area, high capacitance, and potential catalytic activity make them ideal components for SPES.

As demonstrated in recent works, these attributes allow MXenes to enhance charge transfer in biofuel cells (BFCs) and photoelectrochemical cells (PECs), serve as robust scaffolds for high-load immobilization of enzymes and aptamers, boost sensitivity via capacitive signal amplification (particularly $Ti_3C_2T_x$), and contribute synergistically when combined in composites and heterojunctions alongside materials like AuNPs, semiconductors, LDHs, and polymers. This has enabled the development of diverse MXene-enabled SPES platforms, including enzymatic and enzyme-free BFCs, PECs/PFCs, impedimetric sensors, and zinc–air battery-based systems, capable of detecting a wide range of analytes from heavy metals and environmental toxins to critical clinical biomarkers and physiological ions, utilizing various MXene types such as $Ti_3C_2T_x$, $V_2CT_x$, and $Nb_4C_3T_x$. Furthermore, these platforms often incorporate sophisticated strategies like DNA nanotechnology, matched capacitors for signal boosting, microfluidics, and wireless communication, pushing toward fully integrated, intelligent sensing systems.

Despite this rapid progress and demonstrated potential, several challenges impede the widespread practical application of MXene-based SPES. Ensuring the long-term operational stability and resistance to degradation or oxidation, particularly in complex biological or environmental media, remains a critical concern. Achieving consistent high selectivity against diverse interferents in real-world samples often requires sophisticated recognition elements and anti-fouling strategies. Reproducible, scalable, and cost-effective synthesis of MXenes with precisely controlled properties is essential for commercial viability but currently presents difficulties. A deeper fundamental understanding of charge transfer mechanisms and synergistic effects in complex MXene-based systems is often needed for rational design optimization. Moreover, the complex integration of multiple components into miniaturized, robust devices poses significant engineering hurdles, and the absolute power output generated by some BFC or PEC configurations may still be insufficient for certain high-demand applications without effective energy management solutions. Nevertheless, the future trajectory for MXene-based SPES is exceptionally bright, focused on translating laboratory successes into impactful real-world applications. Key directions include validated point-of-care diagnostics, continuous environmental and food safety monitoring networks, and advanced wearable or even implantable sensors for personalized health tracking, leveraging MXenes' flexibility and potential biocompatibility. Driving this translation requires ongoing advancements in the precise surface engineering of MXenes to tailor functional groups and properties, the rational design of novel heterostructures (with quantum dots, MOFs, and other 2D materials) to maximize synergistic effects, and the integration of new, highly stable, and selective biorecognition elements. Improving energy conversion efficiencies, developing integrated MXene-based energy storage (e.g., supercapacitors) for better power management, exploring hybrid energy harvesting modes, and advancing scalable micro/nanofabrication techniques for device miniaturization and integration (including microfluidics and wireless communication like NFC for internet of things (IoT) connectivity) are vital. Furthermore, developing multiplexed platforms for simultaneous detection of multiple analytes and utilizing computational modeling to accelerate materials discovery and understand fundamental mechanisms will likely define the next generation of these powerful sensors. By addressing the current limitations through these concerted research and development efforts, the field can unlock the full potential of MXenes to create truly transformative autonomous sensing technologies for diverse critical applications.

## REFERENCES

Arjun, Ajith. Moham., Menon Ankitha, Neermunda. Shabana, Pushpalatha Vijayakumar Vaishag, Fathima Shamsheera, Mundummal Mufeeda, and Pathath Abdul Rasheed. 2023. "An overview on surface modification of niobium MXenes for diagnostic and prognostic applications." *FlatChem* 41:100538. doi: 10.1016/j.flatc.2023.100538.

Doglioni, Maria., Matteo. Nardello, and Davide. Brunelli. 2024. "Plant microbial fuel cells: Energy sources and biosensors for battery-free smart agriculture." *IEEE Transactions on AgriFood Electronics* 2 (2):460–470. doi: 10.1109/TAFE.2024.3417644.

Duan, Fenghe, Chuanpan Guo, Mengyao Hu, Yingpan Song, Minghua Wang, Linghao He, Zhihong Zhang, Riccardo Pettinari, and Liming Zhou. 2020. "Construction of the 0D/2D heterojunction of $Ti_3C_2T_x$ MXene nanosheets and iron phthalocyanine quantum dots for the impedimetric aptasensing of microRNA-155." *Sensors and Actuators B: Chemical* 310:127844. doi: 10.1016/j.snb.2020.127844.

Hou, Yang-Yang, Jing Xu, Wan-Zhen Xie, Ke-Jing Huang, Xuecai Tan, Bing-Rui Zhao, Si-Qi Zhang, and Meng-Tian Gao. 2023. "3D DNA walker recognition-driven homogeneous dual-mode sensing strategy based on enzyme biofuel cell for ultrasensitive detection of HER2." *Sensors and Actuators B: Chemical* 376:132998. doi: 10.1016/j.snb.2022.132998.

Ji, Kaixiang, Peilin Wang, Yameng Li, Qiang Ma, and Xingguang Su. 2025. "$Ti_3C_2T_x$/Au NPs/PPy ternary heterostructure-based intra-capacitive self-powered sensor for DEHP detection." *Journal of Hazardous Materials* 488:137311. doi: 10.1016/j.jhazmat.2025.137311.

Ji, Kaixiang, Zihui Liang, Peilin Wang, Zhenrun Li, Qiang Ma, and Xingguang Su. 2024. "Mxene-based capacitive enzyme-free biofuel cell self-powered sensor for lead ion detection in human plasma." *Chemical Engineering Journal* 495:153598. doi: 10.1016/j.cej.2024.153598.

Ji, Xingyu, Xiaojiao Du, Jun Sun, Bing Zhang, Yuebin Lian, Min Zheng, Rongquan Zhu, and Long Geng. 2023. "Persulfate mediated single-channel bimodal self-checking light-assisted self-powered aptasensor for highly accurate assay." *Sensors and Actuators B: Chemical* 397:134647. doi: 10.1016/j.snb.2023.134647.

Jin, Yunxia, Yang Luan, Zhen Wu, Wei Wen, Xiuhua Zhang, and Shengfu Wang. 2021. "Photocatalytic fuel cell-assisted molecularly imprinted self-powered sensor: A flexible and sensitive tool for detecting aflatoxin B1." *Analytical Chemistry* 93 (39):13204–13211. doi: 10.1021/acs.analchem.1c02074.

Liu, Jiameng, Mengfei Wang, Zheng Tao, Linghao He, Chuanpan Guo, Baozhong Liu, and Zhihong Zhang. 2023. "Photo-assisted Zn-air battery-driven self-powered aptasensor based on the 2D/2D Schottky heterojunction of cadmium-doped molybdenum disulfide and $Ti_3C_2T_x$ nanosheets for the sensitive detection of penicillin G." *Analytica Chimica Acta* 1270:341396. doi: 10.1016/j.aca.2023.341396.

Liu, Jiameng, Zheng Tao, Yinpeng Zhang, Tianjun Ni, Baozhong Liu, and Zhihong Zhang. 2024. "Dual-modal flexible zinc-air battery-driven self-powered and impedimetric aptasensor based on the hybrid of $V_2CT_x$ MXene and bimetallic layered double hydroxide for the detection of cortisol." *Microchemical Journal* 197:109757. doi: 10.1016/j.microc.2023.109757.

Luo, Dan, Jinfei Yi, Yongju Wu, Yan Luo, Yanli Zhang, Xue Men, Hongbin Wang, Wenrong Yang, and Pengfei Pang. 2024. "Biofuel cell–based self-powered immunosensor for detection of 17β-estradiol by integrating the target-induced biofuel release and biogate immunoassay." *Microchimica Acta* 191 (8):477. doi: 10.1007/s00604-024-06553-4.

Ou-Yang, Wei, Liqiang Liu, Mingjun Xie, Siqian Zhou, Xiaowei Hu, Han Wu, Zhiyu Tian, Xucong Chen, Yirui Zhu, and Jun Li. 2024. "Recent advances in triboelectric nanogenerator-based self-powered sensors for monitoring human body signals." *Nano Energy* 120:109151. doi: 10.1016/j.nanoen.2023.109151.

Qiu, Zhenli, Xianghang Xue, Yufen Lei, Xintong Lin, Dianping Tang, and Yiting Chen. 2023. "MXene-$TiO_2$-based photocatalytic fuel cell with bioresponsive controlled glucose release system: An innovative mode for Ochratoxin A detection." *Analytica Chimica Acta* 1257:341126. doi: 10.1016/j.aca.2023.341126.

San Nah, Joong, Sharat C. Barman, Md Abu Zahed, Md Sharifuzzaman, Hyosang Yoon, Chani Park, Sanghyuk Yoon, Shipeng Zhang, and Jae Y. Park. 2021. "A wearable microfluidics-integrated impedimetric immunosensor based on $Ti_3C_2T_x$ MXene incorporated laser-burned graphene for noninvasive sweat cortisol detection." *Sensors Actuators B: Chemical* 329:129206.

Sarode, Amit, Sri R. Torati, Md Faruk Hossain, and Gymama Slaughter. 2024. "A photo-driven bioanode based on MXene-decorated graphene." *Electrochimica Acta* 498:144637. doi: 10.1016/j.electacta.2024.144637.

Sun, Jun, Rongquan Zhu, Xiaojiao Du, Bing Zhang, Min Zheng, Xingyu Ji, and Long Geng. 2023. "An ultrasensitive photo-driven self-powered aptasensor for microcystin-RR assay based on $ZnIn_2S_4/Ti_3C_2$ MXenes integrated with a matching capacitor for multiple signal amplification." *Analyst* 148 (20):5060–5069. doi: 10.1039/D3AN00914A.

Sun, Shanshan, Meng Su, Han Xiao, Xiaoshuang Yin, Ying Liu, Wenzhong Yang, and Yun Chen. 2024. "Self-powered biosensing platform for highly sensitive detection of soluble CD44 protein." *Talanta* 272:125824. doi: 10.1016/j.talanta.2024.125824.

Wang, Shi-Hang, Jian-Wei Wang, Li-Ting Zhao, Syed Z. Abbas, Zhugen Yang, and Yang-Chun Yong. 2023. "Soil microbial fuel cell based self-powered cathodic biosensor for sensitive detection of heavy metals." *Biosensors* 13 (1-12). doi:10.3390/bios13010145.

Wei, Jie, Junling Sha, Kezuo Di, Sheng Chen, Wenhui Liu, Lingliang Long, Lijun Ding, Yubo Zhou, Xu Wang, and Kun Wang. 2025. "Reusable self-powered electrochromic sensor patch based on enzymatic biofuel cells for on-site visualized monitoring of lactic acid." *Analytical Chemistry* 97 (5):2604–2609. doi: 10.1021/acs.analchem.4c06000.

Wu, Minying, Qian Zhang, Yuanyuan Fang, Chao Deng, Fangzhou Zhou, Yi Zhang, Xingdong Wang, Yi Tang, and Yajun Wang. 2021. "Polylysine-modified MXene nanosheets with highly loaded glucose oxidase as cascade nanoreactor for glucose decomposition and electrochemical sensing." *Journal of Colloid and Interface Science* 586:20–29. doi: 10.1016/j.jcis.2020.10.065.

Xiao, Xinxin, Hong-qi Xia, Ranran Wu, Lu Bai, Lu Yan, Edmond Magner, Serge Cosnier, Elisabeth Lojou, Zhiguang Zhu, and Aihua Liu. 2019. "Tackling the challenges of enzymatic (bio)fuel cells." *Chemical Reviews* 119 (16):9509–9558. doi: 10.1021/acs.chemrev.9b00115.

Yi, Jinfei, Dan Luo, Hua Meng, Yongju Wu, Lihong Ma, Yanli Zhang, Hongbin Wang, Wenrong Yang, and Pengfei Pang. 2024. "Enzymatic biofuel cells-based self-powered electrochemical biosensor for detection of 17β-estradiol based on dual-enzyme cascade amplification and bioconjugate." *Sensors and Actuators B: Chemical* 418:136269. doi: 10.1016/j.snb.2024.136269.

Zeng, Hui, Kang Liang, Lei Jiang, Dongyuan Zhao, and Biao Kong. 2025. "Electrochemical sensing mechanisms and interfacial design strategies of mesoporous nanochannel membranes in biosensing applications." *Accounts of Chemical Research* 58 (5):732–745. doi: 10.1021/acs.accounts.4c00764.

Zhang, Shipeng, Md Abu Zahed, Md Sharifuzzaman, Sanghyuk Yoon, Xue Hui, Sharat C. Barman, Sudeep Sharma, Hyo S. Yoon, Chani Park, and Jae Y. Park. 2021. "A wearable battery-free wireless and skin-interfaced microfluidics integrated electrochemical sensing patch for on-site biomarkers monitoring in human perspiration." *Biosensors and Bioelectronics* 175:112844. doi: 10.1016/j.bios.2020.112844.

Zhu, Junlun, Dong Shao, Wei Wen, Zhengfang Tian, Xiuhua Zhang, and Shengfu Wang. 2024a. "Self-powered electrochemical sensor based on photoelectrode: An up-to-date review." *Coordination Chemistry Reviews* 518:216095. doi: 10.1016/j.ccr.2024.216095.

Zhu, Junlun, Yao Xiao, Wei Hu, Qian Cui, Yuying Yuan, Xu Peng, Wei Wen, Xiuhua Zhang, and Shengfu Wang. 2024b. "A portable self-powered electrochemical sensor based on zinc-air battery for detection of hydrogen sulfide." *Analytical Chemistry* 96 (5):1852–1860. doi: 10.1021/acs.analchem.3c03423.

# 14 Microfluidic Sensors in the Palette of MXene

*Bartholomew Richard*

## 14.1 INTRODUCTION

Microfluidics is a multidisciplinary field that deals with the behavior, control, and manipulation of fluids in microscale environments. It involves the precise handling of small volumes of liquids, typically in the range of microliters ($10^{-6}$ L) to picoliters ($10^{-12}$ L), within channels and chambers that are often on the order of micrometers ($10^{-6}$ m) in size (Nielsen et al. 2020). The field of microfluidics emerged from advancements in microfabrication techniques, such as in the fields of biomedical, energy, and environment. Microfluidic devices, often referred to as lab-on-a-chip systems, are constructed using materials such as glass, silicon, or polymers (Padash, Enz, and Carrara 2020). These devices contain a network of channels, valves, pumps, and sensors that enable the manipulation and analysis of fluids. One of the key advantages of microfluidics is its ability to offer precise control over fluid flow and manipulation, with the ability to integrate multiple functions onto a single chip. Microfluidics in the realm of electrochemical sensing has paved a new dimension in the field of wearable devices for the precise detection of bioanalytes (Vinoth et al. 2021). In the field of biology, microfluidics has enabled the miniaturization and automation of laboratory processes, such as DNA analysis (Hopwood et al. 2010), cell sorting (Shen, Yalikun, and Tanaka 2019), and drug discovery (Lombardi and Dittrich 2010). It has also facilitated the study of cellular and molecular interactions at the single-cell level, offering insights into fundamental biological processes (Delamarche et al. 2013).

Microfluidic systems have been utilized in chemistry for chemical synthesis, reaction optimization, and high throughput screening (Liu and Jiang 2017). The small dimensions and rapid mixing capabilities of microfluidic devices provide enhanced control over reaction kinetics and allow for the efficient utilization of reagents. In medicine, microfluidics has the potential to revolutionize diagnostics and point-of-care testing (Pandey et al. 2018). Microfluidic devices can be portable, requiring only small sample volumes, and can provide real-time results. Overall, microfluidics offers a powerful platform for conducting experiments and performing analyses at the microscale, providing advantages such as reduced sample and reagent consumption, enhanced sensitivity, and improved automation (Richard et al. 2023). As the field continues to advance, it holds great promise for numerous applications and has the potential to revolutionize various industries. The advantages of microfluidics over other fluidic techniques are schematically depicted in Figure 14.1.

Continuous-flow microfluidics is a field of study and technology that focuses on manipulating and controlling small volumes of fluids within microscale channels

DOI: 10.1201/9781003736318-14

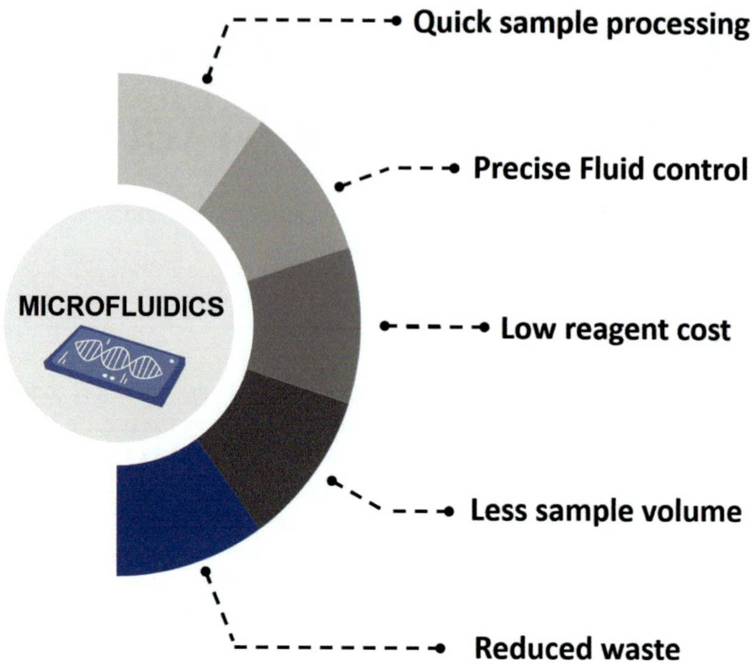


**FIGURE 14.1** Schematic illustration of the advantages of microfluidics compared to other techniques.

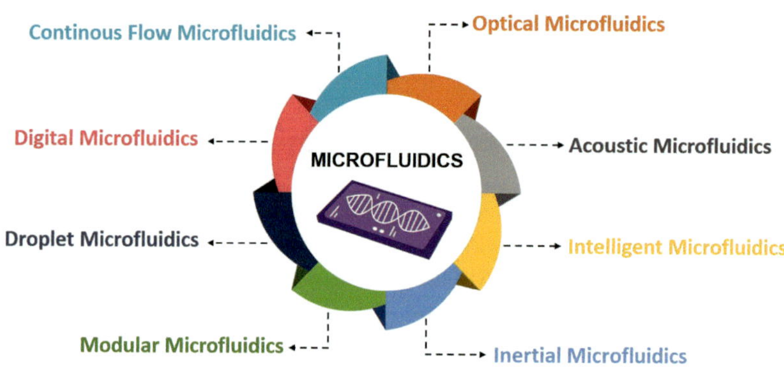


**FIGURE 14.2** Schematic illustration of different types of microfluidics.

(Jahn et al. 2008). In continuous-flow microfluidics, fluids are transported and manipulated through microchannels using precise fluidic control systems. These microchannels are typically fabricated on a chip using photolithography, soft lithography, or microfabrication techniques. The channels are often designed to have specific geometries and features to enable precise fluid flow control and manipulation. One of the key advantages of continuous-flow microfluidics is the ability to perform highly controlled experiments and analyses with small amounts of samples and reagents (Zhang and Jiang 2016). This technology enables high-throughput and parallel processing, as multiple reactions or analyses can be carried out simultaneously within separate microchannels on the same chip (https://www.asiga.com/types-of-microfluidic-devices/). Based on the platform on which they work, there are several types of microfluidics, and Figure 14.2 illustrates the classification of different types of microfluidics.

## 14.2 DIFFERENT TYPES OF MICROFLUIDICS

### 14.2.1 Digital Microfluidics

Digital microfluidics, also known as electrowetting-on-dielectric (EWOD), is a technique that enables the manipulation and control of discrete droplets of fluid on a two-dimensional surface. Unlike continuous-flow microfluidics, which focuses on the movement of fluids within microchannels, digital microfluidics operates on an open surface. The three digital microfluidics platforms are liquid droplets (Kaminski, Scheler, and Garstecki 2016), liquid marbles (Ooi et al. 2021), and core-shell beads (Gorgannezhad et al. 2020). In digital microfluidics, a microscale array of electrodes is patterned on a substrate, typically made of glass or silicon. The substrate is coated with a hydrophobic material, such as Teflon or a polymer, which creates a dielectric layer. By applying voltage to specific electrodes, the wetting properties of the surface can be modified, allowing droplets to be moved, merged, split, and mixed on the surface (Coudron et al. 2019). The basic principle of digital microfluidics relies on the phenomenon of electrowetting. When a voltage is applied to an electrode, an electric field is created, which modifies the interfacial tension between the droplet and the surface. By adjusting the voltage, the contact angle between the droplet and the surface can be controlled, allowing the droplet to be manipulated (Tong et al. 2023).

Digital microfluidics offers several advantages:

1. Reconfigurability: The surface can be reconfigured dynamically by applying different voltage patterns to the electrodes, allowing for flexible and programmable fluid manipulations (Xu et al. 2023).
2. Precise control: Digital microfluidics enables precise control over droplet movements, including transportation, splitting, merging, and mixing. This level of control is essential for performing complex biochemical and chemical reactions (Xu et al. 2023).
3. Small sample volumes: Since digital microfluidics operates on discrete droplets, it requires smaller sample volumes compared to other microfluidic techniques. This can be advantageous when dealing with precious or limited samples (Xu et al. 2023).
4. Compatibility with various applications: Digital microfluidics has been applied in various fields, including chemical synthesis, biological assays, diagnostics, drug discovery, and point-of-care testing.

Applications of digital microfluidics include:

1. Lab-on-a-chip systems: Digital microfluidics provides a platform for integrating various laboratory processes, such as sample preparation, mixing, and analysis, on a single chip (Srinivasan, Pamula, and Fair 2004).
2. Biochemical assays: It allows for the manipulation of small volumes of reagents, cells, and biomolecules for applications such as DNA amplification, protein analysis, and immunoassays (Malic et al. 2010).
3. Drug screening: Digital microfluidics can be used for high-throughput screening of drug candidates, allowing for efficient testing of their effects on cells or biochemical reactions (Zhai et al. 2024).
4. Point-of-care diagnostics: The precise control and small sample volumes of digital microfluidics make it suitable for developing portable diagnostic devices for rapid and accurate testing in resource-limited settings (Sista et al. 2008).

### 14.2.2 Modular Microfluidics

Modular microfluidics refers to the concept of constructing microfluidic systems using modular components that can be easily interconnected and reconfigured to perform different functions. It involves the design and fabrication of individual building blocks or modules that can be combined in a flexible manner to create complex microfluidic architectures (Fan et al. 2018). The idea behind modular microfluidics is to enable rapid prototyping and customization of microfluidic systems by assembling pre-fabricated modules rather than starting from scratch for each application. These modules can include channels, valves, mixers, pumps, sensors, and other functional components. By arranging and connecting these modules in various configurations, different fluidic functionalities can be achieved (Yuen 2008).

The advantages of modular microfluidics are numerous. First, it simplifies the design and fabrication process by eliminating the need to create complex, monolithic microfluidic devices from scratch. Instead, designers can select and combine modular components based on their specific requirements, reducing time, cost, and fabrication complexities. Modularity also enables greater flexibility and scalability. Microfluidic systems can be easily reconfigured or modified by rearranging or replacing modules, allowing for rapid iterations and adaptation to changing experimental needs (Wu et al. 2010). This flexibility is particularly beneficial in research and development settings, where experimentation and optimization are crucial. Moreover, modular microfluidics promotes standardization and compatibility. If modules are designed with specific interconnection methods or standardized dimensions, they can be easily integrated into a larger ecosystem of microfluidic components, facilitating collaboration, knowledge sharing, and the exchange of modules among different laboratories or researchers (Wu et al. 2023).

Another advantage is the potential for automation and high-throughput applications. By incorporating active components like valves and pumps into modular microfluidic systems, it becomes possible to automate fluid control, sample handling, and analysis processes. This opens up opportunities for applications such as high-throughput screening, point-of-care diagnostics, and lab-on-a-chip systems. However, modular microfluidics also has some challenges. Ensuring precise alignment and sealing between modules can be critical for maintaining fluidic integrity and preventing leaks. Compatibility issues between modules from different sources or manufacturers may also arise, requiring standardized interfaces and protocols to ensure seamless integration (Liou et al. 2011).

In summary, modular microfluidics offers a flexible and scalable approach to designing and building microfluidic systems. By using interchangeable and reconfigurable modules, researchers and developers can quickly prototype, customize, and adapt microfluidic architectures for various applications. With ongoing advancements in modular component design and standardization, modular microfluidics has the potential to accelerate the development and adoption of microfluidic technology across multiple fields.

### 14.2.3 Inertial Microfluidics

Inertial microfluidics is a powerful technique that exploits hydrodynamic forces to manipulate and separate particles or cells in microscale fluidic systems. It offers advantages such as label-free operation, continuous processing, and compatibility with various particles. It allows for label-free and continuous particle separation without the need for complex external equipment or reagents. The technique is compatible with various types of particles, including rigid and deformable particles, as well as biological cells. Unlike conventional microfluidic technologies, where fluid inertia is negligible and flow remains within the Stokes flow region with very low Reynolds numbers ($Re \ll 1$), inertial microfluidics operates in the intermediate Reynolds number range ($\sim 1 < Re < \sim 100$). In this range, both inertia and fluid viscosity play significant roles, leading to several interesting phenomena. These include (i) inertial migration and (ii) secondary flow. With its advantages of high throughput, simplicity, precise control, and low cost, inertial microfluidics holds great potential for cellular sample processing, particularly for samples with low-abundance targets (Zhang et al. 2016). Moreover, it enables high-throughput processing and has the potential for integration with other microfluidic functionalities. However, inertial microfluidics also has its limitations. It is primarily effective for particles with sizes in the range of a few micrometers to tens of micrometers. Beyond this range, other techniques may be more suitable, such as dielectrophoresis or acoustic manipulation. Additionally, particle concentration and sample purity can affect the performance of inertial microfluidic systems, requiring careful optimization for specific applications. With further advancements and optimization, inertial microfluidics holds promise for applications in biotechnology, medical diagnostics, and chemical analysis (Chung 2019).

### 14.2.4 Intelligent Microfluidics

Intelligent microfluidics refers to the integration of intelligence and control mechanisms into microfluidic systems, enabling autonomous operation, decision-making, and adaptive behavior. It combines the principles of microfluidics with advanced technologies such as microsensors, actuators, microcontrollers, machine learning, and artificial intelligence (AI) algorithms (Galan et al. 2020). Traditionally, microfluidic systems have been controlled externally by external pumps, valves, and other mechanical components. Intelligent microfluidics, however, aims to incorporate onboard intelligence within the microfluidic device itself, enabling it to perform complex tasks autonomously without constant external control. The integration of intelligence into microfluidic systems offers several benefits. It enables real-time monitoring and feedback, allowing the system to respond and adapt to changing conditions. It also reduces the reliance on bulky external equipment, making the system more compact and portable (Liu et al. 2023).

Intelligent microfluidics can be achieved through different approaches:

1. Sensing and feedback control: Microsensors embedded within the microfluidic device can monitor various parameters such as temperature, pressure, pH, and biochemical concentrations. This information is then fed back to an onboard microcontroller, which processes the data and makes decisions based on pre-programmed algorithms or machine learning techniques.
2. Closed-loop control: The system can employ closed-loop control mechanisms, where the sensors continuously monitor the system's output and adjust the input parameters accordingly. This ensures precise and accurate control over fluid flow, mixing, and other dynamic processes within the microfluidic device.
3. Artificial intelligence and machine learning: Advanced AI algorithms can be used to analyze sensor data, identify patterns, and make intelligent decisions. Machine learning techniques enable the microfluidic system to learn from previous experiences and optimize its performance over time.

Intelligent microfluidics has a wide range of applications. In biology and medicine, it can be utilized for automated cell sorting, drug discovery (Liu et al. 2024), point-of-care diagnostics (Doganay et al. 2024), and personalized medicine (Nosrati and Nosrati 2023). In chemical synthesis and analysis, it enables autonomous reaction optimization, self-regulating processes, and adaptive experiments (Chen and Lv 2022). Intelligent microfluidics also finds applications in environmental monitoring, lab automation, and microscale robotics. By integrating intelligence and control mechanisms, intelligent microfluidics opens up new possibilities for automation, precision, and adaptability in microfluidic systems. It paves the way for more advanced and autonomous lab-on-a-chip devices that can revolutionize fields ranging from healthcare to chemistry and beyond.

### 14.2.5 Acoustic Microfludics (Acoustofluidics)

Acoustic microfluidics, or acoustofluidics, is a field that utilizes acoustic waves to manipulate, control, and analyze fluids and particles at the microscale. It combines principles from acoustics, microfluidics, and materials science to enable precise and contactless handling of samples within microfluidic systems. Acoustic waves are mechanical vibrations that propagate through a medium, such as a fluid (Gao et al. 2020). In acoustic microfluidics, these waves are generated and controlled using piezoelectric transducers or any other acoustic sources. The acoustic waves interact with the fluid and the particles within it, exerting forces that can be used to manipulate and control their behavior (Chen et al. 2021).

There are several key mechanisms employed in acoustic microfluidics:

- Acoustic streaming: Acoustic waves can create streaming flows within a fluid. This phenomenon, known as acoustic streaming, involves the generation of steady streaming vortices and flows near solid boundaries or in regions of acoustic field gradients. Acoustic streaming can be used to mix fluids, enhance mass transport, and create microscale fluidic pumps (Doinikov et al. 2020).
- Acoustic radiation force: Acoustic waves exert forces on particles suspended in a fluid through a phenomenon called acoustic radiation force. The force arises due to the interaction of the particles with the traveling acoustic waves. By controlling the properties of the acoustic field, such as frequency, intensity, and phase, it is possible to manipulate particles, cells, or droplets. This includes tasks such as particle sorting, trapping, aggregation, and separation (Liu et al. 2017).
- Acoustic standing waves: Standing waves can be created by reflecting acoustic waves at boundaries or within microchannels. These standing waves generate pressure nodes and antinodes, forming high and low acoustic pressure regions. Particles or droplets within the standing wave experience forces that depend on their size and position relative to the pressure nodes. This enables tasks such as particle focusing, concentration, and positioning (Mao et al. 2016).

Acoustic microfluidics offers several advantages over other manipulation techniques:

- Non-contact and gentle handling: Acoustic manipulation is non-invasive and does not require physical contact with the samples. This makes it suitable for handling delicate biological samples, such as cells or microorganisms, without causing damage or altering their properties.
- Real-time and rapid manipulation: Acoustic forces act almost instantaneously, allowing for real-time manipulation and control of particles or fluids. Acoustic microfluidics enables high-speed processing and analysis, making it suitable for applications requiring rapid sample handling and analysis.
- Compatibility with diverse samples: Acoustic manipulation is applicable to a wide range of samples, including cells, particles, droplets, and macromolecules. It is compatible with various fluids, such as water, biological media, and complex solutions. This versatility enables applications in biology, medicine, chemistry, and materials science.

While acoustic microfluidics offers many advantages, challenges remain. Designing optimized acoustic fields, mitigating the potential for bubble formation and cavitation, and integrating acoustic components with microfluidic systems can be technically demanding.

Acoustic microfluidics has found applications in several fields:

- Cell and particle manipulation: Acoustic manipulation is used for cell sorting, cell trapping, particle separation, and concentration. It enables high-throughput cell analysis, rare cell isolation, and sample preparation in biology and medicine.
- Lab-on-a-chip systems: Acoustic microfluidics can be integrated into lab-on-a-chip devices to provide on-chip sample handling, mixing, and analysis. It allows for the miniaturization and automation of complex biochemical assays and diagnostics.
- Drug delivery: Acoustic waves can be used to generate localized forces that facilitate controlled drug release from microparticles or targeted drug delivery to specific tissue regions.
- Material synthesis and assembly: Acoustic microfluidics has applications in the fabrication and assembly of micro- and nanoscale materials. It enables the controlled synthesis of particles, emulsions, and composite materials.

### 14.2.6 Optical Microfluidics (Optofluidics)

Optical microfluidics, also known as optofluidics or lab-on-a-chip optical systems, refers to the integration of optical techniques and components with microfluidic platforms. It combines the advantages of microfluidics, such as precise fluid manipulation and miniaturization, with the capabilities of optical analysis, sensing, and manipulation (Myers and Lee 2008). In optical microfluidics, light is used as a tool for various purposes, including detection, imaging, manipulation, and analysis of fluids, particles, or biochemical samples at the microscale. It involves the integration of optical elements such as light sources, lenses, filters, detectors, and waveguides into microfluidic devices to enable optical interrogation and control of the fluids and particles of interest.

There are several key areas of application for optical microfluidics:

- Optical detection and sensing: Optical techniques are employed for label-free detection and sensing of analytes in microfluidic systems. This includes methods such as absorbance, fluorescence, Raman spectroscopy, surface plasmon resonance (SPR), and interferometry. By combining microfluidics with optical detection, it becomes possible to perform rapid and sensitive analysis of various biological and chemical samples (Chiu et al. 2015; Huang et al. 2009).
- Optical trapping and manipulation: Optical microfluidics allow for the precise manipulation and control of particles, cells, or droplets using optical trapping techniques. Optical traps, created using laser beams, can be used to confine and move particles within microchannels, enabling tasks such as sorting, positioning, and aggregation of particles. This is particularly useful in biological and biomedical applications, where single-cell manipulation and analysis are required (Zhao et al. 2020b).

- Optofluidic imaging: Optical microfluidics enables high-resolution imaging of samples in microfluidic channels. It integrates imaging components such as lenses, filters, and detectors into the microfluidic device, allowing for on-chip imaging of cells, particles, or biochemical reactions. This facilitates real-time observation of dynamic processes and reduces the need for external microscopy equipment (Zhao et al. 2020b).
- Optogenetics: Optogenetics is a technique that uses light to control and manipulate genetically modified cells or organisms. By integrating light sources and light-sensitive molecules into microfluidic systems, optical microfluidics provides a platform for precise spatiotemporal control of cellular responses and neural activity. It enables the study of complex biological systems and the development of new therapies (Renault et al. 2015).

The combination of optics and microfluidics offers several advantages. The small dimensions of microfluidic channels allow for reduced sample and reagent consumption, rapid analysis, and integration of multiple functionalities onto a single chip. Optical detection and manipulation techniques provide high sensitivity, specificity, and precision in sample analysis and handling. However, optical microfluidics also presents challenges. The integration of optical components into microfluidic devices requires careful alignment and sealing to prevent light leakage and maintain optical performance (White, Yazdi, and Yu 2012). Fabrication techniques need to be optimized to ensure compatibility between microfluidic and optical elements. Moreover, complex light paths and multi-layer designs may be necessary for certain applications, which can increase the complexity of device fabrication and assembly.

In summary, optical microfluidics combines the benefits of microfluidics and optical techniques to enable advanced analysis, sensing, imaging, and manipulation of fluids and particles at the microscale. It has applications in various fields, including biology, medicine, chemistry, and materials science. With ongoing advancements in microfabrication techniques and optical technologies, optical microfluidics continues to drive innovation in lab-on-a-chip systems and offers new opportunities for research and commercialization.

## 14.3 MICROFLUIDICS IN THE PALETTE OF MXENE

### 14.3.1 Synthesis of MXenes Using Microfluidic Techniques

Two-dimensional materials are a promising candidate for a multitude of applications due to their unique properties, diversity of structural characteristics, and fabrication steps. Apart from other applications, these microfluidic techniques are employed for the synthesis of two-dimensional materials like graphene (Qiu et al. 2020), metal nanoparticles (Abalde-Cela et al. 2018), metal-organic frameworks (Rohra et al. 2022), $MoS_2$ (Marqus et al. 2018), metal oxides (Hao, Zhang, and Zhang 2020), perovskites (Koryakina et al. 2021), etc. Therefore, microfluidic technology has witnessed great interest among chemists, physicists, and materials scientists alike. Among the family of 2D materials, MXenes are the most prominent candidates that gained attraction in the world of material science. MXene is the family comprising two-dimensional

(2D) transition metal carbides (TMCs), nitrides, and carbonitrides with properties and arrangements similar to graphene. MXenes are attractive among other classes of 2D materials due to their structure-wise diversity, as well as rich surface properties. The unique combination of features like hydrophilicity and high conductivity makes MXenes a better candidate among other 2D materials. MXenes are classically synthesized by selective separation of "A" atomic layers from $M_{n+1}AX_n$ (MAX phases), where "M" denotes an early transition metal (Ti, V, Cr, Mo, Nb, V), "A" is an element

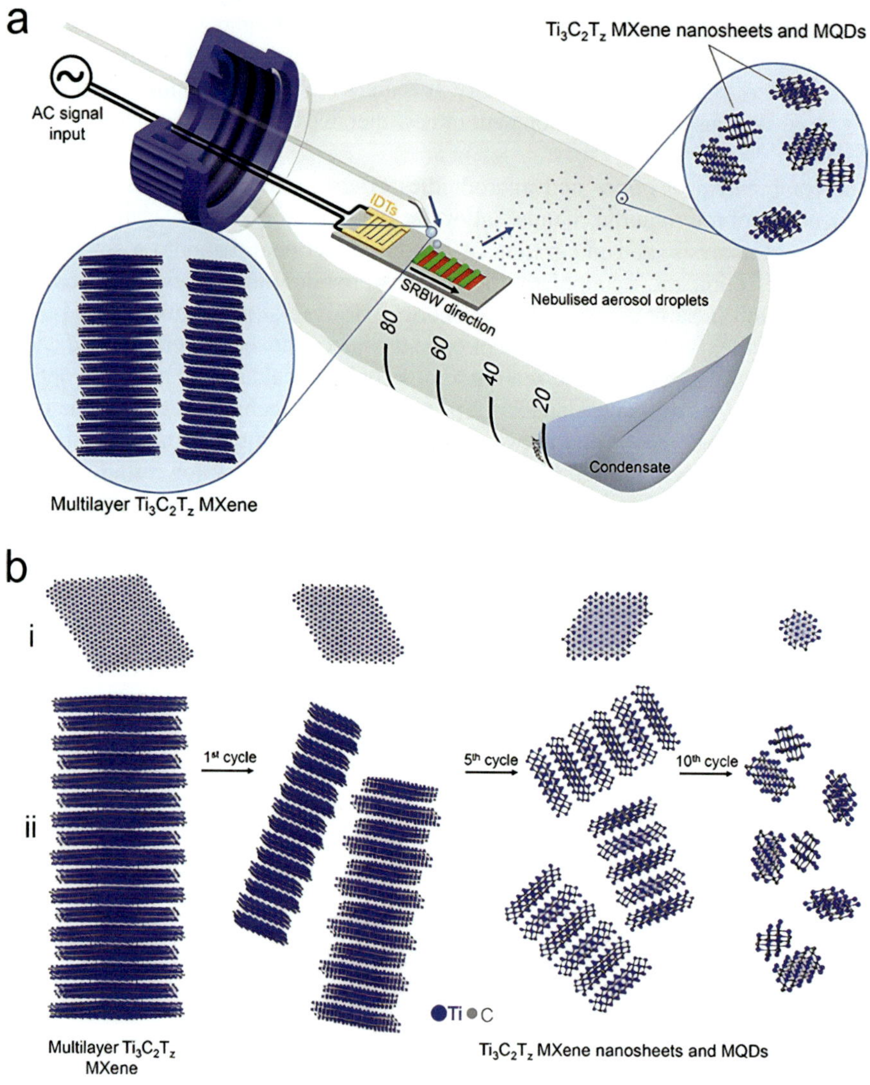


**FIGURE 14.3** (a) Schematic representation of the continuous synthesis of MXene-quantum dots (M-QDs) (b) M-QDs synthesized due to the gradual decrease of the MXene nanosheets' size and thickness. Reprinted with permission from Alijani et al. (2021) (Copyright 2021 ACS).

in group 13, 16 (Al, Si, etc.), and "X" represents carbon and/or nitrogen and $n = 1$–4. MXenes have a unique blend of properties, and surface functional groups that make MXenes hydrophilic and able to bond to various species; a large value of negative zeta-potential, enabling excellent stability enabling colloidal suspension in water; that has led to a wide variety of applications (Gogotsi and Anasori 2019).

Recently, Leslie Yeo and their team reported a novel methodology for the synthesis of MXene nanosheets and MXene-quantum dots (M-QDs) (Alijani et al. 2021). MXenes being one of the prime topics in 2D materials research, this new strategy was considered a breakthrough. The synthetic techniques described in the previous literature to break down MXene nanosheets into M-QDs require severe reaction conditions, such as the use of harsh chemicals, higher temperatures, and sonication, which affect the quality of the M-QDs produced. These drastic conditions can result in harsh oxidative degradation of $Ti_3C_2T_x$ and the production of $TiO_2$ (40%–80%) in M-QDs. To overcome this, single to very few $Ti_3C_2T_x$ was delaminated at ambient temperature using chemical-free acoustomicrofluidics, and the resulting M-QDs maintained the material's purity by reducing oxidation during synthesis (Figure 14.3). In this method, they make use of a class of recently discovered nanometer amplitude high frequency (MHz-order) hybrid sound waves (SRBW), whose extraordinary 10 m/s surface acceleration and 100 v/m evanescent electric field are an incredibly versatile platform for fluids, bulk, and 2D crystals, including exfoliation of bulk 2D crystals, TMDs, and recently etching of $Ti_3C_2T_z$ from $Ti_3AlC_2T_z$ (Ghazaly et al. 2021). In this work, they reported that multi-layer MXenes can be mechanically delaminated into monolayer and few-layer nanosheets and M-QDs without the need for conventional routes. Due to the great purity of synthesized M-QDs, these materials performed better as electrode materials for applications like electrochemical sensing. In particular, they experimented with M-QDs to detect different concentrations of $H_2O_2$ as low as 5 nM, which is the lowest limit to date using $Ti_3C_2T_z$ sensors.

### 14.3.2 Microfluidic MXene Sensors for Analyte Detection

In the field of flexible, portable electronics, including wearable sensors and microfluidic chips, MXenes and their hybrids have demonstrated great potential. MXene might be easily incorporated into microfluidic circuits and wearable sensors due to its adaptability and strong laminar structure. Electrochemical sensing electrodes are essential for the development of wearable sensors. Traditional metal-based electrodes have now been replaced in wearable technology with cutting-edge 2D materials like MXene to improve sensor performance by enlarging the contact area and incorporating necessary electrical and mechanical properties. The multi-layer architecture of MXene provides sufficient surface area for target detection in sensors. Excellent physicochemical properties, good biocompatibility, and conductivity make MXenes an ideal candidate for the fabrication of microfluidic sensors.

MXene as an approach toward the fabrication of fluidic devices has come into the field in recent years only, out of the reports microfluidics and optofluidics-based techniques were explored to date. MXene biosensors using microfluidic technologies have recently only gained attraction in the research world. A portable electrochemical patch sensor for detecting the potassium [$K^+$] concentration level in human sweat was reported by Zhang et al. (2021) (Figure 14.4). The developed sensor was a one-of-a-kind MXene/MWCNT (Multiwalled carbon nanotube) hybrid-based all-solid-state [$K^+$] ion-selective electrode by layer-by-layer deposition on PET film on a PDMS (polydimethylsiloxane) substrate (Figure 14.4(a)). This flexible patch sensor was connected to a battery-free near-field communication (NFC) wireless patch system with the goal of real-time [$K^+$] ion detection. A microfluidic channel was built within that electrochemical patch sensor using 3D printing technology that collects human sweat and minimizes sensor surface fouling (Figure 14.4(b)–(g)). This sensor had a linearity of 1–32 mM and a sensitivity of about 63 mV/dec, which could be

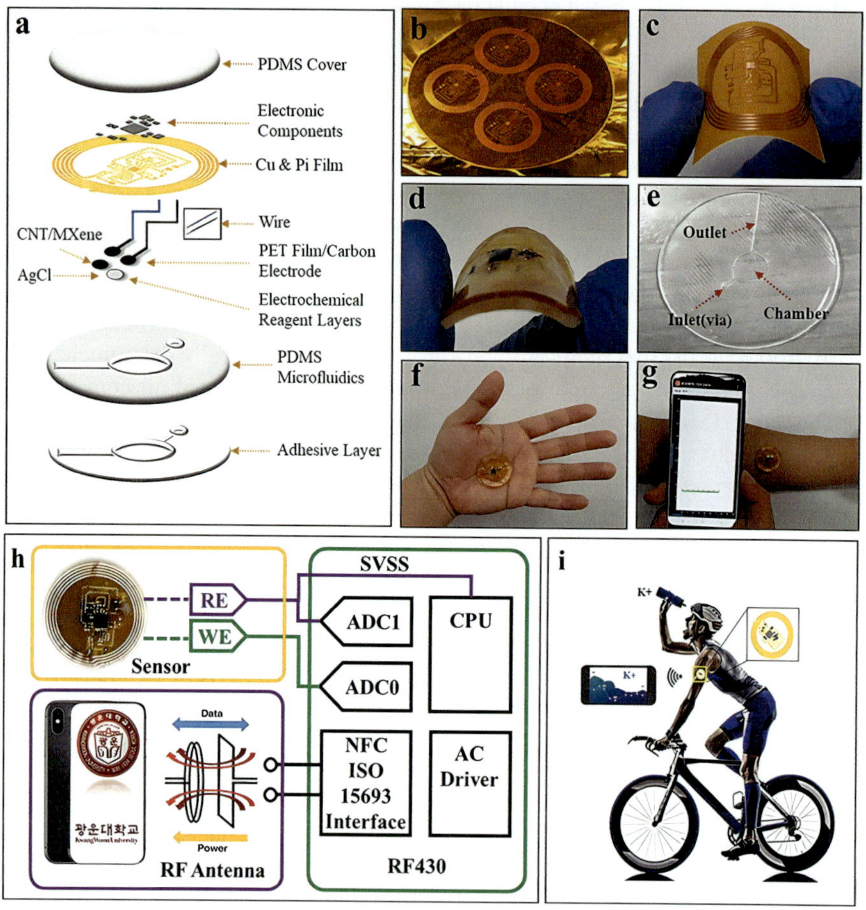


**FIGURE 14.4** (a) Schematic representation of the flexible, wireless, battery-free microfluidic system. (b, c) Lithography process to make flexible circuits and antennae. (d) Fixing of the flexible system on the PDMS substrate. (e) The microfluidic circuit system. (f) Circular (diameter = 3.3 cm) and small-sized patch sensor. (g) Sensor with smartphone connection for real-time analysis. (h) Complete system with a measurement strategy. (i) Measurement of health parameters from athletes' sweat for personalized treatment. Reproduced with permission from Zhang et al. (2021). Copyright (2021) Elsevier B.V.

multiplied by 173 mV/dec using NFC. The complete representation of the measurement strategy was also described (Figure 14.4(h)). This sensor can be used for the personalized treatment of the athlete from sweat to measure the health parameters (Figure 14.4(i)).

Li et al. introduced a novel approach using a cellulose-paper-based substrate to develop a highly integrated sensing (HIS) device for detecting lactate and glucose in sweat (Li et al. 2021a). They created this sensor through a screen-printing process, using MXene combined with methylene blue as the active material on foldable cellulose substrates designed for sweat analysis. The 3D structure of this sensor enhances sweat absorption, minimizing fluid buildup and reducing user discomfort. The device consists of a sweat-absorbing layer with a vortex pattern (the layer that contacts the skin), a 3D electrode layer, and a rapid diffusion layer, where the unique positioning of three electrodes in a 3D arrangement enhances stability. By folding this HIS paper-based substrate into a multi-layer structure, they created a well-structured 3D diffusion channel that connects hydrophilic regions across layers. The MXene-based paper sensors represent a cutting-edge technology, with this dual-channel system being the first of its kind. This electrochemical HIS sensor can simultaneously detect both molecules with sensitivity values of 0.49 μA/mM for lactate and 2.4 nA/μM for glucose. MXene provided a susceptible surface for biomolecule immobilization, while modification with methylene blue facilitates charge transfer, increasing the sensor's sensitivity.

Additionally, MXene-based immunosensors are described for the detection of sweat indicators. Based on $Ti_3C_2T_x$ MXene-laser-burned graphene (LBG) electrodes, a wearable microfluidic-integrated impedimetric immunosensor for cortisol detection in human sweat was developed (Li et al. 2021a). The method of deposition of highly conductive $Ti_3C_2T_x$ MXenes onto the LBG electrode solved the issue of inter-flake disconnection brought on by burning and transferring, consequently enhancing the LBG electrode's electrical characteristics. The fabricated patch immunosensor showed good selectivity and a low detection limit of 3.88 pM when linked with a microfluidic device.

Liu et al. reported another electrochemical microfluidic system that can be used to monitor bodily conditions while undergoing hemodialysis (Liu et al. 2019). This device allows for the detection of urea, creatinine, and uric acid in the blood. A four-layered microfluidic device was developed for the continuous and direct multicomponent analysis of blood. The top layer had a blood flow channel, the second layer contained a dialysis membrane, the third layer had an isotonic solution channel, and the bottom layer occupied a detecting chamber with a sensing electrode (screen-printed MXene). Methylene blue (MB) was deposited over this screen-printed electrode after immobilization with urease with the aid of glutaraldehyde for urea detection to increase the performance of this microfluidic electrochemical sensor. The synergistic interplay of MXene electrocatalysis, the enzymes immobilized on MXene, and probes/ions adsorbed by MXene nanosheets allowed for the simultaneous detection of several analytes. The microfluidic chip was precisely designed to meet the demands of comprehensive blood analysis and to make it simple to replace the electrodes for long-term use. With a conventional biochemical analysis device serving as the gold standard, this chip was used to examine the renal function of 105 healthy individuals. The results showed no statistical difference between the two methods, and the amounts of creatinine, urea, and uric acid determined using the chip method matched those determined using the gold standard approach. For subsequent electrocatalysis, MXene adsorbs uric acid molecules, which create hydrogen bonds with the functional groups of MXene to raise the effective concentration of uric acid at the electrode/electrolyte interface. A brief display of microfluidic sensors based on 2D MXenes is shown in Table 14.1.

The palette of opto-fluidics for bio applications includes lab-on-chip devices, molecular imaging tools, displays, lenses, etc. The potential advantages of merging optofluidics and MXenes for various applications are currently only the subject of very little research. An interesting work exhibited by Miao et al. was the development of an organophosphate pesticide (OP) biosensor using an optofluidic system using photothermal spectroscopy blending with molecular photothermal effect (Figure 14.5) (Li et al. 2021b). They detected an organophosphorus pesticide

**TABLE 14.1**
**A brief display of microfluidic sensors based on 2D MXenes**

| Sensing material | Detection principle | Target/analyte | Substrate | Limit of detection (LOD) | Ref. |
|---|---|---|---|---|---|
| MXene/polystyrene/$SiO_2$ nanospheres | Surface Enhanced Raman Spectroscopy (SERS) | Indole, benzaldehyde, 2,4-dinitrotoluene (DNT) | Polyethylene terephthalate (PET) film | Indole = 10 ppb<br>Benzaldehyde = 100 ppb<br>DNT = 50 ppb | Yang et al. (2021) |
| MXene/methylene blue | Electrochemical studies | Urea, uric acid, and creatinine | PMMA | Creatinine = $1.2 \times 10^{-6}$ M<br>Uric acid = $5 \times 10^{-6}$ M | Liu et al. (2019) |
| MXene/Prussian blue (MXene/PB) | Electrochemical studies | Lactate and glucose | Ecoflex substrate | Glucose = $0.33 \times 10^{-6}$ M<br>Lactate = $0.67 \times 10^{-6}$ M | Li et al. (2021a) |
| MXene/Au nanoparticles | SERS | Nile red | – | – | Chen et al. (2022) |
| MXene/He@CCNT (hemin-coated carboxylic carbon nanotube) | Electrochemical studies | Carcinoembryonic antigen (CEA) | PMMA | 2.88 pg/mL | Zhao et al. (2021) |
| Laccase/MXene/LIG (laser-induced graphene) | Electrochemical studies | Dopamine | Polyamide | 0.47 nM | Wagh et al. (2022) |

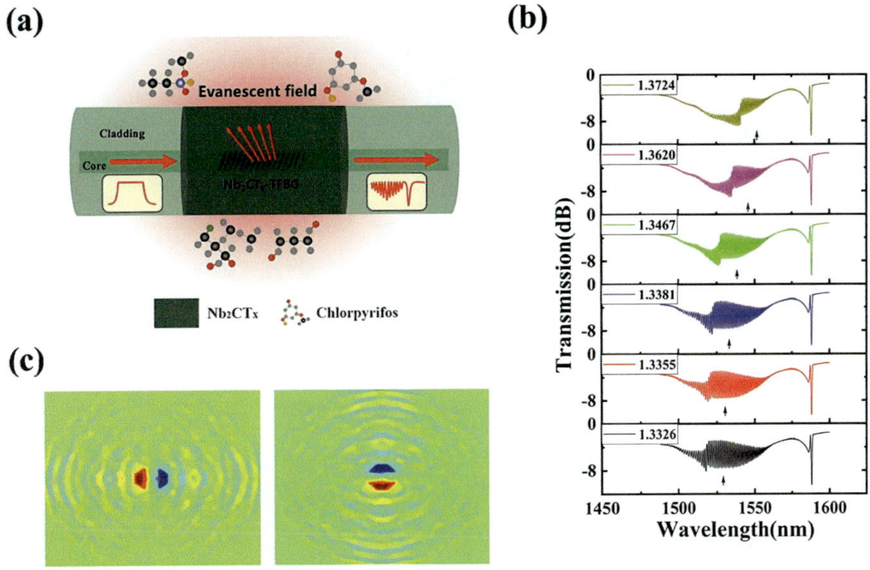


**FIGURE 14.5** (a) Schematic representation of optofluidic MXene-based TFBG biosensor. (b) The transmission spectrum of 16° TFBG using different refractive indices. (c) The performance of the horizontal component of the P-mode and S-mode transverse electric fields was studied using a simulation. Reproduced with permission from Li et al. (2021) Copyright (2021) Optica Publishing Group.

particularly chlorpyrifos using the $Nb_2CT_x$ MXene nanosheets coated on a tilted fiber Bragg grating (TFBG) sensing platform. The functional groups on the surface of MXene nanosheets, like =O, –F, and –OH, boost the sensing of analytes in the solution. To strengthen the interface between light and matter and enhance the sensing performance, integration of MXenes with TFBG is a potential strategy. The photothermal impact of OP causes a visible change in the sample's refractive index when excitation light is applied, thus TFBG can be used to measure the concentration of chlorpyrifos. TFBG was coated with $Nb_2CT_x$ MXene using the optical deposition technique. Chlorpyrifos molecules could adsorb to the surface of MXene owing to their large surface area and biocompatibility, which improves the light-matter interaction. The unification of MXene with TFBG increased the sensitivity to 1.8 pm/ppm and 0.0285 dB/ppm. Also, monitoring the TFBG's core mode would assist in minimizing the impact of the surrounding temperature, and 4.84 ppm/°C was the calculated temperature compensation coefficient. The obtained detection value was 0.35 ppm, and this low limit of detection, as well as specificity, makes this system an excellent candidate for biosensing. Thus, combining the photothermal detection approach with $Nb_2CT_x$ MXene-TFBG results in the real-time label-free analysis of organophosphorus pesticides with high sensitivity, temperature control, and no pre-treatment.

Apart from the above applications, MXene-based fluidic technology is also found in biomedical applications, like in the development of electronic skin, especially MXene-based electronic skin nowadays finding new dimensions in

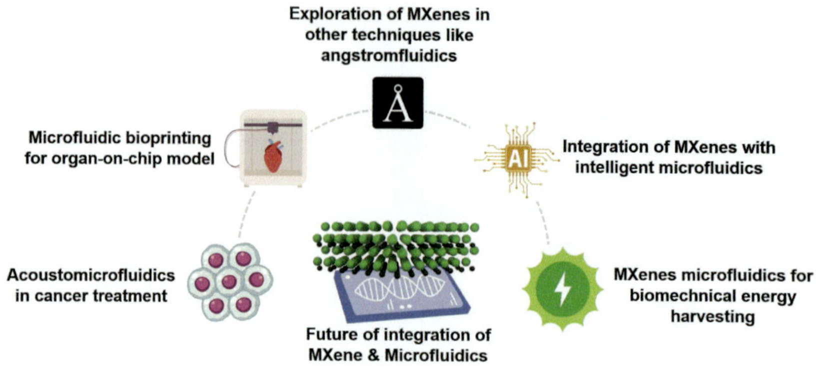


**FIGURE 14.6** Schematic representation of future perspectives on integrating MXene and microfluidics.

biomedical research. In a work, Guo et al. developed morphologically conducting hydrogel microfibers with MXene encapsulated using a multi-injection coflowing glass capillary microfluidic system (Guo et al. 2021). In summation, cross-linked BIS (N,N′-methylenebisacrylamide), monomeric NIPAM (N-isopropyl acrylamide), and photoinitiator HMPP (2-hydroxy-2-methyl-1-phenyl-1–241 propanone) were blended to create a pregel solution. Using the microfluidic system, including an aqueous MXene suspension as the inner phase in the spindle capillary, MXene microfibrils were produced. The MXene microfiber surface was then treated with the pregel solution and the gel was UV-cured. The MXene hydrogel fiber's composition and shape, which showed good photothermal conversion ability and strain capacity, may be controlled by the flow rate. In another work, the group discussed mixing the reaction's necessary ingredients in direct proportions, including MXenes, to create the inner injection fluid as a precursor. The microfibers were preserved in glass vials with the heated $CaCl_2$ solution after being exposed to ultraviolet radiation. In this work, they used cross-linking polyacrylamide and alginate with MXenes to get a network structure with good electrical conductivity. The synthesized hydrogel also showed practical applications in the sensing of joint movements.

## 14.4 CONCLUSION AND FUTURE PERSPECTIVES

MXenes have emerged as a rapidly growing field in biomedical research over the past decade, as shown by recent statistics. Integrating microfluidic technologies with MXenes marked a new direction in materials science. This combination has led to the fabrication of various biosensors, including wearable, electrochemical, and immunosensors. However, several challenges remain before microfluidic sensors can be fully realized, such as optimizing microstructure design, improving electronic functionality, and selecting suitable materials and substrates. One of the most frequently discussed issues in wearable systems is the interface between the sensor, material, and human skin (Radhakrishnan, Mathew, and Rout 2022). Human skin, with its soft, stretchable, viscoelastic outer layer, poses a barrier to fluid exchange, making material compatibility crucial. Wearability depends on materials that align with the skin's physical, chemical, and biological characteristics. Foreign materials that disrupt this interface can cause skin irritation and discomfort, highlighting the need for careful consideration when designing wearable sensors. Though wearable sensors are still in their early stages, significant progress has been made in enhancing their performance in terms of accuracy, resolution, detection limits, and reliability. The challenges posed by individual factors, such as material selection, can be addressed through the introduction of new materials like MXenes. Simulation and modeling techniques, such as molecular dynamics (Manzanares-Palenzuela et al. 2019) and Monte Carlo methods (Zhao et al. 2020a), have already been used to study the adsorption of biomolecules on functionalized MXenes. These techniques can also be applied to investigate MXenes' interactions with biomolecules to better understand their cytotoxicity and long-term stability in vivo. To unlock the full potential of MXenes for biological applications, issues related to storage, thermal stability, and oxidative resistance must be addressed, with simulation studies offering valuable insights. Beyond sensor technologies, MXene-based microfluidics is expected to advance other biomedical applications. For example, digital microfluidic platforms (Wang, Chen, and Sun 2017), such as liquid marble (Ooi et al. 2021) and core-shell beads (Gorgannezhad et al. 2020) integrated with MXenes could revolutionize digital health monitoring. The emerging field of intelligent microfluidics (Galan et al. 2020), which combines microfluidic systems with machine learning, is poised to drive the next generation of biomedical technologies. The synergy of MXenes, microfluidics, and machine learning holds great promise for opening new avenues in chemical and biological research. Also, the exploration of MXene in acoustomicrofluidics for cancer-related studies, microfluidic bioprinting for developing organ-on-chip models, etc. are still in its blooming way, which has a great potential in next-generation healthcare solutions. MXene microfluidics for the biomechanical energy harvesting is one of the unexplored realms in the domain of energy applications which could be tried in the near future in powering wearable devices Lastly, the integration of MXenes in other fluidic techniques like angstrom fluidics and magnetofluidics has not been explored; much thrust to these fields has to be addressed among the material scientists. A brief schematic illustration of the future of integrating MXene and microfluidics is given in Figure 14.6. For MXene microfluidics integration to achieve transformative progress, concerted efforts are required to address the challenges and perspectives mentioned above. This study aimed to highlight innovations and practical examples that illuminate the proposed pathway, helping to shape a technological landscape where MXenes and microfluidics together can overcome complex obstacles.

## REFERENCES

Abalde-Cela, Sara, Patricia Taladriz-Blanco, Marcelo G. de Oliveira, and Chris Abell. 2018. "Droplet microfluidics for the highly controlled synthesis of branched gold nanoparticles." *Scientific Reports* 8 (1):2440. doi: 10.1038/s41598-018-20754-x.

Alijani, Hossein, Amgad R. Rezk, Mohammad M. K. Farsani, Heba Ahmed, Joseph Halim, Philipp Reineck, Billy J. Murdoch, Ahmed El-Ghazaly, Johanna Rosen, and Leslie Y. Yeo. 2021. "Acoustomicrofluidic synthesis of pristine ultrathin $Ti_3C_2T_z$ MXene nanosheets and quantum dots." *ACS Nano* 15 (7):12099–12108. doi: 10.1021/acsnano.1c03428.

Chen, Xueye, and Honglin Lv. 2022. "Intelligent control of nanoparticle synthesis on microfluidic chips with machine learning." *NPG Asia Materials* 14 (1):69. doi: 10.1038/s41427-022-00416-1.

Chen, Zhaoxian, Anping Liu, Xiumei Zhang, Jiawei Jiao, Yuan Yuan, Yingzhou Huang, and Sheng Yan. 2022. "Mxenes–Au NP hybrid plasmonic 2D microplates in microfluidics for SERS detection." *Biosensors*, 12, 505. https://doi.org/10.3390/bios12070505.

Chiu, Tzu-Keng, Kin-Fong Lei, Chia-Hsun Hsieh, Hung-Bo Hsiao, Hung-Ming Wang, and Min-Hsien Wu. 2015. "Development of a microfluidic-based optical sensing device for label-free detection of circulating tumor cells (CTCs) through their lactic acid metabolism." *Sensors* 15 (3):6789–6806.

Chung, Aram J. 2019. "A minireview on inertial microfluidics fundamentals: Inertial particle focusing and secondary flow." *BioChip Journal* 13 (1):53–63.

Coudron, Loïc, Martin B. McDonnell, Ian Munro, Daniel K. McCluskey, Ian D. Johnston, Christabel K. L. Tan, and Mark C. Tracey. 2019. "Fully integrated digital microfluidics platform for automated immunoassay; A versatile tool for rapid, specific detection of a wide range of pathogens." *Biosensors and Bioelectronics* 128:52–60. doi: 10.1016/j.bios.2018.12.014.

Delamarche, Emmanuel, Noemi Tonna, Robert D. Lovchik, Fabio Bianco, and Michela Matteoli. 2013. "Pharmacology on microfluidics: multimodal analysis for studying cell–cell interaction." *Current Opinion in Pharmacology* 13 (5):821–828. doi: 10.1016/j.coph.2013.07.005.

Doganay, Mert T., Purbali Chakraborty, Sri M. Bommakanti, Soujanya Jammalamadaka, Dheerendranath Battalapalli, Anant Madabhushi, and Mohamed S. Draz. 2024. "Artificial intelligence performance in testing microfluidics for point-of-care." *Lab on a Chip* 24 (21):4998–5008. doi: 10.1039/D4LC00671B.

Doinikov, Alexander A., Michael S. Gerlt, Alen Pavlic, and Jürg Dual. 2020. "Acoustic streaming produced by sharp-edge structures in microfluidic devices." *Microfluidics and Nanofluidics* 24 (5):32. doi: 10.1007/s10404-020-02335-5.

Fan, Yi-Qiang, Hong-Liang Wang, Ke-Xin Gao, Jing-Ji Liu, Dong-Ping Chai, and Ya-Jun Zhang. 2018. "Applications of modular microfluidics technology." *Chinese Journal of Analytical Chemistry* 46 (12):1863–1871. doi: 10.1016/S1872-2040(18)61126-0.

Galan, Edgar A., Haoran Zhao, Xukang Wang, Qionghai Dai, Wilhelm T. S. Huck, and Shaohua Ma. 2020. "Intelligent microfluidics: The convergence of machine learning and microfluidics in materials science and biomedicine." *Matter* 3 (6):1893–1922. doi: 10.1016/j.matt.2020.08.034.

Gao, Yuan, Mengren Wu, Yang Lin, and Jie Xu. 2020. "Acoustic microfluidic separation techniques and bioapplications: A review." *Micromachines* 11 (10):921.

Ghazaly, Ahmed El, Heba Ahmed, Amgad R. Rezk, Joseph Halim, Per O. Å. Persson, Leslie Y. Yeo, and Johanna Rosen. 2021. "Ultrafast, one-step, salt-solution-based acoustic synthesis of $Ti_3C_2$ MXene." *ACS Nano* 15 (3):4287–4293. doi: 10.1021/acsnano.0c07242.

Gogotsi, Yury, and Babak Anasori. 2019. "The rise of MXenes." *ACS Nano* 13 (8):8491–8494. doi: 10.1021/acsnano.9b06394.

Gorgannezhad, Lena, Kamalalayam R. Sreejith, Melody Christie, Jing Jin, Chin H. Ooi, Mohammad Katouli, Helen Stratton, and Nam-Trung Nguyen. 2020. "Core-shell beads as microreactors for phylogrouping of E. coli strains." *Micromachines* 11 (8):761.

Guo, Jiahui, Yunru Yu, Dagan Zhang, Han Zhang, and Yuanjin Zhao. 2021. "Morphological hydrogel microfibers with MXene encapsulation for electronic skin." *Research* 2021. doi: 10.34133/2021/7065907.

Hao, Nanjing, Michael Zhang, and John X. J. Zhang. 2020. "Microfluidics for ZnO micro-/nanomaterials development: Rational design, controllable synthesis, and on-chip bioapplications." *Biomaterials Science* 8 (7):1783–1801. doi: 10.1039/C9BM01787A.

Hopwood, Andrew J., Cedric Hurth, Jianing Yang, Zhi Cai, Nina Moran, John G. Lee-Edghill, Alan Nordquist, Ralf Lenigk, Matthew D. Estes, John P. Haley, Colin R. McAlister, Xiaojia Chen, Carla Brooks, Stan Smith, Keith Elliott, Pieris Koumi, Frederic Zenhausern, and Gillian Tully. 2010. "Integrated microfluidic system for rapid forensic DNA analysis: Sample collection to DNA profile." *Analytical Chemistry* 82 (16):6991–6999. doi: 10.1021/ac101355r.

Huang, Chengjun, Kristien Bonroy, Gunter Reekmans, Wim Laureyn, Katarina Verhaegen, Iwijn De Vlaminck, Liesbet Lagae, and Gustaaf Borghs. 2009. "Localized surface plasmon resonance biosensor integrated with microfluidic chip." *Biomedical Microdevices* 11 (4):893–901. https://doi.org/10.1007/s10544-009-9306-8.

Jahn, Andreas, Joseph E. Reiner, Wyatt N. Vreeland, Don L. DeVoe, Laurie E. Locascio, and Michael Gaitan. 2008. "Preparation of nanoparticles by continuous-flow microfluidics." *Journal of Nanoparticle Research* 10:925–934.

Kaminski, Tomasz S., Ott Scheler, and Piotr Garstecki. 2016. "Droplet microfluidics for microbiology: Techniques, applications and challenges." *Lab on a Chip* 16 (12):2168–2187. doi: 10.1039/C6LC00367B.

Koryakina, Irina G., Maksim Naumochkin, Daria I. Markina, Soslan A. Khubezhov, Anatoly P. Pushkarev, Anatoly A. Evstrapov, Sergey V. Makarov, and Mikhail V. Zyuzin. 2021. "Single-step microfluidic synthesis of halide perovskite nanolasers in suspension." *Chemistry of Materials* 33 (8):2777–2784. doi: 10.1021/acs.chemmater.0c04263.

Li, Meng, Lichao Wang, Rui Liu, Jianmin Li, Qinghong Zhang, Guoying Shi, Yaogang Li, Chengyi Hou, and Hongzhi Wang. 2021a. "A highly integrated sensing paper for wearable electrochemical sweat analysis." *Biosensors and Bioelectronics* 174:112828. doi: 10.1016/j.bios.2020.112828.

Li, Wenjie, Yinping Miao, Tuan Guo, Kialiang Zhang, and Jianquan Yao. 2021b. "$Nb_2CT_x$ MXene-tilted fiber Bragg grating optofluidic system based on photothermal spectroscopy for pesticide detection." *Biomedical Optics Express* 12 (11):7051–7063. doi: 10.1364/BOE.442602.

Liou, Dar-Sun, Yi-Fan Hsieh, Long-Sheng Kuo, Chin-Ting Yang, and Ping-Hei Chen. 2011. "Modular component design for portable microfluidic devices." *Microfluidics and Nanofluidics* 10:465–474.

Liu, Huan, Lang Nan, Feng Chen, Yue Zhao, and Yongxi Zhao. 2023. "Functions and applications of artificial intelligence in droplet microfluidics." *Lab on a Chip* 23 (11):2497–2513. doi: 10.1039/D3LC00224A.

Liu, Jiang, Xiantao Jiang, Ruyue Zhang, Yang Zhang, Leiming Wu, Wei Lu, Jianqing Li, Yingchun Li, and Han Zhang. 2019. "MXene-enabled electrochemical microfluidic biosensor: Applications toward multicomponent continuous monitoring in whole blood." *Advanced Functional Materials* 29 (6):1807326. doi: 10.1002/adfm.201807326.

Liu, Junchi, Hanze Du, Lei Huang, Wangni Xie, Kexuan Liu, Xue Zhang, Shi Chen, Yuan Zhang, Daowei Li, and Hui Pan. 2024. "AI-powered microfluidics: Shaping the future of phenotypic drug discovery." *ACS Applied Materials & Interfaces* 16 (30):38832–38851. doi: 10.1021/acsami.4c07665.

Liu, Shilei, Yanye Yang, Zhengyang Ni, Xiasheng Guo, Linjiao Luo, Juan Tu, Dong Zhang, Zhang, and Jie. 2017. "Investigation into the effect of acoustic radiation force and acoustic streaming on particle patterning in acoustic standing wave fields." *Sensors* 17 (7):1664.

Liu, Yong, and Xingyu Jiang. 2017. "Why microfluidics? Merits and trends in chemical synthesis." *Lab on a Chip* 17 (23):3960–3978.

Lombardi, Dario, and Petra S. Dittrich. 2010. "Advances in microfluidics for drug discovery." *Expert Opinion on Drug Discovery* 5 (11):1081–1094. doi: 10.1517/17460441.2010.521149.

Malic, Lidija, Daniel Brassard, Teodor Veres, and Maryam Tabrizian. 2010. "Integration and detection of biochemical assays in digital microfluidic LOC devices." *Lab on a Chip* 10 (4):418–431. doi: 10.1039/B917668C.

Manzanares-Palenzuela, C. L., Amir M. Pourrahimi, Jesus Gonzalez-Julian, Zdenek Sofer, Martin Pykal, Michal Otyepka, and Martin Pumera. 2019. "Interaction of single- and double-stranded DNA with multilayer MXene by fluorescence spectroscopy and molecular dynamics simulations." *Chemical Science* 10 (43):10010–10017. doi: 10.1039/C9SC03049B.

Mao, Zhangming, Yuliang Xie, Feng Guo, Liqiang Ren, Po-Hsun Huang, Yuchao Chen, Joseph Rufo, Francesco Costanzo, and Tony Jun Huang. 2016. "Experimental and numerical studies on standing surface acoustic wave microfluidics." *Lab on a Chip* 16 (3):515–524. doi: 10.1039/C5LC00707K.

Marqus, Susan, Heba Ahmed, Mustafa Ahmed, Chenglong Xu, Amgad R. Rezk, and Leslie Y. Yeo. 2018. "Increasing exfoliation yield in the synthesis of $MoS_2$ quantum dots for optoelectronic and other applications through a continuous multicycle acoustomicrofluidic approach." *ACS Applied Nano Materials* 1 (6):2503–2508. doi: 10.1021/acsanm.8b00559.

Myers, Frank B., and Luke P. Lee. 2008. "Innovations in optical microfluidic technologies for point-of-care diagnostics." *Lab on a Chip* 8 (12):2015–2031. doi: 10.1039/B812343H.

Nielsen, Jacob B., Robert L. Hanson, Haifa M. Almughamsi, Chao Pang, Taylor R. Fish, and Adam T. Woolley. 2020. "Microfluidics: Innovations in materials and their fabrication and functionalization." *Analytical Chemistry* 92 (1):150–168. doi: 10.1021/acs.analchem.9b04986.

Nosrati, Hamed, and Masoud Nosrati. 2023. "Artificial intelligence in regenerative medicine: Applications and implications." *Biomimetics* 8 (5):442.

Ooi, Chin H., Raja Vadivelu, Jing Jin, Kamalalayam R. Sreejith, Pradip Singha, Nhat-Khuong Nguyen, and Nam-Trung Nguyen. 2021. "Liquid marble-based digital microfluidics – Fundamentals and applications." *Lab on a Chip* 21 (7):1199–1216. doi: 10.1039/D0LC01290D.

Padash, Mahshid, Christian Enz, and Sandro Carrara. 2020. "Microfluidics by additive manufacturing for wearable biosensors: A review." *Sensors* 20 (15):4236.

Pandey, Chandra M., Shine Augustine, Saurabh Kumar, Suveen Kumar, Sharda Nara, Saurabh Srivastava, and Bansi D. Malhotra. 2018. "Microfluidics based point-of-care diagnostics." *Biotechnology Journal* 13 (1):1700047. doi: 10.1002/biot.201700047.

Qiu, Hui, Xingjiang Wu, Ri Hong, Guan Wu, and Su Chen. 2020. "Microfluidic-oriented synthesis of graphene oxide nanosheets toward high energy density supercapacitors." *Energy & Fuels* 34 (9):11519–11526. doi: 10.1021/acs.energyfuels.0c02143.

Radhakrishnan, Sithara, Minu Mathew, and Chandra S. Rout. 2022. "Microfluidic sensors based on two-dimensional materials for chemical and biological assessments." *Materials Advances* 3 (4):1874–1904. doi: 10.1039/D1MA00929J.

Renault, Renaud, Nirit Sukenik, Stéphanie Descroix, Laurent Malaquin, Jean-Louis Viovy, Jean-Michel Peyrin, Samuel Bottani, Pascal Monceau, Elisha Moses, and Maéva Vignes. 2015. "Combining microfluidics, optogenetics and calcium imaging to study neuronal communication in vitro." *PloS One* 10 (4):e0120680.

Richard, Bartholomew, Susmi A. Thomas, Amarendar R. M., Mohan R. Pallavolu, and Jayesh Cherusseri. 2023. "Minireview on fluid manipulation techniques for the synthesis and energy applications of two-dimensional MXenes: Advances, challenges, and perspectives." *Energy & Fuels* 37 (10):6999–7013. doi: 10.1021/acs.energyfuels.3c00589.

Rohra, Nanda, Ganesh Gaikwad, Prajakta Dandekar, and Ratnesh Jain. 2022. "Microfluidic synthesis of a bioactive metal–organic framework for glucose-responsive insulin delivery." *ACS Applied Materials & Interfaces* 14 (6):8251–8265. doi: 10.1021/acsami.1c22153.

Shen, Yigang, Yaxiaer Yalikun, and Yo Tanaka. 2019. "Recent advances in microfluidic cell sorting systems." *Sensors and Actuators B: Chemical* 282: 268–281. https://doi.org/10.1016/j.snb.2018.11.025.

Sista, Ramakrishna, Zhishan Hua, Prasanna Thwar, Arjun Sudarsan, Vijay Srinivasan, Allen Eckhardt, Michael Pollack, and Vamsee Pamula. 2008. "Development of a digital microfluidic platform for point of care testing." *Lab on a Chip* 8 (12):2091–2104. doi: 10.1039/B814922D.

Srinivasan, Vijay, Vamsee K. Pamula, and Richard B. Fair. 2004. "An integrated digital microfluidic lab-on-a-chip for clinical diagnostics on human physiological fluids." *Lab on a Chip* 4 (4):310–315. doi: 10.1039/B403341H.

Tong, Zhaoduo, Chuanjie Shen, Qiushi Li, Hao Yin, and Hongju Mao. 2023. "Combining sensors and actuators with electrowetting-on-dielectric (EWOD): Advanced digital microfluidic systems for biomedical applications." *Analyst* 148 (7):1399–1421. doi: 10.1039/D2AN01707E.

Vinoth, Rajendran, Tatsuo Nakagawa, Jayaraman Mathiyarasu, and Vinu Mohan A. M. 2021. "Fully printed wearable microfluidic devices for high-throughput sweat sampling and multiplexed electrochemical analysis." *ACS Sensors* 6 (3):1174–1186. doi: 10.1021/acssensors.0c02446.

Wagh, Mrunali D., Pavar S. Kumar, Khairunnisa Amreen, Subhendu K. Sahoo, and Sanket Goel. 2022. "Integrated microfluidic device with MXene enhanced laser-induced graphene bioelectrode for sensitive and selective electroanalytical detection of dopamine." *IEEE Sensors Journal* 22 (14):14620–14627.

Wang, He, Liguo Chen, and Lining Sun. 2017. "Digital microfluidics: A promising technique for biochemical applications." *Frontiers of Mechanical Engineering* 12 (4):510–525. doi: 10.1007/s11465-017-0460-z.

White, Ian M., Soroush H. Yazdi, and Wei W. Yu. 2012. "Optofluidic SERS: Synergizing photonics and microfluidics for chemical and biological analysis." *Microfluidics and Nanofluidics* 13 (2):205–216. doi: 10.1007/s10404-012-0962-2.

Wu, Amy, Lisen Wang, Erik Jensen, Richard Mathies, and Bernhard Boser. 2010. "Modular integration of electronics and microfluidic systems using flexible printed circuit boards." *Lab on a Chip* 10 (4):519–521. doi: 10.1039/B922830F.

Wu, Jialin, Hui Fang, Jun Zhang, and Sheng Yan. 2023. "Modular microfluidics for life sciences." *Journal of Nanobiotechnology* 21 (1):85. doi: 10.1186/s12951-023-01846-x.

Xu, Xing, Linfeng Cai, Shanshan Liang, Qiannan Zhang, Shiyan Lin, Mingying Li, Qizheng Yang, Chong Li, Ziyan Han, and Chaoyong Yang. 2023. "Digital microfluidics for biological analysis and applications." *Lab on a Chip* 23 (5):1169–1191. doi: 10.1039/D2LC00756H.

Yang, Kuo, Kai Zhu, Yuanzhe Wang, Ziting Qian, Yizhi Zhang, Zhaoyan Yang, Zhuyuan Wang, Lei Wu, Shenfei Zong, and Yiping Cui. 2021. "$Ti_3C_2T_x$ MXene-loaded 3D substrate toward on-chip multi-gas sensing with surface-enhanced raman spectroscopy (SERS) barcode readout." *ACS Nano* 15 (8):12996–13006. doi: 10.1021/acsnano.1c01890.

Yuen, Po K. 2008. "SmartBuild–A truly plug-n-play modular microfluidic system." *Lab on a Chip* 8 (8):1374–1378. doi: 10.1039/B805086D.

Zhai, Jiao, Yingying Liu, Weiqing Ji, Xinru Huang, Ping Wang, Yunyi Li, Haoran Li, Ada Hang-Heng Wong, Xiong Zhou, Ping Chen, Lianhong Wang, Ning Yang, Chi Chen, Haitian Chen, Pui-In Mak, Chu-Xia Deng, Rui Martins, Mengsu Yang, Tsung-Yi Ho, Shuhong Yi, Hailong Yao, and Yanwei Jia. 2024. "Drug screening on digital microfluidics for cancer precision medicine." *Nature Communications* 15 (1):4363. doi: 10.1038/s41467-024-48616-3.

Zhang, Jun, Sheng Yan, Dan Yuan, Gursel Alici, Nam-Trung Nguyen, Majid E. Warkiani, and Weihua Li. 2016. "Fundamentals and applications of inertial microfluidics: A review." *Lab on a Chip* 16 (1):10–34. doi: 10.1039/C5LC01159K.

Zhang, Shipeng, Md Abu Zahed, Md Sharifuzzaman, Sanghyuk Yoon, Xue Hui, Sharat C. Barman, Sudeep Sharma, Hyo S. Yoon, Chani Park, and Jae Y. Park. 2021. "A wearable battery-free wireless and skin-interfaced microfluidics integrated electrochemical sensing patch for on-site biomarkers monitoring in human perspiration." *Biosensors and Bioelectronics* 175:112844. doi: 10.1016/j.bios.2020.112844.

Zhang, Yonghao, and Hui-Rong Jiang. 2016. "A review on continuous-flow microfluidic PCR in droplets: Advances, challenges and future." *Analytica Chimica Acta* 914:7–16. doi: 10.1016/j.aca.2016.02.006.

Zhao, Daohui, Yuqing Wang, Qianwen Su, Libo Li, and Jian Zhou. 2020a. "Lysozyme adsorption on porous organic cages: A molecular simulation study." *Langmuir* 36 (41):12299–12308. doi: 10.1021/acs.langmuir.0c02233.

Zhao, Peng, Jilin Zheng, Yi Liang, Fengchun Tian, Lan Peng, Danqun Huo, and Changjun Hou. 2021. "Functionalized carbon nanotube-decorated MXene nanosheet-enabled microfluidic electrochemical aptasensor for carcinoembryonic antigen determination." *ACS Sustainable Chemistry & Engineering* 9 (46):15386–15393. doi: 10.1021/acssuschemeng.1c03597.

Zhao, Xiaoting, Nan Zhao, Yang Shi, Hongbao Xin, and Baojun Li. 2020b. "Optical fiber tweezers: A versatile tool for optical trapping and manipulation." *Micromachines* 11 (2):114. https://doi.org/10.3390/mi11020114.

# 15 Strategies to Mitigate Bottlenecks in the Commercialization of MXenes

*Chiranjit Das and Menon Ankitha*

## 15.1 INTRODUCTION

MXenes are an innovative class of two-dimensional (2D) materials composed of transition metal carbides, nitrides, or carbonitrides (VahidMohammadi, Rosen, and Gogotsi 2021). They have garnered immense interest due to their exceptional electrical conductivity (Shayesteh Zeraati et al. 2021), excellent hydrophilicity (Zeng, Peng, and Lang 2017), enhanced specific surface area, good biocompatibility (Gayathri et al. 2025), and outstanding mechanical properties (Wyatt, Rosenkranz, and Anasori 2021). Since their discovery in 2011 by Naguib et al. (2011), MXenes have rapidly emerged as a leading candidate for diverse technological applications including energy storage, catalysis, selective adsorption/filtration, electronics, sensing, electromagnetic interference shielding (EMI), antennas, and biomedical applications (Kumar, Kumari, and Seo 2024; Zhang, Ji, and Ma 2023; Xu and Gogotsi 2020). The foundational synthesis of MXenes involves the selective removal of the A-layer (typically a group 13 or 14 element) from a parent MAX phase, where M represents an early transition metal, A is an A-group element, and X is carbon and/or nitrogen. The origin of MXenes can be traced back to the discovery of MAX phases, a family of layered ternary carbides and nitrides first studied in the 1960s. MAX phases, with the general formula $M_{n+1}AX_n$ ($n$ = 1, 2, or 3), are characterized by their unique combination of metallic and ceramic properties. In 2011, Naguib et al. demonstrated that etching of the A-layer selectively (A=Al, In, Ga, Ge, Tl, Sn, Si, As, S, and Pb) from MAX phases using hydrofluoric acid resulted in the formation of two-dimensional materials, which they named MXenes (Naguib et al. 2011). This breakthrough opened new avenues for the development of 2D materials beyond graphene, expanding the horizon of materials science. The etching process, typically achieved using fluoride-containing chemicals, results in layered materials with properties that can be fine-tuned for specific applications, including storage of energy, water purification, electromagnetic shielding, catalysis, and biomedical devices (Gogotsi and Anasori 2019). MXenes are synthesized by selective etching process, where certain atomic layers of their precursors, mostly the MAX phases are removed. MXenes are represented by the chemical formula $M_{n+1}X_nT_x$, where $T_x$

DOI: 10.1201/9781003736318-15

stands for surface functional groups like –F, –OH, and –O–, –Cl, –Br, –NH, –Te and n is an integer from 1 to 4 (Gogotsi 2023; Ezzah Ab Latif et al. 2022). This structural arrangement is further enriched by surface terminations that impart chemical tunability (Figure 15.1).

MXenes are often compared to graphene due to their exceptionally high conductivity (~20,000 S/cm). However, unlike graphene, which requires chemical modification for functionalization, MXenes naturally exhibit surface terminations that enhance their chemical reactivity and compatibility with other materials. This inherent advantage positions MXenes as a versatile platform for applications ranging from flexible electronics to water purification systems. One of the most significant drivers of research into MXenes is their superior performance in energy storage devices. For example, their high volumetric capacitance and excellent rate capability make them ideal candidates for supercapacitors and batteries. Additionally, their hydrophilic surfaces enable uniform dispersion in aqueous and non-aqueous media, facilitating composite formation for multifunctional applications (Lukatskaya et al. 2013).

Despite the extensive potential of MXenes, their large-scale adoption faces several hurdles. Among these, stability-related issues are paramount. MXenes are susceptible to oxidation, hydrolysis, and thermal degradation, which compromise their performance and shelf life. Furthermore, the production of MXenes often involves hazardous chemicals, such as hydrofluoric acid (HF), raising environmental and safety

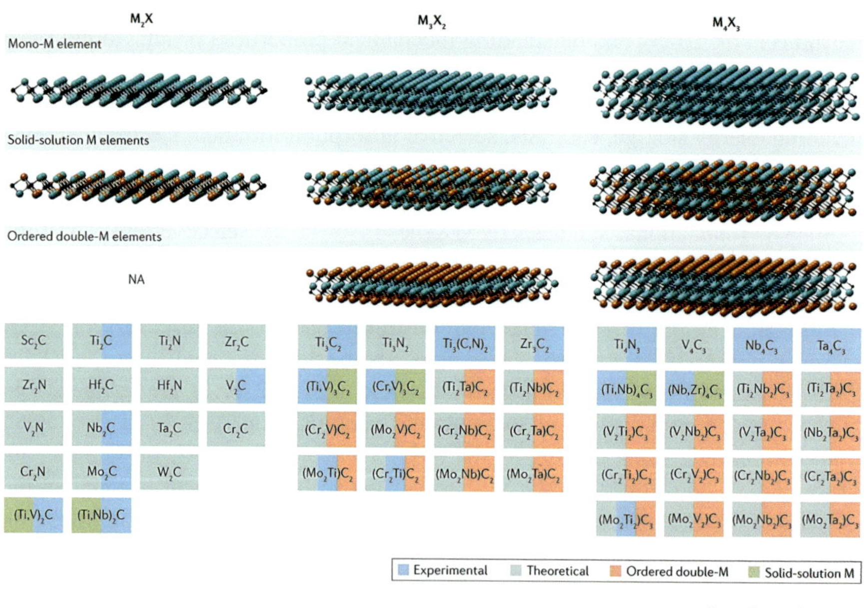


**FIGURE 15.1** A comprehensive overview of both experimental findings and theoretical predictions developed MXene materials with the general composition of $M_{n+1}X_n$ (where $n$ ranges from 1 to 3), encompassing systems based on single metallic components as well as those derived from multi-metal solid solutions. Reproduced with permission from Anasori, Lukatskaya, and Gogotsi (2017). Copyright 2017, Springer Nature.

concerns (Alhabeb et al. 2017). Addressing these challenges requires a multidisciplinary approach that encompasses materials science, chemistry, and engineering.

The handling and disposal of HF require stringent safety protocols to prevent accidents and environmental contamination. Moreover, the potential toxicity of MXenes themselves has been a subject of research, with studies indicating that certain MXene compositions can exhibit cytotoxic effects, endangering public health and ecological balance (Lakhe et al. 2019). Also, developing alternative, less hazardous synthesis methods is crucial. For example, employing alkaline solutions instead of HF for etching can reduce environmental impact (Mim et al. 2024). Enhancing the stability of MXenes through surface modifications and protective coatings can mitigate degradation issues. Additionally, comprehensive safety assessments and the development of standardized protocols for the handling and disposal of MXenes are vital to ensure environmental and human health.

In summary, while MXenes hold significant promise for various applications, overcoming the hurdles related to their stability, environmental impact, and safety is imperative for their successful commercialization. Ongoing research and development efforts are essential to address these challenges and facilitate the transition of MXenes from laboratory-scale studies to industrial-scale applications. This part of this chapter seeks to provide an in-depth understanding of the stability challenges faced by MXenes, emphasizing the intrinsic instability, environmental sensitivity, and structural and chemical factors that contribute to their degradation. Tackling these challenges is crucial to fully harnessing the potential of MXenes and accelerating their integration into industrial and commercial applications. By systematically analyzing the factors influencing their stability and proposing mitigation strategies. Through this discussion, we aim to lay the groundwork for developing strategies to enhance MXene stability and facilitate their transition from the laboratory to industrial-scale applications.

## 15.2 THERMODYNAMIC STABILITY OF MXENES

MXenes exhibit significant thermodynamic instability due to their high surface energy, which makes them prone to degradation even under mild conditions. Surface terminations such as –OH, –O, or –F, commonly present in MXenes, further aggravate this issue by increasing their reactivity with environmental agents. For example, $Ti_3C_2T_x$ MXenes are particularly prone to oxidation in ambient conditions, leading to the formation of titanium dioxide ($TiO_2$) and a loss of electrical conductivity. Studies have shown that even brief exposure to air can result in measurable degradation, with oxidation rates heavily influenced by the density and type of surface terminations (Kumar et al. 2024). The metastable nature of MXenes contributes to their propensity for phase transformation into more stable compounds, such as oxides or non-layered structures. Elevated temperatures accelerate these transformations, with thermal decomposition often leading to the conversion of MXenes into metal oxides completely, and this phenomenon is particularly problematic for high-temperature applications. For instance, $Ti_3C_2T_x$ MXenes, when subjected to temperatures above 400°C, decompose into anatase or rutile $TiO_2$, as reported in thermal stability studies (Adomaviciute-Grabusove et al. 2024).

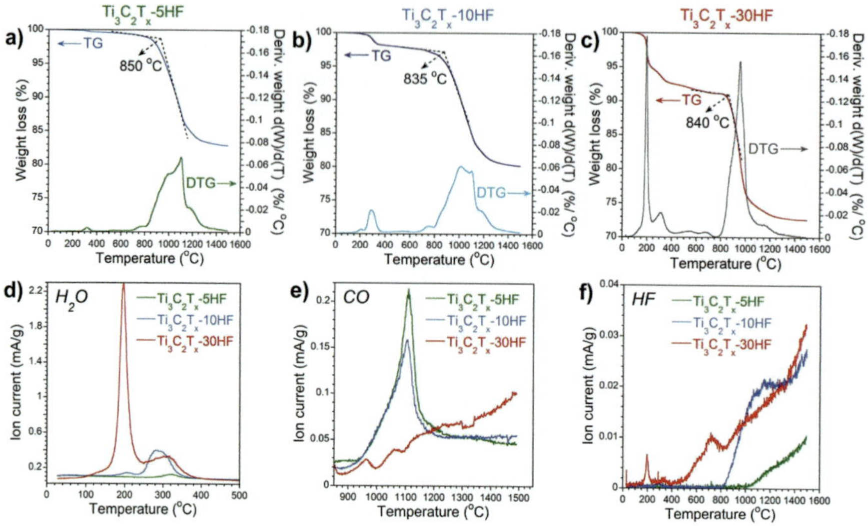


**FIGURE 15.2** Thermal gravimetric (TG) curves with derivatives of the weight losses (DTG) for $Ti_3C_2T_x$ MXene using three different HF concentrations of 5, 10, and 30 wt%: (a) 5 wt% HF for etching, (b) 10 wt% HF for etching, and (c) 30 wt% HF for etching. Mass spectrometry analysis for the atomic mass unit (amu) of (d) 18/$H_2O$, (e) 28/CO, and (f) 20/HF (Seredych et al. 2019). Copyright 2019, American Chemical Society.

To better understand the nature of surface functionalities in carbide-based MXenes, viz., $Ti_3C_2T_x$, $Mo_2CT_x$, and $Nb_2CT_x$. An in-depth thermal analysis was conducted using thermogravimetric techniques combined with mass spectrometry (TGA–MS) (Figure 15.2) under an inert helium environment, with temperatures reaching up to 1500°C. This method enabled the identification of surface-bound groups such as fluorides, hydroxyls, and oxo species, as well as interlayer components like structural water and residual salts. The degree of hydrophilicity observed in these materials was found to depend heavily on both the etching medium, whether a single acid like HF or a combination such as $H_2SO_4$ and HF or HCl and HF. Interestingly, $Ti_3C_2T_x$ exhibited a greater retention of structural water when fewer oxygen-based terminations were present on the surface. Furthermore, employing milder etching conditions or mixed acid systems led to improved thermal robustness by minimizing defect sites during synthesis. The thermal durability of MXenes is closely linked to their intrinsic structure and elemental makeup. Among them, $Ti_3C_2T_x$ exhibits superior heat resistance compared to $Mo_2CT_x$ and $Nb_2CT_x$, both of which possess fewer atomic layers. Between the latter two, $Mo_2CT_x$ is thermally more robust than $Nb_2CT_x$ (Seredych et al. 2019).

## 15.3 OXIDATIVE STABILITY OF MXENES

Recent research has revealed that MXene nanosheets degrade rapidly due to oxidation, significantly compromising their electrochemical performance. This susceptibility is largely attributed to intrinsic atomic defects and the conditions under which they are stored. Lotfi et al. employed reactive force field molecular dynamics (ReaxFF) to simulate the oxidative behavior of $Ti_3C_2$-based MXenes (Lotfi et al. 2018). The study observed that Ti atoms migrated from the inner layers to the surface of the sheet, promoting the formation and compaction of $TiO_2$ nanoparticles. Simulations at 100 ps and 1000 K revealed that MXenes remained structurally intact in dry air. In contrast, exposure to humid conditions prompted the transfer of titanium atoms to the surface, resulting in surface functionalization with oxygen-containing groups such as O and OH. Among the tested environments, hydrogen peroxide induced the most aggressive oxidation, producing a high density of surface functional groups. Similar degradation behavior was noted at temperatures exceeding 1500 K. These findings shed light on the oxidative stability and possible degradation pathways of MXenes under varying environmental conditions. In a $CO_2$ atmosphere, Li et al. investigated the oxidative behavior of $Ti_2CT_x$ MXene, as depicted in Figure 15.3(a)–(h) (Li et al. 2017). At lower temperatures, $CO_2$ interacts with the surface and internal titanium layers, leading to the formation of $TiO_2$ and resulting in a $Ti_2CT_x/TiO_2$ nanocomposite. This transformation causes the initially smooth MXene nanosheets to develop a roughened surface morphology (Figure 15.3(c) and (d)). The oxidation process is initially limited by both thermodynamic feasibility and kinetic barriers, allowing only partial oxidation. However, as temperature increases, $CO_2$ begins to penetrate through the newly formed $TiO_2$ layer, progressively driving the system

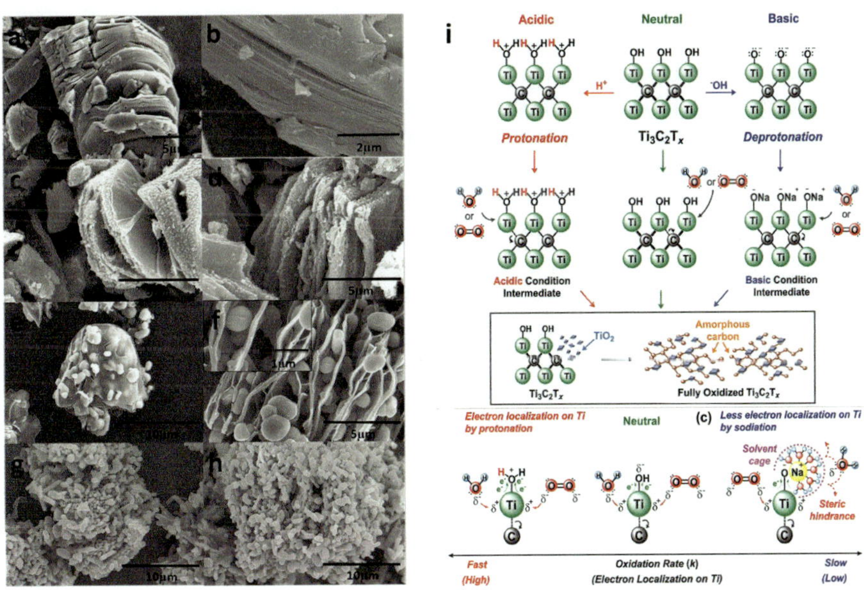


**FIGURE 15.3** SEM images of (a and b) $Ti_2CTx$, (c and d) calcined in $CO_2$ atmosphere at 500°C, (e and f) calcined in $CO_2$ atmosphere at 800°C, and (g and h) calcined in $CO_2$ atmosphere at 900°C. Reproduced with permission from (Li et al. 2017). Copyright 2017, Royal Society of Chemistry. (i) Proposed oxidation mechanism of $Ti_3C_2T_x$ colloids at different pH conditions. Reproduced with permission from Doo et al. (2021). Copyright 2021, American Chemical Society.

toward complete oxidation. In the case of $Ti_3C_2$ MXenes, once the Ti content is fully consumed and converted into $TiO_2$ nanoparticles, the remaining carbon reorganizes to form a distinct carbon layer. This sequential reaction results in the emergence of a $C/TiO_2$ composite structure (Figure 15.3(e) and (f)). With continued exposure to $CO_2$, the carbon layer undergoes complete oxidation, eventually leaving behind only the rutile $TiO_2$ phase (Figure 15.3(g) and (h)). These findings offer a strategic route for the controlled, in situ development of novel nanohybrid architectures that retain the unique 2D framework of MXenes.

During the oxidation of MXenes, oxide formation initiates across the nanosheet surfaces in a fairly uniform manner. As oxidation advances, nanocrystals of oxides have been found to grow both at the interlayer spaces and along the sheet edges (Zhang et al. 2016). Some reports suggest that the oxidation process starts at the more chemically reactive sides of the nanosheets, rather than on their basal surfaces (Lipatov et al. 2016). Although the complete oxidation pathway of MXenes is still under investigation, evidence indicates that the edges serve as the primary sites for initial degradation. In dispersions exposed to atmospheric conditions, the combined presence of dissolved oxygen and water accelerates this process by selectively attacking these susceptible regions, leading to the formation of titanium dioxide. Controlling the oxidative degradation may be achievable by removing dissolved oxygen from aqueous environments, such as through the introduction of compressed inert gases.

Doo et al. (2021) explored how $Ti_3C_2T_x$ suspensions undergo oxidation under different pH and temperatures. A detailed examination concluded that oxidation generates both solid residues like $TiO_2$ and amorphous carbon, along with gaseous byproducts such as carbon di/mono oxide, methane, and HF. Kinetic analysis indicates that the degradation in acidic media follows a proton-driven mechanism: protonation of surface hydroxyl groups on $Ti_3C_2T_x$ promotes electron accumulation on Ti atoms, accelerating oxidation. Conversely, in alkaline environments, the formation of bulky, alkali-stabilized hydroxyl complexes reduces electron concentration on Ti centers, thus slowing down the reaction rate (Figure 15.3(i)).

## 15.4 FACTORS AFFECTING THE OXIDATION OF MXENES

Numerous research ideas have highlighted that the structural integrity of MXenes is governed by a combination of external chemical conditions and inherent material properties, such as atmospheric exposure, temperature, compositional variations, microstructural features, and interactions with aqueous environments. Furthermore, even MXenes sharing the same chemical composition can display markedly different oxidation behaviors, primarily due to variations in their synthesis techniques and processing conditions.

### 15.4.1 Atmospheric Conditions

Extensive research has explored the structural and chemical stability of MXenes under various atmospheric conditions, including vacuum, inert, oxidizing, and reducing environments. Studies have shown that $Ti_3C_2$ MXenes maintain their stability

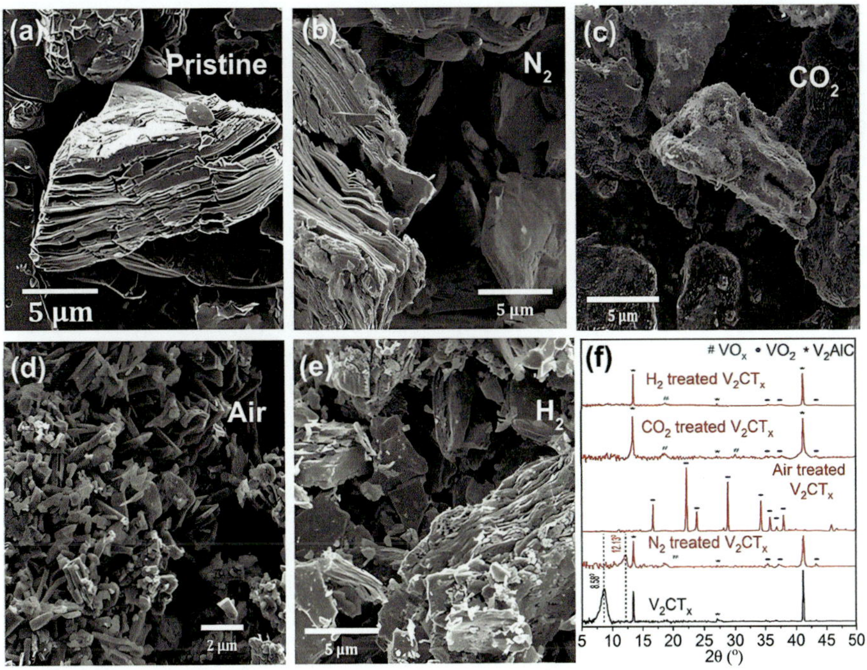


**FIGURE 15.4** SEM images of (a) $V_2CT_x$ MXenes and $V_2CT_x$ MXenes after heat treatment in (b) $N_2$, (c) $CO_2$, (d) air, and (e) $H_2$ at 600°C, with (f) showing the corresponding XRD patterns. Reproduced with permission from Thakur et al. (2019) Copyright 2019, Royal Society of Chemistry.

in an argon gas atmosphere even at the temperature of 800°C. In contrast, exposure to an oxygen-rich environment initiates oxidation at a low temperature (~200°C), which causes the formation of anatase uniformly across the surface. Upon further heating to 1000°C, this anatase phase undergoes a complete transformation into the rutile form (Li et al. 2015). The thermal behavior of $Ti_2CT_x$ MXenes under different atmospheric conditions was systematically investigated by Rakhi et al. (2015). Their findings revealed that heat treatment in air leads to a complete structural transformation, producing numerous $TiO_2$ nanocrystals distributed over ultrathin graphitic nanosheets. In contrast, when annealed in inert (Ar, $N_2$) or reducing ($N_2/H_2$) environments, $Ti_2CT_x$ maintained its inherent morphology and chemical structure, exhibiting minimal alteration. Thakur et al. (2019) conducted a thorough investigation into the stability of $V_2CT_x$ MXenes across various environments, including air, $CO_2$, $H_2$, and $N_2$, at temperatures up to 600°C (Figure 15.4). They observed that at temperatures above 300°C, $V_2CT_x$ MXenes were converted into $V_2O_5$ or various $VO_x$ phases. when exposed to air or $CO_2$, accompanied by a complete loss of the initial layered structure. However, in a reducing $H_2$ atmosphere, the layered structure was preserved to a certain extent. When exposed to $N_2$, the surface groups of $V_2CT_x$ MXenes underwent slight oxidation due to reactions with trace $H_2O$ molecules present in the as-fabricated material.

### 15.4.2 Temperature Effects

At lower temperatures, MXenes preserve their phase integrity and layered structure; however, their oxidation behavior evolves markedly at elevated temperatures. Oxidation of the metallic layers and elimination of the central carbon layers lead to the formation of carbon/$TiO_2$ composites with a well-defined multilayered architecture. As the temperature increases, both the $TiO_2$ grain size and content progressively grow, accompanied by oxidation-driven degradation of the carbon framework. $Ti_3C_2$ MXenes maintain structural integrity up to approximately 600°C under vacuum, with improved electrical conductivity arising from the removal of surface terminations and shortened conduction pathways. Beyond this threshold, significant structural collapse occurs, characterized by the predominant formation of rutile $TiO_2$ and minor anatase phases. This transformation is attributed to oxidation reactions with residual oxygen impurities or decomposition of oxygenated surface functional groups such as carbonyl (C=O) and hydroxyl (–OH) functionalities (Wang et al. 2015) (Figure 15.5). However, Han and co-workers demonstrated that $Ti_3C_2T_x$ MXenes, when annealed in an Ar atmosphere, retained their original layered architecture without detectable formation of $TiO_2$ or carbon phases up to 800°C. In contrast, annealing at 1000°C resulted in pronounced structural disintegration of the nanosheets (Han et al. 2016). Interestingly, the disintegration of MXenes at elevated temperatures can occur even in the absence of an external oxidizing agent. This behavior is attributed to the combined influence of surface terminations and molecular hydrogen trapped between the nanosheets. Consequently, improving the high-temperature stability of MXenes requires careful control over both the surface functional groups and the entrapped molecular hydrogen present in the initial material.

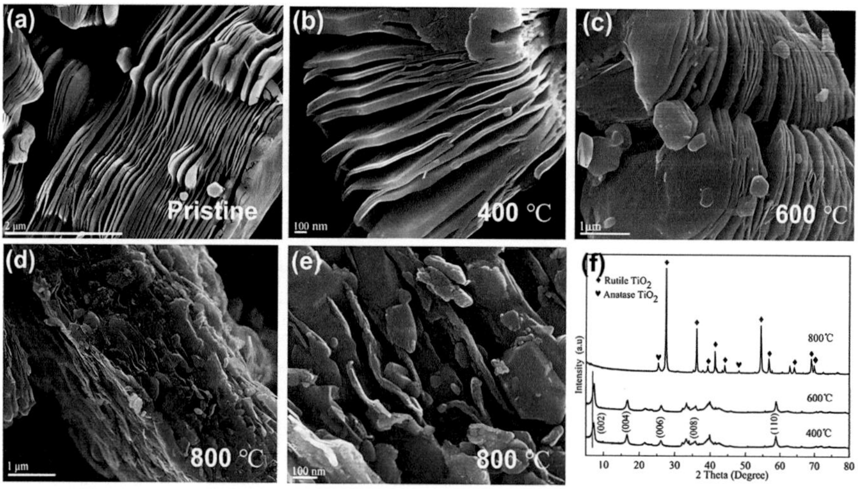


**FIGURE 15.5** SEM images of (a) MXene $Ti_3C_2$ NSs, MXene $Ti_3C_2$ NSs after calcination at (b) 400°C, (c) 600°C, (d) 800°C with (e) high-magnification image of (d), and (f) XRD patterns of MXene $Ti_3C_2$ NSs after calcinations at various temperatures. Reproduced with permission from Wang et al. (2015). Copyright 2015, Elsevier.

### 15.4.3 Effects of Humidity and Water Exposure

The oxidation of MXenes is accelerated when they are dispersed in aqueous solutions, progressing at a much faster rate than in their freestanding form. Previous studies indicated that dissolved $O_2$ in water was a key factor in the oxidation of MXene dispersions. However, more recent findings reveal that MXenes are more reactive toward $H_2O$ molecules than toward dissolved oxygen (Habib et al. 2019). Research has also focused on solutions beyond neutral ones, such as alkaline and oxidizing solutions. Doo et al. proposed a distinct oxidation process for aqueous $Ti_3C_2$ dispersions under different pH values, emphasizing the role of pH in influencing the formation of reaction intermediates between the –OH terminal groups and $H^+$ or $OH^-$ ions, as illustrated in Figure 15.3(i) (Doo et al. 2021). However, Zhao et al. (2020a) confirmed that in basic media, the reaction with hydroxyl ions enhances the oxidation of MXenes. This conclusion was supported by the changes in atomic percentage of the Ti(IV) component, as determined through X-ray photoelectron spectroscopy. Cao et al. (2017) investigated the oxidation behavior of $Ti_3C_2$ in both water and NaOH solutions under ambient conditions. Their observations indicated distinct differences in the formation of $TiO_2$. In aqueous solutions, $TiO_2$ nanoparticles predominantly formed on the surface and within the interlayers of the $Ti_3C_2$ nanosheets. Conversely, in NaOH solutions, $TiO_2$ nanowires were found to develop specifically at the edges of the $Ti_3C_2$ MXene. Additionally, both the size and the quantity of $TiO_2$ loading on the $Ti_3C_2$ surface were found to increase with prolonged reaction times or higher concentrations of NaOH.

### 15.4.4 Intrinsic Composition of MXenes

The intrinsic chemical composition and microstructural characteristics of MXenes, such as the nature of the M layers and surface terminations, play an important role in determining their stability. Zhang et al. (2016) investigated the oxidation behavior of $Nb_4C_3T_x$ MXenes in a $CO_2$ atmosphere at 850°C, observing the formation of a multilayer structure with $Nb_2O_5$ nanoparticles uniformly distributed on the surface of the MXene nanosheets. Their findings indicated that $Nb_4C_3T_x$ exhibits greater resistance to oxidation compared to $Ti_3C_2T_x$, which is more prone to oxidation in $CO_2$ flow at a low temperature of 500°C. Zhou et al. studied the stability of $Zr_3C_2T_x$ and $Ti_3C_2T_x$ MXenes, which were examined at higher temperatures in vacuum and argon environments as well (Zhou et al. 2016). The results showed that $Zr_3C_2T_x$ demonstrated a significantly better ability to preserve its 2D structure and overall integrity compared to $Ti_3C_2T_x$ under the same conditions.

The chemical termination of MXene surfaces plays a critical role in determining the material's properties and performance. Due to their inherently high surface-to-volume ratio, the adsorbed molecules or atomic groups on the MXene surface are key factors in regulating stoichiometry, stability, and other material characteristics. Following the topochemical deintercalation from MAX phases to MXenes, surface terminations such as =O and –OH, which naturally form on the MXene surface, can also function as reactive oxygen species (Naguib and Gogotsi 2015). MXenes synthesized with –Cl terminations have demonstrated greater stability compared to those terminated with –F and =O, maintaining thermal stability up to 750°C (Li et al. 2019). The synthesis methods for high-quality MXenes, which ensure both chemical and structural stability, have recently been recognized as essential for their various applications in nanotechnology. Nevertheless, MXene nanosheets with HF-etched multilayers typically display considerable structural defects (Seredych et al. 2019). Degradation studies of MXene colloidal solutions also reveal that their stability is influenced by size, with smaller MXenes generally exhibiting lower stability. This size-dependent instability is an important consideration alongside the structural defects present in the material (Zhang et al. 2017). To conclude, the behavior and durability of MXenes are governed by a range of interrelated factors, including synthesis methods, surface terminations, structural integrity, and size, all of which must be carefully optimized for their effective use in various applications.

## 15.5 STRATEGIES TO MITIGATE OXIDATION OF MXENES

The environmental instability of MXenes in both air and water significantly limits their practical applications. To address this, there has been a concerted effort to enhance the stability of MXenes or to develop devices that operate in non-aqueous environments. This growing need has catalyzed extensive research aimed at mitigating the oxidation and degradation of MXenes, thereby expanding their potential for use in various applications. Efforts to reduce oxidation have led to various strategies, including the careful management of storage conditions and complete isolation of MXenes from both water and oxygen. Furthermore, antioxidants like polyphosphoric anions and ascorbates are also utilized to effectively counteract oxidation.

### 15.5.1 Optimizing the Synthetic Conditions

MXenes with minimal structural defects are typically considered more chemically stable, as defect sites tend to facilitate the initiation of oxidation. By controlling defect density during synthesis, the quality of MXenes can be enhanced, which in turn allows for modulation of oxidation behavior and suppression of oxidation progression. Enhanced chemical stability in MXenes can be achieved through synthesis routes that limit defect formation, such as employing mixed acid etchants (HCl/HF, LiF/HCl) or bifluorides like $NH_4HF_2$, instead of relying solely on HF (Feng et al. 2017). By leveraging electrostatic interactions between alkali metal cations (such as $Li^+$, $Na^+$, and $Mg^{2+}$) and negatively charged $V_2CT_x$ nanosheets, VahidMohammadi et al. (2019) developed a cation-driven assembly approach that promotes the formation of ordered multilayered structures. This configuration helps retain the nanosheets in a more reduced state, thereby markedly suppressing their oxidation compared to delaminated $V_2CT_x$. By incorporating excess aluminum during synthesis, Mathis et al. (2021) achieved $Ti_3AlC_2$ grains with superior structural integrity and refined morphology. Continual advancements in the synthesis optimization of both MXenes and their MAX phase precursors are crucial for realizing new compositions and achieving high-quality MXenes. Optimizing the synthesis processes of MXenes and their corresponding MAX precursors remains essential for obtaining high-quality materials and expanding the range of accessible compositions.

### 15.5.2 Thermal Processing

Oxidation reactions are most likely to initiate at vulnerable functional groups, particularly hydroxyl sites, which are highly prone to oxidative attack (Xia et al. 2019). Eliminating surface terminations via thermal treatment under vacuum or inert conditions is a proven method for improving the inherent thermal stability of MXenes, thereby expanding their suitability for diverse applications (Persson et al. 2019). Wang et al. (2016) demonstrated that heat-treated, bare MXenes without surface terminations can withstand temperatures up to 1200°C in an argon atmosphere. Through thermal annealing under inert Ar gas, Zhao et al. (2020b) demonstrated that the formation of a protective $TiO_2$ layer on the surface of an MXene film significantly improves its stability. The protective $TiO_2$ layer formed on films with denser nanosheet stacking hinders water infiltration and preserves the internal structure, significantly enhancing the chemical stability of the surface and extending its shelf life in aqueous environments. Compared to untreated $Ti_3C_2T_x$ films, the $Ti_3C_2T_x$ films annealed at 600°C remained stable in water for over ten months, thereby broadening the potential applications of MXenes in diverse environments.

### 15.5.3 Impact of Storage Conditions

Optimizing storage conditions is another key approach to safeguarding MXenes against oxidation. Numerous research studies have demonstrated that chemical stability can be significantly enhanced by carefully controlling the storage temperature. Zhang et al. (2017) suggested an optimal approach for prolonging the shelf life of MXene solutions by preparing concentrated nanosheet solutions with large lateral dimensions and storing them in sealed Ar-filled containers at low temperatures. Storing MXenes in a dioxygen-free atmosphere and refrigerating them can significantly enhance stability. Xia et al. (2019) demonstrated that MXenes stored in organic solvents without moisture had improved stability. They also noted that water presence encourages the local diffusion of $Ti^{4+}$ and $O^{2-}$ ions, which are involved in oxidation. Maleski et al. (2017) compared the stability of $Ti_3C_2T_x$ dispersions in 12 different solvents, as depicted in Figure 15.6. Their results show that dispersing $Ti_3C_2T_x$ in organic solvents, such as DMF, NMP, PC, and ethanol, effectively alleviates or inhibits oxidation, thereby increasing the storage duration time of MXene colloidal solutions. Additionally, MXene powders can be isolated from the colloidal solution through vacuum filtration or freeze-drying, effectively preventing degradation caused by the aqueous environment.

### 15.5.4 Surface Modifications

To prevent oxidation, antioxidants are used in 2D MXenes colloidal dispersions, similar to their use in food preservation. $Ti_3C_2T_x$, when exposed to air and water, rapidly oxidizes and degrades, causing its 2D structure to break down into $TiO_2$ and carbon. To address this, Zhao et al. (2019) introduced sodium L-ascorbate (NaAsc), a common preservative and antioxidant in the food and pharmaceutical industries, to create a protective barrier around the MXenes. This addition effectively halts

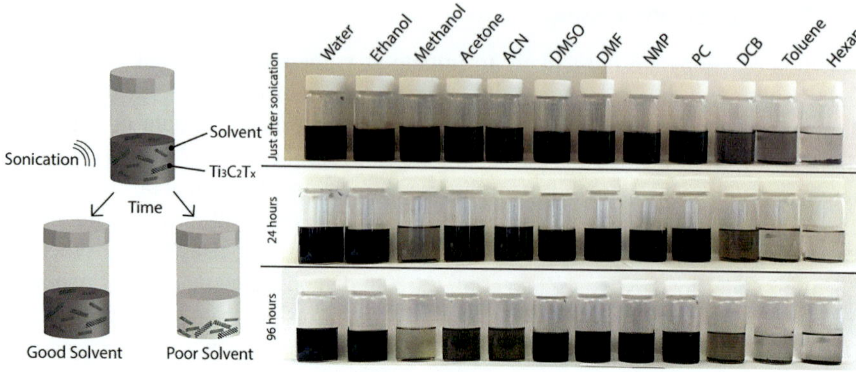


**FIGURE 15.6** Dispersions of 50% HF-etched $Ti_3C_2T_x$ in the 12 solvents mentioned above were evaluated for stability at three time points: immediately after sonication (top), 24 hours after sonication (middle), and 96 hours after sonication (bottom). Reproduced with permission from (Maleski, Mochalin, and Gogotsi 2017). Copyright 2017, American Society of Chemistry.

the oxidation of $Ti_3C_2T_x$ nanosheets, allowing them to remain stable and be stored in a closed container at room temperature for up to 21 days. In the absence of antioxidants, $Ti_3C_2T_x$ nanosheets would quickly degrade into $TiO_2$. However, when dispersed in a NaAsc solution, the $Ti_3C_2T_x$ nanosheets demonstrated prolonged stability, effectively retaining their physicochemical properties such as crystal structure, morphology, chemical composition, and conductivity. To mitigate oxidation, which typically starts at the edges, Natu et al. (2019) incorporated inorganic polyanionic salts, such as polyphosphates, polyborates, and polysilicates, into MXenes colloidal dispersions. MXenes are known to have negative charges on their surfaces and positive charges on their edges. Their findings revealed that adding just 0.1 $mol{\cdot}L^{-1}$ of polyanionic salt was enough to prevent the oxidation of MXenes in aerated water at room temperature for up to three weeks. A protective barrier was established to block edge-initiated oxidation and extend the storage lifespan of MXenes. Functionalization at the edges by introducing suitable antioxidants or polyanions to create a protective layer offers an effective strategy for minimizing their oxidation.

To enhance the performance of MXenes, surface chemical modification has been widely explored. Ji et al. (2019), for example, improved the surface characteristics and effectively suppressed the oxidation and degradation of MXenes by modifying their surface hydrophilicity using silanes. Due to the abundance of active terminal groups on their surfaces, MXenes can form strong covalent bonds with a broad range of polymers, leading to the creation of a protective shielding layer through in situ blending or polymerization. The interaction between the polymer chains and MXene sheets is further reinforced by van der Waals forces, electrostatic interactions, and hydrogen bonding (Gao et al. 2020). Krecker et al. (2020) introduced a strategy to enhance the stability of MXenes by co-assembling MXene flakes with silk fibroin proteins in an aqueous suspension, forming thin, silk-wrapped MXene sheets. The interaction between the silk proteins and MXene flakes, driven by electrostatic, van der Waals, and hydrophobic forces, strengthened the interfacial bonding and created a protective layer around the MXenes. This encapsulation effectively preserved the Ti–C bonds and suppressed the formation of titanium oxides, thereby significantly improving the resistance to oxidation and extending the shelf life of MXenes in aqueous environments. Applying a nano-scale carbon coating is a useful technique for preventing oxidation and structural deterioration of MXenes during their synthesis. Wu et al. (2017) demonstrated this by applying carbon nanoplating to create a thin physical barrier that inhibits the instant oxidation of MXenes and blocks the continuous diffusion of oxygen into the material. This process involved hydrothermally carbonizing glucose at 160°C, where glucose molecules preferentially adsorbed onto the surface of MXenes, lowering the surface free energy. Through intermolecular polymerization, the glucose molecules were converted into hydrothermal carbon, which further developed into a conductive carbon layer via thermal carbonization. With the protective carbon coating, $Ti_3C_2$ MXenes remained structurally intact throughout both the hydrothermal reactions and annealing at 500°C, showing minimal damage to their morphology and texture. In contrast, $Ti_3C_2$ MXenes without the carbon coating were quickly converted into $TiO_2$.

In summary, surface functionalization strategies such as hydrophobic silylation, polymerization, and carbon coating offer effective protection against the oxidation of MXenes while largely preserving their original structure and properties. Moreover, the protective layers formed through these functionalization approaches can substantially enhance the electrochemical performance of MXene-based materials.

## 15.6 CONCLUSIONS AND FUTURE PERSPECTIVES

MXenes, with their hydrophilic nature, excellent electrical conductivity, abundant surface functional groups, ease of processing, and remarkable energy storage properties, have garnered significant interest in cutting-edge fields since their discovery. However, their application is hindered by a major challenge: oxidation and degradation when exposed to oxygen-dissolved water. This leads to a loss in their unique properties, such as electrical conductivity and surface reactivity, thus limiting their practical use. As a result, creating effective methods to prevent or slow down the oxidation of MXenes is vital for preserving their 2D structure and functionality. Several strategies have been proposed to increase the stability of MXene-based materials, including optimizing the MAX phase synthesis, modifying MXene preparation, and regulating storage conditions, such as frozen storage under an inert gas atmosphere. Additionally, surface and edge shielding techniques, including the introduction of suitable antioxidants, silane coupling agents, and biopolymers, have been explored. Addressing the oxidation issue is crucial for improving the design, preparation, and application of MXenes, and future research should focus on developing innovative solutions to further enhance their stability and expand their applications. It is anticipated that with continued global collaboration, novel methods for the high-efficiency, safe, and very large-scale production of stable MXene-based materials will be realized in the near future.

## REFERENCES


Adomaviciute-Grabusove, Sonata, Anton Popov, Simonas Ramanavicius, Valdas Sablinskas, Kateryna Shevchuk, Oleksiy Gogotsi, Ivan Baginskiy, Yury Gogotsi, and Arunas Ramanavicius. 2024. "Monitoring $Ti_3C_2T_x$ MXene degradation pathways using Raman spectroscopy." *ACS Nano* 18 (20):13184–13195. doi: 10.1021/acsnano.4c02150.

Alhabeb, Mohamed, Kathleen Maleski, Babak Anasori, Pavel Lelyukh, Leah Clark, Saleesha Sin, and Yury Gogotsi. 2017. "Guidelines for synthesis and processing of two-dimensional titanium carbide ($Ti_3C_2T_x$ MXene)." *Chemistry of Materials* 29 (18):7633–7644. doi: 10.1021/acs.chemmater.7b02847.

Cao, Minjuan, Fen Wang, Lei Wang, Wenling Wu, Wenjing Lv, and Jianfeng Zhu. 2017. "Room temperature oxidation of $Ti_3C_2$ MXene for supercapacitor electrodes." *Journal of The Electrochemical Society* 164 (14):A3933.

Doo, Sehyun, Ari Chae, Daesin Kim, Taegon Oh, Tae Y. Ko, Seon J. Kim, Dong-Yeun Koh, and Chong M. Koo. 2021. "Mechanism and kinetics of oxidation reaction of aqueous $Ti_3C_2T_x$ suspensions at different pHs and temperatures." *ACS Applied Materials & Interfaces* 13 (19):22855–22865. doi: 10.1021/acsami.1c04663.

Ezzah Ab Latif, Farah, Arshid Numan, Nabisab M. Mubarak, Mohammad Khalid, Ezzat C. Abdullah, Norhuda A. Manaf, and Rashmi Walvekar. 2022. "Evolution of MXene and its 2D heterostructure in electrochemical sensor applications." *Coordination Chemistry Reviews* 471:214755. doi: 10.1016/j.ccr.2022.214755.

Feng, Aihu, Yun Yu, Feng Jiang, Yong Wang, Le Mi, Yang Yu, and Lixin Song. 2017. "Fabrication and thermal stability of $NH_4HF_2$-etched $Ti_3C_2$ MXene." *Ceramics International* 43 (8):6322–6328.

Gao, Lingfeng, Chao Li, Weichun Huang, Shan Mei, Han Lin, Qi Ou, Ye Zhang, Jia Guo, Feng Zhang, and Shixiang Xu. 2020. "MXene/polymer membranes: synthesis, properties, and emerging applications." *Chemistry of Materials* 32 (5):1703–1747.

Gayathri, Vijayakumar G., Richard Bartholomew, Jithin T. Chacko, Jagadeesh Bayry, and Pathath A. Rasheed. 2025. "Non-Ti MXenes: new biocompatible and biodegradable candidates for biomedical applications." *Journal of Materials Chemistry B* 13:1212–1228.

Gogotsi, Yury 2023. "The future of MXenes." *Chemistry of Materials* 35 (21):8767–8770.

Gogotsi, Yury, and Babak Anasori. 2019. "The rise of MXenes." *ACS Nano* 13 (8):8491–8494. doi: 10.1021/acsnano.9b06394.

Habib, Touseef, Xiaofei Zhao, Smit A. Shah, Yexiao Chen, Wanmei Sun, Hyosung An, Jodie L. Lutkenhaus, Miladin Radovic, and Micah J. Green. 2019. "Oxidation stability of $Ti_3C_2T_x$ MXene nanosheets in solvents and composite films." *npj 2D Materials and Applications* 3 (1):8. doi: 10.1038/s41699-019-0089-3.

Han, Meikang, Xiaowei Yin, Heng Wu, Zexin Hou, Changqing Song, Xinliang Li, Litong Zhang, and Laifei Cheng. 2016. "$Ti_3C_2$ MXenes with modified surface for high-performance electromagnetic absorption and shielding in the X-band." *ACS applied materials interfaces* 8 (32):21011–21019.

Ji, Jingjing, Lufang Zhao, Yanfei Shen, Songqin Liu, and Yuanjian Zhang. 2019. "Covalent stabilization and functionalization of MXene via silylation reactions with improved surface properties." *FlatChem* 17:100128.

Krecker, Michelle C., Daria Bukharina, Christine B. Hatter, Yury Gogotsi, and Vladimir V. Tsukruk. 2020. "Bioencapsulated MXene flakes for enhanced stability and composite precursors." *Advanced Functional Materials* 30 (43):2004554.

Kumar, Sunil, Nitu Kumari, Tej Singh, and Yongho Seo. 2024. "Shielding 2D MXenes against oxidative degradation: recent advances, factors and preventive measures." *Journal of Materials Chemistry C* 12 (23):8243–8281. doi: 10.1039/D4TC00884G.

Kumar, Sunil, Nitu Kumari, and Yongho Seo. 2024. "MXenes: versatile 2D materials with tailored surface chemistry and diverse applications." *Journal of Energy Chemistry* 90:253–293. doi: 10.1016/j.jechem.2023.11.031.

Lakhe, Pritishma, Evan M. Prehn, Touseef Habib, Jodie L. Lutkenhaus, Miladin Radovic, M. Sam Mannan, and Micah J. Green. 2019. "Process safety analysis for $Ti_3C_2T_x$ MXene synthesis and processing." *Industrial & Engineering Chemistry Research* 58 (4):1570–1579. doi: 10.1021/acs.iecr.8b05416.

Li, Mian, Jun Lu, Kan Luo, Youbing Li, Keke Chang, Ke Chen, Jie Zhou, Johanna Rosen, Lars Hultman, and Per Eklund. 2019. "Element replacement approach by reaction with Lewis acidic molten salts to synthesize nanolaminated MAX phases and MXenes." *Journal of the American Chemical Society* 141 (11):4730–4737.

Li, Xinliang, Xiaowei Yin, Meikang Han, Changqing Song, Xinnan Sun, Hailong Xu, Laifei Cheng, and Litong Zhang. 2017. "A controllable heterogeneous structure and electromagnetic wave absorption properties of $Ti_2CT_x$ MXene." *Journal of Materials Chemistry C* 5 (30):7621–7628. doi: 10.1039/C7TC01991B.

Li, Zhengyang, Libo Wang, Dandan Sun, Yude Zhang, Baozhong Liu, Qianku Hu, and Aiguo Zhou. 2015. "Synthesis and thermal stability of two-dimensional carbide MXene $Ti_3C_2$." *Materials Science Engineering: B* 191:33–40.

Lipatov, Alexey, Mohamed Alhabeb, Maria R. Lukatskaya, Alex Boson, Yury Gogotsi, and Alexander Sinitskii. 2016. "Effect of synthesis on quality, electronic properties and environmental stability of individual monolayer $Ti_3C_2$ MXene flakes." *Advanced Electronic Materials* 2 (12):1600255.

Lotfi, Roghayyeh, Michael Naguib, Dundar E. Yilmaz, Jagjit Nanda, and Adri C. T. van Duin. 2018. "A comparative study on the oxidation of two-dimensional $Ti_3C_2$ MXene structures in different environments." *Journal of Materials Chemistry A* 6 (26):12733–12743. doi: 10.1039/C8TA01468J.

Lukatskaya, Maria R., Olha Mashtalir, Chang E. Ren, Yohan Dall'Agnese, Patrick Rozier, Pierre L. Taberna, Michael Naguib, Patrice Simon, Michel W. Barsoum, and Yury Gogotsi. 2013. "Cation intercalation and high volumetric capacitance of two-dimensional titanium carbide." *Science* 341 (6153):1502–1505. doi: 10.1126/science.1241488.

Maleski, Kathleen, Vadym N. Mochalin, and Yury Gogotsi. 2017. "Dispersions of two-dimensional titanium carbide MXene in organic solvents." *Chemistry of Materials* 29 (4):1632–1640.

Mathis, Tyler S., Kathleen Maleski, Adam Goad, Asia Sarycheva, Mark Anayee, Alexandre C. Foucher, Kanit Hantanasirisakul, Christopher E. Shuck, Eric A. Stach, and Yury Gogotsi. 2021. "Modified MAX phase synthesis for environmentally stable and highly conductive $Ti_3C_2$ MXene." *ACS Nano* 15 (4):6420–6429.

Mim, Mumtahina, Khairul Habib, Sazratul N. Farabi, Syed A. Ali, Md Abu Zaed, Mohammad Younas, and Saidur Rahman. 2024. "MXene: a roadmap to sustainable energy management, synthesis routes, stabilization, and economic assessment." *ACS Omega* 9 (30):32350–32393. doi: 10.1021/acsomega.4c04849.

Naguib, Michael, Murat Kurtoglu, Volker Presser, Jun Lu, Junjie Niu, Min Heon, Lars Hultman, Yury Gogotsi, and Michel W. Barsoum. 2011. "Two-dimensional nanocrystals produced by exfoliation of $Ti_3AlC_2$." *Advanced Materials* 23 (37):4248–4253. doi: 10.1002/adma.201102306.

Naguib, Michael, and Yury Gogotsi. 2015. "Synthesis of two-dimensional materials by selective extraction." *Accounts of Chemical Research* 48 (1):128–135.

Natu, Varun, James L. Hart, Maxim Sokol, Helen Chiang, Mitra L. Taheri, and Michel W. Barsoum. 2019. "Edge capping of 2D-MXene sheets with polyanionic salts to mitigate oxidation in aqueous colloidal suspensions." *Angewandte Chemie* 131 (36):12785–12790.

Persson, Ingemar, Joseph Halim, Hans Lind, Thomas W. Hansen, Jakob B. Wagner, Lars-Åke Näslund, Vanya Darakchieva, Justinas Palisaitis, Johanna Rosen, and Per O. Å. Persson. 2019. "2D transition metal carbides (MXenes) for carbon capture." *Advanced Materials* 31 (2):1805472.

Rakhi, Raghavan B., Bilal Ahmed, Mohamed N. Hedhili, Dalaver H. Anjum, and Husam N. Alshareef. 2015. "Effect of postetch annealing gas composition on the structural and electrochemical properties of $Ti_2CT_x$ MXene electrodes for supercapacitor applications." *Chemistry of Materials* 27 (15):5314–5323.

Seredych, Mykola, Christopher E. Shuck, David Pinto, Mohamed Alhabeb, Eliot Precetti, Grayson Deysher, Babak Anasori, Narendra Kurra, and Yury Gogotsi. 2019. "High-temperature behavior and surface chemistry of carbide MXenes studied by thermal analysis." *Chemistry of Materials* 31 (9):3324–3332. doi: 10.1021/acs.chemmater.9b00397.

Shayesteh Zeraati, Ali, Seyyed Alireza Mirkhani, Pengcheng Sun, Michael Naguib, Paul V. Braun, and Uttandaraman Sundararaj. 2021. "Improved synthesis of $Ti_3C_2T_x$ MXenes resulting in exceptional electrical conductivity, high synthesis yield, and enhanced capacitance." *Nanoscale* 13 (6):3572–3580. doi: 10.1039/D0NR06671K.

Thakur, Raj, Armin VahidMohammadi, Jorge Moncada, William R. Adams, Mingyang Chi, Bruce Tatarchuk, Majid Beidaghi, and Carlos A. Carrero. 2019. "Insights into the thermal and chemical stability of multilayered $V_2CT_x$ MXene." *Nanoscale* 11 (22):10716–10726.

VahidMohammadi, Armin, Johanna Rosen, and Yury Gogotsi. 2021. "The world of two-dimensional carbides and nitrides (MXenes)." *Science* 372 (6547):eabf1581. doi: 10.1126/science.abf1581.

VahidMohammadi, Armin, Mehrnaz Mojtabavi, Nuala M. Caffrey, Meni Wanunu, and Majid Beidaghi. 2019. "Assembling 2D MXenes into highly stable pseudocapacitive electrodes with high power and energy densities." *Advanced Materials* 31 (8):1806931.

Wang, Hongbing, Yuping Wu, Jianfeng Zhang, Gaiye Li, Huajie Huang, Xin Zhang, and Quanguo Jiang. 2015. "Enhancement of the electrical properties of MXene $Ti_3C_2$ nanosheets by post-treatments of alkalization and calcination." *Materials Letters* 160:537–540.

Wang, Kun, Youfu Zhou, Wentao Xu, Decai Huang, Zhiguang Wang, and Maochun Hong. 2016. "Fabrication and thermal stability of two-dimensional carbide $Ti_3C_2$ nanosheets." *Ceramics International* 42 (7):8419–8424.

Wu, Xianhong, Zhiyu Wang, Mengzhou Yu, Luyang Xiu, and Jieshan Qiu. 2017. "Stabilizing the MXenes by carbon nanoplating for developing hierarchical nanohybrids with efficient lithium storage and hydrogen evolution capability." *Advanced Materials* 29 (24):1607017.

Wyatt, Brian C., Andreas Rosenkranz, and Babak Anasori. 2021. "2D MXenes: tunable mechanical and tribological properties." *Advanced Materials* 33 (17):2007973.

Xia, Fanjie, Junchao Lao, Ruohan Yu, Xiahan Sang, Jiayan Luo, Yu Li, and Jinsong Wu. 2019. "Ambient oxidation of $Ti_3C_2$ MXene initialized by atomic defects." *Nanoscale* 11 (48):23330–23337.

Xu, Bin, and Yury Gogotsi. 2020. "MXenes: from discovery to applications." *Advanced Functional Materials* 30 (47), 2007011.

Zeng, Xingzhong, Yitian Peng, and Haojie Lang. 2017. "A novel approach to decrease friction of graphene." *Carbon* 118:233–240. doi: 10.1016/j.carbon.2017.03.042.

Zhang, Chuanfang, Seon J. Kim, Michael Ghidiu, Meng-Qiang Zhao, Michel W. Barsoum, Valeria Nicolosi, and Yury Gogotsi. 2016. "Layered orthorhombic $Nb_2O_5@Nb_4C_3T_x$ and $TiO_2@Ti_3C_2T_x$ hierarchical composites for high performance Li-ion batteries." *Advanced Functional Materials* 26 (23):4143–4151.

Zhang, Chuanfang J., Sergio Pinilla, Niall McEvoy, Conor P. Cullen, Babak Anasori, Edmund Long, Sang-Hoon Park, Andrés Seral-Ascaso, Aleksey Shmeliov, and Dileep Krishnan. 2017. "Oxidation stability of colloidal two-dimensional titanium carbides (MXenes)." *Chemistry of Materials* 29 (11):4848–4856.

Zhang, Wei, Xing-Xiang Ji, and Ming-Guo Ma. 2023. "Emerging MXene/cellulose composites: design strategies and diverse applications." *Chemical Engineering Journal* 458:141402.

Zhao, Xiaofei, Aniruddh Vashisth, Evan Prehn, Wanmei Sun, Smit A. Shah, Touseef Habib, Yexiao Chen, Zeyi Tan, Jodie L. Lutkenhaus, and Miladin Radovic. 2019. "Antioxidants unlock shelf-stable $Ti_3C_2T_x$ (MXene) nanosheet dispersions." *Matter* 1 (2):513–526.

Zhao, Xiaofei, Aniruddh Vashisth, Jackson W. Blivin, Zeyi Tan, Dustin E. Holta, Vrushali Kotasthane, Smit A. Shah, Touseef Habib, Shuhao Liu, Jodie L. Lutkenhaus, Miladin Radovic, and Micah J. Green. 2020a. "pH, nanosheet concentration, and antioxidant affect the oxidation of $Ti_3C_2T$ and $Ti_2CT$ MXene dispersions." *Advanced Materials* 7 (20):2000845. doi: 10.1002/admi.202000845.

Zhao, Xiaofei, Dustin E. Holta, Zeyi Tan, Ju-Hyun Oh, Ian J. Echols, Muhammad Anas, Huaixuan Cao, Jodie L. Lutkenhaus, Miladin Radovic, and Micah J. Green. 2020b. "Annealed $Ti_3C_2T_z$ MXene films for oxidation-resistant functional coatings." *ACS Applied Nano Materials* 3 (11):10578–10585.

Zhou, Jie, Xianhu Zha, Fan Y. Chen, Qun Ye, Per Eklund, Shiyu Du, and Qing Huang. 2016. "A two-dimensional zirconium carbide by selective etching of $Al_3C_3$ from nanolaminated $Zr_3Al_3C_5$." *Angewandte Chemie International Edition* 55 (16):5008–5013.

# Index

## E



## F



## G



## H



## I

## L



## M



## N



## O



## P



## R

## S



## T



## U



## W



## X



## Z